# Sturm-Liouville Theory

## Past and Present

Werner O. Amrein
Andreas M. Hinz
David B. Pearson
(Editors)

Birkhäuser Verlag
Basel · Boston · Berlin

Editors:

Werner O. Amrein
Section de Physique
Université de Genève
24, quai Ernest-Ansermet
1211 Genève 4
Switzerland
Werner.Amrein@physics.unige.ch

Andreas M. Hinz
Mathematisches Institut
Universität München
Theresienstrasse 39
D-80333 München
Germany
Andreas.Hinz@mathematik.uni-muenchen.de

David P. Pearson
Department of Mathematics
University of Hull
Cottingham Road
Hull HU6 7RX
United Kingdom
D.B.Pearson@hull.ac.uk

2000 Mathematical Subject Classification 34B24, 34C10, 34L05, 34L10, 01A55, 01A10

A CIP catalogue record for this book is available from the
Library of Congress, Washington D.C., USA

Bibliographic information published by Die Deutsche Bibliothek
Die Deutsche Bibliothek lists this publication in the Deutsche Nationalbibliografie;
detailed bibliographic data is available in the Internet at <http://dnb.ddb.de>.

ISBN 3-7643-7066-1 Birkhäuser Verlag, Basel – Boston – Berlin


Part of Springer Science+Business Media
Cover design: Micha Lotrovsky, CH-4106 Therwil, Switzerland
Printed on acid-free paper produced of chlorine-free pulp. TCF ∞
Printed in Germany
ISBN-10: 3-7643-7066-1
ISBN-13: 978-3-7643-7066-4

9 8 7 6 5 4 3 2 1

www.birkhauser.ch

# Contents

# Preface

Charles François Sturm, through his papers published in the 1830's, is considered to be the founder of Sturm-Liouville theory. He was born in Geneva in September 1803. To commemorate the 200th anniversary of his birth, an international colloquium in recognition of Sturm's major contributions to science took place at the University of Geneva, Switzerland, following a proposal by Andreas Hinz. The colloquium was held from 15 to 19 September 2003 and attended by more than 60 participants from 16 countries. It was organized by Werner Amrein of the Department of Theoretical Physics and Jean-Claude Pont, leader of the History of Science group of the University of Geneva. The meeting was divided into two parts. In the first part, historians of science discussed the many contributions of Charles Sturm to mathematics and physics, including his pedagogical work. The second part of the colloquium was then devoted to Sturm-Liouville theory. The impact and development of this theory, from the death of Sturm to the present day, was the subject of a series of general presentations by leading experts in the field, and the colloquium concluded with a workshop covering recent research in this highly active area.

This drawing together of historical presentations with seminars on current mathematical research left participants in no doubt of the degree to which Sturm's original ideas are continuing to have an impact on the mathematics of our own times. The format of the conference provided many opportunities for exchange of ideas and collaboration and might serve as a model for other multidisciplinary meetings.

The organizers had decided not to publish proceedings of the meeting in the usual form (a complete list of scientific talks is appended, however). Instead it was planned to prepare, in conjunction with the colloquium, a volume containing a complete collection of Sturm's published articles and a volume presenting the various aspects of Sturm-Liouville theory at a rather general level, accessible to the non-specialist. Thus Jean-Claude Pont will edit a volume[1] containing the collected works of Sturm accompanied by a biographical review as well as abundant historical and technical comments provided by the contributors to the first part of the meeting.

The present volume is a collection of twelve refereed articles relating to the second part of the colloquium. It contains, in somewhat extended form, the survey lectures on Sturm-Liouville theory given by the invited speakers; these are the first

[1] *The Collected Works of Charles François Sturm*, J.-C. Pont, editor (in preparation).

six papers of the book. To complement this range of topics, the editors invited a few participants in the colloquium to provide a review or other contribution in an area related to their presentation and which should cover some important aspects of current interest. The volume ends with a comprehensive catalogue of Sturm-Liouville differential equations. At the conclusion of the Introduction is a brief description of the articles in the book, placing them in the context of the developing theory of Sturm-Liouville differential equations. We hope that these articles, besides being a tribute to Charles François Sturm, will be a useful resource for researchers, graduate students and others looking for an overview of the field.

We have refrained from presenting details of Sturm's life and his other scientific work in this volume. As regards Sturm-Liouville theory, some aspects of Sturm's original approach are presented in the contributions to the present book, and a more detailed discussion will be given in the article by Jesper Lützen and Angelo Mingarelli in the companion volume. Of course, the more recent literature concerned with this theory and its applications is strikingly vast (on the day of writing, MathSciNet yields 1835 entries having the term "Sturm-Liouville" in their title); it is therefore unavoidable that there may be certain aspects of the theory which are not sufficiently covered here.

The articles in this volume can be read essentially independently. The authors have included cross-references to other contributions. In order to respect the style and habits of the authors, the editors did not ask them to use a uniform standard for notations and conventions of terminology. For example, the reader should take note that, according to author, inner products may be anti-linear in the first or in the second argument, and deficiency indices are either single natural numbers or pairs of numbers. Moreover, there are some differences in terminology as regards spectral theory.

The colloquium would not have been possible without support from numerous individuals and organizations. Financial contributions were received from various divisions of the University of Geneva (Commission administrative du Rectorat, Faculté des Lettres, Faculté des Sciences, Histoire et Philosophie des Sciences, Section de Physique), from the History of Science Museum and the City of Geneva, the Société Académique de Genève, the Société de Physique et d'Histoire Naturelle de Genève, the Swiss Academy of Sciences and the Swiss National Science Foundation. To all these sponsors we express our sincere gratitude. We also thank the various persons who volunteered to take care of numerous organizational tasks in relation with the colloquium, in particular Francine Gennai-Nicole who undertook most of the secretarial work, Jan Lacki and Andreas Malaspinas for technical support, Danièle Chevalier, Laurent Freland, Serge Richard and Rafael Tiedra de Aldecoa for attending to the needs of the speakers and other participants. Special thanks are due to Jean-Claude Pont for his enthusiastic collaboration over a period of more than three years in the entire project, as well as to all the speakers of the meeting for their stimulating contributions.

As regards the present volume, we are grateful to our authors for all the efforts they have put into the project, as well as to our referees for generously giving of their time. We thank Norrie Everitt, Hubert Kalf, Karl Michael Schmidt, Charles Stuart and Peter Wittwer who freely gave their scientific advice, Serge Richard who undertook the immense task of preparing manuscripts for the publishers, and Christian Clason for further technical help. We are much indebted to Thomas Hempfling from Birkhäuser Verlag for continuing support in a fruitful and rewarding partnership.

The cover of this book displays, in Liouville's handwriting, the original formulation by Sturm and Liouville, in the manuscript of their joint 1837 paper, of the regular second-order boundary value problem on a finite interval. The paper, which is discussed here by W.N. Everitt on pages 47–50, was presented to the Paris Académie des sciences on 8 May 1837 and published in Comptes rendus de l'Académie des sciences, Vol. IV (1837), 675–677, as well as in Journal de Mathématiques Pures et Appliquées, Vol. 2 (1837), 220–223. The original manuscript, with the title "Analyse d'un Mémoire sur le développement des fonctions en séries, dont les différents termes sont assujettis à satisfaire à une même équation différentielle linéaire contenant un paramètre variable", is preserved in the archives of the Académie des sciences to whom we are much indebted for kind permission to reproduce an extract.

Geneva, September 2004

Werner Amrein
Andreas Hinz
David Pearson

## Scientific Lectures given at the Sturm Colloquium

*J. Dhombres*
Charles Sturm et la Géométrie

*H. Sinaceur*
Charles Sturm et l'Algèbre

*J. Lützen*
The history of Sturm-Liouville theory, in particular its early part

*A. Mingarelli*
Two papers by Sturm (1829 and 1833) are considered
in the light of their impact on his famous 1836 Memoir

*P. Radelet*
Charles Sturm et la Mécanique

*E.J. Atzema*
Charles Sturm et l'Optique

*J.-C. Pont*
Charles Sturm, Daniel Colladon et la compressibilité de l'eau

---

*D. Hinton*
Sturm's 1836 Oscillation Results: Evolution of the Theory

*B. Simon*
Sturm Oscillation and Comparison Theorems and Some Applications

*W.N. Everitt*
The Development of Sturm-Liouville Theory in the Years 1900 to 1950

*J. Weidmann*
Spectral Theory of Sturm-Liouville Operators;
Approximation of Singular Problems by Regular Problems

*Y. Last*
Spectral Theory of Sturm-Liouville Operators on Infinite Intervals:
A Review of Recent Developments

*D. Gilbert*
Asymptotic Methods in the Spectral Analysis of Sturm-Liouville Operators

*E. Sanchez Palencia*
Singular Perturbations with Limit Essential Spectrum
and Complexification of the Solutions

---

*H. Behncke*
Asymptotics and Oscillation Theory of Eigenfunctions of Fourth Order Operators

*C.-N. Chen*
Nonlinear Sturm-Liouville Equations

*H.O. Cordes*
The Split of the Dirac Hamiltonian into Precisely Predictable Kinetic and Potential Energy

*R. del Río*
Boundary Conditions of Sturm-Liouville Operators with Mixed Spectra

*L.V. Eppelbaum and V.R. Kardashov*
On one Strongly Nonlinear Generalization of the Sturm-Liouville Problem

*S.E. Guseinov*
A Simple Method for Solving a Class of Inverse Problems

*P. Harwin*
On Evolution Completeness of Eigenfunctions for Nonlinear Diffusion Operators: Application of Sturm's Theorem

*A. Kerouanton*
Some Properties of the $m$-Function associated with a Non-selfadjoint Sturm-Liouville Type Operator

*A. Makin*
An Eigenvalue Problem for Nonlinear Sturm-Liouville Operators

*M. Malamud*
Inverse Spectral Problems for Matrix Sturm-Liouville Equations

*D.B. Pearson*
Three Results for $m$-Functions

*A. Ronveaux*
Extension of Rodrigues' Formula for Second Kind Solution of the Hypergeometric Equation : History – Developments – Generalization

*Y. Yakubov*
General Sturm-Liouville Problems with a Spectral Parameter in Boundary Conditions

*P. Zhidkov*
Basis Properties of Eigenfunctions of Nonlinear Sturm-Liouville Problems

# Introduction

David Pearson

Charles François Sturm was born in Geneva on 29 September 1803[1]. He received his scientific education in this city, in which science has traditionally been of such great importance. Though he was later drawn to Paris, where he settled permanently in 1825 and carried out most of his scientific work, he has left his mark also on the city of Geneva, where his name is commemorated by the Place Sturm and the Rue Charles-Sturm. On the first floor of the Museum of History of Science, in its beautiful setting with magnificent views over Lake Geneva, you can see some of the equipment with which his friend and collaborator Daniel Colladon pursued his research on the lake into the propagation of sound through water[2].

Sturm's family came to Geneva from Strasbourg a few decades before his birth. He frequently moved house, and at least two of the addresses where he spent some of his early years can still be found in Geneva's old town[3,4].

Not only did Charles Sturm leave his mark on Geneva, but his rich scientific legacy is recognized by mathematicians and scientists the world over, and continues to influence the direction of mathematical development in our own times[5]. In

[1]This corresponds to the sixth day of the month of Vendémiaire in year XII of the French revolutionary calendar then in use in the Département du Léman.

[2]Colladon was the physicist and experimentalist of this partnership, while Sturm played an important role as theoretician. Their joint work on sound propagation and compressibility of fluids was recognized in 1827 by the award of the Grand Prix of the Paris Academy of Sciences.

[3]The address 29, Place du Bourg-de-Four was home to ancestors of Charles Sturm in 1798. The present building appears on J.-M. Billon's map of Geneva, dated 1726, which is the earliest extant cadastral map of the city. The home of Charles Sturm in 1806, with his parents and first sister, was 11, Rue de l'Hôtel-de-Ville. The building now on this site was constructed in 1840. The two houses are in close proximity.

[4]For details on Sturm's life, see the biographical notice by J.-C. Pont and I. Benguigui in *The Collected Works of Charles François Sturm*, J.-C. Pont, editor (in preparation), as well as Chapter 21 of the book by P. Speziali, *Physica Genevensis, La vie et l'oeuvre de 33 physiciens genevois*, Georg, Chêne-Bourg (1997).

[5]Sturm was already judged by his contemporaries to be an outstanding theoretician. Of the numerous honours which he received during his lifetime, special mention might be made of the Grand Prix in Mathematics of the Paris Academy, in 1834, and membership of the Royal Society of London as well as the Copley Medal, in 1840. The citation for membership of the Royal Society was as follows: "Jacques Charles François Sturm, of Paris, a Gentleman eminently distinguished for his original investigations in mathematical science, is recommended by us as a proper person

bringing together leading experts in the scientific history of Sturm's work with some of the major contributors to recent and contemporary mathematical developments in related fields, the Sturm Colloquium provided a unique opportunity for the sharing of knowledge and exchange of new ideas.

Interactions of this kind between individuals from different academic backgrounds can be of great value. There is, of course, a powerful argument for mathematics to take note of its history. Mathematical results, concepts and methods do not spring from nowhere. Often new results are motivated by existing or potential applications. Some of Sturm's early work on sound propagation in fluids is a good example of this, as are his fundamental contributions to the theory of differential equations, which were partly motivated by problems of heat flow. Some of the later developments in areas that Sturm had initiated proceeded in parallel with one of the revolutions in twentieth century physics, namely quantum mechanics. New ideas in mathematics need to be considered in the light of the mathematical and cultural environment of their time.

Sturm's mathematical publications covered diverse areas of geometry, algebra, analysis, mechanics and optics. He published textbooks in analysis and mechanics, both of which were still in use as late as the twentieth century[6].

To most mathematicians today, Sturm's best-known contributions, and those which are usually considered to have had the greatest influence on mathematics since Sturm's day, have been in two main areas.

The first of Sturm's major contributions to mathematics was his remarkable solution, presented to the Paris Academy of Sciences in 1829 and later elaborated in a memoir of 1835[7], of the problem of determining the number of roots, on a given interval, of a real polynomial equation of arbitrary degree. Sturm found a complete solution of this problem, which had been open since the seventeenth century. His solution is algorithmic; a sequence of auxiliary polynomials (now called Sturm

---

to be placed on the list of Foreign members of the Royal Society". The Copley Medal was in recognition of his seminal work on the roots of real polynomial equations and was the second medal awarded that year, the first having gone to the chemist J. Liebig. The citation for the Medal was: "Resolved, by ballot. – That another Copley Medal be awarded to M. C. Sturm, for his "Mémoire sur la Résolution des Equations Numériques," published in the Mémoires des Savans Etrangers for 1835". Sturm is also one of the few mathematicians commemorated in the series of plaques at the Eiffel tower in Paris.

[6]Both of these books were published posthumously, Sturm having died on 18 December 1855. The analysis text went through 15 editions, of which the last printing was as late as 1929. A reference for the first edition is: *Cours d'analyse de l'École polytechnique* (2 vols.), published by E. Prouhet (Paris, 1857–59). The text was translated into German by T. Fischer as: *Lehrbuch der Analysis* (Berlin, 1897–98). The first edition of the mechanics text was: *Cours de mécanique de l'École polytechnique* (2 vols.), published by E. Prouhet (Paris, 1861). The fifth and last edition, revised and annotated by A. de Saint-Germain, was in print at least until 1925.

[7]The full text of Sturm's resolution of this problem is to be found in: *Mémoire sur la résolution des équations numériques*, in the journal Mémoires présentés par divers savans à l'Académie Royale des Sciences de l'Institut de France, sciences mathématiques et physiques **6** (1835), 271–318 (also cited as Mémoires Savants Étrangers). See also *The Collected works of Charles François Sturm*, J.-C. Pont, editor (in preparation) for further discussion of this work.

functions), is calculated, and the number of roots on an interval is determined by the signs of the Sturm functions at the ends of the intervals. Sturms work on zeros of polynomials undoubtedly influenced his work on related problems for solutions of differential equations, which was to follow.

His second major mathematical contribution, or rather a whole series of contributions, was to the theory of second-order linear ordinary differential equations. In 1833 he read a paper to the Academy of Sciences on this subject, to be followed in 1836 by a long and detailed memoir in the Journal de Mathématiques Pures et Appliquées. This memoir was one of the first to appear in the journal, which had recently been founded by Joseph Liouville, who was to become a collaborator and one of Sturm's closest friends in Paris. It contained the first full treatment of the oscillation, comparison and separation theorems which were to bear Sturm's name, and was succeeded the following year by a remarkable short paper, in the same journal and in collaboration with Liouville, which established the basic principles of what was to become known as Sturm-Liouville theory[8]. The problems treated in this paper would be described today as Sturm-Liouville boundary value problems (second-order linear differential equations, with linear dependence on a parameter) on a finite interval, with separated boundary conditions. Sturm's earlier work had shown that such problems led to an infinity of possible values of the parameter. The collaboration between Sturm and Liouville took the theory some way forward by proving the expansion theorem, namely that a large class of functions could be represented by a Fourier-type expansion in terms of the family of solutions to the boundary value problem. In modern terminology, the solutions would later be known as eigenfunctions and the corresponding values of the parameter as eigenvalues.

The 1837 memoir, published jointly by Sturm and Liouville, was to become the foundation of a whole new branch of mathematics, namely the spectral theory of differential operators. Sturm-Liouville theory is central to a large part of modern analysis. The theory has been successively generalized in a number of directions, with applications to Mathematical Physics and other branches of modern science. This volume provides the reader with an account of the evolution of Sturm-Liouville theory since the pioneering work of its two founders, and presents some of the most recent research. The companion volume will treat aspects of the work of Sturm and his successors as a branch of the history of scientific ideas. We believe that the two volumes together will provide a perspective which will help to make clear the significant position of Sturm-Liouville theory in modern mathematics.

Sturm-Liouville theory, as originally conceived by its founders, may be regarded, from a modern standpoint, as a first, tentative step towards the development of a spectral theory for a class of second-order ordinary differential operators.

---

[8]For a more extended treatment of the early development of Sturm-Liouville theory, with detailed references, see the paper on Sturm and differential equations by J. Lützen and A. Mingarelli in the companion volume, as well as the first contribution by Everitt to this volume.

Liouville had already covered in some detail the case of a finite interval with two regular endpoints and boundary conditions at each endpoint. He regarded the resulting expansion theorem in terms of orthogonal eigenfunctions[9] as an extension of corresponding results for Fourier series, and the analysis was applicable only to cases for which, in modern terminology, the spectrum could be shown to be pure point. In fact the term "spectrum" itself, in a sense close to its current meaning, only began to emerge at the end of the nineteenth and the beginning of the twentieth century, and is usually attributed to David Hilbert.

The first decade of the twentieth century was a period of rapid and highly significant development in the concepts of spectral theory. A number of mathematicians were at that time groping towards an understanding of the idea of continuous spectrum. Among these was Hilbert himself, in Göttingen. Hilbert was concerned not with differential equations (though his work was to have a profound impact on the spectral analysis of second-order differential equations) but with what today we would describe as quadratic forms in the infinite-dimensional space $l^2$. Within this framework, he was able to construct the equivalent of a spectral function for the quadratic form, in terms of which both discrete and continuous spectrum could be defined. Examples of both types of spectrum could be found, and from these examples emerged the branch of mathematics known as spectral analysis. For the first time, spectral theory began to make sense even in cases where the point spectrum was empty. The time was ripe for such developments, and the theory rapidly began to incorporate advances in integration and measure theory coming from the work of Lebesgue, Borel, Stieltjes and others.

As far as Sturm-Liouville theory itself is concerned, the most significant progress during this first decade of the twentieth century was undoubtedly due to the work of the young Hermann Weyl. Weyl had been a student of Hilbert in Göttingen, graduating in 1908. (He was later, in 1930, to become professor at the same university.) His 1910 paper[10] did much to revolutionise the spectral theory of second-order linear ordinary differential equations. Weyl's spectrum is close to the modern definition via resolvent operators, and his analysis of endpoints based on limit point/limit circle criteria anticipates later ideas in functional analysis in which deficiency indices play the central role. For Weyl, continuous spectrum was not only to be tolerated, but was totally absorbed into the new theory. The expansion theorem, from 1910 onwards, was to cover contributions from both discrete and continuous parts of the spectrum. Weyl's example of continuous spectrum, corresponding to the differential equation $-d^2f(x)/dx^2 - xf(x) = \lambda f(x)$ on the

[9]Liouville's proof of the expansion theorem was not quite complete in that it depended on assumptions involving some additional regularity of eigenfunctions. Later extensions of this theory, as well as a full and original proof of completeness of eigenfunctions, can be found in the article by Bennewitz and Everitt in this volume.

[10]A full discussion of Weyl's paper and its impact on Sturm-Liouville theory is to be found in the first contribution by Everitt to this volume.

half line $[0, \infty)$ , could hardly have been simpler[11]. And, perhaps most importantly, with Weyl's 1910 paper complex function theory began to move to the center stage in spectral analysis.

The year 1913 saw a further advance through the publication of a research monograph by the Hungarian mathematician Frigyes Riesz[12], in which he continued the ideas of Hilbert, with the new point of view that it was the linear operator associated with a given quadratic form, rather than the form itself, which was to be the focus of analysis. In other words, Riesz shifted attention towards the spectral theory of linear operators. In doing so he was able to arrive at the definition of spectrum in terms of the resolvent operator, to define a functional calculus for linear operators, and to explore the idea of what was to become the resolution of the identity for bounded self-adjoint operators. An important consequence of these results was that it became possible to incorporate many of Weyl's results on Sturm-Liouville problems into the developing theory of functional analysis. Thus, for example, the role of boundary conditions in determining self-adjoint extensions of differential operators could then be fully appreciated.

The modern theory of Sturm-Liouville differential equations, which grew from these beginnings, was profoundly influenced by the emergence of quantum mechanics, which also had its birth in the early years of the twentieth century. At the heart of the development of a mathematical theory to meet the demands of the new physics was John von Neumann[13].

Von Neumann joined Hilbert as assistant in Göttingen in 1926, the very year that Schrödinger first published his fundamental wave equation. The Schrödinger equation is, in fact, a partial differential equation, but, in the case of spherically symmetric potentials such as the Coulomb potential, the standard technique of separation of variables reduces the equation to a sequence of ordinary differential equations, one for each pair of angular momentum quantum numbers. In this way, under the assumption of spherical symmetry, Sturm-Liouville theory can be applied to the Schrödinger equation.

Von Neumann found in functional analysis the perfect medium for understanding the foundations of quantum mechanics. Quantum theory led in a natural way to a close correspondence (one could almost say identification, though that would not quite be true) of the physical objects of the theory with mathematical objects drawn from the theory of linear operators (usually differential operators) in Hilbert space. The state of a quantum system could be described by a normalized element (or vector, or wave function) in the Hilbert space. Corresponding to each

[11] Later it was to emerge that examples of this kind could be interpreted physically in terms of a quantum mechanical charged particle moving in a uniform electric field.

[12] F. Riesz, *Les systèmes d'équations linéaires à une infinité d'inconnues*, Gauthier-Villars, Paris (1913). See also J. Dieudonné, *History of functional analysis*, North-Holland, Amsterdam (1981). With Riesz we begin to see the development of an "abstract" operator theory, in which the special example of Sturm-Liouville differential operators was to play a central role.

[13] Von Neumann established a mathematical framework for quantum theory in his book *Mathematische Grundlagen der Quantenmechanik*, Springer, Berlin (1932). An English translation appeared as *Mathematical Foundations of Quantum Mechanics*, Princeton University Press (1955).

quantum observable was a self-adjoint operator, the spectrum of which represented the range of physically realizable values of the observable. Both point spectrum and continuous spectrum were important – in the case of the hydrogen atom the energy spectrum had both discrete and continuous components, the discrete points (eigenvalues of the corresponding Schrödinger operator) agreeing closely with observed energy levels of hydrogen, and the continuous spectrum corresponding to states of positive energy.

Von Neumann quickly saw the implications for quantum mechanics of the new theory, and played a major part in developing the correspondence between physical theory and the analysis of operators and operator algebras. Physics and mathematical theory were able to develop in close parallel for many years, greatly to the advantage of both. He developed to a high art the spectral theory of self-adjoint and normal operators in abstract Hilbert space. A complete spectral analysis of self-adjoint operators in Hilbert space, generalizing the earlier results of Riesz, was just one outcome of this work, and a highly significant one for quantum theory. Similar results were independently discovered by Marshall Stone, who expounded the theory in his book published in 1932. (See the first article by Everitt.)

Of central importance for the future development of applications to mathematical physics, particularly in scattering theory which existed already in embryonic form in the work of Heisenberg, was the realization that the Lebesgue decomposition of measures into its singular and absolutely continuous (with respect to Lebesgue measure) components led to an analogous decomposition of the Hilbert space into singular and absolutely continuous subspaces for a given self-adjoint operator. Moreover, these two subspaces are mutually orthogonal. The singular subspace may itself be decomposed into two orthogonal components, namely the subspace of discontinuity, spanned by eigenvectors, and the subspace of singular continuity. Physical interpretations have been found for all of these subspaces, though in most applications only the discontinuous and absolutely continuous subspaces are non-trivial. In the case of the Hamiltonian (energy operator) for a quantum particle subject to a Coulomb force, the discontinuous subspace is the subspace of negative energy states and describes bound states of the system, whereas the absolutely continuous subspace corresponds to scattering states, which have positive energy.

The influence of the work of Charles Sturm and his close friend and collaborator Joseph Liouville may be found in the numerous modern developments of the theory which bears their names. A principal aim of this volume is to follow in detail the evolution of the theory since its early days, and to present an overview of the most important aspects of the theory as it stands today at the beginning of the twenty-first century.

We are grateful indeed to Norrie Everitt for his contributions to this volume, as author of two articles and coauthor of another. Over a long mathematical career, he has played an important role in the continuing progress of Sturm-Liouville theory.

The first of Norrie's articles in this volume deals with the development of Sturm-Liouville theory up to the year 1950, and covers in particular the work of

Weyl, Stone and Titchmarsh, of whom Norrie was himself a one-time student. (He also had the good fortune, on one occasion, to have encountered Weyl, who was visiting Titchmarsh at the time.)

Don Hinton's article is concerned with a series of results which follow from Sturm's original oscillation theorems developed in 1836 for second-order equations. Criteria are obtained for the oscillatory nature of solutions of the differential equation, and implications for the point spectrum are derived. Extensions of the theory to systems of equations and to higher-order equations are described.

Joachim Weidmann's contribution considers the impact of functional analysis on the spectral theory of Sturm-Liouville operators. Starting from ideas of resolvent convergence, it is shown how spectral behavior for singular problems may in appropriate cases be derived through limiting arguments from an analysis of regular problems. Conditions are obtained for the existence (or non-existence) of absolutely continuous spectrum in an interval.

Spectral properties of Sturm-Liouville operators are often derived, directly or indirectly, as a consequence of an established link between large distance asymptotic behavior of solutions of the associated differential equation and spectral properties of the corresponding differential operator. In the case of complex spectral parameter, the existence of solutions which are square-integrable at infinity may be described by the values of an analytic function, known as the Weyl-Titchmarsh $m$-function or $m$-coefficient, and spectral properties of Sturm-Liouville operators may be correlated with the boundary behavior of the $m$-function close to the real axis. The article by Daphne Gilbert explores further the link between asymptotics and spectral properties, particularly through the concept of subordinacy of solutions, an area of spectral analysis to which she has made important contributions.

A useful resource for readers of this volume, particularly those with an interest in numerical approaches to spectral analysis, will be the catalogue of Sturm-Liouville equations, compiled by Norrie Everitt with the help of colleagues. More than 50 examples are described, with details of their Weyl limit point/limit circle endpoint classification, the location of eigenvalues, other spectral information, and some background on applications. This collection of examples from an extensive literature should also provide a reference to some of the sources in which the interested reader can find further details of the theory and its applications, as well as numerical data on spectral properties.

In collaboration with Christer Bennewitz, Everitt has contributed a new version of the proof of the expansion theorem for general Sturm-Liouville operators, incorporating both continuous and discontinuous spectra.

The article by Barry Simon presents some recent results related to Sturm's oscillation theory for second-order equations. The cases of both Schrödinger operators and Jacobi matrices (which may be regarded as a discrete analogue of Schrödinger operators) are considered. A focus of this work is the establishment of a connection between the dimension of spectral projections and the number of zeros of appropriate functions defined in terms of solutions of the Schrödinger equation. Some deep results in spectral theory follow from this analysis, and there are links with the theory of orthogonal polynomials on the unit circle.

Yoram Last has provided a review of progress over recent years in spectral theory for discrete and continuous Schrödinger operators. Of particular interest has been the progress in analysis of spectral types, with a finer decomposition of spectral measures than hitherto, and the development of new ways of characterizing absolutely continuous and singular continuous spectrum.

Rafael del Río's article is an exposition of recent results relating to the influence of boundary conditions on spectral behavior. For Schrödinger operators, a change of boundary condition will not affect the location of absolutely continuous spectrum, whereas the nature of singular spectrum may be profoundly influenced by choice of boundary conditions.

In view of the major influence that Sturm-Liouville theory has had over the years on the development of spectral theory for linear differential equations, it is not surprising that there have been many attempts to extend the ideas and methods to nonlinear equations. Chao-Nien Chen describes some recent results in the nonlinear theory, with particular emphasis on the characterization of nodal sets, an area related to Sturm's original ideas on oscillation criteria in the linear case.

Another productive area of research into Sturm-Liouville theory is the extension of the theory to partial differential equations. Sturm had himself published results on zero sets for parabolic linear partial differential equations in a paper of 1836. In their contribution to this volume, Victor Galaktionov and Petra Harwin survey recent progress in this area, including extensions to some quasilinear equations.

A continuing and flourishing branch of spectral theory, with applications in many areas, is that of inverse spectral theory. The aim of inverse theory is to derive the Sturm-Liouville equation from its spectral properties. An early example of this kind of result was the proof, due originally to Borg in 1946, that for the Schrödinger equation with potential function $q$ over a finite interval and subject to boundary conditions at both endpoints, the spectrum for the associated Schrödinger operator for two distinct boundary conditions at one endpoint (and given fixed boundary condition at the other endpoint) is sufficient to determine $q$ uniquely. This result has been greatly extended over recent years, for example to systems of differential equations, and some of the more recent developments are treated in the survey by Mark Malamud.

We believe that the contents of this book will confirm that Sturm-Liouville theory has, indeed, a very rich Past and a most active and influential Present. It is our hope, too, that the book will help to contribute to a continuing productive Future for this fundamental branch of mathematics and its applications.

*W.O. Amrein, A.M. Hinz, D.B. Pearson*
Sturm-Liouville Theory: Past and Present, 1–27


# Sturm's 1836 Oscillation Results Evolution of the Theory

Don Hinton

*This paper is dedicated to the memory of Charles François Sturm*

**Abstract.** We examine how Sturm's oscillation theorems on comparison, separation, and indexing the number of zeros of eigenfunctions have evolved. It was Bôcher who first put the proofs on a rigorous basis, and major tools of analysis where introduced by Picone, Prüfer, Morse, Reid, and others. Some basic oscillation and disconjugacy results are given for the second-order case. We show how the definitions of oscillation and disconjugacy have more than one interpretation for higher-order equations and systems, but it is the definitions from the calculus of variations that provide the most fruitful concepts; they also have application to the spectral theory of differential equations. The comparison and separation theorems are given for systems, and it is shown how they apply to scalar equations to give a natural extension of Sturm's second-order case. Finally we return to the second-order case to show how the indexing of zeros of eigenfunctions changes when there is a parameter in the boundary condition or if the weight function changes sign.

**Mathematics Subject Classification (2000).** Primary 34C10; Secondary 34C11, 34C20.

**Keywords.** separation, comparison, disconjugate, oscillatory, conjugate point.

## 1. Introduction

In a series of papers in the 1830's, Charles Sturm and Joseph Liouville studied the qualitative properties of the differential equation

$$\frac{d}{dx}\left(K\frac{dV}{dx}\right) + GV = 0, \quad \text{for} \quad x \geq \alpha \tag{1.1}$$

where $K$, $G$, and $V$ are real functions of the two variables $x$, $r$. Their work began research into the qualitative theory of differential equations, i.e., the deduction of properties of solutions of the differential equation directly from the equation and without benefit of knowing the solutions. However, it was half a century before significant interest in the qualitative theory took hold. In (1.1) and elsewhere, we consider only real solutions unless otherwise indicated.

In more modern notation (for spectral theory it is convenient to have the leading coefficient negative; for the oscillation results of Sections 2 and 3, we return to the convention of positive leading coefficient), (1.1) would be written as

$$-(py')' + qy = 0, \quad x \in I, \tag{1.2}$$

or as (when eigenvalue problems are studied )

$$-(py')' + qy = \lambda wy, \quad x \in I, \tag{1.3}$$

where the real functions $p, q, w$ satisfy

$$p(x),\, w(x) > 0 \text{ on } I\ , 1/p, q, w \in L_{\text{loc}}(I), \tag{1.4}$$

where $L_{\text{loc}}(I)$ denotes the locally Lebesgue integrable functions on $I$. These are the minimal conditions the coefficients must satisfy for the initial value problem,

$$-(py')' + qy = 0, \quad x \in I, \quad y(a) = y_0, \quad y'(a) = y_1,$$

to have a unique solution. Sturm imposed no conditions on his coefficients, but was perhaps thinking of continuous coefficients. It is fair to say that thousands of papers have been written concerning the properties of solutions of (1.2), and hundreds more are published each year. Tony Zettl has called (1.2) the world's most popular differential equation. A recent check in math reviews shows 8178 entries for the word "oscillatory", 3284 entries for "disconjugacy", 1412 entries for "non-oscillatory", and even 62 for "Picone identity". The applications of (1.2) and (1.3) are ubiquitous. Their appearance in problems of heat flow and vibrations were well known since the work of Fourier. They play an important role in quantum mechanics where the problems are singular in the sense that $I$ is an interval of infinite extent or where at a finite endpoint a coefficient fails to satisfy certain integrability conditions. Today we can find numerically with computers the solutions of (1.2) or the eigenvalues and eigenfunctions associated with (1.3). However, even with current technology, there are still problems which give computational difficulty such as computing two eigenvalues which are close together. Codes such as SLEIGN2 [9] (developed by Bailey, Everitt, and Zettl) or the NAG routines give quickly and accurately the eigenvalues and eigenfunctions of large classes of Sturm-Liouville problems. The recent text by Pryce [85] is devoted to the numerical solution of Sturm-Liouville problems.

For (1.1), Sturm imposed a condition ($h(r)$ is a given function),

$$\frac{K(\alpha,r)}{V(\alpha,r)}\frac{\partial V(\alpha,r)}{\partial x} = h(r), \tag{1.5}$$

and obtained the following central result [94] (after noting that when the values of $V(\alpha,r)$, $\partial V(\alpha,r)/\partial x$ are given, the solution $V(x,r)$ is uniquely determined). We have also used Lützen's translation [74].

**Theorem A.** *If $V$ is a nontrivial solution of* (1.1) *and* (1.5), *and if for all $x \in [\alpha, \beta]$,*

1. *$K > 0$ for all $r$ and $K$ is a decreasing function of $r$,*
2. *$G$ is an increasing function of $r$,*
3. *$h(r)$ is a decreasing function of $r$,*

*then $\left(\frac{K}{V}\frac{\partial V}{\partial x}\right)$ is a decreasing function of $r$ for all $x \in [\alpha, \beta]$.*

Here decreasing or increasing means strictly. If $V(\alpha, r) = 0$, then $h(r)$ decreasing means $\partial V/\partial x \cdot \partial V/\partial r < 0$ at $x = \alpha$. Sturm's method of proof of Theorem A was to differentiate (1.1) with respect to $r$, multiply this by $V$, and then subtract this from $\partial V/\partial r$ times (1.1). After an integration by parts over $[\alpha, x]$, the resulting equation obtained is

$$\left(-V^2 \frac{\partial}{\partial r}\left(\frac{K}{V}\frac{\partial V}{\partial x}\right)\right)(x) = \left(-V^2(\alpha, r)\frac{dh}{dr}\right) + \int_\alpha^x \left[\frac{\partial G}{\partial r}V^2 - \frac{\partial K}{\partial r}\left(\frac{\partial V}{\partial r}\right)^2\right], \tag{1.6}$$

where we have used

$$-V^2 \frac{\partial}{\partial r}\left(\frac{K}{V}\frac{\partial V}{\partial x}\right) = K\frac{\partial V}{\partial x}\frac{\partial V}{\partial r} - V\frac{\partial}{\partial r}\left(K\frac{\partial V}{\partial x}\right). \tag{1.7}$$

If we solve this equation for the term $\frac{\partial}{\partial r}\left(\frac{K}{V}\frac{\partial V}{\partial x}\right)(x)$, then we get

$$\frac{\partial}{\partial r}\left(\frac{K}{V}\frac{\partial V}{\partial x}\right)(x, r) < 0, \tag{1.8}$$

which completes the proof.

An examination of the above proof shows that the same conclusion can be reached with less restrictive hypotheses. With $K > 0$, an examination of the right-hand side of (1.6) shows that it is positive, and hence (1.8) holds under any one of the following three conditions.

$$\frac{\partial G}{\partial r} > 0, \ \frac{\partial K}{\partial r} \le 0, \ \frac{dh}{dr} \le 0, \tag{1.9}$$

$$\frac{\partial G}{\partial r} \ge 0, \ \frac{\partial K}{\partial r} \le 0, \ \frac{dh}{dr} < 0, \tag{1.10}$$

$$\frac{\partial G}{\partial r} \ge 0, \ \frac{\partial K}{\partial r} < 0, \ \frac{dh}{dr} \le 0, \ V \text{ is not constant.} \tag{1.11}$$

Theorem A has immediate consequences. The first is that if $x(r)$ denotes a solution of $V(x, r) = 0$, then by implicit differentiation, we get from (1.7) and (1.8) that

$$\frac{dr}{dx} = -\frac{\partial V}{\partial x}/\frac{\partial V}{\partial r} < 0. \tag{1.12}$$

Note that this implies under the conditions of Theorem A, that the roots $x(r)$ of $V(x, r)$ are decreasing with respect to $r$. With $K > 0$ the same conclusion may be reached by replacing the hypothesis of Theorem A with (1.9), (1.10), or (1.11).

By considering two equations, $(K_iV_i')' + G_iV_i = 0$, $i = 1, 2$, with $G_2(x) \geq G_1(x)$, $K_2(x) \leq K_1(x)$ and embedding the functions $h_1$, $h_2$, $G_1$, $G_2$ and $K_1$, $K_2$ into a continuous family, e.g., one can define

$$\hat{G}(r, x) = rG_2(x) + (1 - r)G_1(x),\ 0 \leq r \leq 1,$$

and similarly for $K$, Sturm was able to prove comparison theorems. In particular he proved

**Theorem B (Sturm's Comparison Theorem).** *For $i = 1, 2$ let $V_i$ be a nontrivial solution of $(K_iV_i')' + G_iV_i = 0$. Suppose further that with $h_i = (K_iV_i'/V_i)(\alpha)$,*

$$h_2 < h_1, \quad G_2(x) \geq G_1(x), \quad K_2(x) \leq K_1(x), \quad x \in [\alpha, \beta].$$

*Then if $\alpha, \beta$ are two consecutive zeros of $V_1$, the open interval $(\alpha, \beta)$ will contain at least one zero of $V_2$.*

In case $V_i(\alpha) = 0$, the proper interpretation of infinity must be made.

This version of comparison corresponds to using the hypothesis (1.10). Other versions may be proved by using either (1.9) or (1.11). Perhaps the most widely stated version of Sturm's comparison theorem (not the version he proved) may be stated as follows.

**Theorem B*.** *For $i = 1, 2$ let $V_i$ be a nontrivial solution of $(K_iV_i')' + G_iV_i = 0$ on $\alpha \leq x \leq \beta$. Suppose further that the coefficients are continuous and for $x \in [\alpha, \beta]$,*

$$G_2(x) \geq G_1(x), \text{ with } G_2(x_0) > G_1(x_0) \text{ for some } x_0, \quad K_2(x) \leq K_1(x).$$

*Then if $\alpha, \beta$ are two consecutive zeros of $V_1$, the open interval $(\alpha, \beta)$ will contain at least one zero of $V_2$.*

Sturm's methods also yielded (in modern terminology):

**Theorem C (Sturm's Separation Theorem).** *If $V_1$, $V_2$ are two linearly independent solutions of $(KV')' + GV = 0$ and a,b are two consecutive zeros of $V_1$, then $V_2$ has a zero on the open interval $(a, b)$.*

The final result of Sturm that we wish to quote concerns the zeros of eigenfunctions and is proved in his second memoir [95]. Here he considered the eigenvalue problem,

$$(k(x)V'(x))' + [\lambda g(x) - l(x)]V(x) = 0, \quad \alpha \leq x \leq \beta, \tag{1.13}$$

with separated boundary conditions,

$$k(\alpha)V'(\alpha) - hV(\alpha) = 0, \quad k(\beta)V'(\beta) + HV(\beta) = 0. \tag{1.14}$$

Further the functions $k, g$, and $l$ are assumed positive. Some properties he established are:

**Theorem D.** *There are infinitely many real simple eigenvalues $\lambda_1, \lambda_2, \dots$ of (1.13) and (1.14), and if $V_1, V_2, \dots$ are the corresponding eigenfunctions, then for $n = 1, 2, \dots,$*

1. *$V_n$ has exactly $n - 1$ zeros in the open interval $(\alpha, \beta)$,*
2. *between two consecutive zeros of $V_{n+1}$ there is exactly one zero of $V_n$.*

Theorem D relates to the spectral theory of the operator associated with (1.13) and (1.14). For (1.2) considered on an infinite interval $I = [a, \infty)$, an eigenvalue problem, in order to define a self-adjoint operator, may only require one boundary condition at $a$ (limit point case at infinity), or it may require two boundary conditions involving both $a$ and infinity (limit circle case at infinity). This dichotomy was discovered by Weyl. In the limit point case with $w \equiv 1$, a self-adjoint operator is defined in the Hilbert space $L^2(a, \infty)$ of Lebesgue square integrable functions by

$$L_\alpha[y] = -(py')' + qy, \quad y \in \mathcal{D},$$

where

$$\mathcal{D} = \{y \in L^2(a, \infty) : y,\, py' \in AC_{loc}, L_\alpha[y] \in L^2(a, \infty), \\ y(a) \sin \alpha - (py')(a) \cos \alpha = 0\}, \quad (1.15)$$

and $AC_{loc}$ denotes the locally absolutely continuous functions.

Unlike the case (1.13) and (1.14) for the compact interval, the spectrum for the infinite interval may contain essential spectrum, i.e., numbers $\lambda$ such that $L_\alpha - \lambda I$ has a range that is not closed, and Theorem D does not apply. However in the case of a purely discrete spectrum bounded below, a version of Theorem D carries over to the operator $L_\alpha$ above in the relation of the index of the eigenvalue to the number of zeros of the eigenfunction in $(a, \infty)$ [22]. In general, one can say that the number of points in the spectrum of $L_\alpha$ below a real number $\lambda_0$ is infinite if and only if the equation $-(py')' + qy = \lambda_0 y$ is oscillatory, i.e., the solutions have infinitely many zeros on $[a, \infty)$. This same result carries over to self-adjoint equations of arbitrary order if the definition of oscillation in Section 4 is used [80, 99]. This basic connection has been used extensively in spectral theory. Note that if $-(py')' + qy = \lambda_0 y$ is non-oscillatory for every $\lambda_0$, then the spectrum of $L_\alpha$ consists only of a sequence of eigenvalues tending to infinity. Theorem D and its generalizations have also important numerical consequences. When an eigenvalue is computed, it allows one to be sure which eigenvalue it is, i.e., just count the zeros of the eigenfunction. It also allows the calculation of an eigenvalue without first calculating the eigenvalues that precede it. This feature is built into some eigenvalue codes.

A number of monographs deal almost exclusively with the oscillation theory of linear differential equations and systems. The books of Coppel [24] and Reid [88] emphasize linear Hamiltonian systems, but also contain substantial material on the second-order case. Coppel contains perhaps the most concise treatment of Hamiltonian systems; Reid is the most comprehensive development of Sturm theory. The book of Elias [29] is based on the oscillation and boundary value problem theory for two term ordinary differential equations, while Greguš [38] deals entirely with third-order equations. The text by Kreith [62] includes abstract oscillation theory as well as oscillation theory for partial differential equations. Finally the classic book by Swanson [96] has special chapters on second, third, fourth-order ordinary differential equations as well as results for partial differential equations. The reader is also referred to the survey papers of Barrett [10] and Willett [100]. The books by Atkinson [8], Glazman [37], Hartman [44], Ince [53], Kratz [61], Müller-Pfeiffer [80], and Reid [86] contain many results on oscillation theory.

As noted, the literature on the Sturm theory is voluminous. There are extensive results on difference equations, delay and functional differential equations, and partial differential equations. The Sturm theory for difference equations is similar to that of ordinary differential equations, but contains many new twists. The book by Ahlbrandt and Peterson [6] details this theory (see also the text by B. Simon in the present volume). Oscillation results for delay and functional equations as well as further work on difference equations can be found in the books by Agarwal, Grace, and O'Regan [1, 2], I. Gyori and G. Ladas [39], and L. Erbe, Q. Kong, and B. Zhang [31]. We confine ourselves to the case of ordinary differential equations and at that we are only able to pursue a few themes.

The comparison and oscillation theorems of Sturm have remained a topic of considerable interest. While the extensions and generalizations have much intrinsic interest, we believe their continued relevance is due in no small part to their intimate connection with problems of physical origin. Particularly the connections with the minimization problems of the calculus of variations and optimal control as well as the spectral theory of differential operators are important. We will discuss some of these connections below. We will trace some of the developments that have occurred with respect to the comparison and separation theorems as well as other developments related to Theorem D. The tools introduced by Picone, Prüfer, and the variational methods will be discussed and their applications to second-order equations as well as to higher-order equations and systems. Sample results will be stated and a few short and elegant proofs will be given. The problem of extending Sturm's results to systems was only considered about one hundred years after Sturm; the work of Morse was fundamental in this development. It is interesting that it was variational theory which gave the most natural and fruitful generalization of the definitions of oscillation. In a very loose way, we show that the theme of largeness of the coefficient $q$ in $(py')' + qy = 0$ leads to oscillation in not only the second-order, but also higher-order equations, while $q \leq 0$, or $|q|$ small leads to disconjugacy.

## 2. Extensions and more rigor

Sturm's proofs of course do not meet the standards of modern rigor. They meet the standards of his time, and are in fact correct in method and can without too much trouble be made rigorous. The first efforts to do this are due to Bôcher in a series of papers in the Bulletin of the AMS [17] and are also contained in his book [18]. Bôcher [17] remarks that "the work of Sturm may, however, be made perfectly rigorous without serious trouble and with no real modification of method". The conditions placed on the coefficients were to make them piecewise continuous. Bôcher used Riccati equation techniques in some of his proofs; we note that Sturm mentions the Riccati equation, but does not employ it in his proofs. Riccati equation techniques in variational theory go back at least to Legendre who in 1786 gave a flawed proof of his necessary condition for a minimizer of an integral functional. A correct proof of Legendre's condition using Riccati equations can be found in Bolza's 1904 lecture notes [19]. Bolza attributes this proof to Weierstrass.

Bôcher was also motivated by the oscillation theorem of Klein [58] which is a multiparameter version of Sturm's existence proof for eigenvalues. Bôcher [17] noted that Klein "had given rough geometrical proofs which however made no pretence at rigor". The general form of Klein's problem may be stated as follows, see Ince [53, p. 248]. Suppose in (1.2), $q$ is of the form

$$q(x) = -l(x) + [\lambda_0 + \lambda_1 x + \cdots + \lambda_n x^n]g(x),$$

where $p$, $l$, $g$ are continuous with $p(x), g(x) > 0$. Further let there be $n+1$ intervals $[a_0, b_0], \ldots, [a_n, b_n]$ with $a_0 < b_0 < a_1 < \cdots < a_n < b_n$. Suppose $m_s$, $s = 0, \ldots, n$ are given nonnegative integers and on each interval $[a_s, b_s]$, separated boundary conditions of the form (1.14) are given. Then there exist a set of simultaneous characteristic numbers $\lambda_0, \ldots, \lambda_n$ and corresponding functions $y_0, \ldots, y_n$ such that on each $[a_s, b_s]$, $y_s$ has $m_s$ zeros in $(a_s, b_s)$ and satisfies the boundary conditions for $[a_s, b_s]$. Klein was interested in the two parameter Lamé equation

$$y'' + \frac{1}{2}\left[\frac{1}{x-e_1} + \frac{1}{x-e_2} + \frac{1}{x-e_3}\right]y' - \frac{Ax+B}{4(x-e_1)(x-e_2)(x-e_3)}y = 0$$

because of its application to physics. The text by Halvorsen and Mingarelli [40] deals with the oscillation theory of the two parameter case.

The proofs of Sturm's theorems depend on existence-uniqueness results for (1.2), and Norrie Everitt has brought to our attention that it was Dixon [25] who first proved that these are valid under only the assumption that the coefficients $1/p$, $q$ are Lebesgue integrable functions. The details of Dixon's work may be found in N. Everitt's text in the present volume. Later Carathéodory generalized the concept of a solution of a system of differential equations to only require the equation hold almost everywhere. When (1.2) is written in system form, the Dixon and Carathéodory conditions are the same. Richardson [89, 90] extended the results of counting zeros of eigenfunctions further by allowing the weight $g(x)$ in (1.13) to not be of constant sign and called this the non-definite case. We will return to his case in Section 5. Part (1) of Theorem D, which is for the separated boundary conditions (1.14), was extended by Birkhoff [16] to the case of arbitrary self-adjoint boundary conditions.

To simplify our discussion, we will henceforth assume that all coefficients and matrix components are real and piecewise continuous unless otherwise stated.

Thinking of examples like $y'' + ky = 0, k > 0$, whose solutions are sines and cosines or the Euler equation $y'' + kx^{-2}y = 0$ which has oscillatory solutions if and only if $k > 1/4$, it is natural to pose the problem:

$$\text{When are all solutions of } (py')' + qy = 0 \text{ oscillatory on } I? \tag{2.1}$$

We use the term *oscillatory* (*non-oscillatory*) here in the sense of infinitely (finitely) many zeros for all nontrivial solutions. Because of the Sturm separation theorem, if one nontrivial solution has infinitely many zeros, then all do, but this property fails for nonlinear equations. A second problem, not quite so obvious, but which arose naturally from the calculus of variations, is

$$\text{When is the equation } (py')' + qy = 0 \text{ disconjugate on } I? \tag{2.2}$$

The term *disconjugate* is used here to mean that no nontrivial solution has more than one zero on $I$. If a nontrivial solution of $(py')' + qy = 0$ has a zero at $a$, then the first zero of $y$ to the right of $a$ is called the first right *conjugate point* of $a$; if there are no zeros to the right of $a$, then we say the equation is right disconjugate. Successive zeros are isolated and hence yield a counting of conjugate points. If $y$ satisfies $y'(a) = 0$, then the first zero of $y$ to the right of $a$ is called the first right *focal point* of $a$. If $y$ has no zeros to the right of $a$, then $(py')' + qy = 0$ is called right *disfocal*. Similar definitions are made to the left. The simplest criterion for both right disconjugate and disfocal is for $q(x) \le 0$, for then an easy argument shows $y$ is monotone if $y(a) \ge 0$, $y'(a) \ge 0$. On a compact or open interval $I$ disconjugacy is equivalent to there being a solution of $(py')' + qy = 0$ with no zeros on $I$ [24, p.5]. For a half-open interval $(py')' + qy = 0$ can be disconjugate without there being a solution with no zeros as is shown by the equation $y'' + y = 0$ on $[0, \pi)$ which is disconjugate, but every solution has a zero in $[0, \pi)$.

A major advance was made by Picone [83] in his 1909 thesis. He discovered the identity

$$\left[\frac{u}{v}\left(vpu' - uPv'\right)\right]' = u(pu')' - \frac{u^2}{v}(Pv')' + (p - P)\,u'^2 + P\left(u' - \frac{u}{v}v'\right)^2 \tag{2.3}$$

which holds when $u$, $v$, $pu'$, and $Pv'$ are differentiable and $v(x) \neq 0$. In case $u$, $v$ are solutions of the differential equations

$$(pu')' + qu = 0, \qquad (Pv')' + Qv = 0,$$

(2.3) reduces to

$$\left[\frac{u}{v}\left(vpu' - uPv'\right)\right]' = (Q - q)\,u^2 + (p - P)\,u'^2 + P\left(u' - \frac{u}{v}v'\right)^2. \tag{2.4}$$

With this identity one can give an elementary proof of Sturm's comparison Theorem B* which we now give. Suppose $p(x) \ge P(x)$, $Q(x) \ge q(x)$ with $Q(x_0) > q(x_0)$ at some $x_0$, $\alpha$, $\beta$ are consecutive zeros of a nontrivial solution $u$ of $(pu')' + qu = 0$, and that $v$ is a solution of $(Pv')' + Qv = 0$ with no zeros in the open interval $(\alpha, \beta)$. Note the quotient $u(x)/v(x)$ has a limit at the endpoints. For example the limit at $\alpha$ is zero if $v(\alpha) \neq 0$, and the limit is $u'(\alpha)/v'(\alpha)$ if $v(\alpha) = 0$. Integration of (2.4) over $[\alpha, \beta]$ yields that the left-hand side integrates to zero while the right-hand side integrates to a positive number. This contradiction proves the theorem.

Another major advance was made by Prüfer [84] with the use of trigonometric substitution. In the equation $(pu')' + (q + \lambda w)u = 0$, he made the substitution

$$u = \rho \sin\theta, \qquad pu' = \rho\cos\theta,$$

and then proved that $\rho, \theta$ satisfy the differential equations

$$\theta' = \frac{1}{p}\cos^2\theta + (q + \lambda w)\sin^2\theta, \quad \rho' = (\frac{1}{p} - q - \lambda w)(\sin\theta\cos\theta)\rho.$$

The zeros of the solution $u$ are given by the values of $x$ such that $\theta(x) = n\pi$ for some integer $n$. The equation for $\theta$ is independent of $\rho$, and by using a first-order comparison theorem for nonlinear equations, it is possible to establish Sturm's comparison theorem. Prüfer used the equation for $\theta$ to establish the link stated in Theorem D between the number of zeros of an eigenfunction and the corresponding eigenvalue. These equations can also be used to prove the existence of infinitely many eigenvalues. This is the method used in most textbooks today for the proof of Theorem D.

Note that with Prüfer's transformation, the equation $(py')' + qy = 0$, $a \leq x < \infty$, is oscillatory if and only if $\theta(x) \to \infty$ as $x \to \infty$. It also follows easily from this transformation that

$$\int_a^\infty \left[\frac{1}{p} + |q|\right] dx < \infty \Rightarrow \text{ non-oscillation,}$$

$$\int_a^\infty \left[\frac{1}{p} + |q|\right] dx < \pi \Rightarrow \text{ disconjugacy.}$$

Kamke [56] used the trigonometric substitution technique to prove a Sturm type comparison theorem for a system of first-order equations

$$y' = Py + Qz, \quad z' = Ry + Sz$$

where the coefficients are continuous functions.

Klaus and Shaw [57] used the Prüfer transformation to study the eigenvalues of a Zakharov-Shabat system. One of their results shows that the first-order system

$$v_1' = sv_1 + q(t)v_2, \quad v_2' = -sv_2 - q(t)v_1,$$

is (in our terminology ) right disfocal on $-d \leq t \leq d$ if $\int_{-d}^{d} |q(t)|dt \leq \pi/2$; moreover the constant $\pi/2$ is sharp. Extension is then made to the interval $(-\infty, \infty)$ and for complex-valued $q$. Application is made to the nonexistence of eigenvalues (s is the eigenparameter) of the Zakharov-Shabat system, and hence to the nonexistence of soliton solutions of an associated nonlinear Schrödinger equation.

Sturm's comparison Theorem B* has been generalized to include integral comparisons of the coefficients. Consider the two equations, for $a \leq x < \infty$,

$$y'' + q_1(x)y = 0, \tag{2.5}$$

$$y'' + q_2(x)y = 0. \tag{2.6}$$

Then we may phrase Sturm's comparison theorem by:

If $q_1(x) \leq q_2(x)$, $a \leq x < \infty$, then (2.6) disconjugate $\Rightarrow$ (2.5) disconjugate.

This result was extended by Hille [50] (as generalized by Hartman [44, p. 369]) to read:

$$\text{If } \int_t^\infty q_1(x)dx \le \int_t^\infty q_2(x)dx,\ a \le t < \infty,$$

$$\text{then (2.6) disconjugate} \Rightarrow \text{ (2.5) disconjugate.}$$

Further results of this nature were given by Levin [67] and Stafford and Heidel [92].

## 3. Some basic oscillation results

The first major attack on problem (2.1) seems to have been made in 1883 by Kneser [59] who studied the higher-order equation $y^{(n)} + qy = 0$, and proved that all solutions oscillate an infinite number of times provided that $x^m q(x) > k > 0$ for all sufficiently large values of $x$, where $n \ge 2m > 0$ and $n$ is even. Of course for $n = 2$, this follows immediately from the Sturm comparison theorem applied to the oscillatory Euler equation $y'' + kx^{-2}y = 0$, $k > 1/4$, since $k/x^2 \le k/x$ for $x \ge 1$. Hubert Kalf has noted that Weber [98] refined Kneser's result to decide on oscillation or non-oscillation in the case where $x^2q(x)$ tends to a limit as $x$ tends to infinity. The Kneser criterion has recently been extended by Gesztesy and Ünal [36].

A result which subsequently received a lot of attention was proved by Fite [33] in studying the equation $y^{(n)} + py^{(n-1)} + qy = 0$ on a ray $x \ge x_1$. Fite's result was if $q \ge 0$, $\int_{x_1}^\infty q dx = \infty$ and $y$ is a solution of $y^{(n)} + qy = 0$, then $y$ must change sign an infinite number of times in case $n$ is even, and in case $n$ is odd such a solution must either change sign an infinite number of times or not vanish at all for $x \ge x_1$. For $n = 2$ we then have a sufficient condition for (2.1), i.e.,

$$q(x) \ge 0, \quad \int_{x_1}^\infty q(x)dx = \infty \Rightarrow y'' + qy = 0 \text{ is oscillatory.}$$

This theme of $q(x)$ being sufficiently large has reoccurred in oscillation theory in many situations. The first improvement of the Fite result was due to Wintner [101] who removed the sign restriction on $q(x)$ and proved the stronger result

$$t^{-1}\int^t q(x)(t-x)dx \to \infty \text{ as } t \to \infty \Rightarrow y'' + qy = 0 \text{ is oscillatory.}$$

Independently Leighton [64] proved, for $(py')' + qy = 0$, that

$$\int^\infty \frac{dx}{p(x)} = \infty, \quad \int^\infty q(x)dx = \infty \Rightarrow (py')' + qy = 0 \text{ is oscillatory.}$$

Again there is no sign restriction on $q(x)$.

An elegant proof of this Fite-Wintner-Leighton result has been given by Coles [23]. We give this proof since it a good illustration of Riccati equation techniques.

Suppose that $\int^\infty p^{-1}\,dx = \infty$, $\int^\infty q\,dx = \infty$, and that $u$ is a non-oscillatory solution of $(pu')' + qu = 0$, say $u(x) > 0$ on $[b,\infty)$. Define $r = pu'/u$. Then a calculation shows that $r' = -q - r^2/p$, and hence for large $x$, say $x \geq c$,

$$r(x) + \int_b^x \frac{r^2}{p}\,dt = r(b) - \int_b^x q\,dx < 0.$$

This implies that $r(x) < -\int_b^x p^{-1}r^2\,dt$. Thus defining $R(x) = \int_b^x p^{-1}r^2\,dt$, one has that for $x \geq c$, $R' = r^2/p \geq R^2/p$. Integration of this inequality gives

$$\int_c^x \frac{1}{p}\,dt \leq \int_c^x \frac{R'}{R^2}\,dt = \frac{1}{R(c)} - \frac{1}{R(x)} \leq \frac{1}{R(c)}$$

which is contrary to $\int^\infty p^{-1}\,dx = \infty$.

Related to the above result of Wintner is that of Kamenev [55] who showed that if for some positive integer $m > 2$,

$$\limsup_{t\to\infty} \frac{1}{t^{m-1}} \int_a^t (t-s)^{m-1} q(s)\,ds = \infty,$$

then the equation $y'' + qy = 0$ is oscillatory on $[a,\infty)$. The Kamenev type results have been extended to operators with matrix coefficients and Hamiltonian systems by Erbe, Kong, and Ruan [30], Meng and Mingarelli [75], and others.

The mid-twentieth century saw a large number of papers written on problems (2.1) and (2.2). We mention a small sampling of these results.

**Theorem 3.1 (Hille, 1948).** *If $q(x) \geq 0$ is a continuous function on $I = [a,\infty)$, such that $\int_a^\infty q < \infty$, and*

$$g_* := \liminf_{x\to\infty} x \int_x^\infty q(t)dt, \quad g^* := \limsup_{x\to\infty} x \int_x^\infty q(t)dt,$$

*then $g^* > 1$ or $g_* > 1/4$ implies $y'' + qy = 0$ is oscillatory, and $g^* < 1/4$ implies $y'' + qy = 0$ is non-oscillatory.*

Hille's results have been extended to equations with matrix coefficients and linear Hamiltonian systems by Sternberg [93] and Ahlbrandt [3].

**Theorem 3.2 (Hartman, 1948).** *If $y'' + qy = 0$ is non-oscillatory on $[a,\infty)$, then there are solutions $u$, $v$ of $y'' + qy = 0$ such that*

$$\int^\infty u^{-2}(t)\,dt < \infty \quad \text{and} \quad \int^\infty v^{-2}(t)\,dt = \infty.$$

**Theorem 3.3 (Wintner, 1951).** *The equation $y'' + qy$ is non-oscillatory on $[a,\infty)$ if $\int_x^\infty q(t)\,dt$ converges and either $-3/4 \leq x\int_x^\infty q(t)\,dt \leq 1/4$ or $[\int_x^\infty q(t)\,dt]^2 \leq q(x)/4$.*

**Theorem 3.4 (Nehari, 1954).** *If $I = [a,\infty)$ and $\lambda_0(b)$ is the smallest eigenvalue of*

$$-y'' = \lambda c(x) y, \quad y(a) = y'(b) = 0,$$

*where $c(x) > 0$ is continuous on $I$, then $y'' + c(x)y = 0$ is non-oscillatory on $I$ iff $\lambda_0(b) > 1$ for all $b > a$.*

**Theorem 3.5 (Hartman-Wintner, 1954).** *The equation $y''+qy=0$ is non-oscillatory on $[a,\infty)$ if $f(x)=\int_x^\infty q(t)dt$ converges and the differential equation $v''+4f^2(x)v=0$ is non-oscillatory.*

**Theorem 3.6 (Hawking-Penrose, 1970).** *If $I=(-\infty,\infty)$ and $q(x)\geq 0$ is a continuous function on $I$ such that $q(x_0)>0$ for some $x_0$, then $y''+q(x)y=0$ is not disconjugate on $I$.*

A particularly simple proof of this result has been given by Tipler [97] which we now present. Suppose $y$ is the unique solution of $y''+q(x)y=0$ with the initial conditions $y(x_0)=1$, $y'(x_0)=0$. Then $y''(x_0)=-q(x_0)y(x_0)<0$, and further $y''(x)\leq 0$ as long as $y(x)\geq 0$. Since $y'(x_0)=0$, this concavity of $y$ implies that $y$ eventually has a zero both to the right and to the left of $x_0$.

Many results on oscillation can be expanded by making a change of independent and dependent variables of the form $y(x)=\mu(x)z(t)$, $t=f(x)$, where $\mu(x)$ and $f'(x)$ are nonzero on the interval $I$. In the case of $(py')'+qy$, this leads to

$$(py')'+qy=(\gamma/\mu)[\dot{w}+Qz], \quad w=P\dot{z}, \quad \gamma(x)=f'(x),$$

where $\dot{z}=dz/dt$ and

$$P(t)=p(x)\mu^2(x)\gamma(x), \quad Q(t)=\frac{\mu(x)}{\gamma(x)}\left[(p\mu')'+q\mu\right].$$

Applications of these ideas can be found in Ahlbrandt, Hinton, and Lewis [5].

To return to the concept of disconjugacy and the link to the calculus of variations, it was in 1837 that Jacobi [54] gave his sufficient condition for the existence for a (weak) minimum of the functional

$$J[y]=\int_a^b f(x,y,y')dx \tag{3.1}$$

over the class of admissible functions $y$ defined as those sufficiently smooth $y$ satisfying the endpoint conditions $y(a)=A$, $y(b)=B$. A necessary condition for an extremal is the vanishing of the first variation, $dJ(y+\epsilon\eta)/d\epsilon\rfloor_{\epsilon=0}$, for sufficiently smooth variations $\eta$ satisfying $\eta(a)=\eta(b)=0$. This leads to the Euler-Lagrange equation $f_y-d(f_{y'})/dx=0$ for $y$. A sufficient condition for a weak minimum is that the second variation

$$\delta^2 J(\eta)=\int_a^b\left[p\eta'^2+q\eta^2\right]dx \tag{3.2}$$

be positive for all nontrivial admissible $\eta$ where $p=f_{y'y}$ and $q=f_{yy}-d(f_{y'y})/dx$. Jacobi discovered that the positivity of (3.2) was related to the oscillation properties of $-(py')'+qy=0$. In particular he discovered (3.2) is positive if $-(py')'+qy=0$ has a solution $y$ which is positive on $[a,b]$. The condition of (3.2) being positive is equivalent to $-(py')'+qy=0$ being disconjugate on $[a,b]$. This is the principal connection of oscillation theory to the calculus of variations. This connection may be proved with Picone's identity as we now demonstrate.

First suppose (1.2) is disconjugate on $[a, b]$; hence there is a solution $v$ of (1.2) which is positive on [a,b]. Then (2.3) with $p = P$ yields for the variation $\eta$,

$$\left[\eta(p\eta') - \frac{\eta^2}{v}(Pv')\right]' = \eta(p\eta')' - \frac{\eta^2}{v}(Pv')' + p\left(\eta' - \frac{\eta}{v}v'\right)^2.$$

Simplifying this expression yields that

$$p\eta'^2 - \left[\frac{\eta^2}{v}(pv')\right]' = -q\eta^2 + p\left(\eta' - \frac{\eta}{v}v'\right)^2,$$

which one can verify directly using only one derivative for $\eta$. An integration and applying $\eta(a) = \eta(b) = 0$ gives that

$$0 \leq \int_a^b p\left(\eta' - \frac{\eta}{v}v'\right)^2 dx = \int_a^b \left(p\eta'^2 + q\eta^2\right) dx$$

with equality if and only if $\eta' = \eta v'/v$. But $\eta' = \eta v'/v$ implies $(\eta/v)' = 0$ or $\eta/v$ is constant. This is contrary to $\eta(a) = 0$, $v(a) \neq 0$. Hence $\delta^2 J(\eta)$ is positive. On the other hand if (1.2) is not disconjugate, there is a nontrivial solution $u$ with $u(c) = u(d) = 0$, $a \leq c < d \leq b$. By defining $\eta(x) = u(x)$, $c \leq x \leq d$, and $\eta(x) = 0$ otherwise, it follows that $\delta^2 J(\eta) = 0$ so that $\delta^2 J$ fails to be positive for all nontrivial admissible functions.

Leighton was able to exploit this equivalence to obtain comparison theorems, e.g., as in [65]. One of his results is that if there is a nontrivial solution $u$ in $[a, b]$ of $(pu')' + qu = 0$ such that $u(a) = u(b) = 0$, and

$$\int_a^b \left[(p - P)u'^2 + (Q - q)u^2\right] dx > 0,$$

then every solution of $(Pv')' + Qv = 0$ has at least one zero in (a,b). This has as a corollary Sturm's Comparison Theorem B*. Angelo Mingarelli has pointed out that the monotonicity condition on the $G$ coefficient in Sturm's comparison theorem has been replaced by a convexity condition by Hartman [45].

When the equivalence of disconjugacy of (1.2) to positivity of (3.2) is used to show oscillation, it is frequently done by a construction. That is, if (1.2) is considered on $I = [a, \infty)$, and it can be shown that for each $b > a$ there is a function $\eta_b$ with compact support in $[b, \infty)$ such that $\delta^2 J(\eta_b) \leq 0$, then (1.2) is oscillatory. When the equivalence is used to show disconjugacy, it is usually done by the use of inequalities which bound the integral $\int_a^b q\eta^2\, dx$ in terms of the integral $\int_a^b p\eta'^2\, dx$. For example, the Hardy inequality $\int_a^b x^{-2}u^2\, dx \leq (1/4)\int_a^b u'^2\, dx$ for functions $u$ satisfying $u(a) = u(b) = 0$, $a > 0$, can be used to show that $u'' + qu = 0$ is disconjugate on $[a, \infty)$, $a > 0$, if $|x \int_x^\infty q(t)dt| \leq 1/4$.

Oscillation theory in the complex domain, i.e., for an equation of the form

$$w''(z) + G(z)w(z) = 0, \quad z \in \mathcal{D}, \tag{3.3}$$

where $w(z)$ is a function analytic in the domain $\mathcal{D}$, did not begin until the end of the nineteenth century. The earliest work dealt with special functions which are themselves solutions of second-order linear differential equations. Hurwitz [52] in 1889 investigated the zeros of Bessel functions in the complex plane. Work soon followed on other special functions. The definitions of disconjugate and nonoscillatory are the same as in the real case although now there is no simple ordering of the zeros. The location of complex zeros has found recent application in the quantum mechanical problem of locating resonances and anti-bound states as in Brown and Eastham [20], Eastham [28], or Simon [91]. A fairly extensive analytic oscillation theory has been developed by Hille [49], Beesack [11], London [73], Nehari [81], and others. We state two such results.

**Theorem 3.7 (Nehari, 1954).** *If $G(z)$ is analytic in $|z| < 1$, then (3.3) is disconjugate in $|z| < 1$ if $|G(z)| \leq (1 - |z|^2)^{-2}$ in $|z| < 1$.*

**Theorem 3.8 (London, 1962).** *If $G(z)$ is analytic in $|z| < 1$, then (3.3) is disconjugate in $|z| < 1$ if*

$$\iint_{|z|<1} |G(z)| dx dy \leq \pi.$$

It is surprising that the oscillation theory on the real axis, especially the comparison theory, plays an important role in the analytic oscillation theory, cf., Beesack [11]. Analytic oscillation theory is also connected with the theory of univalent functions. If $f(z)$ is analytic in $\mathcal{D}$ and $G(z) = \{f(z), z\}/2$ where $\{f(z), z\}$ is the Schwarzian derivative of $f$, then the univalence of $f$ in $\mathcal{D}$ is equivalent to the disconjugacy of (3.3) in $\mathcal{D}$ [11]. A summary of the analytic oscillation theory can be found in the books by Hille [51] and Swanson [96].

A notable result on disconjugacy was given by Lyapunov in 1893 [71].

**Theorem 3.9 (Lyapunov).** *The equation $y'' + q(x)y = 0$ is disconjugate on $[a, b]$ if $(b - a) \int_a^b |q(x)| dx \leq 4$.*

Extensions of Lyapunov's theorem to systems in the Stieltjes integral setting have been made by Brown, Clark, and Hinton [21]; further the $L[a, b]$ norm on $q$ has been replaced by an $L_p[a, b]$ norm for $1 \leq p \leq 2$.

Disconjugacy theorems play an important role in the stability of differential equations with periodic coefficients. For $-y'' + qy = \mu y$ on $[0, \infty)$ with $q(t)$ periodic of period $T$, the equation is called stable if all solutions are bounded. This occurs if $\lambda_0 < \mu < \lambda_0^*$, where $\lambda_0$ is the first eigenvalue of $-y'' + qy = \lambda y$ with periodic boundary conditions, and where $\lambda_0^*$ is the first eigenvalue of $-y'' + qy = \lambda y$ with semi-periodic boundary conditions. The criterion of Krein/Borg [103, II, p. 729] (see also Eastham [27, p. 49]) states that $-y'' + qy = 0$ is stable if $\int_0^T q \leq 0$, $q \neq 0$, and $T \int_0^T q_- \leq 4$, where $q_-(t) = \max\{-q(t), 0\}$. The proof of this uses the fact that $T \int_0^T q_- \leq 4$ and $q$ periodic implies the spacing of zeros of solutions of $-y'' + qy = 0$ is greater than $T$. Much of the work on stability of solutions of periodic equations and systems can be found in the Russian literature; in particular, see Yakubovich and Starzhinskii [103].

Thus we see from these theorems that $q$ sufficiently large in $(py')' + qy = 0$ will give oscillation, and that $q \le 0$ or $|q|$ sufficiently small will give disconjugacy.

## 4. Higher-order equations and systems

For higher-order differential equations, what is the "correct" extension of the definition of oscillatory? of disconjugate? Consider for example the two-term fourth-order equation $y^{(iv)} + q(x)y = 0$. For the distribution of four zeros of a nontrivial solution, there are seven possibilities, 3-1 (meaning $y(a) = y'(a) = y''(a) = y(b) = 0$ for some $a < b$), and with similar meanings the distributions 2-2, 1-3, 2-1-1, 1-2-1, 1-1-2, 1-1-1-1. Hence one could define seven different kinds of disconjugacy.

A widely studied point of view is that an $n$th-order linear ordinary differential equation is disconjugate if no nontrivial solution has $n$ zeros containing multiplicities. This was the definition used by Levin [68, 69] and others. For the differential expression

$$l[y] = y^{(n)} + \sum_{i=1}^{n} a_i(x) y^{(n-i)} = 0, \quad \alpha \le x \le \beta, \tag{4.1}$$

one defines the first conjugate point $\delta(\alpha)$ as the supremum of all $\gamma$ such that no nontrivial solution of (4.1) has more than $n-1$ zeros, counting multiplicities, on $[\alpha, \gamma]$. One result of Levin is that if $\delta(\alpha) < \infty$, then there is a nontrivial solution of (4.1) which is positive on $(\alpha, \delta(\alpha))$, and for some $k$, $1 \le k \le n-1$, it has a zero of order not less than $k$ at $\alpha$ and a zero of order not less than $n-k$ at $\delta(\alpha)$. Green's functions are useful in establishing disconjugacy criteria in this sense. One such result by Levin is that $y^{(4)} + q(x)y = 0$ is disconjugate on $[\alpha, \beta]$ if $q(x) \ge 0$, and $\int_\alpha^\beta q(x)\,dx \le 384(\beta - \alpha)^{-3}$.

For oscillation one could again say that the equation is oscillatory if all nontrivial solutions have infinitely many zeros. However for the equation $y^{(iv)} - y = 0$ some solutions have infinitely many zeros and others have none, so some modification of the definition is required. There has been much research on the structure of somewhat special equations. In a classic paper on fourth-order equations, Leighton and Nehari [66] studied the oscillatory structure of the equations

$$(ry'')'' + qy = 0, \tag{4.2}$$

$$(ry'')'' - qy = 0, \tag{4.3}$$

where $r, q$ are positive continuous functions on an interval $I = [a, \infty)$. Typical of their results are:

(1) If $u$ and $v$ are linearly independent solutions of (4.3) on $[a, \infty)$ such that $u(a) = u'(a) = v(a) = v'(a) = 0$, then the zeros of $u$ and $v$ separate each other in $(a, \infty)$.

(2) If $u$ and $v$ are nontrivial solutions of (4.2), the number of zeros of $u$ on any closed interval $[\alpha, \beta]$ cannot differ by more than 4 from the number of zeros of $v$ on $[\alpha, \beta]$. In particular the nontrivial solutions of (4.2) all have infinitely many zeros on $[a, \infty)$ or none do.

(3) Suppose that $r(x) \geq R(x)$ and $q(x) \leq Q(x)$ in (4.3) and in $(Ry'')'' - Qy = 0$. Let $u$ and $v$ be nontrivial solutions of (4.3) and $(Ry'')'' - Qy = 0$, respectively, such that $u(\alpha) = v(\alpha) = u(\beta) = v(\beta) = 0$. If $n$, $m$ denote the number of zeros of $u$, $v$ respectively on $[\alpha, \beta]$ $(n \geq 4)$, then $m \geq n - 1$.

This type of separation, where the zeros of one solution of a scalar equation have interlacing properties with another solution, has been developed by Hanan [41] for third-order equations.

However, we will concentrate here on the definition of oscillation and disconjugacy that comes from the calculus of variations and has other applications such as in optimal control and spectral theory of differential equations. If in (3.1) the functions $f$ and $y$ are $n$-vector-valued, the Euler-Lagrange equation is a coupled system of $n$ second-order differential equations. The quadratic form of the second variation is (where * indicates transpose)

$$J[\eta, \xi] = \int_a^b (\xi^* [R(x)\xi + Q(x)\eta] + \eta^* [Q^*(x)\xi + P(x)\eta])dx$$

which arises from the vector equation

$$(Ru' + Qu)' - (Q^* u' + Pu) = 0 \tag{4.4}$$

with $R(x)$, $P(x)$ hermitian and $R(x)$ nonsingular. The functions $P(x)$, $R(x)$, $Q(x)$ are expressed in terms of the partial derivatives of the components of $f$. Equation (4.4) can be written in the linear Hamiltonian system form

$$u' = Au + Bv, \quad v' = Cu - A^* v \tag{4.5}$$

with $A = -R^{-1}Q$, $B = R^{-1}$, and $C = P - Q^* R^{-1} Q$.

Symmetric scalar differential equations can also be put in the form (4.5). For example, the equation

$$(ry'')'' + (py')' + q(x)y = 0, \tag{4.6}$$

has the system form (4.5) with

$$u = \begin{bmatrix} y \\ y' \end{bmatrix}, \; v = \begin{bmatrix} -(ry'')' - py' \\ ry'' \end{bmatrix}, \; A = \begin{bmatrix} 0 & 1 \\ 0 & 0 \end{bmatrix}, \; B = \begin{bmatrix} 0 & 0 \\ 0 & 1/r \end{bmatrix}, \; C = \begin{bmatrix} q & 0 \\ 0 & -p \end{bmatrix}.$$

In analogy to the scalar case, the vector minimization problem with fixed endpoints leads to admissible perturbations with $\eta(a) = \eta(b) = 0$. Thus we say that a solution $u$, $v$ of (4.5) has a zero at $a$ provided $u(a) = 0$, and we say $b > a$ is *conjugate* to $a$ if there is a nontrivial solution $u$, $v$ of (4.5) such that $u(a) = u(b) = 0$. Note that as applied to the scalar equation (4.6), $u(a) = 0$ is equivalent to $y(a) = y'(a) = 0$. We will say the system (4.5) is *disconjugate* on $[a, b]$ provided that there do not exist $c < d$ in $[a, b]$ such that $d$ is conjugate to $c$. Otherwise we say (4.5) is *oscillatory* on [a,b]. The definition of oscillatory on a ray $I = [a, \infty)$ that turns out to be useful for spectral theory is that (4.5) is *oscillatory* on $[a, \infty)$ if for every $b > a$ there exist $b \leq c < d$ such that $d$ is conjugate to $c$. Analogous to problems (2.1) and (2.2) are the questions of when (4.5) is oscillatory or disconjugate on an interval. The definitions of disfocal are similar to those in the second-order case. The system (4.5) is called *identically normal* on an interval $I$ if $u \equiv 0$ on a subinterval of $I$ implies also $v \equiv 0$ on the subinterval. This is a controllability condition, cf. [24].

The theme of a sufficiently large coefficient that is in the Fite-Leighton-Wintner Theorem of Section 3 has continued in the case of scalar equations of order greater than 2 and for Hamiltonian systems. Some of these results are described below.

**Theorem 4.1 (Byers, Harris, Kwong, 1986).** *If $Q(x)$ is a continuous symmetric $n \times n$ matrix function on $I = [a, \infty)$, and*

$$\max \textit{ eigenvalue } \int_a^x Q(t)dt \longrightarrow \infty \textit{ as } x \longrightarrow \infty,$$

*then the equation $y'' + Q(x)y = 0$ is oscillatory on $[a, \infty)$.*

Note that the scalar condition $\int_a^\infty q(x)dx = \infty$ has been replaced by the maximum eigenvalue condition.

Glazman [37] proved that the scalar equation $(-1)^{n+1}y^{(2n)} + q(x)y = 0$ is oscillatory on $[a, \infty)$ if $\int_a^\infty q(x)dx = \infty$. Various extensions of this have been made. In particular we quote the result:

**Theorem 4.2 (Müller-Pfeiffer, 1982).** *The equation $(-1)^{n+1}(p(x)y^{(n)})^{(n)} + q(x)y = 0$ is oscillatory on $[a, \infty)$ if*

1. *$p(x) > 0$ and for some $m$, $0 \le m \le n-1$, $\int_a^\infty x^{2m}[p(x)]^{-1}\, dx = \infty$,*
2. *$\int_a^\infty q(x)Q^2(x)dx = \infty$ for some polynomial $Q$ of degree $\le n - m - 1$.*

For two-term equations, the theory of reciprocal equations has been fruitful. Using the results of Ahlbrandt [4], it follows that the equation $(-1)^n(r^{-1}y^{(n)})^{(n)} - py = 0$ is non-oscillatory on $[a, \infty)$ if and only if $(-1)^n(p^{-1}y^{(n)})^{(n)} - ry = 0$ is non-oscillatory on $[a, \infty)$. Using these ideas, Lewis [70] was able to answer affirmatively an open question posed by Glazman that the condition $\lim_{x\to\infty} x^{2n-1}\int_x^\infty 1/r = 0$ was a necessary condition for the equation $(-1)^n(ry^{(n)})^{(n)} = \lambda y$ to be non-oscillatory on $[a, \infty)$ for all $\lambda$. The condition was known to be sufficient.

As noted in the theorems for second-order equations, the equation is disconjugate if the coefficient of $y$ is sufficiently small. A theorem of this type for scalar equations is

**Theorem 4.3 (Ashbaugh, Brown, Hinton, 1992).** *The scalar equation $(x^\delta y^{(n)})^{(n)} + q(x)y = 0$, $\delta$ not in $\{-1, 1, \ldots, 2n-1\}$, is non-oscillatory on $I = [a, \infty)$, $a > 0$, if there is an $s$, $1 \le s < \infty$, such that $\int_a^\infty x^{2n-\delta-1/s}|q(x)|^s dx < \infty$.*

Associated with the system (4.5) is the matrix system

$$U' = AU + BV, \quad V' = CU - A^*V \tag{4.7}$$

where $U, V$ are $n \times n$ matrix functions. When $U$ is nonsingular, the function $W = VU^{-1}$ satisfies the Riccati equation

$$W' = C - WA - A^*W - WBW. \tag{4.8}$$

A solution of (4.7) is called *conjoined or isotropic* if $U^*V = V^*U$. When $U$ is nonsingular, it is easy to show $W = W^*$ if and only if the solution $U$, $V$ is conjoined. All of these concepts can be brought together in what Calvin Ahlbrandt calls the Reid Roundabout Theorem [88, p. 285].

**Theorem 4.4.** *Suppose on $I = [a, b]$ the coefficients $A$, $B$, $C$ are Lebesgue integrable with $C$, $B$ hermitian and $B$ positive semi-definite and the system (4.5) is identically normal on $I$. Define $\mathcal{D}_0[a, b]$ to be the set of all $n$-dimensional vector functions $\eta$ on $[a, b]$ which are absolutely continuous, satisfy $\eta(a) = \eta(b) = 0$, and for which there is an essentially bounded function $\xi$ such that $\eta'(x) = A(x)\eta(x) + B(x)\xi(x)$ a.e. on $[a, b]$. For $\eta \in \mathcal{D}_0[a, b]$ define*

$$J(\eta, a, b) = \int_a^b [\xi^*(x)B(x)\xi(x) + \eta^*(x)C(x)\eta(x)]\, dx. \tag{4.9}$$

*Then the following statements are equivalent.*

1. *There is a conjoined solution $U$, $V$ of (4.7) such that $U$ is nonsingular on $[a, b]$.*
2. *If $\eta \in \mathcal{D}_0[a, b]$ and $\eta$ is not the zero function, then $J(\eta, a, b) > 0$.*
3. *The system (4.5) is disconjugate on $[a, b]$.*
4. *The equation (4.8) has a hermitian solution on $[a, b]$.*

The proof of Theorem 4.4 is greatly facilitated by the Legendre or Clebsch transformation of the functional (4.9) which we now state. Suppose $U$, $V$ are $n \times n$ matrix solutions of (4.7) on an interval [a,b] and $U$ is nonsingular on $[a, b]$. If $\eta \in \mathcal{D}_0[a, b]$ with corresponding function $\xi$, and $W = VU^{-1}$, then

$$[\eta^* W\eta]' + [\xi - W\eta]^*\, B\, [\xi - W\eta] = \eta^* C\eta + \xi^* B\xi.$$

This follows by differentiation and substitution from (4.8).

A general Picone identity for the system (4.7) may be stated. Suppose for i=1,2 we have on an interval matrix solutions $U_i$, $V_i$ of

$$U_i' = A_i U_i + B_i V_i, \quad V_i' = C_i U_i - A_i^* V_i,$$

where $U_1$, $V_1$ are $n \times r$ matrices and $U_2$, $V_2$ are $n \times n$ matrices with $U_2$ nonsingular. Define $W = V_2 U_2^{-1}$. Then if $A_1 = A_2$ and $B_1 = B_2$,

$$[U_1^* W U_1 - U_1^* V_1]' = U_1^* (C_2 - C_1) U_1 - [V_1 - WU_1]^*\, B_2\, [V_1 - WU_1].$$

The general result can be found in [88, p. 354].

Calvin Ahlbrandt has pointed out that prior to Weierstrass it was thought that, as for point functions, if an admissible arc satisfied the Euler equation, the strengthened Legendre condition and the strengthened Jacobi condition (condition (3) in Theorem 4.4), then it would provide a local minimum. This was true for weak local minimums, but not for strong local minimums. Thus the theory of the second variation was discredited as having the analogous utility as the second derivative for point functions.

Theorem 4.4 gives immediately a comparison theorem. If $B_1(x) \geq B(x)$ and $C_1(x) \geq C(x)$, and $J_1$ is the functional corresponding to (4.9), then $J_1(\eta, a, b) \geq J(\eta, a, b)$ so that disconjugacy of (4.5) implies disconjugacy of

$$U' = AU + B_1 V, \quad V' = C_1 U - A^* V.$$

For (4.6), $J$ is given by

$$J(\eta, a, b) = J(y, a, b) = \int_a^b [ry''^2 - py'^2 + qy^2]dx$$

over those sufficiently smooth $y$ satisfying $y(a) = y'(a) = y(b) = y'(b) = 0$. Hence the comparison reads $r_1(x) \geq r(x), p_1(x) \leq p(x), q_1(x) \geq q(x)$ and disconjugacy of (4.6) implies disconjugacy of

$$(r_1 y'')'' + (p_1 y')' + q_1(x)y = 0.$$

Similar comparisons are immediate for the $2n$th-order symmetric differential expression $l[y] = \sum_{i=0}^{n}(p_i y^{(i)})^{(i)}$.

It was Morse [78] who gave the first generalizations of the Sturm theorems of separation and comparison to self-adjoint second-order linear differential systems. Morse proved the number of points on an interval $(a, b)$ which are conjugate to the point $a$ is the same as the number of negative eigenvalues of a quadratic form defined on a certain finite-dimensional space. This quadratic form is constructed from the form (4.9) of Theorem 4.4. This development can be found in [79] and [86]. In particular, it establishes a comparison between conjugate points for two systems of the form (4.5).

A solution of the problem of extending Sturm's separation Theorem C may be stated as follows [86, p. 307]. If for (4.5) there are $q$ points conjugate to $a$ on $(a, b]$, then for any conjoined basis of (4.5) there are at most $q+n$ points conjugate to $a$ on $(a, b]$ and at least $q - n$ points conjugate to $a$ on $(a, b]$. Thus if we take $U$, $V$ to be the solution of (4.7) with initial conditions $U(a) = 0$, $V(a) = I$, and suppose det $U(x)$ is zero exactly $n+1$ times in $(a, b]$, then for any other conjoined solution $U_1$, $V_1$, det $U_1(x) = 0$ at least once. For $n = 1$ this is Sturm's theorem. Note also if det $U(x) = 0$ infinitely many times on $[a, \infty)$, then det $U_1(x) = 0$ infinitely many times on $[a, \infty)$.

## 5. Parameter dependent boundary conditions and indefinite weights

A large class of physical problems have the eigenparameter in the boundary conditions. Examples are vibration problems under various loads such as a vibrating string with a tip mass or heat conduction through a liquid solid interface. See [34] for a list of references. With the boundary condition at one endpoint containing the eigenparameter, the eigenvalue problem on $[a, b]$ takes the form of (1.3) with boundary conditions

$$y(a)\cos\alpha - (py')(a)\sin\alpha = 0, \tag{5.1}$$

$$[\beta_1\lambda + \beta_1']y(b) = [\beta_2\lambda + \beta_2'](py')(b). \tag{5.2}$$

It was noted independently by several authors, Binding, Browne, and Seddighi [14], Harrington [42], and Linden [72] that there is a skip in the counting of the zeros of the eigenfunction compared to the index of the eigenvalue. The development in [14] is the most comprehensive and also shows how the eigenvalues of (5.1)-(5.2) interlace with those of a standard Sturm-Liouville problem. We quote here Linden's theorem.

**Theorem 5.1 (Linden, 1991).** *For the eigenvalue problem* (1.3), (5.1), *and* (5.2), *suppose that* $\beta_1'\beta_2 - \beta_2'\beta_1 > 0$. *Then there is a countable sequence* $\lambda_1 < \lambda_2 < \cdots$ *of real simple eigenvalues with* $\lambda_k \to \infty$ *for* $k \to \infty$. *Let* $y_k$ *denote the eigenfunction corresponding to the eigenvalue* $\lambda_k$. *If* $\beta_2' = 0$, *then* $y_k$ *has exactly* $(k-1)$ *zeros in* $(a,b)$. *If* $\beta_2' \neq 0$, *then for* $\lambda_k < -\beta_2/\beta_2'$, $y_k$ *has exactly* $(k-1)$ *zeros in* $(a,b)$, *and for* $\lambda_k \geq -\beta_2/\beta_2'$, $y_k$ *has exactly* $(k-2)$ *zeros in* $(a,b)$.

In the case of a parameter in the boundary condition at both endpoints there is in general a skip of two zeros in the indexing of the eigenfunctions [14, p. 65]. The case of the eigenparameter occurring rationally in the boundary conditions has been considered by Binding [12].

Everitt, Kwong, and Zettl [32] considered (1.3) with the separated boundary conditions

$$y(a)\cos\alpha - (py')(a)\sin\alpha = 0, \quad y(b)\cos\beta + (py')(b)\sin\beta = 0, \tag{5.3}$$

where the conditions on $p$ and $w$ were relaxed to

$$p(x),\ w(x) \geq 0, \quad \int_a^b w(x)\,dx > 0.$$

Under these conditions they were able to prove that there is a sequence $\lambda_0 < \lambda_1 < \cdots$ of simple eigenvalues tending to infinity with associated eigenfunctions $\psi_0, \psi_1, \ldots$, where each $\psi_n$ has only a finite number $m_n$ of zeros in the open interval (a,b) and such that

(i) $m_{n+1} = m_n + 1$,
(ii) Given any integer $r \geq 0$ there exist $p$, $q$, and $w$ such that $m_0 = r$ and so $m_n = m_0 + n = n + r$ for $n$=1,2, ....

Of course $m_0 = 0$ in the standard case where $p(x),\ w(x) > 0$. Property (i) may also be deduced from Theorem IV of [90] (see also Section 6 of [90]).

We turn now to the case where $w$ may change sign. This occurs in some physical problems, e.g., the equation

$$-((1-x^2)y')' = \lambda xy, \quad -1 < x < 1,$$

occurs in electron transport theory. We associate with the differential expression $L[y] = -(py')' + qy$ and boundary conditions (5.3) the quadratic forms

$$Q[y,y] = \langle L[y], y\rangle = |y(a)|^2 \cot\alpha + |y(b)|^2 \cot\beta + \int_a^b [p|y'|^2 + q|y|^2]dx \quad (5.4)$$

and

$$W[y,y] = \int_a^b w|y|^2 dx. \quad (5.5)$$

Then the equation (1.3) is called *left definite* (polar by Hilbert and his school) if $Q[y,y] > 0$ for all $y \neq 0$ in the domain of $Q$ which consists of all absolutely continuous y such that $\int_a^b p|y'|^2dx < \infty$. It is called *right definite* if $W[y,y] > 0$ for all $y \neq 0$ such that $\int_a^b |w||y|^2dx < \infty$. It is called *indefinite* (non-definite by Richardson) if $\int_a^b w_+dx > 0$ and $\int_a^b w_-dx > 0$ where $w_+ = \max\{w,0\}$, $w_- = \max\{-w,0\}$. In his survey article, Mingarelli [76] attributes the first investigations of the general indefinite case to Haupt [47] and Richardson [89]. The indefinite equations have been studied in Krein and Pontrjagin spaces where the indefinite metric is given by $\int_a^b w|y|^2dx$, but more for questions of completeness of eigenfunction expansions and operator theory. The indefinite problems may have complex eigenvalues, but can have only finitely many.

An early result (see Mingarelli [76]) of Haupt [47] and Richardson [90] is that in the indefinite case there exists an integer $n_R > 0$ such that for each $n > n_R$ there are at least two solutions of (1.3) and (5.3) having exactly $n$ zeros in (a,b) while for $n < n_R$ there are no real solutions having $n$ zeros in (a,b). Furthermore there exists a possibly different integer $n_H \geq n_R$ such that for each $n \geq n_H$ there are precisely two solutions of (1.3) and (5.3) having exactly $n$ zeros in (a,b). It has been shown by Mingarelli that both cases $n_R = n_H$ and $n_R < n_H$ may occur.

However, in the left definite indefinite case things are more orderly and we quote the following result from Ince [53, p. 237]. If in (1.3) and (5.3), $q(x) \geq 0, 0 \leq \alpha, \beta \leq \pi/2$, and the problem is indefinite, then there are eigenvalues

$$\cdots < \lambda_1^- < \lambda_0^- < 0 < \lambda_0^+ < \lambda_1^+ < \cdots$$

with corresponding eigenfunctions $y_n^-$, $y_n^+$ such that both $y_n^-$, $y_n^+$ have exactly $n$ zeros in $a < x < b$ for $n = 0, 1, \ldots$.

Further work on left-definite and indefinite problems may be found in Binding and Browne [13], Binding and Volkmer [15], and Kong, Wu, and Zettl [60]. In [15] and in Möller [77] the coefficient $p$ is also allowed to change sign. Again the eigenvalues are unbounded above and below.

It is clear that the work of Sturm on oscillation theory has had an enduring impact in mathematics. We have only discussed a few ways in which the theory has been extended. It has been necessary to omit many important topics such as the theory of principal solutions and the renormalization theory of Gesztesy, Simon and Teschl [35] (for the latter see the text of B. Simon in the present volume). Important work on the constants of oscillation theory (as in Hille's 1948 theorem) has been done by O. Došlý [26] and others. We have just touched on the Riccati equations which arise in diverse applications and are a research area by themselves, see Reid [87]. Oscillation theory is a subject in its own right, and theorems such as Theorem 4.4 show it can be pursued independently. In his remark "Le principe sur lequel reposent les théorèmes que je développe, n'a jamais, si je ne me trompe, été employé dans l'analyse, et il ne me paraît pas susceptible de s'étendre à d'autres équations différentielles", Sturm [94] was too pessimistic that his methods could not be applied to other differential equations.

**Acknowledgement**

I express my appreciation to Calvin Ahlbrandt, Richard Brown, Steve Clark, Hubert Kalf and Angelo Mingarelli for many helpful comments which have improved this article. Thanks are also due to Werner Amrein and the organizing committee for the opportunity to participate in this conference celebrating the work of Sturm and for their excellent organization of the conference.

## References

[1] R. Agarwal, S. Grace and D. O'Regan, *Oscillation theory for difference and functional differential equations*, Kluwer, Dordrecht, 2000.

[2] ———, *Oscillation theory for second-order dynamic equations*, Taylor and Francis, London, 2003.

[3] C. Ahlbrandt, Disconjugacy criteria for self-adjoint differential systems, J. Differential Eqs. **6** (1969), 271–295.

[4] ———, Equivalent boundary value problems for self-adjoint differential systems, J. Differential Eqs. **9** (1971), 420–435.

[5] C. Ahlbrandt, D. Hinton and R. Lewis, The effect of variable change on oscillation and disconjugacy criteria with applications to spectral theory and asymptotic theory, J. Math. Anal. Appl. **81** (1981), 234–277.

[6] C. Ahlbrandt and A. Peterson, *Discrete Hamiltionian Systems*, Kluwer Academic, Dordrecht, 1996.

[7] M. Ashbaugh, R. Brown and D. Hinton, Interpolation inequalities and nonoscillatory differential equations, p. 243–255, International Series of Numerical Mathematics **103**, Birkhäuser, Basel, 1992.

[8] F. Atkinson, *Discrete and continuous boundary problems*, Academic Press, New York, 1964.

[9] P. Bailey, N. Everitt and A. Zettl, The SLEIGN2 Sturm-Liouville code, ACM Trans. Math. Software **27** (2001), 143–192.

[10] J.H. Barrett, Oscillation theory of ordinary linear differential equations, Adv. in Math. **3** (1969), 415–509.

[11] P.R. Beesack, Nonoscillation and disconjugacy in the complex domain, Trans. Amer. Math. Soc. **81** (1956), 211–242.

[12] P. Binding, A hierarchy of Sturm-Liouville problems, Math. Methods in the Appl. Sci. **26** (2003), 349–357.

[13] P. Binding and P. Browne, Oscillation theory for indefinite Sturm-Liouville problems with eigenparameter-dependent boundary conditions, Proc. Royal Soc. Edinburgh **127A** (1997), 1123–1136.

[14] P. Binding, P. Browne and K. Seddighi, Sturm-Liouville problems with eigenparameter dependent boundary conditions, Proc. Edinburgh Math. Soc. **37** (1993), 57–72.

[15] P. Binding and H. Volkmer, Oscillation theory for Sturm-Liouville problems with indefinite coefficients, Proc. Royal Soc. Edinburgh **131A** (2001), 989–1002.

[16] G. Birkhoff, Existence and oscillation theorem for a certain boundary value problem, Trans. Amer. Math. Soc. **10** (1909), 259–270.

[17] M. Bôcher, The theorems of oscillation of Sturm and Klein, Bull. Amer. Math. Soc. **4** (1897–1898), 295–313, 365–376.

[18] ———, *Leçons sur les méthodes de Sturm dans la théorie des équations différentielles linéaires, et leurs développements modernes*, Gauthier-Villars, Paris, 1917.

[19] O. Bolza, *Lectures on the Calculus of Variations*, Dover, New York, 1961.

[20] M. Brown and M. Eastham, The Hurwitz theorem for Bessel functions and antibound states in spectral theory, R. Soc. Lond. Proc. Ser. A Math. Phys. Eng. Sci. **459** (2003), 2431–2448.

[21] R. Brown, S. Clark and D. Hinton, Some function space inequalities and their application to oscillation and stability problems in differential equations, pp. 19–41, *Analysis and Applications*, Narosa, New Delhi, 2002.

[22] E. Coddington and N. Levinson, *Theory of Ordinary Differential Equations*, McGraw-Hill, New York, 1955.

[23] W.J. Coles, A simple proof of a well-known oscillation theorem, Proc. Amer. Math. Soc. **19** (1968), 507.

[24] W.A. Coppel, *Disconjugacy*, Lecture Notes in Mathematics **220**, Springer, Berlin, 1971.

[25] A. Dixon, On the series of Sturm and Liouville as derived from a pair of fundamental integral equations instead of a differential equation, Phil. Trans. Royal Soc. London Ser. A **211** (1912), 411–432.

[26] O. Došlý, Constants in the oscillation theory of higher-order Sturm-Liouville differential equations, Elect. J. Diff. Eqs. **2002** (2002), 1–12.

[27] M. Eastham, *The Spectral Theory of Periodic Differential Equations*, Scottish Academic Press, Edinburgh, 1973.

[28] ———, Antibound states and exponentially decaying Sturm Liouville potentials, J. London Math. Soc. (2) **65** (2002), 624–638.

[29] U. Elias, *Oscillation Theory of Two-Term Differential Equations*, Kluwer, Dordrecht, 1997.

[30] L. Erbe, Q. Kong and S. Ruan, Kamenev type theorems for second-order matrix differential systems, Proc. Amer. Math. Soc. **117** (1993), 957–962.

[31] L. Erbe, Q. Kong and B. Zhang, *Oscillation theory for functional differential equations*, Monographs in Pure and Applied Mathematics **190**, Marcel Dekker, New York, 1995.

[32] N. Everitt, M. Kwong and A. Zettl, Oscillation of eigenfunctions of weighted regular Sturm-Liouville problems, J. London Math. Soc. (2) **27** (1983), 106–120.

[33] W.B. Fite, Concerning the zeros of the solutions of certain differential equations, Trans. Amer. Math. Soc. **19** (1918), 341–352.

[34] C. Fulton, Two-point boundary value problems with eigenvalue parameter contained in the boundary conditions, Proc. Royal Soc. Edinburgh **77A** (1977), 293–308.

[35] F. Gesztesy, B. Simon and G. Teschl, Zeros of the Wronskian and renormalized oscillation theory, Amer. J. Math. **118** (1996), 571–594.

[36] F. Gesztesy and M. Ünal, Perturbative oscillation criteria and Hardy-type inequalities, Math. Nachr. **189** (1998), 121–144.

[37] I.M. Glazman, *Direct Methods for the Qualitative Spectral Analysis of Singular Differential Operators*, Israel Program for Scientific Translations, Jerusalem, 1965.

[38] M. Greguš, *Third-Order Linear Differential Equations*, Reidel, Dordrecht, 1987.

[39] I. Gyori and G. Ladas, *Oscillation theory of delay differential equations with applications*, Oxford Press, New York, 1991.

[40] G. Halvorsen and A. Mingarelli, *Non-oscillation Domains of Differential Equations with Two Parameters*, Lecture Notes in Mathematics **1338**, Springer, Berlin, 1988.

[41] M. Hanan, Oscillation criteria for third-order differential equations, Pacific J. Math. **11** (1961), 919–944.

[42] B. Harrington, *Qualitative and Quantitative Properties of Eigenvalue Problems with Eigenparameter in the Boundary Conditions*, M.S. thesis, U. Tennessee, 1988.

[43] P. Hartman, Differential equations with non-oscillatory eigenfunctions, Duke Math. J. **15** (1948), 697–709.

[44] ———, *Ordinary Differential Equations*, Wiley, New York, 1964.

[45] ———, Comparison theorems for self-adjoint second-order systems and uniqueness of eigenvalues of scalar boundary value problems, pp. 1–22, *Contributions to analysis and geometry*, Johns Hopkins Univ. Press, Baltimore, 1980.

[46] P. Hartman and A. Wintner, On non-oscillatory linear differential equations with monotone coefficients, Amer. J. Math. **76** (1954), 207–219.

[47] O. Haupt, *Untersuchungen über Oszillationstheoreme*, Teubner, Leipzig, 1911.

[48] S.W. Hawking and R. Penrose, The singularities of gravity collapse and cosmology, Proc. Roy. Soc. London Ser. A **314** (1970), 529–548.

[49] E. Hille, Oscillation theorems in the complex domain, Trans. Amer. Math. Soc. **23** (1922), 350–385.

[50] ———, Nonoscillation Theorems, Trans. Amer. Math. Soc. **64** (1948), 234–252.

[51] ———, *Lectures on Ordinary Differential Equations*, Addison-Wesley, Reading, 1969.

[52] A. Hurwitz Über die Nullstellen der Bessel'schen Funktion, Math. Annalen **33** (1889), 246–266.

[53] E. Ince, *Ordinary Differential Equations*, Dover, New York, 1956.

[54] C.G.J. Jacobi, Zur Theorie der Variations-Rechnung und der Differential-Gleichungen, J. Reine Angew. Math. **17** (1837), 68–82.

[55] I. Kamenev, Integral criterion for oscillations of linear differential equations of second-order, Mat. Zametki **23** (1978), 249–251.

[56] E. Kamke, A new proof of Sturm's comparison theorem, Amer. Math. Monthly **46** (1939), 417–421.

[57] M. Klaus and J. Shaw, On the eigenvalues of Zakharov-Shabat systems, SIAM J. Math. Analysis **34** (2003), 759–773.

[58] F. Klein, Über Körper, welche von confocalen Flächen zweiten Grades begränzt sind, Math. Ann. **18** (1881), 410–427. In: *Felix Klein Gesammelte Abhandlungen*, vol II, p. 521–539, Springer, Berlin, 1922.

[59] A. Kneser, Untersuchungen über die reellen Nullstellen der Integrale linearer Differentialgleichungen, Math. Ann. **42** (1893), 409–435.

[60] Q. Kong, H. Wu and A. Zettl, Left-definite Sturm-Liouville problems, J. Differential Eqs. **177** (2001), 1–26.

[61] W. Kratz, *Quadratic Functionals in Variational Analysis and Control Theory*, Akademie Verlag, Berlin, 1995.

[62] K. Kreith, *Oscillation Theory*, Lecture Notes in Mathematics **324**, Springer, Berlin, 1973.

[63] ———, PDE generalization of the Sturm comparison theorem, Mem. Amer. Math. Soc. **48** (1984), 31–45.

[64] W. Leighton, On the detection of the oscillation of solutions of a second-order linear differential equation, Duke Math. J. **17** (1950), 57–62.

[65] ———, Comparison theorems for linear differential equations of second-order, Proc. Amer. Math. Soc. **13** (1962), 603–610.

[66] W. Leighton and Z. Nehari, On the oscillation of solutions of self-adjoint linear differential equations of the fourth order, Trans. Amer. Math. Soc. **89** (1958), 325–377.

[67] A.J. Levin, A comparison principle for second-order differential equations, Soviet Math. Dokl. **1** (1960), 1313–1316.

[68] ———, Some properties bearing on the oscillation of linear differential equations, Soviet Math. Dokl. **4** (1963), 121–124.

[69] ———, Distribution of the zeros of a linear differential equation, Soviet Math. Dokl. **5** (1964), 818–822.

[70] R. Lewis, The discreteness of the spectrum of self-adjoint, even order, one-term, differential operators, Proc. Amer. Math Soc. **42** (1974), 480–482.

[71] A. Liapunov, Problème Général de la Stabilité du Mouvement, (French translation of a Russian paper dated 1893), Ann. Fac. Sci. Univ. Toulouse **2** (1907), 27–247; reprinted as Ann. Math. Studies **17**, Princeton, 1949.

[72] H. Linden, Leighton's bounds for Sturm-Liouville eigenvalues with eigenparameter in the boundary conditions, J. Math. Anal. Appl. **156** (1991), 444–456.

[73] D. London, On the zeros of $w''(z) + p(z)w(z) = 0$, Pacific J. Math. **12** (1962), 979–991.

[74] J. Lützen, *Joseph Liouville 1809–1882: Master of Pure and Applied Mathematics*, Springer, Berlin, 1990.

[75] F. Meng and A. Mingarelli, Oscillation of linear Hamiltonian systems, Proc. Amer. Math. Soc. **131** (2003), 897–904.

[76] A. Mingarelli, A survey of the regular weighted Sturm-Liouville problem – the non-definite case, in *International Workshop on Applied Differential Equations*, ed. S. Xiao and F. Pu, p. 109–137, World Scientific, Singapore, 1986.

[77] M. Möller, On the unboundedness below of the Sturm-Liouville operator, Proc. Roy. Soc. Edinburgh **129A** (1999), 1011–1015.

[78] M. Morse, A generalization of Sturm separation and comparison theorems, Math. Ann. **103** (1930), 52–69.

[79] ———, *Variational Analysis: Critical Extremals and Sturmian Extensions*, Wiley-Interscience, New York, 1973.

[80] E. Müller-Pfeiffer, *Spectral Theory of Ordinary Differential Equations*, Ellis Horwood, Chichester, 1981.

[81] Z. Nehari, On the zeros of solutions of second-order linear differential equations, Amer. J. Math. **76** (1954), 689–697.

[82] ———, Oscillation criteria for second-order linear differential equations, Trans. Amer. Math. Soc. **85** (1957), 428–445.

[83] M. Picone, Sui valori eccezionali di un parametro da cui dipende un'equazione differenziale lineare ordinaria del second'ordine, Ann. Scuola Norm. Pisa **11** (1909), 1–141.

[84] H. Prüfer, Neue Herleitung der Sturm-Liouvilleschen Reihenentwicklung stetiger Funktionen, Math. Ann. **95** (1926), 499–518.

[85] J. Pryce, *Numerical Solutions of Sturm-Liouville Problems*, Oxford University Press, Oxford, 1993.

[86] W.T. Reid, *Ordinary Differential Equations*, Wiley, New York, 1971.

[87] ———, *Riccati Differential Equations*, Academic Press, New York, 1972.

[88] ———, *Sturmian Theory for Ordinary Differential Equations*, Applied Mathematical Sciences **31**, Springer, Berlin, 1980.

[89] R. Richardson, Theorems of oscillation for two linear differential equations of the second order with two parameters, Trans. Amer. Math. Soc. **13** (1912), 22–34.

[90] ———, Contributions to the study of oscillation properties of the solutions of linear differential equations of the second order, Amer. J. Math. **40** (1918), 283–316.

[91] B. Simon, Resonances in one dimension and Fredholm determinants, J. Funct. Anal. **178** (2000), 396–420.

[92] R. Stafford and J. Heidel, A new comparison theorem for scalar Riccati equations, Bull. Amer. Math. Soc. **80** (1974), 754–757.

[93] R. Sternberg, Variational methods and non-oscillation theorems for systems of differential equations, Duke Math. J. **19** (1952), 311–322.

[94] C. Sturm, Mémoire sur les Équations différentielles linéaires du second ordre, J. Math. Pures Appl. **1** (1836), 106–186.

[95] ———, Mémoire sur une classe d'Équations à différences partielles, J. Math. Pures Appl. **1** (1836), 373–444.

[96] C.A. Swanson, *Comparison and Oscillation Theory of Linear Differential Equations*, Academic Press, New York, 1968.

[97] F.J. Tipler, General relativity and conjugate ordinary differential equations, J. Differential Eqs. **30** (1978), 165–174.

[98] H. Weber, *Die partiellen Differentialgleichungen der mathematischen Physik*, Band 2, 5. Auflage, Vieweg, Braunschweig, 1912.

[99] J. Weidman, *Spectral Theory of Ordinary Differential Operators*, Lecture Notes in Mathematics **1258**, Springer, Berlin, 1987.

[100] D. Willett, Classification of second-order linear differential equations with respect to oscillation, Adv. Math. **3** (1969), 594–623.

[101] A. Wintner, A criterion of oscillatory stability, Quart. Appl. Math. **7** (1949), 115–117.

[102] ———, On the non-existence of conjugate points, Amer. J. Math. **73** (1951), 368–380.

[103] Y. Yakubovich and V. Starzhinskii, *Linear Differential Equations with Periodic Coefficients, vol. I, II*, Wiley, New York, 1975.

Don Hinton
Mathematics Department
University of Tennessee
Knoxville, TN 37996-1300, USA
e-mail: hinton@math.utk.edu

*W.O. Amrein, A.M. Hinz, D.B. Pearson*
Sturm-Liouville Theory: Past and Present, 29–43


# Sturm Oscillation and Comparison Theorems

Barry Simon

**Abstract.** This is a celebratory and pedagogical discussion of Sturm oscillation theory. Included is the discussion of the difference equation case via determinants and a renormalized oscillation theorem of Gesztesy, Teschl, and the author.

## 1. Introduction

Sturm's greatest contribution is undoubtedly the introduction and focus on Sturm-Liouville operators. But his mathematically deepest results are clearly the oscillation and comparison theorems. In [26, 27], he discussed these results for Sturm-Liouville operators. There has been speculation that in his unpublished papers he had the result also for difference equations, since shortly before his work on Sturm-Liouville operators, he was writing about zeros of polynomials, and there is a brief note referring to a never published manuscript that suggests he had a result for difference equations [17]. Indeed, the Sturm oscillation theorems for difference equations written in terms of orthogonal polynomials are clearly related to Descartes' theorem on zeros and sign changes of coefficients [31].

In any event, the oscillation theorems for difference equations seem to have appeared in print only in 1898 [3], and the usual proof given these days is by linear interpolation and reduction to the ODE result. One of our purposes here is to make propaganda for the approach via determinants and orthogonal polynomials (see Section 2). Our discussion in Sections 3 and 4 is more standard ODE theory [4] – put here to have a brief pedagogical discussion in one place. Section 5 makes propaganda for what I regard as some interesting ideas in a paper written by Gesztesy, Teschl, and me [10]. Section 6 has three applications to illustrate the scope of applicability.

Our purpose here is celebratory and pedagogical, so we make simplifying assumptions, such as only discussing bounded and continuous perturbations. Standard modern techniques allow one to discuss much more general perturbations, but

Supported in part by NSF grant DMS-0140592.

this is not the place to make that precise. And we look at Schrödinger operators, rather than the more general Sturm-Liouville operators.

We study the ODE

$$Hu = -\frac{d^2u}{dx^2} + Vu = Eu \tag{1.1}$$

typically on $[0, a]$ with $u(0) = u(a) = 0$ boundary conditions or on $[0, \infty)$ with $u(0) = 0$ boundary condition. The discrete analog is

$$(hu)_n = a_n u_{n+1} + b_n u_n + a_{n-1} u_{n-1} = Eu \tag{1.2}$$

for $n = 1, 2, \dots$ with $u_0 \equiv 0$.

For discussions of Sturm-Liouville theory and its history, see [4, 28, 12, 8].

It is a pleasure to thank W. Amrein for the invitation to give this talk and for organizing an interesting conference, Y. Last and G. Kilai for the hospitality of Hebrew University where this paper was written, and F. Gesztesy for useful comments.

## 2. Determinants, orthogonal polynomials, and Sturm theory for difference equations

Given a sequence of parameters of positive reals $a_1, a_2, \dots$ and a sequence of reals $b_1, b_2, \dots$ for the difference equation (1.2), we look at the fundamental solution, $u_n(E)$, defined recursively by $u_1(E) = 1$ and

$$a_n u_{n+1}(E) + (b_n - E)u_n(E) + a_{n-1}u_{n-1}(E) = 0 \tag{2.1}$$

with $u_0 \equiv 0$, so

$$u_{n+1}(E) = a_n^{-1}(E - b_n)u_n(E) - a_n^{-1}a_{n-1}u_{n-1}(E). \tag{2.2}$$

Clearly, (2.2) implies, by induction, that $u_{n+1}$ is a polynomial of degree $n$ with leading term $(a_n \dots a_1)^{-1}E^n$. Thus, we define for $n = 0, 1, 2, \dots$

$$p_n(E) = u_{n+1}(E) \qquad P_n(E) = (a_1 \dots a_n)p_n(E). \tag{2.3}$$

Then (2.1) becomes

$$a_{n+1}p_{n+1}(E) + (b_{n+1} - E)p_n(E) + a_n p_{n-1}(E) = 0 \tag{2.4}$$

for $n = 0, 1, 2, \dots$. One also sees that

$$EP_n(E) = P_{n+1}(E) + b_{n+1}(E)P_n(E) + a_n^2 P_{n-1}(E). \tag{2.5}$$

We will eventually see $p_n$ are orthonormal polynomials for a suitable measure on $\mathbb{R}$ and the $P_n$ are what are known as monic orthogonal polynomials.

Let $J_n$ be the finite $n \times n$ tridiagonal matrix

$$J_n = \begin{pmatrix} b_1 & a_1 & 0 & & & \\ a_1 & b_2 & a_2 & & & \\ 0 & a_2 & b_3 & \ddots & & \\ & & \ddots & \ddots & \ddots & \\ & & & \ddots & b_{n-1} & a_{n-1} \\ & & & & a_{n-1} & b_n \end{pmatrix}$$

**Proposition 2.1.** *The eigenvalues of $J_n$ are precisely the zeros of $p_n(E)$. We have*

$$P_n(E) = \det(E - J_n). \tag{2.6}$$

*Proof.* Let $\varphi(E)$ be the vector $\varphi_j(E) = p_{j-1}(E)$, $j = 1, \dots, n$. Then (2.1) implies

$$(J_n - E)\varphi(E) = -a_n p_n(E)\delta_n \tag{2.7}$$

where $\delta_n$ is the vector $(0, 0, \dots, 0, 1)^T$. Thus every zero of $p_n$ is an eigenvalue of $J_n$. Conversely, if $\tilde{\varphi}$ is an eigenvector of $J_n$, then both $\tilde{\varphi}_j$ and $\varphi_j$ solve (2.2), so $\tilde{\varphi}_j = \tilde{\varphi}_1 \varphi_j(E)$. This implies that $E$ is an eigenvalue only if $p_n(E)$ is zero and that eigenvalues are simple.

Since $J_n$ is real symmetric and eigenvalues are simple, $p_n(E)$ has $n$ distinct eigenvalues $E_j^{(n)}$, $j = 1, \dots, n$ with $E_{j-1}^{(n)} < E_j^{(n)}$. Thus, since $p_n$ and $P_n$ have the same zeros,

$$P_n(E) = \prod_{j=1}^{n} (E - E_j^{(n)}) = \det(E - J_n).$$

□

**Proposition 2.2.** (i) *The eigenvalues of $J_n$ and $J_{n+1}$ strictly interlace, that is,*

$$E_1^{(n+1)} < E_1^{(n)} < E_2^{(n+1)} < \cdots < E_n^{(n)} < E_{n+1}^{(n+1)}. \tag{2.8}$$

(ii) *The zeros of $p_n(E)$ are simple, all real, and strictly interlace those of $p_{n+1}(E)$.*

*Proof.* (i) $J_n$ is obtained from $J_{n+1}$ by restricting the quadratic form $u \to \langle u, J_{n+1} u \rangle$ to $\mathbb{C}^n$, a subspace. It follows that

$$E_1^{(n+1)} = \min_{u, \|u\|=1} \langle u, J_{n+1} u \rangle \leq \min_{u \in \mathbb{C}^n, \|u\|=1} \langle u, J_{n+1} u \rangle = E_1^{(n)}.$$

More generally, using the min-max principle

$$E_j^{(n+1)} = \max_{\varphi_1, \dots, \varphi_{j-1}} \min_{\substack{\|u\|=1 \\ u \perp \varphi_1, \dots, \varphi_{j-1}}} \langle u, J_{n+1} u \rangle$$

one sees that

$$E_j^{(n)} \geq E_j^{(n+1)}.$$

By replacing min's with max's,

$$E_j^{(n)} \leq E_{j+1}^{(n+1)}.$$

All that remains is to show that equality is impossible. If $E_0 = E_j^{(n)} = E_j^{(n+1)}$ or $E_0 = E_j^{(n)} = E_j^{(n+1)}$, then $p_{n+1}(E_0) = p_n(E_0) = 0$. By (2.4), this implies $p_{n-1}(E_0) = 0$ so, by induction, $p_0(E) = 0$. But $p_0 \equiv 1$. Thus equality is impossible.

(ii) Given (2.6), this is a restatement of what we have proven about the eigenvalues of $J_n$. □

Here is our first version of Sturm oscillation theorems:

**Theorem 2.3.** *Suppose $E_0$ is not an eigenvalue of $J_k$ for $k = 1, 2, \dots, n$. Then*

$$\#\{j \mid E_j^{(n)} > E_0\} = \#\{\ell = 1, \dots, n \mid \operatorname{sgn}(P_{\ell-1}(E_0)) \neq \operatorname{sgn}(P_\ell(E_0))\}, \tag{2.9}$$

$$\#\{j \mid E_j^{(n)} < E_0\} = \#\{\ell = 1, \dots, n \mid \operatorname{sgn}(P_{\ell-1}(E_0) = \operatorname{sgn}(P_\ell(E_0))\}. \tag{2.10}$$

*Proof.* (2.9) clearly implies (2.10) since the sum of both sides of the equalities is $n$. Thus we need only prove (2.9).

Suppose that $E_1^{(\ell)} < \dots < E_k^{(\ell)} < E_0 < E_{k+1}^{(\ell)} < E_\ell^{(\ell)}$. By eigenvalue interlacing, $J_{\ell+1}$ has $k$ eigenvalues in $(-\infty, E_k^{(\ell)})$ and $n-k$ eigenvalues in $(E_{k+1}^{(\ell)}, \infty)$. The question is whether the eigenvalue in $(E_k^{(\ell)}, E_{k+1}^{(\ell)})$ lies above $E_0$ or below. Since $\operatorname{sgn}\det(E - J_\ell) = (-1)^{\#\{j \mid E_j^{(\ell)} > E_0\}}$, and similarly for $J_{\ell+1}$, and there is at most one extra eigenvalue above $E_0$, we see

$$\operatorname{sgn} P_\ell(E_0) = \operatorname{sgn} P_{\ell+1}(E_0) \Leftrightarrow \#\{j \mid E_j^{(\ell)} > E_0\} = \#\{j \mid E_j^{(\ell+1)} > E_0\},$$

$$\operatorname{sgn} P_\ell(E_0) \neq \operatorname{sgn} P_{\ell+1}(E_0) \Leftrightarrow \#\{j \mid E_j^{(\ell)} > E_0\} + 1 = \#\{j \mid E_j^{(\ell+1)} > E_0\}.$$

(2.9) follows from this by induction. □

We want to extend this in two ways. First, we can allow $P_k(E_0) = 0$ for some $k < n$. In that case, by eigenvalue interlacing, it is easy to see $J_{k+1}$ has one more eigenvalue than $J_{k-1}$ in $(E_0, \infty)$ and also in $(-\infty, E_0)$, so $\operatorname{sgn}(P_{k-1}(E_0)) = -\operatorname{sgn}(P_{k+1}(E_0))$ (also evident from (2.5) and $P_k(E_0) = 0$). Thus we need to be sure to count the change of sign corresponding to three successive values of $P$ which are, respectively, negative, zero and positive, as just one change of sign. We therefore have

**Proposition 2.4.** (2.9) *and* (2.10) *remain true so long as $P_n(E_0) \neq 0$ if we define* $\operatorname{sgn}(0) = 1$. *If $P_n(E_0) = 0$, they remain true so long as $\ell = n$ is dropped from the right side.*

One can summarize this result as follows: for $x \in [0, n]$, define $y(x)$ by linear interpolation of $P$, that is,

$$x = [x] + (x) \Rightarrow y(x) = P_{[x]} + (x)(P_{[x]+1} - P_{[x]}).$$

Then the number of eigenvalues of $J_n$ above $E$ is the number of zeros of $y(x, E)$ in $[0, n)$. If we do the same for $\tilde{y}$ with $P_{[x]}$ replaced by $(-1)^{[x]} P_{[x]}$, then the number of eigenvalues below $E$ is the number of zeros of $\tilde{y}$ in $[0, n)$. Some proofs (see [6]) of oscillation theory for difference equations use $y$ and mimic the continuum proof of the next section.

The second extension involves infinite Jacobi matrices. In discussing eigenvalues of an infinite $J$, domain issues arise if $J$ is not bounded (if the moment problem is not determinate, these are complicated issues; see Akhiezer [1] or Simon [25]). Thus, let us suppose

$$\sup_n \left(|a_n| + |b_n|\right) < \infty. \tag{2.11}$$

If $J$ is bounded, the quadratic form of $J_n$ is a restriction of $J$ to $\mathbb{C}^n$. As in the argument about eigenvalues interlacing, one shows that if $J$ has only $N_0 < \infty$ eigenvalues in $(E_0, \infty)$, then $J_n$ has at most $N_0$ eigenvalues there. Put differently, if $E_1^{(\infty)} > E_2^{(\infty)} > \cdots$ are the eigenvalues of $J$, $E_j^{(\infty)} \geq E_j^{(n)}$. Thus, if $N_n(E)$ is the number of eigenvalues of $J_n$ in $(E, \infty)$ and $N_\infty$ the dimension of $\operatorname{Ran} P_{(E,\infty)}(J)$, the spectral projection, then

$$N_n(E) \leq N_{n+1}(E) \leq \cdots \leq N_\infty(E). \tag{2.12}$$

On the other hand, suppose we can find an orthonormal set $\{\varphi_j\}_{j=1}^N$ with $M_{jk}^{(\infty)} = \langle \varphi_j, J\varphi_k \rangle = e_j \delta_{jk}$ and $\min(e_j) = e_0 > E_0$. If $M_{jk}^{(n)} = \langle \varphi_j, J_n \varphi_k \rangle$, $M^{(n)} \to M^{(\infty)}$, so for $n$ large, $M^{(n)} \geq \min(e_j) + \frac{1}{2}(e_0 - E_0) > E_0$. Thus $N_n(E_0) \geq N$ for $n$ large. It follows that $\lim N_n \geq N_\infty$, that is, we have shown that $N_\infty(E_0) = \lim_{n\to\infty} N_n(E_0)$. Thus,

**Theorem 2.5.** *Let $J$ be an infinite Jacobi matrix with* (2.11). *Then* (*with* $\operatorname{sgn}(0) = 1$) *we have*

$$N_\infty(E_0) = \#\{\ell = 1, 2, \dots \mid \operatorname{sgn}(P_{\ell-1}(E_0)) \neq \operatorname{sgn}(P_\ell(E_0))\}, \tag{2.13}$$

$$\dim P_{(-\infty, E_0)}(J) = \#\{\ell = 1, 2, \dots \mid \operatorname{sgn}(P_{\ell-1}(E_0)) = \operatorname{sgn}(P_\ell(E_0))\}. \tag{2.14}$$

**Corollary 2.6.** $a_- \leq J \leq a_+$ *if and only if for all* $\ell$,

$$P_\ell(a_+) > 0 \qquad \textit{and} \qquad (-1)^\ell P_\ell(a_-) > 0. \tag{2.15}$$

While on the subject of determinants and Jacobi matrices, I would be remiss if I did not make two further remarks.

Given (2.6), (2.5) is an interesting relation among determinants, and you should not be surprised it has a determinantal proof. The matrix $J_{n+1}$ has $b_{n+1}$ and $a_n$ in its bottom row. The minor of $E - b_{n+1}$ in $E - J_{n+1}$ is clearly $\det(E - J_n)$. A little thought shows the minor of $-a_n$ is $-a_n \det(E - J_{n-1})$. Thus

$$\det(E - J_{n+1}) = (E - b_{n+1}) \det(E - J_n) - a_n^2 \det(E - J_{n-1}), \tag{2.16}$$

which is just (2.5).

Secondly, one can look at determinants where we peel off the top and left rather than the right and bottom. Let $J^{(1)}, J^{(2)}$ be the Jacobi matrices obtained from $J$ by removing the first row and column, the first two, .... Making the $J$-dependence of $P_n(\,\cdot\,)$ explicit, Cramer's rule implies

$$[(z - J_n)^{-1}]_{11} = \frac{P_{n-1}(z, J^{(1)})}{P_n(z, J)}. \tag{2.17}$$

In the OP literature, $a_1^{-1} p_n(z, J^{(1)})$ are called the second kind polynomials.

The analog of (2.16) is

$$P_n(z, J) = (z - b_1) P_{n-1}(z, J^{(1)}) - a_1^2 P_{n-2}(z, J^{(2)})$$

which, by (2.17), becomes

$$[(z - J_n)^{-1}]_{11} = \frac{1}{(z - b_1) - a_1^2 [(z - J_{n-1}^{(1)})^{-1}]_{11}}. \tag{2.18}$$

In particular, if $d\gamma$ is the spectral measure [20] of $J$ for the vector $\delta_1$, we have

$$[(z - J)^{-1}]_{11} = \int \frac{d\gamma(x)}{z - x} \equiv -m(z, J) \tag{2.19}$$

and (2.18) becomes in the limit with $[(z - J_{n-1}^{(1)})^{-1}]_{11} \to -m(z, J^{(1)})$

$$m(z; J) = \frac{1}{b_1 - z - a_1^2 m(z; J^{(1)})}. \tag{2.20}$$

(2.18) leads to a finite continued fraction expansion of $[(z - J_n)^{-1}]_{11}$ due to Jacobi, and (2.20) to the Stieltjes continued fraction. Sturm's celebrated paper on zeros of polynomials is essentially also a continued fraction expansion. It would be interesting to know how much Sturm and Jacobi knew of each other's work. Jacobi visited Paris in 1829 (see James [13]), but I have no idea if he and Sturm met at that time.

## 3. Sturm theory on the real line

We will suppose $V$ is a real bounded function on $[0, \infty)$. We are interested in solutions of

$$-u'' + Vu = Eu \tag{3.1}$$

for $E$ real.

**Theorem 3.1 (Sturm Comparison Theorem).** *For $j = 1, 2$, let $u_j$ be not identically zero and solve $-u_j'' + Vu_j = E_j u_j$. Suppose $a < b$, $u_1(a) = u_1(b) = 0$ and $E_2 > E_1$. Then $u_2$ has a zero in $(a, b)$. If $E_2 = E_1$ and $u_2(a) \neq 0$, then $u_2$ has a zero in $(a, b)$.*

*Proof.* Define the Wronskian

$$W(x) = u_1'(x) u_2(x) - u_1(x) u_2'(x). \tag{3.2}$$

Then

$$W'(x) = (E_2 - E_1) u_1(x) u_2(x). \tag{3.3}$$

Without loss, suppose $a$ and $b$ are successive zeros of $u_1$. By changing signs of $u$ if need be, we can suppose $u_1 > 0$ on $(a, b)$ and $u_2 > 0$ on $(a, a + \varepsilon)$ for some $\varepsilon$. Thus $W(a) = u_1'(a)u_2(a) \geq 0$ (and, in case $E_1 = E_2$ and $u_2(a) \neq 0$, $W(a) > 0$). If $u_2$ is non-vanishing in $(a, b)$, then $u_2 \geq 0$ there, so $W(b) > 0$ (if $E_2 > E_1$, $(E_2 - E_1) \int_a^b u_1 u_2 \, dx > 0$, and if $E_2 = E_1$ but $u_2(a) \neq 0$, $W(a) > 0$). Since $W(b) = u_1'(b)u_2(b)$ with $u_1'(b) < 0$ and $u_2(b) \geq 0$, this is impossible. Thus we have the result by contradiction. □

**Corollary 3.2.** *Let $u(x, E)$ be the solution of* (3.1) *with $u(0, E) = 0$, $u'(0, E) = 1$. Let $N(a, E)$ be the number of zeros of $u(x, E)$ in $(0, a)$. Then, if $E_2 > E_1$, we have $N(a, E_2) \geq N(a, E_1)$ for all $a$.*

*Proof.* If $n = N(a, E_1)$ and $0 < x_1 < \cdots < x_n < a$ are the zeros of $u(x, E_1)$, then, by the theorem, $u(x, E_2)$ has zeros in $(0, x_1), (x_1, x_2), \ldots, (x_{n-1}, x_n)$. □

This gives us the first version of the Sturm oscillation theorem:

**Theorem 3.3.** *Let $E_0 < E_1 < \cdots$ be the eigenvalues of $H \equiv -d^2/dx^2 + V(x)$ on $L^2(0, a)$ with boundary conditions $u(0) = u(a) = 0$. Then $u(x, E_n)$ has exactly $n$ zeros in $(0, a)$.*

*Proof.* If $u_k \equiv u(\,\cdot\,, E_k)$ has $m$ zeros $x_1 < x_2 < \cdots < x_m$ in $(0, a)$, then for any $E > E_k$, $u(\,\cdot\,, E)$ has zeros in $(0, x_1), \ldots, (x_{m-1}, x_m), (x_m, a)$ and so, $u_{k+1}$ has at least $m + 1$ zeros. It follows by induction that $u_n$ has at least $n$ zeros, that is, $m \geq n$.

Suppose $u_n$ has $m$ zeros $x_1 < \cdots < x_m$ in $(0, a)$. Let $v_0, \ldots, v_m$ be the function $u_n$ restricted successively to $(0, x_1), (x_1, x_2), \ldots, (x_m, a)$. The $v$'s are continuous and piecewise $C^1$ with $v_\ell(0) = v_\ell(a) = 0$. Thus they lie in the quadratic form domain of $H$ (see [20, 21] for discussions of quadratic forms) and

$$\begin{aligned} \langle v_j, H v_k \rangle &= \int_0^a v_j' v_k' + \int_0^a V v_j v_k \\ &= \delta_{jk} E \int_0^a v_j^2 \, dx \end{aligned} \tag{3.4}$$

since if $j = k$, we can integrate by parts and use $-u'' + Vu = Eu$.

It follows that for any $v$ in the span of $v_j$'s, $\langle v, Hv \rangle = E\|v\|^2$, so by the variational principle, $H$ has at least $m+1$ eigenvalues in $(-\infty, E_n]$, that is, $n+1 \geq m + 1$. □

**Remark 1.** The second half of this argument is due to Courant-Hilbert [5].

If we combine this result with Corollary 3.2, we immediately have:

**Theorem 3.4 (Sturm Oscillation Theorem).** *The number of eigenvalues of $H$ strictly below $E$ is exactly the number of zeros of $u(x, E)$ in $(0, a)$.*

As in the discrete case, if $H_a$ is $-d^2/dx^2 + V(x)$ on $[0, a]$ with $u(0) = u(a) = 0$ boundary conditions and $H_\infty$ is the operator on $L^2(0, \infty)$ with $u(0) = 0$ boundary conditions, and if $N_a(E) = \dim P_{(-\infty, E)}(H_a)$, then $N_a(E) \to N_\infty(E)$, so

**Theorem 3.5.** *The number of eigenvalues of $H_\infty$ strictly below $E$, more generally $\dim P_{(-\infty,E)}(H)$, is exactly the number of zeros of $u(x,E)$ in $(0,\infty)$.*

There is another distinct approach, essentially Sturm's approach in [26], to Sturm theory on the real line that we should mention. Consider zeros of $u(x,E)$, that is, solutions of

$$u(x(E),E) = 0. \tag{3.5}$$

$u$ is a jointly $C^1$ function of $x$ and $E$, and if $u(x_0,E_0) = 0$, then $u'(x_0,E_0) \neq 0$ (since $u$ obeys a second-order ODE). Thus, by the implicit function theorem, for $E$ near $E_0$, there is a unique solution, $x(E)$, of (3.4) near $x_0$, and it obeys

$$\left.\frac{dx}{dE}\right|_{E_0} = -\left.\frac{\partial u/\partial E}{\partial u/\partial x}\right|_{x=x_0, E=E_0}. \tag{3.6}$$

Now, $v \equiv \partial u/\partial E$ obeys the equation

$$-v'' + Vv = Ev + u \tag{3.7}$$

by taking the derivative of $-u'' + Vu = Eu$. Multiply (3.7) by $u$ and integrate by parts from 0 to $x_0$. Since $v(0) = 0$, there is no boundary term at 0, but there is at $x_0$, and we find

$$v(x_0)u'(x_0) = \int_0^{x_0} |u(x)|^2\,dx.$$

Thus (3.6) becomes

$$\frac{dx_0}{dE} = -|u'(x_0,E)|^{-2}\int_0^{x_0} |u(x,E)|^2\,dx < 0. \tag{3.8}$$

Thus, as $E$ increases, zeros of $u$ move towards zero. This immediately implies the comparison theorem. Moreover, starting with $u_n$, the $(n+1)$th eigenfunction at energy $E_n$, if it has $m$ zeros in $(0,a)$ as $E$ decreases from $E_n$ to a value, $E'$, below $-\|V\|_\infty$ (where $u(x,E') > 0$ has no zeros in $(0,\infty)$), the $m$ zeros move out continuously, and so $u(a,E) = 0$ exactly $m$ times, that is, $m = n$. This proves the oscillation theorem.

## 4. Rotation numbers and oscillations

Take the solution $u(x,E)$ of the last section and look at the point

$$\Pi(x,E) = \begin{pmatrix} u'(x,E) \\ u(x,E) \end{pmatrix}$$

in $\mathbb{R}^2$. $\Pi$ is never zero since $u$ and $u'$ have no common zeros. At most points in $\mathbb{R}^2$, the argument of $\Pi$, that is, the angle $\pi$ makes with $\binom{1}{0}$, can increase or decrease. $u$ can wander around and around. But not at points where $u = 0$. If $u' > 0$ at such a point, $\Pi$ moves from the lower right quadrant to the upper right, and similarly, if $u' < 0$, it moves from the upper left to the lower left. Thus, since $\pi$ starts at $\binom{1}{0}$, we see

**Theorem 4.1.** *If $u(x,E)$ has $m$ zeros in $(0,a)$, then $\operatorname{Arg}\pi(a,E)$ (defined by continuity and $\operatorname{Arg}\pi(0,E)=0$) lies in $(m\pi,(m+1)\pi]$.*

If $u$ and $v$ are two solutions of $-u''+Vu=Eu$ with $u(0)=0$, $v(0)\neq 0$, we can look at

$$\tilde{\Pi}(x,E)=\begin{pmatrix} v \\ u \end{pmatrix}.$$

$\tilde{\Pi}$ is never zero since $u$ and $v$ are linear independent. $W(x)=u'v-v'u$ is a constant, say $c$. $c\neq 0$ since $u$ and $v$ are linear independent. Suppose $c>0$. Then if $u(x_0)=0$, $u'(x_0)=c/v(x_0)$ has the same sign as $v(x_0)$. So the above argument applies (if $c<0$, there is winding in the $(u,v)$-plane in the opposite direction). Rather than look at $\tilde{\Pi}$, we can look at $\varphi=u+iv$. Then $uv'-vu'=\operatorname{Im}(\bar{\varphi}\varphi')$. Thus we have

**Theorem 4.2.** *Let $\varphi(x,E)$ obey $-\varphi''+V\varphi=E\varphi$ and be complex-valued with*

$$\operatorname{Im}(\bar{\varphi}(0)\varphi'(0))>0. \tag{4.1}$$

*Suppose $\operatorname{Re}\varphi(0)=0$ and $\operatorname{Im}\varphi(0)<0$. Then, if $\operatorname{Re}\varphi$ has $m$ zeros in $(0,a)$, then $\operatorname{Arg}(\varphi(a))$ is in $((m-\frac{1}{2})\pi,(m+\frac{1}{2})\pi]$.*

The ideas of this section are the basis of the relation of rotation numbers and density of states used by Johnson-Moser [15] (see also [14]). We will use them as the starting point of the next section.

## 5. Renormalized oscillation theory

Consider $H=-d^2/dx^2+V$ on $[0,\infty)$ with $u(0)=0$ boundary conditions where, as usual, for simplicity, we suppose that $V$ is bounded.

By Theorem 3.5, $\dim P_{(-\infty,E)}(H)$ is the number of zeros of $u(x,E)$ in $(0,\infty)$. If we want to know $\dim P_{[E_1,E_2)}(H)$, we can just subtract the number of zeros of $u(x,E_1)$ on $(0,\infty)$ from those of $u(x,E_2)$. At least, if $\dim P_{(-\infty,E_2)}(H)$ is finite, one can count just by subtracting. But if $\dim P_{(-\infty,E_1)}(H)=\infty$ while $\dim P_{[E_1,E_2)}$ is finite, both $u(x,E_2)$ and $u(x,E_1)$ have infinitely many zeros, and so subtraction requires regularization.

One might hope that

$$\dim P_{[E_1,E_2)}(H)=\lim_{a\to\infty}(N(E_2,a)-N(E_1,a)) \tag{5.1}$$

where $N(E,a)$ is the number of zeros of $u(x,E)$ in $(0,a)$. This is an approach of Hartman [11]. (5.1) cannot literally be true since $N(E_2,a)-N(E_1,a)$ is an integer which clearly keeps changing when one passes through a zero of $u(x,E_2)$ that is not also a zero of $u(x,E_1)$. One can show that for $a$ large, the absolute value of the difference of the two sides of (5.1) is at most one, but it is not obvious when one has reached the asymptotic region.

Instead, we will describe an approach of Gesztesy, Simon, and Teschl [10]; see Schmidt [23] for further discussion. Here it is for the half-line (the theorem is true in much greater generality than $V$ bounded and there are whole-line results).

**Theorem 5.1.** *Let $V$ be bounded and let $H = -d^2/dx^2 + V(x)$ on $[0,\infty)$ with $u(0) = 0$ boundary condition. Fix $E_1 < E_2$. Let*

$$W(x) = u(x,E_2)u'(x,E_1) - u'(x,E_2)u(x,E_1) \tag{5.2}$$

*and let $N$ be the number of zeros of $W$ in $(0,\infty)$. Then*

$$\dim P_{(E_1,E_2)}(H) = N. \tag{5.3}$$

The rest of this section will sketch the proof of this theorem under the assumption that $\dim P_{(-\infty,E_2)}(H) = \infty$. This will allow a simplification of the argument and covers cases of greatest interest. Following [10], we will prove this in three steps:

(1) Prove the result in a finite interval $[0,a]$ in case $u(a,E_2) = 0$.
(2) Prove $\dim P_{(E_1,E_2)}(H) \leq N$ by limits from (1) when $\dim P_{(-\infty,E_2)}(H) = \infty$.
(3) Prove $\dim P_{(E_1,E_2)}(H) \geq N$ by a variational argument.

**Step 1.** We use the rotation number picture of the last section. Define the Prüfer angle $\theta(x,E)$ by

$$\tan(\theta(x,E)) = \frac{u(x,E)}{u'(x,E)} \tag{5.4}$$

with $\theta(0,E) = 0$ and $\theta$ continuous at points, $x_0$, where $u'(x_0,E) = 0$. Using $\frac{d}{dy}\tan y = 1 + \tan^2 y$, we get

$$\frac{d\theta}{dx} = \frac{(u')^2 - uu''}{u^2 + (u')^2}. \tag{5.5}$$

Let $\theta_1, \theta_2$ be the Prüfer angles for $u_1(x) \equiv u(x,E_1)$ and $u_2(x) \equiv u(x,E_2)$. Suppose $W(x_0) = 0$. This happens if and only if $u(x_0,E_1)/u'(x_0,E_1) = u(x_0,E_2)/u'(x_0,E_2)$, that is, $\theta_2 = \theta_1 + k\pi$ with $k \in \mathbb{Z}$. If it happens, we can multiply $u_2$ by a constant so $u_1(x_0) = u_2(x_0)$, $u_1'(x_0) = u_2'(x_0)$. Once we do that, (5.5) says

$$\frac{d}{dx}(\theta_2 - \theta_1) = \frac{(E_2 - E_1)u_1^2(x_0)}{u_1'(x_0)^2 + u_1^2(x_0)} > 0.$$

Thus

$$\theta_1 = \theta_2 \mod \pi \Rightarrow \theta_2' > \theta_1'. \tag{5.6}$$

Think of $\theta_2$ as a hare and $\theta_1$ as a tortoise running around a track of length $\pi$. There are two rules in their race. They can each run in either direction, except they can only pass the starting gate going forward (i.e., $\theta_j = 0 \mod \pi \Rightarrow \theta_j' > 0$), and the hare can pass the tortoise, not vice-versa (i.e., (5.6) holds).

Suppose $H_a$, the operator on $(0,a)$ with $u(0) = u(a) = 0$ boundary condition, has $m$ eigenvalues below $E_2$ and $n$ below $E_1$. Since $u(a,E_2) = 0$, $\theta_2(a) = (m+1)\pi$, that is, at $x = a$, the hare makes exactly $m+1$ loops of the track. At $x = a$, the tortoise has made $n$ loops plus part, perhaps all, of an additional one. Since

$\theta_2' - \theta_1' > 0$ at $x = 0$, the hare starts out ahead. Thus, the hare must overtake the tortoise exactly $m - n$ times between 0 and $a$ (if $\theta_1(a) = (n+1)\pi$, since then $\theta_2' - \theta_1' > 0$ at $x = 0$, $\theta_2 - (m+1)\pi < \theta_1 - (n-1)\pi$, and $x = a$; so it is still true that there are exactly $m - n$ crossings). Thus

$$\dim P_{(E_1,E_2)}(H_a) = \#\{x_0 \in (0,a) \mid W(x_0) = 0\}. \tag{5.7}$$

**Step 2.** Since $\dim P_{(-\infty,E_2)}(H) = \infty$, there is, by Theorem 3.5, an infinite sequence $a_1 < a_2 < \cdots \to \infty$ so that $u(a_j, E_2) = 0$. $H_{a_j} \to H$ in strong resolvent sense, so by a simple argument,

$$\begin{aligned} \dim P_{(E_1,E_2)}(H) &\leq \liminf \dim P_{(E_1,E_2)}(H_a) \\ &= N \end{aligned} \tag{5.8}$$

with $N$ the number of zeros of $W$ in $(0,\infty)$. (5.8) comes from (5.7).

**Step 3.** Suppose $N < \infty$. Let $0 < x_1 < \cdots < x_N$ be the zeros of $W$. Define

$$\eta_j(x) = \begin{cases} u_1(x) - \gamma_j u_2(x) & 0 < x \leq x_j \\ 0 & x \geq x_j \end{cases} \tag{5.9}$$

$$\tilde{\eta}_j(x) = \begin{cases} u_1(x) + \gamma_j u_2(x) & 0 < x < x_j \\ 0 & x > x_j \end{cases} \tag{5.10}$$

where $u_j(x) = u(x, E_j)$ and $\gamma_j$ is chosen by

$$\gamma_j = \begin{cases} u_1(x_j)/u_2(x_j) & \text{if } u_2(x_j) \neq 0 \\ u_1'(x_j)/u_2'(x_j) & \text{if } u_2(x_j) = 0. \end{cases} \tag{5.11}$$

Since $W(x_j) = 0$, $\eta_j$ is a $C^1$ function of compact support and piecewise $C^2$, and so in $D(H)$. But $\tilde{\eta}_j$ is discontinuous.

We claim that if $\eta$ is in the span of $\{\eta_j\}_{j=1}^N$, then

$$\left\| \left( H - \frac{E_2 + E_1}{2} \right) \eta \right\| = \frac{|E_2 - E_1|}{2} \|\eta\|. \tag{5.12}$$

Moreover, such $\eta$'s are never a finite linear combination of eigenfunctions of $H$. Accepting these two facts, we note that since the $\eta_j$ are obviously linear independent, (5.12) implies $\dim P_{(E_1,E_2)}(H) \geq N$. This, together with (5.8), proves the result.

To prove (5.12), we note that

$$\left( H - \frac{E_2 + E_1}{2} \right) \eta_j = -\frac{|E_2 - E_1|}{2} \tilde{\eta}_j. \tag{5.13}$$

Since $\tilde{\eta}_j$ is not $C^1$ at $x_j$, $\tilde{\eta}$ is not in $D(H)$, hence $\eta$ cannot be in $D(H^2)$ (so we get control of $\dim P_{(E_1,E_2)}(H)$, not just $\dim P_{[E_1,E_2]}(H)$).

Next, note that since $W'(x) = (E_2 - E_1)u_2u_1$, we have if $W(x_i) = W(x_{i+1}) = 0$ that

$$\int_{x_i}^{x_{i+1}} u_1(x)u_2(x)\,dx = 0$$

for $i = 0, 1, 2, \ldots, N$ where $x_0 = 0$. Thus

$$\langle \eta_i, \eta_j \rangle = \langle \tilde{\eta}_i, \tilde{\eta}_j \rangle \tag{5.14}$$

since if $i < j$, the difference of the two sides is $2(\gamma_i + \gamma_j) \int_{x_i}^{x_j} u_1(x)u_2(x) = 0$. (5.14) and (5.13) imply (5.12). That completes the proof if $N < \infty$.

If $N$ is infinite, pick successive zeros $0 < x_1 < \cdots < x_L$ and deduce $\dim P_{(E_1,E_2)}(H) \geq L$ for all $L$. $\square$

## 6. Some applications

We will consider three typical applications in this section: one classical (i.e., fifty years old!), one recent to difference equations, and one of Theorem 5.1.

**Application 1: Bargmann's Bound.** Let $u$ obey $-u'' + Vu = 0$ on $[0, \infty)$ with $u(0) = 0$. Then, if $V$ is bounded, $u(x)/x$ has a finite limit as $x \downarrow 0$. Also suppose $V \leq 0$.

Define $\tilde{m} = -u'/u$ so

$$\tilde{m}' = |V| + \tilde{m}^2 \tag{6.1}$$

since $-V = |V|$. Thus $\tilde{m}$ is monotone increasing. It has a pole at each zero, $x_0 = 0$, $x_1, x_2, \ldots, x_\ell, \ldots$ of $u$. Define

$$b(x) = -\frac{xu'(x)}{u(x)} = x\tilde{m}(x). \tag{6.2}$$

Then $b(x)$ has limit $-1$ as $x \downarrow 0$ and

$$b'(x) = x|V(x)| + \frac{b(x) + b^2(x)}{x}. \tag{6.3}$$

In particular,

$$-1 \leq b \leq 0 \Rightarrow b'(x) \leq x|V(x)|. \tag{6.4}$$

By the monotonicity of $\tilde{m}$, there are unique points $0 < z_1 < x_1 < \cdots < x_{\ell-1} < z_\ell < x_\ell$ where $b_\ell = 0$, and since $b \to -\infty$ as $x \downarrow x_j$, there are last points $y_j \subset [x_{j-1}, z_j]$ where $b(y) = -1$ for $j = 2, 3, \ldots, \ell$ and at $y_1 = 0$, $b(0) = -1$. Integrating $b'$ from $y_j$ to $z_j$, using (6.4), we find

$$\int_{y_j}^{z_j} x|V(x)|\,dx \geq 1$$

so

$$\int_0^{x_\ell} x|V(x)|\,dx \geq \ell.$$

By the oscillation theorem, if $N(V) = \dim P_{(-\infty,0)}(H)$, then

$$N(V) \leq \int_0^\infty x|V(x)|\,dx. \tag{6.5}$$

This is Bargmann's bound [2]. For further discussion, see Schmidt [24].

**Application 2: Denisov-Rakhmanov Theorem.** Rakhmanov [18, 19] (see also [16]) proved a deep theorem about orthogonal polynomials on the unit circle that translates to

**Rakhmanov's Theorem.** *If $J$ is an infinite Jacobi matrix with its spectral measure, $d\mu = f\,dx + d\mu_{\rm s}$ and $f(x) > 0$ for all $x \in [-2,2]$ and* $\operatorname{supp}(d\mu_{\rm s}) \subset [-2,2]$ *(i.e., $\sigma(J) \subset [-2,2]$), then $a_n \to 1$, $b_n \to 0$.*

From the 1990's, there was some interest in extending this to the more general result, where $\sigma(J) \subset [-2,2]$ is replaced by $\sigma_{\rm ess}(J) \subset [-2,2]$. By using the ideas of the proof of Rakhmanov's theorem, one can prove:

**Extended Rakhmanov Theorem.** *There exist $C(\varepsilon) \to 0$ as $\varepsilon \downarrow 0$ so that if $d\mu = f\,dx + du$ and $f(x) > 0$ for a.e. $x$ in $[-2,2]$ and $\sigma(J) \subset [-2-\varepsilon, 2+\varepsilon]$, then*

$$\limsup_{n\to\infty} (|a_n - 1| + |b_n|) \leq C(\varepsilon).$$

Here is how Denisov [7] used this to prove

**Denisov-Rakhmanov Theorem.** *If $d\mu = f(x)\,dx + d\mu_0$, $f(x) > 0$ for a.e. $x \in [-2,2]$ and $\sigma_{\rm ess}(J) \subset [-2,2]$, then $a_n \to 1$ and $b_n \to 0$.*

His proof goes as follows. Fix $\varepsilon$. Since $J$ has only finitely many eigenvalues in $[2+\varepsilon,\infty)$, $P_n(2+\varepsilon)$ has only finitely many sign changes. Similarly, $(-1)^n P_n(-2-\varepsilon)$ has only finitely many sign changes. Thus, we can find $N_0$ so $P_n(2+\varepsilon)$ and $(-1)^n P_n(-2-\varepsilon)$ both have fixed signs if $n > N_0$. Let $\tilde a, \tilde b$ be given by

$$\tilde a_n = a_{N_0+n} \qquad \tilde b_n = b_{N_0+n}.$$

By a use of the comparison and oscillation theorems, $\tilde J$ has no eigenvalues in $(-\infty,-2-\varepsilon) \cup (2+\varepsilon,\infty)$. Thus, by the Extended Rakhmanov Theorem,

$$\limsup_{n\to\infty} (|a_n - 1| + |b_n|) = \limsup_{n\to\infty} (|\tilde a_n - 1| + |\tilde b_n|) \leq C(\varepsilon).$$

Since $\varepsilon$ is arbitrary, the theorem is proven.

**Application 3: Teschl's Proof of the Rofe-Beketov Theorem.** Let $V_0(x)$ be periodic and continuous. Let $H_0 = -d^2/dx^2 + V_0$ on $L^2(0,\infty)$ with $u(0) = 0$ boundary condition. Then

$$\sigma_{\rm ess}(H_0) = \bigcup_{j=1}^\infty [a_j, b_j]$$

with $b_j < a_{j+1}$. (In some special cases, there is only a finite union with one infinite interval.) $(b_j, a_{j+1})$ are called the gaps. In each gap, $H_0$ has either zero or one eigenvalue. Suppose $X(x) \to 0$ as $x \to \infty$, and let $H = H_0 + X$. Since $\sigma_{\rm ess}(H) =$

$\sigma_{\text{ess}}(H_0)$, $H$ also has gaps in its spectrum. When is it true that each gap has at most finitely many eigenvalues? Teschl [29, 30] has proven that if $\int_0^\infty x|X(x)|\,dx < \infty$, then for each $j$, the Wronskian, $w(x)$, of $u(x,b_j)$ and $u(x,a_{j+1})$ has only finitely many zeros. He does this by showing for $H_0$ that $|X(x)| \to \infty$ as $x \to \infty$ and by an ODE perturbation argument, this implies $|w(x)| \to \infty$ for $H$. Thus, by the results of Section 5, there are finitely many eigenvalues in each gap.

It is easy to go from half-line results to whole-line results, so Teschl proves if $\int |x|\,|X(x)|\,dx < \infty$, each gap has only finitely many eigenvalues.

This result was first proven by Rofe-Beketov [22] with another simple proof in Gesztesy-Simon [9]; see that later paper for additional references. Teschl's results are stated for the discrete (Jacobi) case (and may be the first proof for the finite difference situation), but his argument translates to the one above for Schrödinger operators.

## References


[1] N.I. Akhiezer, *Theory of Approximation*, Dover, New York, 1992; Russian original, 1947.

[2] V. Bargmann, On the number of bound states in a central field of force, Proc. Nat. Acad. Sci. U.S.A. **38** (1952), 961–966.

[3] M. Bôcher, The theorems of oscillation of Sturm and Klein, Bull. Amer. Math. Soc. **4** (1897–1898), 295–313, 365–376.

[4] E.A. Coddington and N. Levinson, *Theory of Ordinary Differential Equations*, McGraw-Hill, New York-Toronto-London, 1955.

[5] R. Courant and D. Hilbert, *Methods of Mathematical Physics, Vol. I*, Interscience Publishers, New York, N.Y., 1953.

[6] F. Delyon and B. Souillard, The rotation number for finite difference operators and its properties, Comm. Math. Phys. **89** (1983), 415–426.

[7] S.A. Denisov, On Rakhmanov's theorem for Jacobi matrices, Proc. Amer. Math. Soc. **132** (2004), 847–852.

[8] V.A. Galaktionov and P.J. Harwin, Sturm's Theorems on Zero Sets in Nonlinear Parabolic Equations, in this volume.

[9] F. Gesztesy and B. Simon, A short proof of Zheludev's theorem, Trans. Amer. Math. Soc. **335** (1993), 329–340.

[10] F. Gesztesy, B. Simon and G. Teschl, Zeros of the Wronskian and renormalized oscillation theory, Amer. J. Math. **118** (1996), 571–594.

[11] P. Hartman, Uniqueness of principal values, complete monotonicity of logarithmic derivatives of principal solutions, and oscillation theorems, Math. Ann. **241** (1979), 257–281.

[12] D. Hinton, Sturm's 1836 Oscillation Results. Evolution of the Theory, in this volume.

[13] I. James, *Remarkable Mathematicians. From Euler to von Neumann*, Mathematical Association of America, Washington DC, Cambridge University Press, Cambridge, 2002.

[14] R. Johnson, Oscillation theory and the density of states for the Schrödinger operator in odd dimension, J. Differential Equations **92** (1991), 145–162.

[15] R. Johnson and J. Moser, The rotation number for almost periodic potentials, Comm. Math. Phys. **84** (1982), 403–438.

[16] A. Máté, P. Nevai and V. Totik, Asymptotics for the ratio of leading coefficients of orthonormal polynomials on the unit circle, Constr. Approx. **1** (1985), 63–69.

[17] E. Neuenschwander, The unpublished papers of Joseph Liouville in Bordeaux, Historia Math. **16** (1989), 334–342.

[18] E.A. Rakhmanov, On the asymptotics of the ratio of orthogonal polynomials, Math. USSR Sb. **32** (1977), 199–213.

[19] E.A. Rakhmanov, Asymptotic properties of polynomials orthogonal on the circle with weights not satisfying the Szegő condition, Math. USSR-Sb. **58** (1987), 149–167; Russian original in Mat. Sb. (N.S.) **130(172)** (1986), 151–169, 284.

[20] M. Reed and B. Simon, *Methods of Modern Mathematical Physics, Vol. I: Functional Analysis*, Academic Press, New York, 1972.

[21] M. Reed and B. Simon, *Methods of Modern Mathematical Physics, Vol. II: Fourier Analysis, Self-Adjointness*, Academic Press, New York, 1975.

[22] F.S. Rofe-Beketov, Perturbation of a Hill operator having a first moment and nonzero integral creates one discrete level in distant spectral gaps, Mat. Fizika i Funkts. Analiz. (Kharkov) **19** (1973), 158–159 (Russian).

[23] K.M. Schmidt, An application of the Gesztesy-Simon-Teschl oscillation theory to a problem in differential geometry, J. Math. Anal. Appl. **261** (2001), 61–71.

[24] K.M. Schmidt, A short proof for Bargmann-type inequalities, Proc. Roy. Soc. London. Ser. A **458** (2002), 2829–2832.

[25] B. Simon, The classical moment problem as a self-adjoint finite difference operator, Adv. Math. **137** (1998), 82–203.

[26] C. Sturm, Mémoire sur les Équations différentielles linéaires du second ordre, J. Math. Pures Appl. **1** (1836), 106–186.

[27] C. Sturm, Mémoire sur une classe d'Équations à différences partielles, J. Math. Pures Appl. **1** (1836), 373–444.

[28] C.A. Swanson, *Comparison and Oscillation Theory of Linear Differential Equation*, Academic Press, New York-London, 1968.

[29] G. Teschl, Oscillation theory and renormalized oscillation theory for Jacobi operators, J. Differential Equations **129** (1996), 532–558.

[30] G. Teschl, *Jacobi Operators and Completely Integrable Nonlinear Lattices*, Mathematical Surveys and Monographs **72**, American Mathematical Society, Providence, R.I., 2000.

[31] `http://mathworld.wolfram.com/PolynomialRoots.html`

Barry Simon
Mathematics 253-37
California Institute of Technology
Pasadena, CA 91125, USA
e-mail: `bsimon@caltech.edu`

*W.O. Amrein, A.M. Hinz, D.B. Pearson*
Sturm-Liouville Theory: Past and Present, 45–74


# Charles Sturm and the Development of Sturm-Liouville Theory in the Years 1900 to 1950

W. Norrie Everitt

*This paper is dedicated to the achievements and memory*
*of*
*Charles François Sturm*
*1803 to 1855*

**Abstract.** The first joint publication by Sturm and Liouville in 1837 introduced the general theory of Sturm-Liouville differential equations.

This present paper is concerned with the remarkable development in the theory of Sturm-Liouville boundary value problems, which took place during the years from 1900 to 1950.

Whilst many mathematicians contributed to Sturm-Liouville theory in this period, this manuscript is concerned with the early work of Sturm and Liouville (1837) and then the contributions of Hermann Weyl (1910), A.C. Dixon (1912), M.H. Stone (1932) and E.C. Titchmarsh (1940 to 1950).

The results of Weyl and Titchmarsh are essentially derived within classical, real and complex mathematical analysis. The results of Stone apply to examples of self-adjoint operators in the abstract theory of Hilbert spaces and in the theory of ordinary linear differential equations.

In addition to giving some details of these varied contributions an attempt is made to show the interaction between these two different methods of studying Sturm-Liouville theory.

**Mathematics Subject Classification (2000).** Primary; 34B24, 01A55: Secondary; 34B20, 34L10.

**Keywords.** Sturm-Liouville, ordinary linear differential equations, history of mathematics.

## Contents



## 1. References

The early papers of Sturm on ordinary linear differential equations, and their initial value and boundary value problems, date from 1829 to 1836; see [21], [22] and [23].

The first joint paper by Sturm and Liouville on boundary problems is given in [24]; the results from this remarkable paper are discussed in Section 2 below.

The place and significance of these Sturm and Liouville results in the history of mathematics in the 19th century are considered in detail in the paper of Lützen [16].

In this present paper there are detailed discussions of contributions to Sturm-Liouville theory from: Weyl [32], [33] and [34]; Dixon [5]; Stone [20]; Titchmarsh [26], [27], [28], [29] and [30].

There are later accounts of the Titchmarsh-Weyl theories in the works of Coddington and Levinson [4]; Everitt [6]; Hellwig [9]; Hilb [10]; Hille [11]; Jörgens [12]; Kodaira [14]; Titchmarsh [30]; Yoshida [36].

The theory of Sturm-Liouville differential operators in Hilbert function spaces is developed in the works of Akhiezer and Glazman [1]; Hellwig [9]; Jörgens and Rellich [13]; Naimark [17]; von Neumann [18]; Stone [20].

In view of the significance for Sturm-Liouville theory of the 1910 paper [33] by Weyl special mention is made of the M.Sc. thesis of Race, see [19]. This thesis contains a translation from the German into English of the major part of the Weyl paper; in particular, there is a complete translation of Chapters I and II, together with the translation of the more significant results and remarks from the remaining Chapters, III and IV.

The numerical treatment of Sturm-Liouville boundary value problems has been extensively developed; a summary of results, references together with information on the SLEIGN2 computer code, is given in [2].

Finally there is an epilogue from Weyl in his paper of 1950 [35].

This work is not to be counted as a history of Sturm-Liouville theory for the period 1900 to 1950; such a history should contain reference to many more individual contributions from mathematicians, other than those named at the end of this paper. This paper is an attempt to view the development of Sturm-Liouville theory in the light of advancing techniques in mathematical analysis over this period: the theories of real and complex functions, measure and integration, and linear operators in function spaces.

## 2. Sturm and Liouville and the paper of 1837

As mentioned in the previous section the history of Sturm-Liouville theory is presented in detail in the scholarly paper of Lützen [16]. The main Sturm and Liouville contributions listed in the references are:

(i) The Sturm papers [21], [22] and [23].
(ii) The Sturm and Liouville paper [24].

For a discussion of the results in the three papers listed in (i) see [16].

The Sturm and Liouville paper [24] in (ii) is totally remarkable; it is four pages long but, in almost modern notation, presents the essentials of a Sturm-Liouville boundary value problem on a compact interval, with separated boundary conditions.

The boundary value problem studied by Sturm and Liouville in [24], see also [16, Introduction], is, in their notation,

$$-\frac{d}{dx}\left(k\frac{dV}{dx}\right)+lV=rgV \text{ on the interval } [\mathbf{x},\mathbf{X}] \tag{2.1}$$

with the imposed separated boundary conditions

$$\frac{dV}{dx}-hV=0 \text{ for } x=\mathbf{x} \tag{2.2}$$

$$\frac{dV}{dx}+HV=0 \text{ for } x=\mathbf{X}. \tag{2.3}$$

Here the coefficients $k, l, g$ are positive on the interval $[\mathbf{x}, \mathbf{X}]$, $h$ and $H$ are given positive numbers and $r$ is a real-valued parameter.

It is shown that the initial value problem at the endpoint $\mathbf{x}$, determined by the differential equation (2.1) and the initial boundary condition (2.2), has a non-null solution $V$ for all values of the parameter $r$.

This boundary value problem (2.1), (2.2) and (2.3) only allows for non-null solutions (now called eigenfunctions) for certain values (now called eigenvalues) of the parameter $r$ in (2.1); these values are determined in [24, Page 221] as roots of a transcendental equation, involving the solutions of the equation,

$$\omega(r) = 0; \tag{2.4}$$

namely the equation obtained by inserting the general solution $V$ of (2.1) and (2.2) into the remaining boundary condition (2.3). In the earlier Sturm papers [21], [22] and [23] it is shown that the transcendental equation (2.4) has an infinity of real simple roots which are positive and denoted in [24, Page 221] by $r_1, r_2, \ldots, r_n, \ldots$ arranged in increasing order of magnitude. These are the eigenvalues of the boundary value problem; likewise the associated solution functions (eigenfunctions) are denoted by $V_1, V_2, \ldots, V_n, \ldots$

It is remarked that the transcendental function $\omega$ in (2.4) has the property that $\omega'(r_n) \neq 0$ for all $n \in \mathbb{N}$; in fact it is shown that

$$\int_{\mathbf{x}}^{\mathbf{X}} g(x)V_n^{\,2}(x)\ dx = -k(\mathbf{X})V_n(\mathbf{X})\omega'(r_n) \tag{2.5}$$

where the numbers $k(\mathbf{X})$ and $V_n(\mathbf{X})$ are both non-zero; it is this result that yields the proof that the zeros of the transcendental function $\omega$ are all simple. (Note the use of the prime $'$ notation in (2.5) for the derivative of the function $\omega$; this is the Lagrange notation for the derivative, see [16, Introduction, Page 310].)

The formulae given also show, in effect, that the solution functions

$$V_1, V_2, \ldots, V_n, \ldots$$

have the orthogonality properties

$$\int_{\mathbf{x}}^{\mathbf{X}} g(x)V_m(x)V_n(x)\ dx = 0 \tag{2.6}$$

for all $m, n \in \mathbb{N}$ with $m \neq n$.

Given a function $f$ defined on the interval $[\mathbf{x}, \mathbf{X}]$ the following formulae are obtained (recall that $V$ is the solution of the initial value problem (2.1) and (2.2))

$$\int_{\mathbf{x}}^{\mathbf{X}} g(x)V(x)f(x)\ dx = \sum_{n=1}^{\infty} \left\{ \frac{\int_{\mathbf{x}}^{\mathbf{X}} g(x)V(x)V_n(x)\ dx \cdot \int_{\mathbf{x}}^{\mathbf{X}} g(x)V_n(x)f(x)\ dx}{\int_{\mathbf{x}}^{\mathbf{X}} g(x)V_n^{\,2}(x)\ dx} \right\}, \tag{2.7}$$

and if $F$ on $[\mathbf{x}, \mathbf{X}]$ is given by

$$F(x) = \sum_{n=1}^{\infty} \left\{ \frac{V_n(x) \int_{\mathbf{x}}^{\mathbf{X}} g(y)V_n(y)f(y)\ dy}{\int_{\mathbf{x}}^{\mathbf{X}} g(y)V_n^{\,2}(y)\ dy} \right\} \tag{2.8}$$

then

$$\int_{\mathbf{x}}^{\mathbf{X}} g(x)V(x)F(x)\,dx = \sum_{n=1}^{\infty} \left\{ \frac{\int_{\mathbf{x}}^{\mathbf{X}} g(x)V(x)V_n(x)\,dx \cdot \int_{\mathbf{x}}^{\mathbf{X}} g(x)V_n(x)f(x)\,dx}{\int_{\mathbf{x}}^{\mathbf{X}} g(x)V_n^2(x)\,dx} \right\}.$$

From these results it follows that

$$\int_{\mathbf{x}}^{\mathbf{X}} g(x)V(x)[F(x) - f(x)]\,dx = 0 \tag{2.9}$$

and leads to the conclusion that

$$F(x) - f(x) = 0 \text{ for all } x \in [\mathbf{x}, \mathbf{X}]. \tag{2.10}$$

Finally then the series expansion is obtained

$$f(x) = \sum_{n=1}^{\infty} \left\{ \frac{V_n(x) \int_{\mathbf{x}}^{\mathbf{X}} g(y)V_n(y)f(y)\,dy}{\int_{\mathbf{x}}^{\mathbf{X}} g(y)V_n^2(y)\,dy} \right\} \text{ for all } x \in [\mathbf{x}, \mathbf{X}]. \tag{2.11}$$

**Remark 2.1.** We enter three remarks:

(a) We have followed the outline details of the proof of the critical result (2.10) from the paper [24]; however there seems to be a difficulty in deducing (2.10) from (2.9), since the function $V$ may not be of one sign on the interval $[\mathbf{x}, \mathbf{X}]$; for clarification on this point see the remarks by Lützen [16, Section 49, Page 348].

(b) At the end of the Sturm and Liouville paper [24, Page 223] there is a footnote written by Liouville indicating that complete details of the analysis of the results announced are to be published in a following mémoire; however, Lützen remarks in his paper [16, Section 49, Page 349, Line 5] that this work has been lost.

(c) In modern terminology this last result (2.11) is then the eigenfunction expansion of a continuous function $f$ in terms of these eigenfunctions, within the weighted Hilbert function space $L^2([\mathbf{x}, \mathbf{X}]; g)$.

Sturm, in his first large paper wrote, see [22, Page 106] and [16, Section II, Page 315],

"La résolution de la plupart des problèmes relatifs à la distribution de la chaleur dans des corps de formes diverses et aux petits mouvements oscillatoires des corps solides élastiques, des corps flexibles, des liquides et des fluides élastiques, conduit à des équations différentielles linéaires du second ordre...".

As an example Sturm discussed heat conduction in an inhomogeneous thin bar; in this case the temperature is governed by the linear partial differential equation

$$g\frac{\partial u}{\partial t} = \frac{\partial}{\partial x}\left(k\frac{\partial u}{\partial x}\right) - lu. \tag{2.12}$$

Applying the method of solution by separation of variables leads to ordinary boundary value problems of the form (2.1), (2.2) and (2.3) with the coefficients $g, k, l$. The expansion in terms of the solution functions of these boundary value problems, as given above, then led to formal solutions of boundary and initial value problems associated with partial differential equations of the form (2.12).

**Remark 2.2.** With the advantage of hindsight the following remarks can be made:

(i) As to be expected, given the period when these early results were obtained there are no continuity or differentiability conditions on the three coefficient functions $k, l, g$.

(ii) Essentially, in the earlier results of Sturm, see [21], [22] and [23], there are underlying assumptions of continuity needed to obtain the existence of the real, simple roots $\{r_n : n \in \mathbb{N}\}$ of the transcendental equation (2.4).

(iii) The sign convention in the differential equation (2.1), in particular the negative sign in the derivative term, comes from consideration of the separation of variables technique applied to such partial differential equations as (2.12).

(iv) The positivity of the coefficients $k, l$ and $g$ in the differential equation (2.1), and the positive values of the boundary numbers $h$ and $H$, are responsible for the parameter values of $r$, obtained from the transcendental equation (2.4), being all non-negative and so bounded below on $\mathbb{R}$.

## 3. Notations

The symbols $\mathbb{N}$ and $\mathbb{N}_0$ represent the positive and the non-negative integers, respectively.

The real and complex number fields are denoted by $\mathbb{R}$ and $\mathbb{C}$ respectively. Lebesgue integration is denoted by $L$; the Hilbert function space $L^2((a, b); w)$, given the interval $(a, b)$ and the weight $w$, is the collection (of equivalence classes) of complex-valued, Lebesgue measurable functions $f$ defined on $(a, b)$ such that

$$\int_a^b w(x) |f(x)|^2 \, dx < +\infty. \tag{3.1}$$

The class of Cauchy entire (integral) complex-valued functions defined on $\mathbb{C}$ is denoted by $\mathbf{H}$.

The Sturm and Liouville differential equation in [24] may be rewritten in the form, in modern notation,

$$-\frac{d}{dx}\left(k(x)\frac{dV(x)}{dx}\right) + l(x)V(x) = rg(x)V(x) \text{ for all } x \in [a, b] \tag{3.2}$$

where:

(a) the coefficients $k, l, g : [a, b] \to \mathbb{R}$, with $l, g \in C[a, b]$ and $k \in C^{(1)}[a, b]$

(b) $k, l, g > 0$ on $[a, b]$

(c) the parameter $r \in \mathbb{R}$

(d) the dependent variable $V : [a, b] \to \mathbb{R}$.

To compare with another widely used modern notation for the Sturm-Liouville differential equation, now involving the concept of quasi-derivative (see [17] and [6]), we have (here the prime $'$ denotes classical differentiation on $\mathbb{R}$)

$$-(p(x)y'(x))' + q(x)y(x) = \lambda w(x)y(x) \text{ for all } x \in (a,b), \qquad (3.3)$$

where a wider class of coefficients is admitted as follows:

($\alpha$) for the interval $(a,b)$ the endpoints, in general, satisfy $-\infty \leq a < b \leq +\infty$
($\beta$) the coefficients $p,q,w : (a,b) \to \mathbb{R}$ and $p^{-1}, q, w \in L^1_{\text{loc}}(a,b)$, where $L^1_{\text{loc}}(a,b)$ is the local Lebesgue integration space
($\gamma$) the weight $w(x) > 0$ for almost all $x \in (a,b)$; there are no sign restrictions on the coefficients $p,q$
($\delta$) the spectral parameter $\lambda \in \mathbb{C}$.

In all the cases which follow, Sturm-Liouville differential equations are given in the form (3.3), with restrictions on the coefficients $p,q,w$ and the interval $(a,b)$.

## 4. Mathematical analysis

The four main areas in mathematical analysis that influenced the development of Sturm-Liouville theory, and were in part influenced by this theory in its own right, are:

(i) The Lebesgue integral
(ii) Integrable-square Hilbert function spaces
(iii) Complex function theory on the plane $\mathbb{C}$
(iv) Spectral theory of unbounded operators in Hilbert spaces.

This influence from within and without Sturm-Liouville theory is to be seen in the sections which follow in this paper.

## 5. Hermann Weyl and the 1910 paper

This paper [33] has now long been regarded as one of the most significant contributions to mathematical analysis in the $20^{\text{th}}$ century; whilst not the first paper to consider the singular case of the Sturm-Liouville differential equation it is the first structured consideration of the analytical properties of the equation. The range of new definitions and results is remarkable and set the stage for the full development of Sturm-Liouville theory in the $20^{\text{th}}$ century, as to be seen in the later theory of differential operators in the work of von Neumann [18] and Stone [20], and in the application of complex variable techniques by Titchmarsh [30].

The paper considers the equation (3.3) with the restrictions:

(i) The interval is $[0,\infty)$ and $p,q : [0,\infty) \to \mathbb{R}$; $w(x) = 1$ for all $x \in [0,\infty)$
(ii) The coefficients satisfy: $p,q \in C[0,\infty)$; $p > 0$ on $[0,\infty)$
(iii) The spectral parameter $\lambda \in \mathbb{C}$.

The differential equation is then

$$-(p(x)y'(x))' + q(x)y(x) = \lambda y(x) \text{ for all } x \in [0, \infty). \tag{5.1}$$

Before listing the main results from this paper there are two comments to be made:

1. At the start of the paper, see [33, Chapter I, Page 221, Footnote †)], Weyl points out that no assumption is made concerning the differentiability of the leading coefficient $p$; it is sufficient to require only the continuity of $p$ on $[0, \infty)$; this assumption has the consequence that any term of the form $py'$ has to be considered as a single symbol; in particular the derivative $y'$ may not exist separately at any point of the interval $[0, \infty)$. The initial conditions at the regular endpoint 0 for the existence theorem in [33, Chapter I, Section 1] are of the form, for numbers $\alpha, \beta \in \mathbb{C}$,

$$y(0) = \alpha \qquad (py')(0) = \beta \tag{5.2}$$

which fits in with the existence result that both $y$ and $(py')$ are continuous on $[0, \infty)$. In this respect Weyl is working with the quasi-derivative $(py')$ in place of the classical derivative $y'$ many years before the general introduction of quasi-derivatives, see [1, Appendix 2, Section 123] and [17, Chapter V, Section 15].

2. Throughout the paper, but not always stated in theorems and other results, Weyl assumes a boundary condition, at the regular endpoint 0, on the solutions of the equation (5.1), of the form

$$\cos(h)y(0) + \sin(h)(py')(0) = 0 \tag{5.3}$$

where $h$ is a given real number; see [33, Chapter I, Section 2, (10)].

The main results from this paper are:

(a) Chapter I, Theorem 1: the introduction of the circle method for the differential equation (5.1), and the definition of the limit-circle and limit-point classification of the equation for any point $\lambda \in \mathbb{C}$.

(b) Chapter I, Theorem 2: for all $\lambda \in \mathbb{C} \setminus \mathbb{R}$ the differential equation (5.1) has at least one non-null solution in the Hilbert function space $L^2(0, \infty)$.

(c) Chapter II, Theorem 5: the limit-circle/limit-point classification of the equation (5.1) is independent of the spectral parameter $\lambda$, and depends only on the choice of the two coefficients $p$ and $q$.

(d) Chapter II, Theorem 5: in the limit-circle case all solutions of the differential equation (5.1) are in the Hilbert function space $L^2(0, \infty)$, for all $\lambda \in \mathbb{C}$.

(e) Chapter II, Theorem 5: in the limit-point case there is at most one non-null solution of the equation in the Hilbert function space $L^2(0, \infty)$, for all $\lambda \in \mathbb{C}$.

(f) Chapter II, Corollary to Theorem 5: if the coefficient $q$ is bounded below on $[0, \infty)$ then for all coefficients $p$ the differential equation is in the limit-point case.

(g) Chapter III: at the start of this chapter there are the Weyl definitions of the point spectrum (Punktspektrum), and the continuous spectrum (Streckenspektrum) involving eigendifferentials; the latter definition agrees with the earlier definition of continuous spectrum given by Hellinger in the paper [8].

(h) Chapter III: the Hellinger/Weyl definition of continuous spectrum introduces essentially the concept of eigenpackets as later studied by Rellich [13, Chapter II, Sections 1 and 2] and Hellwig [9, Chapter 10, Section 10.4].

(i) Chapter II, Theorem 4: this theorem gives the eigenfunction expansion in the limit-circle case for the boundary value problem consisting of the Sturm-Liouville differential equation (5.1), a boundary condition (5.3) at the regular endpoint 0, and a boundary condition [33, Chapter II, (41)] at the endpoint $+\infty$; the spectrum of this problem consists only of the point spectrum of real, simple eigenvalues $\{\lambda_n : n \in \mathbb{N}\}$ with an accumulation point at $+\infty$ on the real axis $\mathbb{R}$ of the complex spectral plane $\mathbb{C}$; the corresponding eigenfunctions $\{\varphi_n : n \in \mathbb{N}\}$ form a complete, orthogonal set in the Hilbert function space $L^2(0,\infty)$; there is a pointwise expansion of a function $f \in L^2(0,\infty)$, subject to additional smoothness and integrability conditions on the function $f$, in terms of the eigenfunctions where the infinite series is absolutely convergent and locally uniformly convergent.

(j) Chapter III, Theorem 7: this theorem gives the eigenfunction expansion in the limit-point case for the boundary value problem consisting of the Sturm-Liouville differential equation (5.1), and a boundary condition (5.3) at the regular endpoint 0; in this case no boundary condition is required at the endpoint $+\infty$; the spectrum of this problem consists of a point spectrum, which may be empty, and a continuous spectrum, which may be empty; the point spectrum gives rise to eigenfunctions and a series expansion; the continuous spectrum gives rise to eigendifferentials and an integral expansion; there is a pointwise expansion of a function $f \in L^2(0,\infty)$, subject to additional smoothness and integrability conditions on $f$, in terms of the eigenfunctions and eigendifferentials, where the series and integrals are, respectively, absolutely convergent and locally uniformly convergent.

(k) Chapter IV, Theorem 8: in the limit-point case the continuous spectrum is independent of the choice of the boundary condition at the regular endpoint 0; the point spectrum is different for each particular boundary condition at the regular endpoint 0.

(l) Chapter IV, Theorem 9: if the coefficient $q$ satisfies the condition

$$\lim_{x \to \infty} q(x) = +\infty \tag{5.4}$$

then, for all coefficients $p$, the limit-point case holds and the spectrum for any boundary value problem consists only of the point spectrum of real, simple eigenvalues $\{\lambda_n : n \in \mathbb{N}\}$ with accumulation only at $+\infty$ on the real axis $\mathbb{R}$ of the complex spectral plane $\mathbb{C}$; the corresponding eigenfunctions $\{\varphi_n : n \in \mathbb{N}\}$ form a complete, orthogonal set in the Hilbert function space $L^2(0,\infty)$; for all $n \in \mathbb{N}$ the eigenfunction $\varphi_n$ has exactly $n$ zeros in the open interval $(0,\infty)$.

(m) Chapter IV, Theorem 11: under the following conditions on the coefficients $p, q$

$$p(x) = 1 \text{ for all } x \in [0, \infty) \qquad \int_0^\infty x\,|q(x)|\; dx < +\infty \tag{5.5}$$

the limit-point case holds and the spectrum for any boundary value problem consists of a finite number of strictly negative eigenvalues for the point spectrum, and the half line $[0, \infty)$ for the continuous spectrum.

(n) Chapter IV, Section 22: here there is a remarkable example which illustrates the effectiveness of the Weyl definition of the continuous spectrum of Sturm-Liouville differential equations. Let the coefficients $p, q$ be given by

$$p(x) = 1 \text{ and } q(x) = -x \text{ for all } x \in [0, \infty). \tag{5.6}$$

The resulting Sturm-Liouville differential equation is

$$-y''(x) - xy(x) = \lambda y(x) \text{ for all } x \in [0, \infty) \tag{5.7}$$

which has solutions that can be expressed in terms of the classical Bessel functions. Weyl gives a proof that this equation is in the limit-point case at the singular endpoint $+\infty$; also that, in terms of his definition of the continuous spectrum, any boundary value problem, determined by a boundary condition at the regular endpoint 0, has no eigenvalues and the whole real line $\mathbb{R}$ as continuous spectrum.

(o) Closing remark: at the end of the paper Weyl remarks that all the main results and theorems can be extended to the case when a weight function $w$ is included in the Sturm-Liouville differential equation; that is the results extend to the general equation (3.3)

$$-(p(x)y'(x))' + q(x)y(x) = \lambda w(x)y(x) \text{ for all } x \in [0, \infty). \tag{5.8}$$

Here the weight function $w$ is positive-valued and continuous on the half line $[0, \infty)$; in these circumstances the Hilbert function space is

$$L^2((0, \infty); w),$$

see Section 3 above.

## 6. A.C. Dixon and the paper of 1912

This paper is significant in the development of the Sturm-Liouville differential equation for one reason; it seems to be the first paper in which the continuity conditions on the coefficients $p, q, w$ are replaced by the Lebesgue integrability conditions; these latter conditions are the minimal conditions to be satisfied by $p, q, w$ within the environment given by the Lebesgue integral, see Section 3 above.

The paper uses the same notation [5, Section 1, (1)] of the Sturm-Liouville differential equation as given in the original paper of Sturm and Liouville, *i.e.*,

$$-\frac{d}{dx}\left(k\frac{dV}{dx}\right) + lV = rgV \text{ on the interval } [a, b]; \tag{6.1}$$

however there is no direct reference to the paper [24].

In the notation of Section 3 this Dixon paper considers the equation (3.3) with the assumptions:

(i) The interval $[a,b]$ is compact and $p,q,w : [a,b] \to \mathbb{R}$.
(ii) The coefficients $p,q,w$ satisfy the Lebesgue minimal conditions $p^{-1},q,w \in L^1[a,b]$, and both $p,w > 0$ almost everywhere on $[a,b]$.

The paper discusses the existence of solutions of this Sturm-Liouville differential equation

$$-(p(x)y'(x))' + q(x)y(x) = \lambda w(x)y(x) \text{ for all } x \in [a,b] \tag{6.2}$$

under these coefficient conditions; the existence proof is based on replacing, formally, the differential equation (6.1) with the two integral equations, see [5, Section 1, (2)],

$$U = \int V(l - gr)\, dx \qquad V = \int \frac{1}{k} U\, dx. \tag{6.3}$$

However, boundary conditions at the endpoints $a$ and $b$, which will determine the associated Sturm-Liouville boundary value problem, are difficult to locate.

Certain expansion theorems are given, see for example [5, Section 19]; however again it is difficult to relate such results to the original series type of expansions associated with regular Sturm-Liouville boundary value problems.

This paper by Dixon raises a number of interesting remarks:

1. In effect, the Dixon existence theorem, see [5, Section 3], is a special case of the existence theorem for linear differential systems, with locally integrable coefficients, see [17, Chapter V, Section 16.1]. Note that there seems to be no reference in the paper [5] to the fact that the quasi-derivative $kV'$ exists for solutions of 6.1 (to see this point differentiate the second term in (6.3)) but that the classical derivative $V'$ may not exist.
2. The paper [5] was published two years after the Weyl paper [33] but no reference is given to these earlier results on Sturm-Liouville differential equations. Nevertheless, the Dixon conditions on the coefficients make for a remarkable advance in the study of such differential equations.
3. The years from 1910 onwards saw the introduction of the Lebesgue integral into mathematics; it is interesting to compare the use of integration in the Weyl paper [33], seemingly generalized Riemann integration, with that of the Dixon paper, Lebesgue integration.

## 7. M.H. Stone and the book of 1932

The general theory of unbounded linear operators in Hilbert spaces was developed by John von Neumann from 1927 onwards, and independently by M.H. Stone from 1929 onwards.

The book [20] appeared in the year 1932; it is a remarkable compendium of results and properties of Hilbert spaces. With regard to Sturm-Liouville theory there is in [20, Chapter X, Section 3] a detailed study of Sturm-Liouville differential operators; this study seems to be the first extended account of the properties of Sturm-Liouville differential operators in Hilbert function spaces, under the Lebesgue minimal conditions on the coefficients of the differential equation.

In respect of the standard form of the Sturm-Liouville differential equation given in Section 3 above, see (3.3), the conditions adopted in [20, Chapter X, Section 3] are:

(i) The open interval $(a,b) \subseteq \mathbb{R}$ is arbitrary, so that $-\infty \leq a < b \leq +\infty$.
(ii) The coefficient $w$ is restricted to $w(x) = 1$ for all $x \in (a,b)$.
(iii) The coefficients $p, q : (a,b) \to \mathbb{R}$ and satisfy $p^{-1}, q \in L^1_{\text{loc}}(a,b)$.
(iv) The Sturm-Liouville differential equation and operators are studied in the Hilbert function space $L^2(a,b)$.

Thus the differential equation studied in [20, Chapter X, Section 3] is

$$-(p(x)y'(x))' + q(x)y(x) = \lambda y(x) \text{ for all } x \in (a,b). \tag{7.1}$$

**Remark 7.1.** Four remarks are important:

1. The general weight coefficient $w$, under the conditions of Section 3, can be included in the differential equation to yield all the results in [20, Chapter X, Section 3], with only some additional technical details to the proofs of the stated lemmas and theorems; thus the Stone theory of Sturm-Liouville differential operators applies to the general differential equation

$$-(p(x)y'(x))' + q(x)y(x) = \lambda w(x)y(x) \text{ for all } x \in (a,b), \tag{7.2}$$

working now in the weighted Hilbert function space $L^2((a,b);w)$.

2. Stone in the 1932 book [20] makes only marginal reference to the results of Weyl given in the 1910 paper [33]; in the following discussion of the Stone results these two contributions to Sturm-Liouville theory are brought closer together.

3. In his paper [33] of 1910 Weyl introduced his classification of the singular endpoint as limit-point or limit-circle; this classification is replaced in the Stone book [20] by the concept of the deficiency index of the minimal closed symmetric operator in the Hilbert function space; reference is made to this connection in some of the statements made below.

4. The spectral theory of self-adjoint operators is considered in detail in [20, Chapter V, Section 5], and is based on the properties of the resolution of the identity of the operator; although not discussed in the book [20] it can be shown that the definition of the continuous spectrum (Streckenspektrum) in the Weyl paper [33] is consistent with the definition of continuous spectrum in the book, see [20, Chapter V, Section 5, Theorem 5.11].

The main results from the Stone book, see [20, Chapter X, Section 3], are:

(a) Lemma 10.1: the existence theorem for solutions of the differential equation (7.2) determined by initial conditions at any point $c \in (a, b)$; this existence result involves the requirement to use the quasi-derivative $py'$ in stating the initial conditions for any solution.

(b) Theorem 10.11: this theorem defines and gives the essential properties of the minimal and maximal differential operators in the Hilbert function space $L^2(a, b)$, generated by the differential expression $-(pf')' + qf$; these definitions involve the use of the bilinear form, from the Green's formula for the differential expression, to determine the domain of the minimal operator; the minimal operator is closed and symmetric in $L^2(a, b)$ and its adjoint operator is the maximal operator; the entries in the deficiency index $(m, m)$ of the minimal operator are equal, since this operator is real in $L^2(a, b)$, and take the values $m = 0, 1$ or $2$.

(c) Theorem 10.15: this theorem considers the special case of Theorem 10.11 when one endpoint of the interval, say $a$, satisfies $a \in \mathbb{R}$ and the coefficients then satisfy $p^{-1}, q \in L^1_{\text{loc}}[a, b)$, *i.e.*, this is the case when one endpoint is regular; the maximal operator is then defined as before; the domain of the minimal operator consists of all elements $f$ of the maximal domain satisfying the boundary conditions

$$f(a) = 0 \text{ and } (pf')(a) = 0, \tag{7.3}$$

in addition to a boundary condition at the endpoint $b$; again the minimal operator is closed and symmetric in $L^2(a, b)$ and its adjoint operator is the maximal operator; the entries in the deficiency index $(m, m)$, in $L^2(a, b)$, of the minimal operator are equal since this operator is real in $L^2(a, b)$, and take the values $m = 1$ or $2$. This case is equivalent to the singular problem considered in the 1910 paper [33] of Weyl; the connection here is that the deficiency index $(1, 1)$ is equivalent to the Weyl limit-point case, and the deficiency index $(2, 2)$ is equivalent to the Weyl limit-circle case.

(d) Theorem 10.16: this theorem considers the properties of the Sturm-Liouville differential operators in the special case of Theorem 10.15, when it is assumed that the deficiency index is $(1, 1)$; this is the limit-point case as defined and considered in the Weyl paper [33].

(e) Theorem 10.17: this theorem considers the properties of the Sturm-Liouville differential operators in the special case of Theorem 10.15, when it is assumed that the deficiency index is $(2, 2)$; this is the limit-circle case as defined and considered in the Weyl paper [33].

(f) Theorem 10.18: this theorem considers properties of self-adjoint Sturm-Liouville differential operators when the interval $(a, b)$ is bounded and the coefficients then satisfy $p^{-1}, q \in L^1_{\text{loc}}[a, b]$, *i.e.*, the so-called regular case of Sturm-Liouville theory; here the deficiency index of the minimal closed symmetric operator is $(2, 2)$; the results of this theorem show how to construct the domains of all self-adjoint extensions of this minimal operator by imposing symmetric, separated or coupled boundary conditions, at the regular endpoints $a$ and $b$, on elements of the domain of the maximal operator.

(g) Theorem 10.19: this theorem returns to the case of the general theorem given above as Theorem 10.11, but when the deficiency index is $(2,2)$; it is remarked that this assumption on the index is equivalent to assuming that for some $\lambda \in \mathbb{C}$ the Sturm-Liouville differential equation (7.1) has all solutions in the space $L^2(a,b)$ (this solution property then holds for all $\lambda \in \mathbb{C}$); this condition is equivalent to assuming that the differential equation is in the limit-circle case at both endpoints $a$ and $b$; the results show how to construct the resolvent operator of any self-adjoint extension of the minimal operator, which resolvent is shown to be a Hilbert-Schmidt integral operator in the space $L^2(a,b)$; the point spectrum of this self-adjoint extension is a denumerable infinite point set, with no finite accumulation point; these points are the eigenvalues of the operator, none of which has multiplicity greater than 2; the continuous spectrum of this self-adjoint operator is empty.

(h) Theorem 10.20: this theorem returns to the case of the general theorem given above as Theorem 10.11, but when the deficiency index is $(1,1)$; if the point $c \in (a,b)$, then this index situation can arise when one only of the following two cases holds:
  1. the deficiency index in the space $L^2(a,c)$ for the interval $(a,c]$ is $(2,2)$, and the deficiency index in the space $L^2(c,b)$ for the interval $[c,b)$ is $(1,1)$
  2. the deficiency index in the space $L^2(a,c)$ for the interval $(a,c]$ is $(1,1)$, and the deficiency index in the space $L^2(c,b)$ for the interval $[c,b)$ is $(2,2)$.

  Self-adjoint extensions of the minimal operator, in these circumstances, may have eigenvalues but only of multiplicity 1; the continuous spectrum of such a self-adjoint operator need not be empty.

(i) Theorem 10.21: this theorem returns to the case of the general theorem given above as Theorem 10.11, but when the deficiency index is $(0,0)$; if the point $c \in (a,b)$, then this index situation can arise only when the deficiency index in the space $L^2(a,c)$ for the interval $(a,c]$ is $(1,1)$, and the deficiency index in the space $L^2(c,b)$ for the interval $[c,b)$ is $(1,1)$, *i.e.*, when the differential equation is in the limit-point case at both endpoints $a$ and $b$. Self-adjoint extensions of the minimal operator, in these circumstances, may have eigenvalues but only of multiplicity 1, and the continuous spectrum may not be empty.

## 8. E.C. Titchmarsh and the papers from 1939

The Titchmarsh contributions to Sturm-Liouville theory began about 1938 and concerned the analytic properties of the differential equation

$$-y''(x) + q(x)y(x) = \lambda y(x) \text{ for all } x \in [0,\infty) \tag{8.1}$$

under the coefficient conditions

(i) $q : [0, \infty) \to \mathbb{R}$
(ii) $q$ is continuous on $[0, \infty)$
(iii) the spectral parameter $\lambda \in \mathbb{C}$.

This is a special case of the general Sturm-Liouville differential equation (3.3); however, as in the work of both Weyl and Stone some, but not all, of the Titchmarsh analysis extends to this general form of the equation, and to the case when the coefficients $p, q, w$ satisfy the local integrability conditions given in Section 3.

Both the regular and singular cases of Sturm-Liouville boundary value problems are considered in the Titchmarsh literature; for the requirements of this present paper the three 1941 contributions [26], [27] and [28] are significant; for the consolidated results from Titchmarsh in Sturm-Liouville theory see the second edition of the volume *Eigenfunction expansions* I, [30], and the relevant chapter in the volume *Eigenfunction expansions* II, [31, Chapter XX].

The main thrust of the Titchmarsh method is to apply the extensive theory of functions of a single complex variable to the study of Sturm-Liouville boundary value problems; in the singular case this method involves the existence proof of the complex analytic form of the Weyl integrable-square solution of the differential equation (8.1). This proof of the basic Titchmarsh result dates from 1941, introduces the $m$-coefficient as a Nevanlinna (Herglotz, Pick, Riesz) analytic function which plays such a significant part in the eigenfunction expansion theory of singular Sturm-Liouville boundary value problems. This structure of the Weyl integrable-square solution enables the definition of the Titchmarsh resolvent function $\Phi$ and this step leads to the classical proof of the eigenfunction expansion theorem by contour integration in the complex $\lambda$-plane.

### 8.1. The regular case

The regular Sturm-Liouville case concerns the differential equation (8.1) when considered on a compact interval $[a, b]$

$$-y''(x) + q(x)y(x) = \lambda y(x) \text{ for all } x \in [a, b] \tag{8.2}$$

see [30, Chapter I, Section 1.5]. The starting point is the existence of a solution $\varphi : [a, b] \times \mathbb{C} \to \mathbb{C}$ determined by the initial conditions, for some $\alpha \in [0, \pi)$,

$$\varphi(a, \lambda) = \sin(\alpha) \qquad \varphi'(a, \lambda) = -\cos(\alpha) \qquad \text{for all } \lambda \in \mathbb{C}; \tag{8.3}$$

it follows that $\varphi(x, \cdot) \in \mathbf{H}$ for all $x \in [a, b]$. Similarly for the solution $\chi$ determined, for some $\beta \in [0, \pi)$, by

$$\chi(b, \lambda) = \sin(\beta) \qquad \chi'(b, \lambda) = -\cos(\beta) \qquad \text{for all } \lambda \in \mathbb{C}. \tag{8.4}$$

Then the boundary value problem determined by the equation (8.2) and the separated boundary conditions, see [30, Chapter V, Section 5.3],

$$y(a)\cos(\alpha) + y'(a)\sin(\alpha) = 0 \tag{8.5}$$

$$y(b)\cos(\beta) + y'(b)\sin(\beta) = 0 \tag{8.6}$$

has a discrete, simple, real spectrum with eigenvalues $\{\lambda_n : n \in \mathbb{N}_0\}$ determined by the zeros of the entire function $\omega \in \mathbf{H}$, where

$$\omega(\lambda) := W(\chi, \varphi)(\lambda). \tag{8.7}$$

Here $W(\chi, \varphi)$ is the Wronskian of the solutions $\varphi$ and $\chi$ which is independent of the variable $x \in [a, b]$.

For any zero $\lambda$ of $\omega$ the solutions $\varphi(\cdot, \lambda)$ and $\chi(\cdot, \lambda)$ are linearly dependent and so there exists a real number $k \neq 0$ such that

$$\chi(x, \lambda) = k\varphi(x, \lambda) \text{ for all } x \in [a, b]. \tag{8.8}$$

The zeros of $\omega$ are all real, see [30, Chapter I, Section 1.8].

It then follows that

$$0 \neq k \int_a^b \varphi(x, \lambda)^2 \, dx = \omega'(\lambda) \tag{8.9}$$

so that all the zeros of $\omega$ are not only real but also simple. (At this stage it is interesting to return to the original paper [24] of Sturm and Liouville; this last result echoes the earlier result given above in Section 2, see (2.5).)

The asymptotic properties of the solutions $\varphi$ and $\chi$, for fixed $x$ and large values of $|\lambda|$, show that $\omega$ is an entire function on $\mathbb{C}$ which is of order $1/2$, see [30, Chapter I, Section 1.7]; this implies that $\omega$ has a denumerable number of zeros, see [25, Chapter VIII, Section 8.6]; let these zeros (eigenvalues) be denoted by $\{\lambda_n : n \in \mathbb{N}_0\}$.

The resolvent function $\Phi : [a, b] \times \mathbb{C} \times L^2(a, b) \to \mathbb{C}$ is defined by

$$\Phi(x, \lambda; f) := \frac{\chi(x, \lambda)}{\omega(\lambda)} \int_a^x \varphi(t, \lambda) f(t) \, dt + \frac{\varphi(x, \lambda)}{\omega(\lambda)} \int_x^b \chi(t, \lambda) f(t) \, dt. \tag{8.10}$$

From this definition it follows that, for almost all $x \in [a, b]$,

$$-\Phi''(x, \lambda; f) + q(x)\Phi(x, \lambda; f) = \lambda\Phi(x, \lambda; f) + f(x) \tag{8.11}$$

and that $\Phi(\cdot, \lambda; f)$ satisfies the boundary conditions (8.5) and (8.6) at the endpoints $a$ and $b$.

For $x \in [a, b]$ and $f \in L^2(a, b)$, the resolvent $\Phi(x, \cdot; f)$ is a Cauchy analytic function, regular on $\mathbb{C}\backslash\{\lambda_n : n \in \mathbb{N}_0\}$, with simple poles at the eigenvalues $\{\lambda_n : n \in \mathbb{N}_0\}$. With the corresponding values of $k$ in (8.8) given by $\{k_n : n \in \mathbb{N}_0\}$, the residues are

$$\frac{k_n}{\omega'(\lambda_n)} \varphi(x, \lambda_n) \int_a^b \varphi(t, \lambda_n) f(t) \, dt. \tag{8.12}$$

If now the function $\Phi(x, \cdot; f)$ is integrated around a closed contour $\Gamma_N$ in the complex plane which avoids any of the zeros of $\omega$ but contains the finite number of eigenvalues $\{\lambda_n : n = 0, 1, 2, \ldots, N\}$, then the Cauchy calculus of residues gives

$$\frac{1}{2\pi i} \int_{\Gamma_N} \Phi(x, \lambda; f) \, d\lambda = \sum_{n=0}^{N} \frac{k_n}{\omega'(\lambda_n)} \varphi(x, \lambda_n) \int_a^b \varphi(t, \lambda_n) f(t) \, dt. \tag{8.13}$$

The sequence of contours $\{\Gamma_N : N \in \mathbb{N}_0\}$ is then chosen so to extend to infinity over the complex plane; an argument based on the asymptotic properties of the solutions $\varphi$ and $\chi$ then shows that, for suitable conditions on the function $f \in L^2(a,b)$ and for certain values of the variable $x \in [a,b]$,

$$\lim_{N\to\infty} \frac{1}{2\pi i} \int_{\Gamma_N} \Phi(x,\lambda;f)\, d\lambda = f(x). \tag{8.14}$$

Formally then this argument gives the classical eigenfunction expansion for regular Sturm-Liouville boundary value problems, see [30, Chapter I, Section 1.6, (1.6.5)],

$$f(x) = \sum_{n=0}^{\infty} \frac{k_n}{\omega'(\lambda_n)} \varphi(x,\lambda_n) \int_a^b \varphi(t,\lambda_n) f(t)\, dt, \tag{8.15}$$

or

$$f(x) = \sum_{n=0}^{\infty} \psi_n(x) \int_a^b \psi_n(t) f(t)\, dt \tag{8.16}$$

where, for each $n \in \mathbb{N}$, $\psi_n$ is the real-valued normalized eigenfunction

$$\left[k_n/\omega'(\lambda_n)\right]^{1/2} \varphi(\cdot,\lambda_n),$$

using (8.9).

In [30, Chapter I, Theorem 1.9] Titchmarsh shows how these formal results can be made rigorous to prove that:

1. The infinite series of eigenfunctions (8.16) converges in the topology of $\mathbb{C}$ to $f(x)$, under Fourier type convergence conditions on the function $f$.
2. The normal orthogonal set of eigenfunctions $\{\psi_n : n \in \mathbb{N}_0\}$ is complete in the Hilbert function space $L^2(a,b)$.
3. If given the element $f \in L^2(a,b)$ the generalized Fourier coefficients $\{c_n : n \in \mathbb{N}_0)$ are defined by

$$c_n := \int_a^b \psi_n(x) f(x)\, dx \text{ for all } n \in \mathbb{N}_0 \tag{8.17}$$

then the Parseval identity holds

$$\int_a^b |f(x)|^2\, dx = \sum_{n=0}^{\infty} |c_n|^2 . \tag{8.18}$$

**Remark 8.1.** Two remarks are important:

1. The use of complex variable techniques in [30, Chapter I] illustrates the use of classical analysis to study this regular Sturm-Liouville boundary value problem, without resource to operator theoretic methods; note that there is no mention of the underlying self-adjoint operator in the Hilbert function space $L^2(a,b)$ although the completeness of the eigenfunctions in $L^2(a,b)$ is established. Moreover the methods used enable a proof of the pointwise eigenfunction expansion on the interval $(a,b)$ of a function $f \in L^2(a,b)$,

subject to $f$ satisfying the same conditions that give direct convergence, *i.e.*, convergence in $\mathbb{C}$, of the classical Fourier series.

2. However, it has to be accepted that these complex variable methods do not extend to the analysis of the general regular Sturm-Liouville differential equation

$$-(p(x)y'(x))' + q(x)y(x) = \lambda w(x)y(x) \text{ for all } x \in [a,b] \tag{8.19}$$

when the minimal coefficient conditions $p^{-1}, q, w \in L^1(a,b)$ only are satisfied. In general, with these conditions, it is impossible to apply the transformation, known as the Liouville transformation, to reduce the equation (8.19) to the Titchmarsh form (8.2); the coefficient $p$ may change sign essentially on the compact interval $[a,b]$, and all three coefficients may be unbounded at endpoints and interior points of this interval.

Moreover, whilst the corresponding solutions $\varphi$ and $\chi$ and their Wronskian $\omega$ have the same holomorphic properties on $\mathbb{C}$, it is impossible, in general, to obtain similar asymptotic properties of these functions for large values of $|\lambda|$; however, the reality of the eigenvalues and the orthogonality of the eigenfunctions can be established. The resolvent function $\Phi$ is defined in the same manner, and formal results such as (8.15) and (8.16) follow as above.

### 8.2. The singular case

For the singular case we return to the differential equation (8.1), to be studied in the Hilbert function space $L^2(0,\infty)$,

$$-y''(x) + q(x)y(x) = \lambda y(x) \text{ for all } x \in [0,\infty) \tag{8.20}$$

with the given conditions on the coefficient $q$. Note that if $y$ is a solution of this equation then $y, y', y''$ are all continuous on the interval $[0,\infty)$.

The main problem of extending the Titchmarsh analysis of the regular case to the singular case is to find the equivalent of the boundary function $\chi$; in general there is no method to use a boundary condition at the singular endpoint $+\infty$ in order to determine such a solution as $\chi$; however such an extension is essential to defining a resolvent function $\Phi$ for the singular case.

This problem was resolved by Titchmarsh by using:

1. The existence of the Weyl integrable-square solution of the equation (8.20) for complex values of the parameter $\lambda$, see [33, Chapter I, Theorem 2] and item $(b)$ of Section 5 above.
2. The definition of the $m$-coefficient to give a complex analytic structure to this Weyl solution.

The relevant Titchmarsh work for this programme is to be found in the three 1941 papers [26], [27] and [28]. In the original notation the $l$-coefficient is introduced in [26, Section 2]; the analytic properties of this coefficient are given in [26, Section 5], noting the use that is made of the Vitali convergence theorem. However this original $l$-coefficient notation was later altered to the present $m$-coefficient notation in the first edition of *Eigenfunction Expansions* I, see [29, Chapters II and III], and the second edition [30, Chapters II and III].

Let the solutions $\theta, \varphi : [0, \infty) \times \mathbb{C} \to \mathbb{C}$ of (8.20) be defined by the initial conditions, for some $\alpha \in [0, \pi)$,

$$\begin{cases} \theta(0,\lambda) = \cos(\alpha) & \theta'(0,\lambda) = \sin(\alpha) \\ \varphi(0,\lambda) = -\sin(\alpha) & \varphi'(0,\lambda) = \cos(\alpha) \end{cases} \tag{8.21}$$

and for all $\lambda \in \mathbb{C}$. Then the pair $\theta, \varphi$ forms a basis for solutions of (8.20), for all $\lambda \in \mathbb{C}$, and $\theta(x,\cdot), \theta'(x,\cdot), \varphi(x,\cdot), \varphi'(x,\cdot)$ are all entire (integral) functions on $\mathbb{C}$, for all $x \in [0,\infty)$. Note that this definition of the initial values of the pair $\theta, \varphi$ at 0 yields the Wronskian condition, for all $x \in [0,\infty)$ and $\lambda \in \mathbb{C}$,

$$W(\theta,\varphi)(x,\lambda) \equiv \theta(x,\lambda)\varphi'(x,\lambda) - \theta'(x,\lambda)\varphi(x,\lambda) = 1; \tag{8.22}$$

this sign convention differs from the Titchmarsh convention of the Wronskian given in [30, Chapter II, Section 2.1, (2.1.4)]; this change is to adopt the now standard sign convention for the $m$-coefficient as a Nevanlinna analytic function, see item (iii) below.

Weyl, see [33], proved that either

(i) in the limit-point case

$$\theta(\cdot,\lambda) \notin L^2(0,\infty) \text{ and } \varphi(\cdot,\lambda) \notin L^2(0,\infty) \text{ for all } \lambda \in \mathbb{C} \setminus \mathbb{R} \tag{8.23}$$

or

(ii) in the limit-circle case

$$\theta(\cdot,\lambda) \in L^2(0,\infty) \text{ and } \varphi(\cdot,\lambda) \in L^2(0,\infty) \text{ for all } \lambda \in \mathbb{C}. \tag{8.24}$$

In both cases Titchmarsh showed, see [26] and later in [30, Chapter II, Sections 2.1 and 2.2], that there exists at least one analytic function (the $m$-coefficient) with the properties (note the sign change that has been effected from the formulae in [30, Chapter II, Sections 2.1 and 2.2], see (8.22) above):

(i) $m$ is regular on $\mathbb{C} \setminus \mathbb{R}$

(ii) $\overline{m}(\lambda) = m(\overline{\lambda})$ for all $\lambda \in \mathbb{C} \setminus \mathbb{R}$

(iii) $\mathrm{Im}(m(\lambda)) > 0$ for all $\lambda$ with $\mathrm{Im}(\lambda) > 0$;
$\mathrm{Im}(m(\lambda)) < 0$ for all $\lambda$ with $\mathrm{Im}(\lambda) < 0$

(iv) the analytic function $m(\cdot)$ considered in the upper half plane

$$\mathbb{C}_+ := \{\lambda \in \mathbb{C} : \mathrm{Im}(\lambda) > 0\}$$

of $\mathbb{C}$ may or may not have a continuation into the lower half plane

$$\mathbb{C}_- := \{\lambda \in \mathbb{C} : \mathrm{Im}(\lambda) < 0\}$$

of $\mathbb{C}$; if it does so continue the continuation may or may not be the analytic function $m(\cdot)$ in the lower half plane $\mathbb{C}_-$

(v) the solution $\psi(\cdot,\lambda)$ of the equation (8.20) defined by

$$\psi(x,\lambda) := \theta(x,\lambda) + m(\lambda)\varphi(x,\lambda) \text{ for all } x \in [0,\infty) \text{ and all } \lambda \in \mathbb{C}\setminus\mathbb{R} \tag{8.25}$$

satisfies

$$\int_0^\infty |\psi(x,\lambda)|^2\, dx = \frac{\operatorname{Im}(m(\lambda))}{\operatorname{Im}(\lambda)} < +\infty \text{ for all } \lambda \in \mathbb{C}\setminus\mathbb{R}. \tag{8.26}$$

The existence of this solution $\psi$, and the analytic $m$-coefficient are fundamental to the Titchmarsh eigenfunction analysis as developed in the text [30]. The existence proofs so concerned involve the introduction of the Weyl circle method, see [33, Chapter I, Theorem 1], but with the additional use of complex function theory, see [30, Chapter II, Sections 2.1 and 2.2].

For the $m$-coefficient the following cases occur, see [30, Chapter II, Section 2.1],

1. If the differential equation (8.20) is in the limit-point case at the singular endpoint $+\infty$ then for each choice of the boundary condition parameter $\alpha \in [0,\pi)$ there is a unique $m$-coefficient, which depends upon $\alpha$, with the above properties; for all $\lambda \in \mathbb{C}\setminus\mathbb{R}$ the unique value $m(\lambda)$ is the limit-point of the circles for that value of $\lambda$.
2. If the differential equation (8.20) is in the limit-circle case at the singular endpoint $+\infty$ then for each choice of the boundary condition parameter $\alpha \in [0,\pi)$ there is a continuum of $m$-coefficients, each continuum depending upon $\alpha$; the determination of any particular $m$-coefficient depends upon the limit-circle process, but see the application of the Vitali convergence theorem in [30, Chapter II, Section 2.2].

Although not part of the Titchmarsh theory it is well to remark that the properties (i), (ii) and (iii) above imply that the analytic coefficient $m(\cdot)$ is a Nevanlinna (Herglotz, Pick, Riesz) function and so has a representation of the form, see [1, Chapter 6, Section 69, Theorem 2], where $\gamma, \delta \in \mathbb{R}$ with $\delta \geq 0$,

$$m(\lambda) = \gamma + \delta\lambda + \int_{-\infty}^{+\infty} \left\{ \frac{1}{t-\lambda} - \frac{t}{t^2+1} \right\} d\rho(t) \text{ for all } \lambda \in \mathbb{C}\setminus\mathbb{R}. \tag{8.27}$$

Here the function $\rho : \mathbb{R} \to \mathbb{R}$ is monotonic non-decreasing on $\mathbb{R}$ and satisfies the growth restriction

$$\int_{-\infty}^{+\infty} \frac{1}{1+t^2}\, d\rho(t) < +\infty; \tag{8.28}$$

this function $\rho$ is the spectral function for the $m$-coefficient. The integrals in (8.27) and (8.28) are best interpreted as Lebesgue-Stieltjes integrals with the symbol $\rho$ representing a Borel measure.

The resolvent function $\Phi : [0,\infty) \times \mathbb{C}\setminus\mathbb{R} \times L^2(0,\infty)$ is now defined by

$$\Phi(x,\lambda;f) := \psi(x,\lambda)\int_0^x \varphi(t,\lambda)f(t)\, dt + \varphi(x,\lambda)\int_x^\infty \psi(t,\lambda)f(t)\, dt. \tag{8.29}$$

The Sturm-Liouville boundary value problem considered by Titchmarsh in the singular case is best formulated by requiring that any solution $y$ of the differential equation (8.20) is to satisfy the following conditions, see [30, Chapter II, Section 2.7, Theorem 2.7 (i)],

$$\begin{cases} \text{(i)} & y \in L^2(0,\infty) \\ \text{(ii)} & W(y,\varphi)(0) \equiv y(0)\cos(\alpha) + y'(0)\sin(\alpha) = 0 \\ \text{(iii)} & \lim_{x\to\infty} W(y,\psi(\cdot,\lambda))(x) = 0 \text{ for all } \lambda \in \mathbb{C}\setminus\mathbb{R}. \end{cases} \tag{8.30}$$

The condition (iii) is the required boundary condition at the singular endpoint $+\infty$; it was introduced by Weyl in 1910, see [33, Chapter II, Section 8, (41)], and later by Titchmarsh in 1941, [26, Section 6, (6.2)]; this form of boundary condition heralded the introduction of structured boundary conditions for classical and quasi-differential operators, see [1, Appendix 2, Section 127, Theorem 2] and [17, Chapter V, Section 18.1, Theorem 4].

As in the regular case, see (8.10), the resolvent function $\Phi$ of (8.29) satisfies the boundary conditions (8.30); see [30, Chapter II, Sections 2.8 and 2.9].

The Titchmarsh eigenfunction expansion for the singular Sturm-Liouville boundary value problem (8.20) and (8.30) is considered in two separate cases; the series case when it is assumed that the $m$-coefficient is meromorphic on $\mathbb{C}$, see [30, Chapter II], and the general case in [30, Chapter III].

These two cases both concern the situation when the interval for the differential equation (8.20) is the closed half-line $[0,\infty)$; for both the series and general case Titchmarsh also considers expansion theorems when the interval is the whole real line $(-\infty,\infty)$, see [30, Chapter II, Section 2.18] and [30, Chapter III, Section 3.8].

**8.2.1. The singular case: series expansion.** In this case there is a significant additional assumption in that, given $\alpha \in [0,\infty)$, the $m$-coefficient is assumed to be a meromorphic analytic function on the complex $\lambda$-plane $\mathbb{C}$; this property for $m$ can arise in the limit-point case (for an example see [30, Chapter IV, Section 4.12]); it is always satisfied in the limit-circle case, see [30, Chapter V, Section 5.12].

Suppose that $m$ has a denumerable set of poles at the points $\{\lambda_n : n \in \mathbb{N}_0\}$; then $\lambda_n \in \mathbb{R}$ for all $n \in \mathbb{N}_0$; it is shown in [30, Chapter II, Section 2.2] that all these poles are simple; let the residue of $m$ at $\lambda_n$ be $r_n$ for all $n \in \mathbb{N}_0$. The analysis in [30, Chapter II, Section 2.5] shows that if the sequence of functions $\{\psi_n : n \in \mathbb{N}_0\}$ is defined by

$$\psi_n(x) := |r_n|^{1/2}\,\varphi(x,\lambda_n) \text{ for all } x \in [0,\infty) \text{ and } n \in \mathbb{N}_0, \tag{8.31}$$

then $\{\psi_n : n \in \mathbb{N}_0\}$ is a normal orthogonal set in the space $L^2(0,\infty)$. From this result it follows that, see [30, Chapter II, Section 2.6], the resolvent function $\Phi(x,\cdot;f)$ is meromorphic on the complex plane $\mathbb{C}$, with simple poles at the points $\{\lambda_n : n \in \mathbb{N}_0\}$; the residue at the pole $\lambda_n$ is

$$r_n\varphi(x,\lambda_n)\int_0^\infty \varphi(t,\lambda_n)f(t)\,dt = \psi_n(x)\int_0^\infty \psi_n(t)f(t)\,dt = c_n\psi_n(x), \tag{8.32}$$

where, given $f \in L^2(0,\infty)$, the generalized Fourier coefficients $\{c_n : n \in \mathbb{N}_0\}$ are defined by

$$c_n := \int_0^\infty \psi_n(t)f(t)\ dt \text{ for all } n \in \mathbb{N}_0.$$

It is now possible to prove, following the analysis in [30, Chapter II, Section 2.6], that the solution $\varphi(\cdot,\lambda_n)$ of the differential equation (8.20), with $\lambda = \lambda_n$, satisfies the boundary conditions (8.30); this $\lambda_n$ is an eigenvalue of the singular Sturm-Liouville boundary value problem (8.20) and (8.30), and $\varphi(\cdot,\lambda_n)$ is the associated eigenfunction.

The Titchmarsh analysis, see [30, Chapter II, Section 2.7], now continues to prove that if $f : [0,\infty) \to \mathbb{C}$ satisfies the conditions

$$\begin{cases} \text{(i)} & f, f' \in AC_{\text{loc}}[0,\infty) \\ \text{(ii)} & f, f'' - qf \in L^2(0,\infty) \\ \text{(iii)} & W(f,\varphi)(0) \equiv f(0)\cos(\alpha) + f'(0)\sin(\alpha) = 0 \\ \text{(iv)} & \lim_{x\to\infty} W(f,\psi(\cdot,\lambda))(x) = 0 \text{ for all } \lambda \in \mathbb{C}\setminus\mathbb{R} \end{cases} \tag{8.33}$$

then

$$f(x) = \sum_{n=0}^{\infty} c_n\psi_n(x) \text{ for all } x \in [0,\infty), \tag{8.34}$$

where the infinite series converges absolutely for all $x \in [0;\infty)$ and is locally uniformly convergent on $[0,\infty)$.

Further analysis then shows that for any element $f \in L^2(0,\infty)$ we have the Parseval identity

$$\int_0^\infty |f(x)|^2\ dx = \sum_{n=0}^{\infty} |c_n|^2. \tag{8.35}$$

These last results represent the classical solution to the singular Sturm-Liouville boundary value problem determined by the differential equation (8.20) and the boundary conditions (8.30).

**Remark 8.2.** The Parseval identity (8.35) shows that the normal orthogonal set $\{\psi_n : n \in \mathbb{N}_0\}$ is complete in the Hilbert function space $L^2(0,\infty)$; this result implies that the meromorphic $m$ does have a denumerable number of poles on the real line $\mathbb{R}$; this property was assumed to hold at the beginning of Section 8.2.1.

**8.2.2. The singular case: the general expansion.** Let all the previous definitions concerning the solutions $\theta$ and $\varphi$ of the equation (8.20) and initial conditions (8.21) hold; let an $m$-coefficient be chosen, which implies that the properties (8.25) and (8.26) are satisfied.

To consider the general singular case, *i.e.*, when no additional assumptions are made on the $m$-coefficient, Titchmarsh introduced the $k$ function; originally this function was defined in the 1941 paper [27, Section 4] but here quoted from [30, Chapter III, Section 3.3, Lemma 3.3].

Let $k : \mathbb{R} \to \mathbb{R}$ be defined by (again there is a sign change from the original definition)

$$k(t) := \lim_{\delta \to 0^+} \int_0^t \operatorname{Im}(m(u + i\delta))\ du \text{ for all } t \in \mathbb{R}. \tag{8.36}$$

The analysis in [30, Chapter III, Section 3.3] shows that this limit exists for all $t \in \mathbb{R}$ and that $k$ is a non-decreasing function on $\mathbb{R}$ which satisfies

$$k(t) = \tfrac{1}{2}\{k(t+0) + k(t-0)\} \text{ for all } t \in \mathbb{R}. \tag{8.37}$$

The function $k$ defines a non-negative Borel measure on the real line $\mathbb{R}$ to give the Lebesgue-Stieltjes integrable-square space $L^2(\mathbb{R}; k(\cdot))$ with elements

$$F : (-\infty, +\infty) \to \mathbb{C}$$

satisfying

$$\int_{(-\infty,+\infty)} |F(t)|^2\ dk(t) < +\infty.$$

To obtain the eigenfunction expansion of any element $f \in L^2(0,\infty)$ Titchmarsh gives the following definitions and properties, see [30, Chapter III, Sections 3.4 to 3.6],

1. Let $\chi : [0,\infty) \times (-\infty,+\infty) \to \mathbb{R}$ be defined by the Lebesgue-Stieltjes integral

$$\chi(x,t) := \int_{[0,t]} \varphi(x,s)\ dk(s) \text{ for all } x \in [0,\infty) \text{ and } t \in (-\infty,+\infty); \tag{8.38}$$

then

$$\chi(\cdot,t) \in L^2(0,\infty) \text{ for all } t \in (-\infty,+\infty). \tag{8.39}$$

2. Given $f \in L^2(0,\infty)$ let $\mathcal{F} : (-\infty,+\infty) \to \mathbb{R}$ be defined by

$$\mathcal{F}(t) := \int_0^\infty \chi(x,t) f(x)\ dx \text{ for all } t \in (-\infty,+\infty); \tag{8.40}$$

then it can be shown that $\mathcal{F} \in BV_{\mathrm{loc}}(-\infty,+\infty)$, the space of complex-valued functions, defined on $\mathbb{R}$, which are of bounded variation on all compact intervals of $\mathbb{R}$.

3. Now let $f$ additionally satisfy the boundary conditions (8.33); then

$$f(x) = \frac{1}{\pi} \int_{(-\infty,+\infty)} \varphi(x,t)\ d\mathcal{F}(t) \text{ for all } x \in [0,\infty) \tag{8.41}$$

where the integral is taken in the sense of Lebesgue-Stieltjes and so is absolutely convergent for all $x \in [0,\infty)$.

The result (8.41) is then the general singular Sturm-Liouville eigenfunction expansion for the Titchmarsh differential equation (8.20) when the function $f$ satisfies the boundary conditions (8.33).

In [30, Chapter III, Section 3.7] Titchmarsh gives the Parseval identity for this eigenfunction expansion:

1. Let $f \in L^2(0, \infty)$; then the sequence of functions $\{F_n : n \in \mathbb{N}_0\}$, where

$$F_n : (-\infty, +\infty) \to \mathbb{C},$$

is defined by

$$F_n(t) := \int_0^n \varphi(x,t) f(x) \; dx \text{ for all } t \in (-\infty, +\infty). \tag{8.42}$$

2. Then it may be shown that $F_n \in L^2(\mathbb{R}; k(\cdot))$ for all $n \in \mathbb{N}_0$, that the sequence $\{F_n : n \in \mathbb{N}_0\}$ converges in mean to, say, $F \in L^2(\mathbb{R}; k(\cdot))$ in this space, and

$$\int_0^\infty |f(x)|^2 \; dx = \int_{(-\infty,+\infty)} |F(t)|^2 \; dk(t). \tag{8.43}$$

## 9. The Titchmarsh-Weyl contributions

In this section we review some aspects of the Weyl and Titchmarsh contributions to the development of Sturm-Liouville boundary value problems in the years 1900 to 1950.

### 9.1. The regular case

From the viewpoint of classical analysis the Titchmarsh theory of the regular case, see [30, Chapter I] is still a significant contribution to Sturm-Liouville theory. The spectrum of the boundary value problem is proved to be discrete with a denumerable number of eigenvalues, the eigenfunctions are complete in the Hilbert space $L^2(a, b)$, and the Parseval identity is established. Of course, all these results also follow from the properties of the associated self-adjoint operators in $L^2(a, b)$, see again [20, Chapter X, Section 3, Theorem 10.18].

However, the additional contribution in [30, Chapter I, Section 1.9, Theorem 1.9] is that of the pointwise convergence result of the eigenfunction expansion under Fourier type conditions on the function $f \in L^2(a, b)$, as in the classical theory of Fourier series. Such results are not in general possible using the operator methods of Sturm-Liouville theory.

In the Titchmarsh theory with the differential equation

$$-y''(x) + q(x)y(x) = \lambda y(x) \text{ for all } x \in [a, b],$$

the coefficient $q$ is required to be continuous; however, this condition can be relaxed to $q \in L^1(a, b)$ to achieve the same results.

### 9.2. The singular case: general remarks

We make the following general remarks on the singular Sturm-Liouville case as to be seen in the work of Weyl and Titchmarsh up to the year 1950.

**9.2.1. Pointwise convergence theorems.** The eigenfunction expansions as envisaged originally by Sturm and Liouville [24], and later developed by Weyl [33], Dixon [5] and Titchmarsh [30] are all modelled on the classical theory of Fourier series.

The pointwise convergence, in $\mathbb{C}$, for a function $f : [0, 2\pi) \to \mathbb{C}$ in Fourier series is given in detail in [25, Chapter XIII, Section 8.2]; in addition to starting with $f \in L^1(0, 2\pi)$ some form of smoothness on $f$ is required. However, if $f \in L^2(0, 2\pi)$ then convergence in this space requires no additional restrictions on $f$; the main tool here is the Bessel inequality; the expansion result is seen in the form of the Parseval identity, see [25, Chapter XIII, Section 13.6].

The original problem of Sturm and Liouville [24] in 1837 was to consider pointwise convergence, as viewed at that time, of the series of solution functions. The Weyl paper of 1910 [33] considers both pointwise and $L^2$ convergence; the Dixon paper of 1912 [5] considers only some form of pointwise convergence; of course, in both these theories the concept of convergence has been made rigorous.

The Titchmarsh theory, as now gathered together in the text [30], is influenced throughout by classical Fourier theory. In both the regular and singular cases we have:

(i) direct or pointwise convergence of the eigenfunction expansion requiring some form of second derivative integrability on the function $f$, in addition to the initial requirement that $f \in L^2(a, b)$ or $L^2(0, \infty)$
(ii) integrable-square convergence involving only that $f \in L^2(a, b)$ or $L^2(0, \infty)$ where the main methods are in obtaining the Bessel inequality and, in particular, the Parseval identity.

There are additional pointwise convergence results under Fourier conditions in [30, Chapter IX]; these results relax the conditions on the function $f$ but require additional constraints on the coefficient $q$.

For the Stone book [20] the only convergence considered is that involved with abstract Hilbert space theory; there are no pointwise convergence results.

**9.2.2. Operator theory.** The theory of Sturm-Liouville differential operators is fully developed in the Stone treatise [20, Chapter X, Section 3]; see Section 7 above.

There are interesting connections between this operator theory and some of the classical convergence results in the work of Weyl [33] and Titchmarsh [30].

In [33] the two main pointwise expansion theorems are [33, Chapter II, Theorem II] and [33, Chapter III, Theorem 7]. In both theorems the domains of functions in the space $L^2(0, \infty)$ for which the expansion results are valid are virtually the domains of the corresponding self-adjoint operators in the Stone theory, see respectively the statements of the theorems [20, Chapter X, Section 3, Theorem 10.17] and [20, Chapter X, Section 3, Theorem 10.16].

These remarks also apply, respectively, to the Titchmarsh expansion results given above in Section 8.2.1 and 8.2.2; the set of functions (8.33) for which these expansions are valid are, in effect, the domains of the corresponding Stone self-adjoint differential operators.

In general, well-posed Sturm-Liouville boundary value problems generate self-adjoint differential operators in $L^2(a,b)$ for which the generalized Parseval identity holds. However, if a pointwise expansion theorem is required for the same boundary value problem, then the function in $L^2(a,b)$, to be expanded, has to satisfy additional smoothness conditions equivalent to the function belonging to the domain of the corresponding self-adjoint differential operator.

In one respect Titchmarsh came much closer to the operator theory than did Weyl. The Titchmarsh $k$ function, see the definition in (8.36), introduces the Lebesgue-Stieltjes Hilbert function space $L^2(\mathbb{R};k(\cdot))$. Now the canonical form of the self-adjoint Stone differential operator in $L^2(a,b)$ is simply the self-adjoint multiplication operator in $L^2(\mathbb{R};k(\cdot))$; these two self-adjoint operators are unitarily equivalent and so the spectrum of the Sturm-Liouville operator can be read off from the jump and continuity properties of the monotonic non-decreasing function $k$. Although, seemingly, Titchmarsh was not aware of this operator theoretic connection, he successfully defines the spectrum of his singular Sturm-Liouville boundary value problem in terms of the $k$ function, see [30, Chapter III, Section 3.9] and Section 9.2.3 below.

**9.2.3. The spectrum.** The definition of the spectrum of singular Sturm-Liouville boundary value problems is best seen from the operator theoretic viewpoint; for self-adjoint operators this definition concerns the resolution of the identity of the operator, see [20, Chapter V, Section 5, Definition 5.2 and Theorem 5.11].

From the classical viewpoint, such as is involved with the results and work of Weyl and Titchmarsh, the definitions are equivalent to the operator theoretic definitions but this statement has to be justified analytically. It should be remembered that Weyl, see [33, Chapter III] and items ($g$) and ($h$) of Section 5 above, formulated his definitions some twenty years before the Hilbert space definitions were in place. In the case of Titchmarsh, seemingly, he framed his definition of the spectrum of his Sturm-Liouville boundary value problems, see [30, Chapter III, Section 3.9], solely in analytical terms of his $k$ function, as derived from the $m$-coefficient.

In the case of the Weyl definition of the spectrum the connection with the operator theoretic definition is given by Hellwig, see [9, Chapter 10, Sections 10.4 and 10.5]

As mentioned above, the Titchmarsh definition of the spectrum is made in terms of the monotonic non-decreasing function $k$; see [30, Chapter III, Section 3.9].

Given a singular Sturm-Liouville boundary value problem as discussed in Section 8.2.2 above, let the function $k$ be defined as in (8.36). If $k$ is constant over any open interval of $\mathbb{R}$ then it follows from the formulae (8.38) and (8.40) that this open interval makes no contribution to the expansion formula (8.41). The spectrum of the boundary value problem is then defined as the complement in the real line $\mathbb{R}$ of the set of all such open intervals of constancy of the function $k$. Thus the spectrum of the boundary value problem is a closed subset of the real line $\mathbb{R}$.

Points of $\mathbb{R}$ being being points of discontinuity of $k$ belong to the spectrum and represent the eigenvalues of the boundary value problem; points of continuity but where $k$ is increasing are in the continuous spectrum; also the limit points of these two sets are in the spectrum.

The connection between the Titchmarsh definition and the operator theory definition of the spectrum is best considered in terms of the self-adjoint multiplication operator in the Hilbert space $L^2(\mathbb{R}; k(\cdot))$; see the remarks in the last paragraph of Section 9.2.2 above.

The Titchmarsh definition of the spectrum can also be made in terms of the properties of the $m$-coefficient on the real line $\mathbb{R}$ of the complex plane $\mathbb{C}$. For the definitions concerned and the connection with the spectrum defined by the $k$ function see the paper by Chaudhuri and Everitt [3].

Finally, the Titchmarsh spectral properties can be determined, or defined, from the spectral function $\rho$ of the $m$-coefficient, see (8.27). There is a connection between the Titchmarsh $k$ function and the Nevanlinna $\rho$ function

$$k(t) = \pi\rho(t) \text{ for all } t \in \mathbb{R};$$

see [30, Chapter VI, Section 6.7, (6.7.5)], so that spectral properties may be deduced equally well from $k$ as from $\rho$.

## 10. Aftermath

From 1950 onwards all these properties and results of the Sturm-Liouville differential equations and boundary value problems formed the basis of the spectral theory of ordinary and quasi-differential equations, and the associated differential operators, of arbitrary integer order with real and complex coefficients.

For some details of the progress made in the study of Sturm-Liouville differential equations and boundary value problems, following soon after the years 1900 to 1950, see the texts of Akhiezer and Glazman [1, Appendix 2], Naimark [17], Glazman [7] and Coddington and Levinson [4], and the papers by Kodaira [14] and [15]. For a new proof of the expansion theorem for Sturm-Liouville equations, see the article by Bennewitz and Everitt in this volume.

For a final word from Hermann Weyl see his epilogue [35] written in 1950.

## 11. Acknowledgements

All those who attended the Sturm Bicentennial meeting at the University of Geneva in September 2003 extend their best thanks to Professor Werner Amrein and his colleagues, and to the University, for the excellent and memorable organization of the conference.

I extend personal thanks to Werner Amrein and to David Pearson for advice and help in attending the meeting and in the preparation of this manuscript.

I thank Clemens Markett for a careful scrutiny of the draft of this paper, and for his list of required additions and corrections.

As on so many previous occasions I thank Professor Hubert Kalf, University of Munich, for help in preparing the list of references for this paper, and for his continuing wise council.

I am grateful to two referees who read the original manuscript with great care and attention to detail; the final form of this paper has benefited from their constructive criticisms.

## 12. Salute

When Charles Sturm died in 1855 Liouville said, at the side of the grave,

"Adieu, Sturm, Adieu".

At my lecture to the Sturm Bicentennial meeting at the University of Geneva, in September 2003, I finished with the words

"Merci, Sturm-Liouville, Merci Bien".

## References

[1] N.I. Akhiezer and I.M. Glazman, *Theory of linear operators in Hilbert space: I and II*, Pitman, London and Scottish Academic Press, Edinburgh, 1981.

[2] P.B. Bailey, W.N. Everitt and A. Zettl, The SLEIGN2 Sturm-Liouville code, ACM Trans. Math. Software **27** (2001), 143–192.

[3] Jyoti Chaudhuri and W.N. Everitt, On the spectrum of ordinary second order differential operators, Proc. Royal Soc. Edinburgh (A) **68** (1969), 95–119.

[4] E.A. Coddington and N. Levinson, *Theory of ordinary differential equations*, McGraw-Hill, New York, 1955.

[5] A.C. Dixon, On the series of Sturm and Liouville, as derived from a pair of fundamental integral equations instead of a differential equation, Phil. Trans. Royal Soc. London Ser. A **211** (1912), 411–432.

[6] W.N. Everitt, Some remarks on the Titchmarsh-Weyl $m$-coefficient and associated differential operators, in *Differential Equations, Dynamical Systems and Control Science; A festschrift in Honor of Lawrence Markus*, 33–53, Lecture Notes in Pure and Applied Mathematics **152**, edited by K.D. Elworthy, W.N. Everitt and E.B. Lee, Marcel Dekker, New York, 1994.

[7] I.M. Glazman, *Direct methods of the qualitative spectral analysis of singular differential operators*, Israel Program for Scientific Translations, Jerusalem, 1965, Daniel Davey and Co, Inc., New York, 1966.

[8] E. Hellinger, Neue Begründung der Theorie quadratischer Formen von unendlich vielen Veränderlichen, J. Reine Angew. Math. **136** (1909), 210–271.

[9] G. Hellwig, *Differential operators of mathematical physics*, Addison-Wesley, London, 1964.

[10] E. Hilb, Über gewöhnliche Differentialgleichungen mit Singularitäten und die dazugehörigen Entwicklungen willkürlicher Funktionen, Math. Ann. **76** (1915), 333–339.

[11] E. Hille, *Lectures on ordinary differential equations*, Addison-Wesley, London, 1969.

[12] K. Jörgens, *Spectral theory of second-order ordinary differential operators*, Matematisk Institut, Universitet Århus, 1962/63.

[13] K. Jörgens and F. Rellich, *Eigenwerttheorie gewöhnlicher Differentialgleichungen*, edited by J. Weidmann, Springer, Heidelberg, 1976.

[14] K. Kodaira, Eigenvalue problems for ordinary differential equations of the second order and Heisenberg's theory of $S$-matrices, Amer. J. Math. **71** (1949), 921–945.

[15] K. Kodaira, On ordinary differential equations of any even order and the corresponding eigenfunction expansions, Amer. J. Math. **72** (1950), 502–544.

[16] J. Lützen, Sturm and Liouville's work on ordinary linear differential equations. The emergence of Sturm-Liouville theory, Arch. Hist. Exact Sci. **29** (1984), 309–376.

[17] M.A. Naimark, *Linear differential operators II*, Ungar Publishing Company, New York, 1968.

[18] J. von Neumann, Allgemeine Eigenwerttheorie Hermitescher Funktionaloperatoren, Math. Ann. **102** (1929), 49–131.

[19] D. Race, *Limit-point and limit-circle: 1910-1970*, M.Sc. thesis, University of Dundee, Scotland, UK, 1976.

[20] M.H. Stone, *Linear transformations in Hilbert space*, American Mathematical Society Colloquium Publications **15**, American Mathematical Society, Providence, Rhode Island, 1932.

[21] C. Sturm, Extrait d'un Mémoire sur l'intégration d'un système d'équations différentielles linéaires, Bull. Sci. Math. Férussac **12** (1829), 313–322.

[22] C. Sturm, Mémoire sur les Équations différentielles linéaires du second ordre, J. Math. Pures Appl. **1** (1836), 106–186.

[23] C. Sturm, Mémoire sur une classe d'Équations à différences partielles, J. Math. Pures Appl. **1** (1836), 373–444.

[24] C. Sturm and J. Liouville, Extrait d'un Mémoire sur le développement des fonctions en séries dont les différents termes sont assujettis à satisfaire à une même équation différentielle linéaire, contenant un paramètre variable, J. Math. Pures Appl. **2** (1837), 220–223.

[25] E.C. Titchmarsh, *Theory of functions*, Oxford University Press, second edition, 1952.

[26] E.C. Titchmarsh, On expansions in eigenfunctions IV, Quart. J. of Math. **12** (1941), 33–50.

[27] E.C. Titchmarsh, On expansions in eigenfunctions V, Quart. J. of Math. **12** (1941), 89–107.

[28] E.C. Titchmarsh, On expansions in eigenfunctions VI, Quart. J. of Math. **12** (1941), 154–166.

[29] E.C. Titchmarsh, *Eigenfunction expansions I*, Oxford University Press, first edition, 1946.

[30] E.C. Titchmarsh, *Eigenfunction expansions I*, Oxford University Press, second edition, 1962.

[31] E.C. Titchmarsh, *Eigenfunction expansions II*, Oxford University Press, 1958.

[32] H. Weyl, Über gewöhnliche lineare Differentialgleichungen mit singulären Stellen und ihre Eigenfunktionen, Göttinger Nachrichten (1909), 37–63.

[33] H. Weyl, Über gewöhnliche Differentialgleichungen mit Singularitäten und die zugehörigen Entwicklungen willkürlicher Funktionen, Math. Ann. **68** (1910), 220–269.

[34] H. Weyl, Über gewöhnliche lineare Differentialgleichungen mit singulären Stellen und ihre Eigenfunktionen, Göttinger Nachrichten (1910), 442–467.

[35] H. Weyl, Ramifications old and new of the eigenvalue problem, Bull. Amer. Math. Soc. **56** (1950), 115–139.

[36] K. Yosida, On Titchmarsh-Kodaira's formula concerning Weyl-Stone's eigenfunction expansion, Nagoya Math. J. **1** (1950), 49–58, and **6** (1953), 187–188.

W. Norrie Everitt
School of Mathematics and Statistics
University of Birmingham
Edgbaston
Birmingham B15 2TT
England, UK
e-mail: `w.n.everitt@bham.ac.uk`

*W.O. Amrein, A.M. Hinz, D.B. Pearson*
Sturm-Liouville Theory: Past and Present, 75–98


# Spectral Theory of Sturm-Liouville Operators Approximation by Regular Problems

Joachim Weidmann

**Abstract.** It is the aim of this article to present a brief overview of the theory of Sturm-Liouville operators, self-adjointness and spectral theory: minimal and maximal operators, Weyl's alternative (limit point/limit circle case), deficiency indices, self-adjoint realizations, spectral representation.

The main part of the lecture will be devoted to the method of proving spectral results by approximating singular problems by regular problems: calculation/approximation of the discrete spectrum as well as the study of the absolutely continuous spectrum. For simplicity, most results will be presented only for the case where one end point is regular, but they can be extended to the general case, as well as to Dirac systems, to discrete operators, and (partially) to ordinary differential operators of arbitrary order.

## 1. Introduction

We study self-adjoint operators generated by Sturm-Liouville differential expressions

$$\tau f(x) = \frac{1}{r(x)}\Big\{ -(pf')'(x) + q(x)f(x)\Big\} \text{ in } (a,b), -\infty \le a < b \le \infty$$

in the Hilbert space $L^2(a,b;r)$ with inner product $\langle f,g\rangle := \int_a^b \overline{f(x)}g(x)r(x)\mathrm{d}x$.
We require the following minimal assumptions on the coefficients of $\tau$:

- $p,q,r$ are real-valued measurable functions on $(a,b)$,
- $p(x), r(x) > 0$ almost everywhere in $(a,b)$ (sometimes $p$ is allowed to change sign, but this changes the type of operators radically; therefore we will ignore this case here),
- $1/p, q, r$ are locally integrable in $(a,b)$.

A very special – but anyhow very interesting – case is $(p = r = 1)$

$$\tau f(x) = -f''(x) + q(x)f(x) \;\text{ in } (a,b),\; -\infty \le a < b \le \infty. \tag{1}$$

The above assumptions reduce in this case to:

- $q$ is a real-valued measurable function on $(a,b)$,
- $q$ is locally integrable on $(a,b)$.

All interesting phenomena do occur in this special case (1); therefore in the literature often only this case is studied.

Many technical problems disappear if we assume in addition that $p$ is continuously differentiable and that $r$ and $q$ are continuous in $(a,b)$. In the special case (1) it is reasonable (and in most cases sufficiently general) to assume that $q$ is locally square integrable on $(a,b)$.

We say that $\tau$ is *regular at* $a$, if $a > -\infty$ and the above assumptions hold in $[a,b)$ instead of $(a,b)$. Similarly one defines *regularity* at $b$. $\tau$ is called *regular* if it is regular at $a$ and at $b$. $\tau$ is said to be *singular at* $a$ (resp. $b$) if it is not regular at $a$ (resp. $b$); it is said to be *singular* if it is singular at $a$ or at $b$.

Since in the general case $p$ is not assumed to be continuous, $\tau$ is not a differential expression in the usual sense: $(pf')'$ cannot be written in the form $p'f'+pf''$. The expression $pf'$ is sometimes called the first *quasi derivative.*

The "differential equation"

$$(\tau - z)u = g \quad (z \in \mathbb{C})$$

is transformed to the linear system of first order

$$Y'(x) = \begin{pmatrix} 0 & 1/p(x) \\ q(x) - zr(x) & 0 \end{pmatrix} Y(x) + G(x),$$

where

$$Y(x) = \begin{pmatrix} u(x) \\ pu'(x) \end{pmatrix} \quad \text{and } G(x) = \begin{pmatrix} 0 \\ -r(x)g(x) \end{pmatrix}.$$

Although the coefficients of this system are not continuous but only locally integrable in $(a,b)$, the usual existence and uniqueness for the corresponding initial value problem hold (cf. J. Weidmann [19], Satz 13.2, or [17], Theorem 2.1).

This implies that, given $g \in L^2_{\text{loc}}(a,b;r)$, a fundamental system $u_1, u_2$ and a $c \in (a,b)$, all solutions $f$ of $(\tau - z)u = g$ have the form

$$\begin{aligned} f(x) &= c_1 u_1(x) + c_2 u_2(x) \\ &\quad + \frac{1}{W(u_1,u_2)} \Big\{ u_1(x) \int_c^x u_2(y) g(y) r(y) \mathrm{d}y - u_2(x) \int_c^x u_1(y) g(y) r(y) \mathrm{d}y \Big\}, \end{aligned}$$

with $c_j \in \mathbb{C}$, where $W(u_1,u_2)$ denotes the *generalized Wronskian*

$$W(u_1,u_2) := u_1(x) p u_2'(x) - p u_1'(x) u_2(x).$$

If $\tau$ is regular at $a$ (resp. $b$) and $g$ is $L^2$ at $a$ (resp. $b$), then $c$ may be chosen to be $a$ (resp. $b$).

In this lecture we give a brief introduction to the spectral theory of self-adjoint realizations of $\tau$ and describe the method of approximation of singular operators by regular operators in order to prove spectral properties.

## 2. The minimal and maximal operator

Since all operators studied here will be defined by the map $f \mapsto \tau f$, we need only to describe their domain of definition. The *maximal operator* $T$ is defined on the domain

$$D(T) := \Big\{ f \in L^2(a,b;r) : f \text{ and } pf' \text{ absolutely continuous}, \tau f \in L^2(a,b;r) \Big\}$$

(apparently this is the maximal domain in $L^2(a,b;r)$ on which the map $f \mapsto \tau f$ operates as a differential operator).

For $f, g : (a,b) \to \mathbb{C}$ with $f, g$ and $pf', pg'$ absolutely continuous, the *Lagrange bracket* is defined by

$$[f,g]_x := \overline{f(x)} pg'(x) - \overline{pf'(x)} g(x)$$

(we always write $pf'(x)$ and not $p(x)f'(x)$ since in general $p$ and $f$ are only measurable while $pf'$ is continuous). For $f, g \in D(T)$ the limits

$$[f,g]_a := \lim_{x \searrow a} [f,g]_x \text{ and } [f,g]_b := \lim_{x \nearrow b} [f,g]_x$$

exist, and the *Lagrange identity* holds:

$$\langle Tf, g \rangle - \langle f, Tg \rangle = [f,g]_a^b := [f,g]_b - [f,g]_a.$$

The *(pre-)minimal operator* $T_0'$ is defined on

$$D(T_0') := \Big\{ f \in D(T) : f \text{ has compact support in } (a,b) \Big\}.$$

$D(T_0')$ is dense in $L^2(a,b;r)$; this is trivial under the additional assumptions mentioned in Section 1 ($p$ continuously differentiable, $r$ and $q$ continuous); for the general case see J. Weidmann [19], Satz 13.1, or [17], Theorem 3.7. Hence $T_0'$ is a real (with respect to the natural conjugation $K : f \mapsto \overline{f}$) symmetric operator. This implies that $T_0'$ is closable and that it allows self-adjoint extensions.

The *minimal operator* $T_0$ is defined to be the closure of $T_0'$, $T_0 := \overline{T_0'}$.

**Theorem 2.1.** a) *For every $z \in \mathbb{C}$ the range of the operator $T_0' - z$ is given by*

$$R(T_0' - z) = \left\{ g \in L_0^2(a,b;r) : \begin{array}{l} \int_a^b \overline{u(x)} g(x) r(x) \mathrm{d}x = 0 \textit{ for every} \\ \textit{solution } u \textit{ of } (\tau - \overline{z})u = 0 \end{array} \right\}$$

*($L_0^2(a,b;r)$ is the subspace of $L^2(a,b;r)$ of functions which vanish almost everywhere close to $a$ and $b$).*

b) *If $\tau$ is regular, then $R(T_0 - z) = N(T - \overline{z})^\perp$, $R(T_0 - z)^\perp = N(T - \overline{z})$ (where $N(A)$ is the null-space of the operator $A$).*

c) *$(T_0')^* = T$.*

*Proof.* a) Denote the set on the right-hand side by $R$. Obviously $R(T_0' - z) \subset R$, since for every $f \in D(T_0')$ the function $(T_0' - z)f = (\tau - z)f$ has compact support in $(a,b)$, and for every solution $u$ of $(\tau - \overline{z})u = 0$

$$\int_a^b \overline{u(x)}(\tau - z)f(x)r(x)\mathrm{d}x = \int_a^b \overline{(\tau - \overline{z})u(x)}f(x)r(x)\mathrm{d}x = 0.$$

Let $g \in R$, $h$ the solution of $(\tau - z)h = g$ given by

$$h(x) = \frac{1}{W(u_1,u_2)}\Big\{u_1(x)\int_a^x u_2(y)g(y)r(y)\mathrm{d}y - u_2(x)\int_a^x u_1(y)g(y)r(y)\mathrm{d}y\Big\},$$

where $u_1, u_2$ is a fundamental system of $(\tau - z)u = 0$. The function $h(x)$ vanishes close to $a$ and $b$ since $g$ does so, and $\int_a^b u_j(y)g(y)r(y)\mathrm{d}y = 0$ for $j = 1,2$ (notice that $\overline{u_j(\cdot)}$ are solutions of $(\tau - \overline{z}) = 0$). Hence $h \in D(T_0')$.

b) $R(T_0 - z) = N(T - \overline{z})^\perp$ follows by essentially the same arguments. Since $N(T - \overline{z})$ is finite-dimensional, the second equation follows from this.

c) $T \subset (T_0')^*$ is obvious.

Let $f \in D((T_0')^*)$, $h$ any solution of $\tau h = (T_0')^* f$. This implies for every $g \in D(T_0')$

$$\int_a^b \overline{(f(x) - h(x))}T_0'g(x)r(x)\mathrm{d}x = \Big\langle (T_0')^* f - \tau h, g\Big\rangle = 0,$$

i.e., $R(T_0') \subset N(F)$, where $F$ is the linear functional

$$F : L_0^2(a,b;r) \to \mathbb{C}, \quad k \mapsto \int_a^b \overline{(f(x) - h(x))}k(x)r(x)\mathrm{d}x.$$

If $\{u_1, u_2\}$ is a fundamental system of $\tau u = 0$, this implies together with part a) that

$$N(F_1) \cap N(F_2) = R(T_0') \subset N(F),$$

where the $F_j$ are the linear functionals

$$F_j : L_0^2(a,b;r) \to \mathbb{C}, \quad k \mapsto \int_a^b \overline{u_j(x)}k(x)r(x)\mathrm{d}x.$$

With [18], Lemma 2.21 this implies that $F$ is a linear combination of $F_1$ and $F_2$, hence $f - h$ is the corresponding linear combination of $u_1$ and $u_2$. Therefore $f$ and $pf'$ are absolutely continuous and $\tau f = \tau h \in L^2(a,b;r)$, i. e. $f \in D(T)$. □

The identity $(T_0')^* = T$ implies that $T_0'$ is minimal in the following sense: If $S$ is defined by $f \mapsto \tau f$ on an "essentially" smaller domain than $T_0'$, then $S^* \supsetneqq T$ is not a differential operator in the above sense.

The domain of the minimal operator $T_0$ is characterized by

$$D(T_0) = \Big\{ f \in D(T) : [f,g]_a = [f,g]_b = 0 \;\text{ for every } g \in D(T) \Big\}.$$

If $\tau$ is regular at $a$, then, as $g$ varies over $D(T)$, $g(a)$ and $pg'(a)$ will assume any value, hence the condition $[f,g]_a = 0$ for all $g \in D(T)$ may be replaced by

$$f(a) = pf'(a) = 0.$$

Similarly if $\tau$ is regular at $b$, the condition $[f,g]_b = 0$ for all $g \in D(T)$ is equivalent to

$$f(b) = pf'(b) = 0.$$

## 3. Self-adjoint realizations

As mentioned above $T_0'$, and therefore $T_0 = \overline{T_0'}$, has self-adjoint extensions; for short we call these just *self-adjoint realizations* of $\tau$. From

$$R(T_0 - z)^{\perp} = N(T_0^* - \overline{z}) = N(T - \overline{z}) = \Big\{ L^2(a,b;r)\text{-solutions of } (\tau - \overline{z})u = 0 \Big\}$$

it follows that the *deficiency indices* of $T_0$ can only be $(0,0)$, $(1,1)$ or $(2,2)$.

From this we conclude immediately:

**Theorem 3.1.** *All self-adjoint realizations of $\tau$ have the same essential spectrum. Their absolutely continuous parts are unitarily equivalent.*

The *proof* follows from abstract results of spectral and scattering theory (cf. J. Weidmann [18], Satz 10.17 and [19], Satz 22.19).

The deficiency indices may be explicitly determined by means of the following Weyl's alternative. Here we use the following convenient notation: a function $f : (a,b) \to \mathbb{C}$ is said to lie *left* (respectively *right*) *in* $L^2(a,b;r)$, if the restriction of $f$ to $(a,c)$ (respectively $(c,b)$) for any $c \in (a,b)$ lies in $L^2(a,c;r)$ (respectively $L^2(c,b;r)$).

**Theorem 3.2 (Weyl's alternative).** *Either*

- *for every $z \in \mathbb{C}$ all solutions of $(\tau - z)u = 0$ lie left (respectively right) in $L^2(a,b;r)$,* limit circle case *(lcc) at $a$ (respectively $b$),*
- *or for every $z \in \mathbb{C} \setminus \mathbb{R}$ there is a unique (up to a factor) solution of $(\tau - z)u = 0$ which lies left (respectively right) in $L^2(a,b;r)$,* limit point case *(lpc) at $a$ (respectively $b$) (in this case for every $z \in \mathbb{C}$ there is at least one solution of $(\tau - z)u = 0$ which lies not left (respectively right) in $L^2(a,b;r)$).*

For a proof see, e.g., J. Weidmann [17], Theorem 5.6, or [19], Satz 13.18.

This alternative makes it very easy to decide if $\tau$ is in lpc or lcc at $a$ (respectively $b$): *if for some $z \in \mathbb{C}$ two linearly independent solutions of $(\tau - z)u = 0$ lie left (respectively right) in $L^2(a,b;r)$, then $\tau$ is lcc at $a$ (respectively $b$). If for some $z \in \mathbb{C}$ at least one solution does not lie left (respectively right) in $L^2(a,b;r)$, then $\tau$ is lpc at $a$ (respectively $b$).*

The deficiency indices are:

- $(0, 0)$ if $\tau$ is lpc at $a$ and at $b$,
- $(1, 1)$ if $\tau$ is lpc at one boundary point and lcc at the other one,
- $(2, 2)$ if $\tau$ is lcc at $a$ and at $b$.

If $\tau$ is lpc at $a$ (respectively $b$), then for arbitrary $f, g \in D(T)$ we have $[f, g]_a = 0$ (respectively $[f, g]_b = 0$).

The self-adjoint realizations $A$ of $\tau$ can be explicitly described as follows:

- Deficiency $(0, 0)$: $A = T_0 = T$ is the only self-adjoint realization of $\tau$, i.e., $T_0'$ is essentially self-adjoint.
- Deficiency $(1, 1)$: All self-adjoint realizations $A$ of $\tau$ are given by

$$D(A) = D(A_g) = \Big\{ f \in D(T) : [f, g]_a^b = 0 \Big\},$$

where $g$ is a real function from $D(T) \setminus D(T_0)$. Notice that here $[f, g]_a = 0$ (resp. $[f, g]_b = 0$) if $\tau$ is lpc at $a$ (resp. $b$). Hence
  - lcc at $a$, lpc at $b$: $D(A) = \{ f \in D(T) : [f, g]_a = 0 \}$,
  - lcc at $b$, lpc at $a$: $D(A) = \{ f \in D(T) : [f, g]_b = 0 \}$.

In this special representation of $D(A)$ we may choose $g$ to be any nontrivial real solution of $(\tau - \lambda)u = 0$ with $\lambda \in \mathbb{R}$.
- Deficiency $(2, 2)$: All self-adjoint realizations $A$ of $\tau$ are given by

$$D(A) = \Big\{ f \in D(T) : [f, g_j]_a^b = 0, j = 1, 2 \Big\},$$

where $g_1, g_2 \in D(T)$ are linearly independent modulo $D(T_0)$ and satisfy $[g_j, g_k]_a^b = 0$ for $j, k = 1, 2$. Without restriction $g_1$ and $g_2$ may be chosen to be equal to solutions of $(\tau - \lambda)g = 0$ in neighborhoods of $a$ and $b$ for some $\lambda \in \mathbb{R}$. Only in this case *coupled boundary* conditions are possible. All self-adjoint realizations with separated boundary conditions are given by

$$D(A) = \Big\{ f \in D(T) : [f, g_a]_a = 0, [f, g_b]_b = 0 \Big\}$$

where $g_a$ and $g_b$ are nontrivial real solutions of $(\tau - \lambda)u = 0$ with $\lambda \in \mathbb{R}$.

There exists a huge number of *limit point – limit circle criteria* in the literature. Here we just mention the most important ones for the special case (1) (for these and some more results see N. Dunford-J. Schwartz [3] Chapter XIII, or J. Weidmann [19], Section 13.4).

**Theorem 3.3.** *Let* $\tau = -\dfrac{\mathrm{d}^2}{\mathrm{d}x^2} + q$ *on* $(a, b)$.

a) i) *If there exists a* $C \in \mathbb{R}$ *such that*

$$q(x) \geq C + \frac{3}{4} \frac{1}{(x-a)^2} \quad \textit{for } x \textit{ close to } a,$$

*then* $\tau$ *is lpc at* $a$ *(similarly for* $b$*).*

ii) *If there exists an* $\varepsilon > 0$ *such that*

$$|q(x)| \leq \left(\frac{3}{4} - \varepsilon\right) \frac{1}{(x-a)^2} \quad \textit{for } x \textit{ close to } a,$$

*then* $\tau$ *is lcc at* $a$ *(similarly for* $b$*).*

b) *Let* $b = \infty$. *If there exists a* $C \geq 0$ *such that*

$$q(x) \geq -C|x|^2 \quad \textit{for } x \textit{ close to } \infty,$$

*then* $\tau$ *is lpc at* $\infty$ *(similarly for* $a = -\infty$*).*

In the case of deficiency $(2,2)$ (i.e., lcc at both end points: *quasi regular case*) the resolvent turns out to be compact (actually Hilbert-Schmidt; cf. the explicit description of the resolvent in Section 6); therefore the spectral theory of these operators is not really interesting (their spectrum is purely discrete, with eigenvalues of multiplicity at most 2). For this reason we essentially restrict ourselves to the cases of deficiency indices $(0,0)$ and $(1,1)$, where at most separated boundary conditions occur.

## 4. Spherically symmetric Schrödinger operators

The *Schrödinger operator* $H = -\Delta + V(\cdot)$ of a charged particle moving in the gradient field of the real-valued potential $V(\cdot)$ in $\mathbb{R}^d$ $(d \geq 2)$ generates an operator in $L^2(\mathbb{R}^d)$ with domain $D \supset C_0^\infty(\mathbb{R}^d)$ if $V$ is locally square integrable. If $V$ is

- uniformly locally square integrable for $d \leq 3$,
- uniformly locally in $L^p$ with $p > d/2$ for $d > 3$,

then $V$ is $-\Delta$-bounded with relative bound 0. This implies that $H$ is essentially self-adjoint on $C_0^\infty(\mathbb{R}^d)$; the self-adjoint closure is bounded from below. If in addition $V(x)$ tends to 0 for $|x| \to \infty$ in a suitable sense, then $V$ is $-\Delta$-compact. In this case the self-adjoint closure has essential spectrum $[0,\infty)$, and the discrete spectrum is bounded below with the possible accumulation point 0.

In the case of a spherically symmetric potential $V$, i.e., $V(x) = v(|x|)$, one adequate technique for the more detailed study of the spectral properties of $H$ is the decomposition of $H$ into an orthogonal sum of one-dimensional operators; these are of Sturm-Liouville type.

The subspaces

$$X_{l,j} := \left\{ f \in L^2(\mathbb{R}^d) : f(x) = Y_{l,j}\left(\frac{x}{|x|}\right) g(|x|) \right\}$$

with the spherical harmonics $Y_{l,j}$ $(j = 1, \ldots, N(d,l)$ [1], $l \in \mathbb{N}_0)$ reduce the operator $H$, and the restriction of $H$ to $X_{l,j}$ is unitarily equivalent to

$$\tau_{d,l} = -\frac{\mathrm{d}^2}{\mathrm{d}r^2} + q_{d,l}(r) \quad \text{on } (0,\infty)$$

[1] with $N(d,l) = \frac{(2l+d-2)(l+d-3)!}{l!(d-2)!}$ for $d \geq 2$.

with

$$q_{d,l}(r) = \left\{ l(l+d-2) + \frac{1}{4}(d-1)(d-3) \right\} \frac{1}{r^2} + v(r).$$

If $v$ has reasonable behavior near 0 and near $\infty$ (e.g., not more singular than $1/r$ near 0, and bounded [2] near $\infty$), then $\tau$ is

- lpc at $\infty$ for every $l \in \mathbb{N}$ and every $d \geq 2$,
- lpc at 0 for $l > 0$, $d \geq 3$ and for $l = 0$, $d > 3$,
- lcc at 0 for $l = 0$, $d = 3$

(for $d = 2$ the situation is more complicated).

In the most interesting case $d = 3$ we have lcc only for $l = 0$ at 0; in this case $q_{3,0}(r) = v(r)$. In order to define self-adjoint realizations of $\tau_{3,l}$ a boundary condition at 0 is needed only for $l = 0$; in all other cases the maximal operator generated by $\tau_{3,l}$ is the only self-adjoint realization of $\tau_{3,l}$ (for $d > 3$ this is true for all $l \geq 0$).

The self-adjoint Schrödinger operator $H = -\Delta + V(\cdot)$ in $L^2(\mathbb{R}^d)$ is unitarily equivalent to a sum of self-adjoint realizations of $\tau_{d,l}$ ($l \in \mathbb{N}_0$; the self-adjoint realization of $\tau_{d,l}$ occurs exactly $N(d,l)$ times in the sum).

Only for $d = 3$, $l = 0$ we have the problem of choosing the suitable boundary condition at 0. For reasonable $v(\cdot)$ this boundary condition is as follows (cf. Weidmann [19], Satz 18.14, or [17], Section 17.F)

$$\lim_{r \to 0} \Big( r g'(r) - g(r) \Big) = 0.$$

Together with methods as they are described in the following sections this allows a detailed study of the spectral properties of $H$: the negative eigenvalues can be calculated numerically, and the absolute continuity of the positive spectrum can be shown in many interesting cases.

## 5. Essential self-adjointness on $C_0^\infty(a,b)$

In the case of deficiency $(0,0)$, where $T_0'$ is essentially self-adjoint, one often asks if the operator $T_{00}$ with $D(T_{00}) = C_0^\infty(a,b)$ is essentially self-adjoint. Actually this seems to be true in most cases where $\tau$ can be defined on $C_0^\infty(a,b)$:

**Theorem 5.1.** *Assume that $p$ is continuously differentiable, $q^2/r$ and $p'^2/r$ are locally integrable and $p^2/r$ is locally bounded and bounded away from 0 on $(a,b)$. Then $T_{00}$ is essentially self-adjoint if and only if $\tau$ is lpc at $a$ and $b$.*

*Proof.* From the assumptions it follows that $T_{00} \subset T_0'$ holds, and therefore $\overline{T_{00}} \subset \overline{T_0'}$. Hence the theorem is proved if we show $T_0' \subset \overline{T_{00}}$. For this let $f \in D(T_0')$, i. e. $f$ has compact support in $(a,b)$, $f$ and $pf'$ are absolutely continuous (which, under

[2]Boundedness in $L^2$-mean for $d = 3$ and $L^p$-mean with $p > d/2$ for $d > 3$ is sufficient.

the above assumptions, actually implies that $f$ is continuously differentiable and $f'$ is absolutely continuous), and

$$\frac{q}{r}f \in L^2(a,b;r),\ \frac{1}{r}(pf')' \in L^2(a,b;r) \text{ and } f'' \in L^2(a,b).$$

Let $f_\varepsilon \in D(T_{00})$ be the convolution of $f$ with a smooth $\delta$-approximation. Then $f_\varepsilon \to f$ and $f'_\varepsilon \to f'$ uniformly for $\varepsilon \to 0$, $f''_\varepsilon \to f''$ in $L^2(a,b;r)$ for $\varepsilon \to 0$; hence

$$\frac{q}{r}f_\varepsilon \to \frac{q}{r}f \quad \text{in } L^2(a,b;r)$$

and

$$\frac{1}{r}(pf'_\varepsilon)' = \frac{p'}{r}f'_\varepsilon + \frac{p}{r}f''_\varepsilon \to \frac{p'}{r}f' + \frac{p}{r}f'' = \frac{1}{r}(pf')' \quad \text{in } L^2(a,b;r).$$

This implies $\tau f_\varepsilon \to \tau f$ and therefore $f \in D(\overline{T_{00}})$. □

## 6. The resolvent of self-adjoint realizations

In all cases the resolvent of every self-adjoint realization $A$ of $\tau$ is an integral operator

$$R_z f(x) = \int_a^b k(x,y;z)f(y)r(y)\mathrm{d}y \quad \text{for } z \in \rho(A), f \in L^2(a,b;r),$$

where $k(\cdot,\cdot;z)$ has the form

$$k(x,y;z) = \begin{cases} \sum_{i,j=1}^2 m_{ij}^+(z)u_i(x)u_j(y) & \text{for } a<y<x<b, \\ \sum_{i,j=1}^2 m_{ij}^-(z)u_i(x)u_j(y) & \text{for } a<x<y<b \end{cases}$$

with a fundamental system $\{u_1,u_2\}$ of $(\tau-\lambda)u=0$ and suitable complex numbers $m_{ij}^\pm(z)$ (cf. J. Weidmann [19], Satz 13.14, 13.20 and 13.21, or [17], Section 7).

In general it is not easy to determine the $m_{ij}^\pm(z)$ explicitly. This is much easier in all cases with separated boundary condition (i.e., for every self-adjoint realization if $\tau$ is in the lpc at least at one boundary point).

The self-adjoint realizations with separated boundary conditions are in all cases given by

$$D(A) = \left\{ f \in D(T) : \begin{array}{l} [f,g_a]_a = 0 \text{ if } \tau \text{ is } lcc \text{ at } a, \\ [f,g_b]_b = 0 \text{ if } \tau \text{ is } lcc \text{ at } b \end{array} \right\}, \tag{2}$$

where $g_a, g_b$ are nontrivial real solutions of $(\tau-\lambda)g=0$ with $\lambda \in \mathbb{R}$. At a regular boundary point $c$ ($=a$ or $=b$) every *self-adjoint boundary condition* (s.a.b.c) has the form

$$f(c)\cos\alpha - pf'(c)\sin\alpha = 0 \quad (\alpha \in [0,\pi)). \tag{3}$$

In this case the resolvent of the self-adjoint realization $A$ for $z \in \rho(A)$ is given by

$$\begin{aligned} R_z g(x) &= (A-z)^{-1} g(x) \\ &= \frac{1}{W(u_b, u_a)} \Big\{ u_b(x) \int_a^x u_a(y) g(y) r(y) \mathrm{d}y \\ &\quad + u_a(x) \int_x^b u_b(y) g(y) r(y) \mathrm{d}y \Big\}, \end{aligned}$$

where $u_a$ and $u_b$ are nontrivial solutions of $(\tau - z)u = 0$ which

- lie left, resp. right, in $L^2(a, b; r)$ in the lpc at $a$, resp. $b$,
- satisfy the boundary condition at $a$, resp. $b$, in the lcc at $a$, resp. $b$,

and $W$ is the (generalized) *Wronskian* (cf. Section 1). If $\tau$ is quasi regular (i.e., lcc at $a$ *and* $b$), then this implies (as we already mentioned above) that the resolvent of every self-adjoint realization $A$ is Hilbert-Schmidt, which implies $\sum_{\lambda_n \neq 0} |\lambda_n|^{-2} < \infty$ for the eigenvalues $\lambda_n$ of $A$. Actually much more holds (cf. J. Weidmann [15]):

**Theorem 6.1.** *For every self-adjoint realization $A$ of a quasi regular Sturm-Liouville expression and every $\varepsilon > 0$ we have $\sum_{\lambda_n \neq 0} |\lambda_n|^{-1-\varepsilon} < \infty$. If $A$ is semi-bounded (which holds either for none or for all self-adjoint realizations), then one has $\sum_{\lambda_n \neq 0} |\lambda_n|^{-1} < \infty$. (Notice that in many explicitly solvable regular problems $\sum_{\lambda_n \neq 0} |\lambda_n|^{-1/2-\varepsilon} < \infty$ for every $\varepsilon > 0$.)*

In the general case (i.e., lpc at least at one boundary point) things are more complicated. Again we consider only the case where $\tau$ is regular at $a$ and lpc at $b$. It seems to be something like a general rule: if for every $\lambda$ from an interval $I$ there exists an $L^2$-solution, then there is a purely discrete spectrum in $I$ (only isolated eigenvalues); if for no $\lambda \in I$ there is an $L^2$-solution, then the spectrum is absolutely continuous in $I$ (obviously it is continuous in this case since there cannot be eigenvalues). Although this is true in all (or at least most) explicitly solvable problems, the following sections will show that we are far from being able to prove this in general.

## 7. Approximation by regular problems

In many cases results about the spectral properties (continuity and discreteness) of singular problems can be proved by means of an "approximation" by regular problems. We shall demonstrate this in several cases. But first: *How can singular problems be approximated by regular problems?* Notice that for every $[c, d] \subset (a, b)$ the differential expression $\tau$ is regular in $(c, d)$; of course one may choose $c = a$ (or $d = b$) if $\tau$ is regular at $a$ (respectively $b$).

For an operator $A = A_{g_a,g_b}$ defined by (2) and $[c,d] \subset (a,b)$ we define operators $A_{c,d}$ in $L^2(c,d;r)$ by

$$D(A_{c,d}) := \left\{ f \in D(T_{c,d}) : \begin{array}{lll} \text{any s. a. b. c. at } c & \text{if} & \tau \text{ is lpc at } a, \\ [f,g_a]_c = 0 & \text{if} & \tau \text{ is lcc at } a, \\ \text{any s. a. b. c. at } d & \text{if} & \tau \text{ is lpc at } b, \\ [f,g_b]_d = 0 & \text{if} & \tau \text{ is lcc at } b. \end{array} \right\},$$

where we denote by $T_{c,d}$ the maximal operator generated by $\tau$ in $L^2(c,d;r)$.

**Theorem 7.1.** *With the above definition $A_{c,d}$ converges to $A$ in the sense of* generalized strong resolvent convergence, *i.e.,*

$$(A_{c,d} - z)^{-1} P_{c,d} \xrightarrow{s} (A - z)^{-1} \quad \textit{for } z \in \mathbb{C} \setminus \mathbb{R},\ c \to a,\ d \to b,$$

*and*

$$E_{c,d}(\lambda) P_{c,d} \xrightarrow{s} E(\lambda) \quad \textit{if } \lambda \ \textit{ is not an eigenvalue of } A,$$

*where $P_{c,d}$ is the orthogonal projection in $L^2(a,b;r)$ onto $L^2(c,d;r)$ (= restriction to $(c,d)$).*

*Proof.* In fact, if $A_{c,d}$ is extended to a self-adjoint operator $\widetilde{A}_{c,d}$ in $L^2(a,b;r)$ by

$$D(\tilde{A}_{c,d}) := D(A_{c,d}) \oplus L^2((a,b) \setminus (c,d);r)$$
$$\tilde{A}_{c,d}(f+g) := A_{c,d} f \quad \text{for } f+g \in D(\tilde{A}_{c,d}),$$

then $(\tilde{A}_{c,d} - z)^{-1} \xrightarrow{s} (A - z)^{-1}$; this is the usual strong resolvent convergence of $\tilde{A}_{c,d}$ to $A$. In order to prove this one uses the fact that

$$\widetilde{D} := \left\{ f \in D(T) : \begin{array}{lll} f = 0 \text{ close to } a, & \text{if} & \tau \text{ is lpc at } a, \\ f \sim g_a \text{ close to } a, & \text{if} & \tau \text{ is lcc at } a, \\ f = 0 \text{ close to } b, & \text{if} & \tau \text{ is lpc at } b, \\ f \sim g_b \text{ close to } b, & \text{if} & \tau \text{ is lcc at } b \end{array} \right\}$$

is a core of $A$ (cf. G. Stolz and J. Weidmann [12], Theorem 6, or J. Weidmann [19], Section 14.3).

The strong resolvent convergence of $\tilde{A}_{c,d}$ to $A$ implies the strong convergence of the spectral resolution $\widetilde{E}_{c,d}(\lambda)$ of $\tilde{A}_{c,d}$ to the spectral resolution $E(\lambda)$ of $A$ for every $\lambda \in \mathbb{R}$ where $E(\cdot)$ is continuous (i.e., for every $\lambda$ which is not an eigenvalue of $A$). Since $I - P_{c,d} \xrightarrow{s} 0$ for $c \to a$, $d \to b$, this implies $E_{c,d}(\lambda)P_{c,d} \xrightarrow{s} E(\lambda)$ if $\lambda$ is not an eigenvalue of $A$. □

For simplicity we consider in the sequel only the case where $\tau$ is regular at $a$ and lpc at $b$. (The case where $\tau$ is regular at $b$ and lpc at $a$ is treated similarly. The general case is somewhat more complicated although some of the results follow from the special case by means of the so-called *decomposition method*, cf. J. Weidmann [17], Section 11, or [19], Aufgabe 13.6.) This means that we consider operators $A = A_\alpha$ ($\alpha \in [0,\pi)$) with

$$D(A_\alpha) := \Big\{ f \in D(T) : f(a)\cos\alpha - pf'(a)\sin\alpha = 0 \Big\}.$$

With $d \in (a,b)$ and $d \to b$ these operators are approximated in the above sense by $A_{\alpha,d}$ with

$$D(A_{\alpha,d}) := \left\{ f \in D(T_{a,d}) : \begin{array}{l} f(a)\cos\alpha - pf'(a)\sin\alpha = 0, \\ \text{any s. a. b. c. at } d \end{array} \right\},$$

where the boundary condition at $d$ may depend on $d$ or not, for example the following situations will occur later:

- boundary condition $(\alpha)$ at $d$, especially Dirichlet condition $(\alpha = 0)$ $f(d) = 0$, or
- $[f,g]_d = 0$ with a fixed real function $g$ (e.g., a nontrivial real solution $g$ of $(\tau - \lambda)g = 0$ with $\lambda \in \mathbb{R}$).

As mentioned above, the quasi regular case (lcc at both endpoints) is not interesting in connection with the questions treated in this paper. Just for completeness we end this section with a short look at this case:

If $\tau$ is quasi regular, and the self-adjoint realization $A$ is defined by

$$D(A) := \{f \in D(T) : [f, g_j]_a^b = 0 \ \text{ for } j = 1,2\}$$

with solutions $g_1, g_2$ of $(\tau - \lambda)g = 0$ $(\lambda \in \mathbb{R})$ satisfying $[g_j, g_k]_a^b = 0$ for $j,k = 1,2$, then the corresponding way of defining approximating regular operators in $L^2(c,d;r)$ is

$$D(A_{c,d}) := \left\{ f \in D(T_{c,d}) : [f, g_j]_c^d = 0 \ \text{ for } j = 1,2 \right\}.$$

With similar arguments as in Theorem 6.1 it can be shown that $A_{c,d}$ converges to $A$ in the sense of generalized strong resolvent convergence. But actually much more holds:

We recall the following result from the paper by P. B. Bailey, W. N. Everitt, J. Weidmann and A. Zettl [2]. Notice that this result holds for every family of self-adjoint realizations $A_{c,d}$ of $\tau$ in $L^2(c,d;r)$ which, for $c \searrow a$ and $d \nearrow b$, converges to $A$ in the sense of generalized strong resolvent convergence, not only for those constructed above in terms of the functions $g_1, g_2$ which occur in the boundary conditions of $A$.

**Theorem 7.2.** *Assume that $\tau$ is lcc at $a$ and $b$. Let $A$ be any self-adjoint realization of $\tau$ in $L^2(a,b;r)$. For $(c,d) \subset (a,b)$ let $A_{c,d}$ be self-adjoint realizations of $\tau$ in $L^2(c,d;r)$ such that $A_{c,d} \to A$ in the sense of generalized strong resolvent convergence for $c \searrow a$ and $d \nearrow b$. Then $(A_{c,d} - z)^{-1}P_{c,d} \to (A - z)^{-1}$ for $z \in \mathbb{C}\setminus\mathbb{R}$ in the sense of Hilbert-Schmidt norm. (This of course implies convergence of eigenvalues including the norm convergence of the corresponding eigenprojections. Notice that in general the eigenvalues of $A_{c,d}$ and $A$ may have multiplicity 2.)*

The *proof* uses the fact that the resolvent kernels $k(x, y; z)$ and $k_{c,d}(x, y; z)$ have the form (cf. Section 6)

$$k(x,y;z) = \begin{cases} \sum_{i,j=1}^{2} m_{ij}^{+} u_i(x) u_j(y) & \text{for } a < y < x < b, \\ \sum_{i,j=1}^{2} m_{ij}^{-} u_i(x) u_j(y) & \text{for } a < x < y < b \end{cases}$$

and

$$k_{c,d}(x,y;z) = \begin{cases} \sum_{i,j=1}^{2} m_{ij}^{+}(c,d) u_i(x) u_j(y) & \text{for } c < y < x < d, \\ \sum_{i,j=1}^{2} m_{ij}^{-}(c,d) u_i(x) u_j(y) & \text{for } c < x < y < d \end{cases}$$

with a fundamental system $\{u_1, u_2\}$ of $(\tau - z)u = 0$ and suitable complex numbers $m_{ij}^{\pm}$ and $m_{ij}^{\pm}(c,d)$. The resolvent convergence implies that

$$m_{ij}^{\pm}(c,d) \to m_{ij}^{\pm} \quad \text{for } c \searrow a, d \nearrow b.$$

From this the convergence of the resolvents in Hilbert-Schmidt norm follows.

## 8. Point spectrum

In this Section we discuss the question: which spectral properties can be deduced from the existence or nonexistence of $L^2$-solutions of $(\tau - \lambda)u = 0$ for some $\lambda \in \mathbb{R}$, respectively for all $\lambda$ from an interval $I$. The following classical result is essentially due to Ph. Hartman and A. Wintner [7]. A similar result for higher-order operators is given in J. Weidmann [17], Section 11.

**Theorem 8.1.** *Let $\tau$ be regular at $a$, lpc at $b$.*

a) *If for some $\lambda \in \mathbb{R}$ the equation $(\tau - \lambda)u = 0$ has no $L^2$-solution, then $\lambda$ belongs to the essential spectrum of every self-adjoint realization $A_\alpha$ of $\tau$ (of course $\lambda$ is not an eigenvalue of $A_\alpha$ for any $\alpha$; if the assumption holds for all $\lambda$ from an interval $I$, then the spectrum is continuous in $I$).*
b) *If for every $\lambda$ from an interval $I$ there exists an $L^2$-solution of $(\tau - \lambda)u = 0$, then for every self-adjoint realization $A_\alpha$ of $\tau$ the continuous spectrum in $I$ is empty and the point spectrum is nowhere dense in $I$.*[3]

*Proof.* a) By assumption $\lambda$ is certainly not an eigenvalue. Since the deficiency indices of $T_0$ are $(1,1)$ and $R(T_0 - \lambda)^\perp = N(T - \lambda) = \{0\}$, $\lambda$ is not in the regularity domain of $T_0$ [4] and therefore not in the resolvent set of $A_\alpha$. Hence $\lambda$ is in the essential spectrum of $A_\alpha$.

---

[3] It seems to be unknown whether in this case the spectrum is purely discrete in $I$.

[4] This is the set of those $z \in \mathbb{C}$ for which there exists a $k(z) > 0$ such that $\|(T_0 - z)f\| \geq k(z)\|f\|$ for every $f \in D(T_0)$ (cf. M. A. Naimark [9] or J. Weidmann [18]).

b) Throughout this part of the proof we use the solutions $u_j(\cdot, s)$ of $(\tau - s)u = 0$ defined by

$$u_1(a,s) = \sin\alpha, \qquad pu_1'(a,s) = \cos\alpha,$$
$$u_2(a,s) = \cos\alpha, \qquad pu_2'(a,s) = -\sin\alpha.$$

By assumption for every $\lambda \in I$ which is not an eigenvalue of $A_\alpha$ there exists a unique $m(\lambda) \in \mathbb{C}$ ($m$-function) such that

$$\psi(\cdot,\lambda) = m(\lambda)u_1(\cdot,\lambda) + u_2(\cdot,\lambda) \in L^2(a,b;r)$$

(notice that $u_1(\cdot,\lambda) \notin L^2(a,b;r)$ since $\lambda$ is not an eigenvalue of $A_\alpha$). Obviously we have for every $\mu \in \mathbb{R}$

$$[\psi(\cdot,\lambda), u_1(\cdot,\mu)]_a = [u_2(\cdot,\lambda), u_1(\cdot,\mu)]_a = 1.$$

At first we show that the point spectrum is nowhere dense in $I$: Let $\{\lambda_n : n \in M\}$ be the eigenvalues of $A_\alpha$ in $I$,

$$v_n(\cdot) := \frac{u_1(\cdot,\lambda_n)}{\|u_1(\cdot,\lambda_n)\|}$$

the corresponding normalized eigenfunctions. Notice that

$$[\psi(\cdot,\lambda), v_n]_a = \frac{1}{\|u_1(\cdot,\lambda)\|}[\psi(\cdot,\lambda), u_1(\cdot,\lambda_n)]_a = \frac{1}{\|u_1(\cdot,\lambda)\|},$$

while, since $\psi$ and $v_n$ are in $L^2(a,b;r)$,

$$[\psi(\cdot,\lambda), v_n]_b = 0,$$

and therefore

$$\langle \psi(\cdot,\lambda), v_n\rangle = \frac{1}{\lambda - \lambda_n}[\psi(\cdot,\lambda), v_n]_a^b = \frac{1}{\|u_1(\cdot,\lambda_n)\|}\frac{-1}{\lambda-\lambda_n}.$$

This implies for $\lambda \in I \setminus \{\lambda_n : n \in M\}$

$$\infty > \|\psi(\cdot,\lambda)\|^2 \geq \sum_{n\in M} \left|\langle \psi(\cdot,\lambda), v_n\rangle\right|^2 = \sum_{n \in M} \frac{1}{\|u_1(\cdot,\lambda_n)\|^2}\left|\frac{1}{\lambda-\lambda_n}\right|^2.$$

Assume now that $\{\lambda_n : n \in \mathbb{N}\}$ is dense in a subinterval $[\lambda', \lambda'']$. Then for every $N \in \mathbb{N}$ the set

$$K_N := \left\{\lambda \in [\lambda',\lambda''] : \sum_{k=1}^{\infty} \frac{1}{\|u_1(\cdot,\lambda_n)\|^2}\left|\frac{1}{\lambda-\lambda_n}\right|^2 < N\right\}$$

is nowhere dense in $[\lambda', \lambda'']$, and therefore

$$[\lambda',\lambda''] \setminus \{\lambda_n : n \in \mathbb{N}\} = \bigcup_{N\in\mathbb{N}} K_N$$

is of first category. Hence $\{\lambda_n : n \in \mathbb{N}\}$ is of second category, in contradiction to the countability.

It remains to show that there is no continuous spectrum in $I$. Without restriction we may assume that $I$ is compact (otherwise take any compact subinterval of $I$).

Let $\lambda$ be any fixed value in $I$ which is not an eigenvalue of $A_\alpha$. For $d \in (a,b)$ let $A_{\alpha,d}$ be the self-adjoint realization in $L^2(a,d;r)$ defined by

$$D(A_{\alpha,d}) := \Big\{ f \in D(T_{a,d}) : \text{b. c. (3) at } a,\ [\psi(\cdot,\lambda), f]_d = 0 \Big\},$$

$\lambda_{d,n}$ the eigenvalues of $A_{\alpha,d}$ in $I$; $\|u_1(\cdot,\lambda_{d,n})\|^{-1} u_1(x,\lambda_{d,n})$ are the normalized eigenfunctions corresponding to the eigenvalues $\lambda_{d,n}$. From

$$\Big[\psi(x,\lambda), u_1(x,\lambda_{d,n})\Big] = \Big[u_2(x,\lambda), u_1(x,\lambda_{d,n})\Big] = \begin{cases} 1 & \text{for } x = a, \\ 0 & \text{for } x = d, \end{cases}$$

it follows that

$$\Big\langle \psi(\cdot,\lambda), u_1(\cdot,\lambda_{d,n}) \Big\rangle_{(a,d)} = \frac{1}{\lambda - \lambda_{d,n}} \Big[\psi(\cdot,\lambda), u_1(\cdot,\lambda_{d,n})\Big]_a^d = \frac{-1}{\lambda - \lambda_{d,n}},$$

and for every $\varepsilon > 0$ with $\lambda \pm \varepsilon \in I$

$$\begin{aligned} \|\psi(\cdot,\lambda)\|^2 &\geq \int_a^d |\psi(x,\lambda)|^2 r(x) \mathrm{d}x \\ &\geq \sum_{|\lambda - \lambda_{d,n}| \leq \varepsilon} \Big| \Big\langle \psi(\cdot,\lambda), u_1(\cdot,\lambda_{d,n}) \Big\rangle_{(a,d)} \Big|^2 \frac{1}{\|u_1(\cdot,\lambda_{d,n})\|^2} \\ &\geq \sum_{|\lambda - \lambda_{d,n}| \leq \varepsilon} \frac{1}{|\lambda - \lambda_{d,n}|^2} \frac{1}{\|u_1(\cdot,\lambda_{d,n})\|^2} \\ &\geq \frac{1}{\varepsilon^2} \sum_{|\lambda - \lambda_{d,n}| \leq \varepsilon} \frac{1}{\|u_1(\cdot,\lambda_{d,n})\|^2}. \end{aligned}$$

This implies, with $c := \|\psi(\cdot,\lambda)\|^2$

$$\sum_{|\lambda - \lambda_{d,n}| \leq \varepsilon} \frac{1}{\|u_1(\cdot,\lambda_{d,n})\|^2} \leq c\varepsilon^2.$$

For an arbitrary $f \in L^2(a,b;r)$ with compact support in $[a,b)$ and $d \in (a,b)$ such that $\operatorname{supp} f \subset [a,d)$ we have (since $|u_1(x,s)|$ is uniformly bounded for $x \in \operatorname{supp} f$ and $s \in I$)

$$\begin{aligned} \Big| \Big\langle \Big(E_d(\lambda \pm \varepsilon) - E_d(\lambda)\Big) f, f \Big\rangle \Big| &\leq \sum_{|\lambda - \lambda_{d,n}| \leq \varepsilon} \Big| \Big\langle f, u_1(\cdot,\lambda_{d,n}) \Big\rangle \Big|^2 \frac{1}{\|u_1(\cdot,\lambda_{d,n})\|^2} \\ &\leq c_1 \sum_{|\lambda - \lambda_{d,n}| \leq \varepsilon} \frac{1}{\|u_1(\cdot,\lambda_{d,n})\|^2} \leq c_2 \varepsilon^2, \end{aligned}$$

where $c_1$ and $c_2$ of course depend on $f$.

For $d \to b$ it follows, if $\lambda \pm \varepsilon$ are also not eigenvalues of $A_\alpha$, that

$$\Big| \Big\langle \Big(E(\lambda \pm \varepsilon) - E(\lambda)\Big) f, f \Big\rangle \Big| \leq c_2 \varepsilon^2.$$

Hence $\langle E(\cdot)f, f\rangle$ is differentiable with derivative 0 in every point $\lambda \in I$ which is not an eigenvalue of $A_\alpha$; therefore the continuous part of the spectral measure of $A_\alpha$ vanishes in $I$. □

## 9. Approximation of the discrete spectrum

In the complement of the essential spectrum there is at most discrete point spectrum, i.e., isolated eigenvalues (notice that the eigenvalues of a Sturm-Liouville-operator have at most multiplicity 2). For regular problems there exist methods to calculate or at least to approximate numerically the eigenvalues. This leads to the question whether it is possible to determine/approximate the isolated eigenvalues by approximating singular problems by regular problems.

Especially simple is the case where $T_0$ (and therefore every self-adjoint realization of $\tau$) is bounded from below; then the spectrum below the essential spectrum consists at most of simple eigenvalues which accumulate at most at the lower bound of the essential spectrum.

**Theorem 9.1.** *Assume that $\tau$ is regular at $a$, lpc at $b$, and that $T_0$ is bounded below. Let $A_\alpha$ be the self-adjoint realization of $\tau$ with boundary condition* (3) *at $a$, $A_{\alpha,d,0}$ the self-adjoint realization of $\tau$ in $L^2(a,d;r)$ defined by*

$$D(A_{\alpha,d,0}) := \Big\{ f \in D(T_{a,d}) : b.\ c.\ (3), f(d) = 0 \Big\}.$$

*Then for every $\lambda$ below the essential spectrum of $A_\alpha$ which is not an eigenvalue of $A_\alpha$ we have*

$$\Big\| E_{\alpha,d,0}(\lambda)P_d - E_\alpha(\lambda) \Big\| \to 0 \quad \textit{for } d \to b,$$

*where $P_d$ is restriction to $(a,d)$ (the orthogonal projection from $L^2(a,b;r)$ onto $L^2(a,d;r)$). This implies that the eigenvalues of $A_\alpha$ below the essential spectrum are exactly the limits of eigenvalues of $T_{\alpha,d,0}$ for $d \to b$; the corresponding eigenprojections converge in norm.*

*Proof.* From Theorem 7.1 (special case $c = a$) it follows that

$$E_{\alpha,d,0}(\lambda) \xrightarrow{s} E_\alpha(\lambda) \quad \text{for } \lambda \notin \sigma_p(A_\alpha),\ d \to b.$$

Therefore the result follows if we show that

$$\dim E_{\alpha,d,0}(\lambda) = \dim E_{\alpha,d,0}(\lambda)P_d \leq \dim E_\alpha(\lambda) \tag{4}$$

(notice that $\dim E_\alpha(\lambda) < \infty$ for $\lambda$ below the essential spectrum). Here we use the following well-known result: *If $P_n, P$ are orthogonal projections with* $\dim P_n \leq \dim P < \infty$ *and* $P_n \xrightarrow{s} P$, *then* $\|P_n - P\| \to 0$. The inequality (4) follows from the fact that $A_{\alpha,d,0}$ is the Friedrichs extension in $L^2(a,d;r)$ of the operator with domain

$$\Big\{ f \in D(A_\alpha) : f(x) = 0 \ \text{ for } x \geq d \Big\} \subset D(A_\alpha),$$

while $A_\alpha$ is self-adjoint on $D(A_\alpha)$ and therefore may be considered as Friedrichs extension of $A_\alpha$. □

If $T_0$ is not bounded from below or if we want to approximate the eigenvalues in a gap of the essential spectrum, things are somewhat more complicated. If we use the same approximation operators as above, then for every $\lambda$ in this gap and every $d \in (a, b)$ the projection $E_{\alpha,d,0}(\lambda)$ is finite-dimensional, while $E_\alpha(\lambda)$ is infinite-dimensional. This implies that the eigenvalues of $A_{\alpha,d,0}$ for $d \to b$ are "falling down" through the gap, which leads to the so called "trapping and cascading" phenomenon (cf. F. Gesztesy, D. Gurarie, H. Holden, M. Klaus, L. Sudan, B. Simon, P. Vogl [5] and G. Stolz, J. Weidmann [12]).

Of course, every eigenvalue of $A_\alpha$ is a limit of eigenvalues of a sequence $A_{\alpha,d_n,0}$ with $d_n \to b$ (this follows from the strong convergence of the spectral resolution); but actually every $\lambda$ in the gap (not only the eigenvalues) may be represented as such a limit. This problem cannot be solved by considering differences of spectral projections $E_{\alpha,d,0}(\mu) - E_{\alpha,d,0}(\lambda)$, since only the inequality $\dim(E_{\alpha,d,0}(\mu) - E_{\alpha,d,0}(\lambda)) \geq \dim(E_\alpha(\mu) - E_\alpha(\lambda))$ can be shown for $d$ sufficiently large, and there exist $d$ arbitrarily close to $b$ for which ">" holds. Choosing other more suitable boundary conditions at $d$ we get again a correct approximation:

**Theorem 9.2.** *Let $A_\alpha$ be the self-adjoint realization of $\tau$ with boundary condition (3) at $a$, $I = [\lambda, \mu] \subset \mathbb{R} \setminus \sigma_e(A_\alpha)$, $v$ a nontrivial real $L^2$-solution of $(\tau - \gamma)v = 0$ with $\gamma \in I$, $A_{\alpha,d,v}$ the self-adjoint realization of $\tau$ in $L^2(a, d; r)$ defined by*

$$D(A_{\alpha,d,v}) := \Big\{f \in D(T_{a,d}) : \text{b. c. (3) at } a, [v, f]_d = 0\Big\}.$$

*Then for $[\lambda_1, \lambda_2] \subset I$ with $\lambda_j$ not eigenvalues of $A_\alpha$ we have*

$$\Big\|\Big(E_{\alpha,d,v}(\lambda_2) - E_{\alpha,d,v}(\lambda_1)\Big)P_d - \Big(E_\alpha(\lambda_2) - E_\alpha(\lambda_1)\Big)\Big\| \to 0 \quad \text{for } d \to b.$$

*This implies that the eigenvalues of $A_\alpha$ in $I$ are exactly the limits of eigenvalues of $A_{\alpha,d,v}$ in $I$, and that the corresponding eigenprojections converge in norm.*

*Proof.* Again the essential part of the proof is to show that

$$k := \dim\Big(E_{\alpha,d,v}(\mu) - E_{\alpha,d,v}(\lambda)\Big) \leq \dim(E_\alpha(\mu) - E_\alpha(\lambda)).$$

This implies the desired norm convergence, and therefore equality of the dimension for $d$ sufficiently close to $b$. Hence for every eigenvalue $\sigma$ in this interval and sufficiently small $\varepsilon > 0$

$$\dim\Big(E_{\alpha,d,v}(\sigma + \varepsilon) - E_{\alpha,d,v}(\sigma - \varepsilon)\Big) = 1 \quad \text{for } d \text{ close to } b.$$

From this the statement of the theorem follows.

In order to prove the above inequality it is sufficient to find a $k$-dimensional subspace $M$ of $D(A_\alpha)$ such that

$$\Big\|\Big(A_\alpha - \frac{\lambda + \mu}{2}\Big)\psi\Big\| \leq \frac{\mu - \lambda}{2}\|\psi\| \text{ for } \psi \in M.$$

Let $\lambda_1, \ldots, \lambda_k$ be the eigenvalues of $A_{\alpha,d,v}$ in $(\lambda, \mu)$, $\varphi_1, \ldots, \varphi_k$ the corresponding normalized eigenfunctions. Since the functions $\varphi_j$ satisfy the boundary condition $[v, \varphi_j]_d = 0$ at $d$, there exist $d_j \in \mathbb{C}$ such that for every $j \in \{1, \ldots, k\}$

$$\psi_j(x) := \begin{cases} \varphi_j(x) & \text{in } [a, d], \\ d_j v(x) & \text{in } [d, b) \end{cases}$$

lies in $D(A_\alpha)$. The linear span $M$ of these functions obviously is $k$-dimensional. Every $\psi \in M$ has the form

$$\psi(x) = \begin{cases} \sum_{j=1}^{k} c_j \varphi_j(x) & \text{in } [a, d], \\ c_\psi v(x) & \text{in } [d, b). \end{cases}$$

From this it follows that

$$\|\psi\|^2 = \sum_{j=1}^{k} |c_j|^2 + c_\psi^2 \|v\|^2_{(a,b)},$$

and therefore

$$\begin{aligned} \left\| \left(A_\alpha - \frac{\lambda+\mu}{2}\right)\psi \right\|^2 &= \sum_{j=1}^{k} \left|\lambda_j - \frac{\lambda+\mu}{2}\right|^2 |c_j|^2 + c_\psi^2 \left\| \left(\gamma - \frac{\lambda+\mu}{2}\right) v \right\|^2_{(d,b)} \\ &\leq \left(\frac{\mu-\lambda}{2}\right) \|\psi\|^2. \end{aligned}$$

□

Of course, it will be difficult in general to find the $L^2$-solution $v$ of $(\tau-\gamma)v = 0$ explicitly. Actually an $L^2$-solution of $(\tilde{\tau} - \gamma)v = 0$ can be used, where $\tilde{\tau}$ "behaves like $\tau$" close to $b$ (cf. G. Stolz, J. Weidmann [12], Corollary 3).

## 10. Absolutely continuous spectrum

In order to study the continuous/absolutely continuous spectrum we first have to say something about the general form of the spectral representation of self-adjoint Sturm-Liouville operators (cf. J. Weidmann [17], Section 8, or [19], Section 14). An ordered spectral representation of a self-adjoint operator $A$ in Hilbert space $H$ is a unitary map $F : H \to \bigoplus_j L^2(\mathbb{R}, \rho_j)$, where $\rho_{j+1}$ is absolutely continuous with respect to $\rho_j$ such that $FAF^{-1}$ is the (maximal) operator of multiplication by id in $\bigoplus_j L^2(\mathbb{R}, \rho_j)$. The number of measures $\rho_j \not\equiv 0$ is called the *spectral multiplicity* of $A$. If $F_j f$ is the $j$th component of $Ff$, it is not difficult to see that for $[\alpha, \beta] \subset (a, b)$ and $M_j$ = multiplication by id in $L^2(\mathbb{R}, \rho_j)$

$$(M_j - z)^{-1} F_j \chi_{[\alpha,\beta]} = F_j (A - z)^{-1} \chi_{[\alpha,\beta]}$$

is a Hilbert-Schmidt operator (here $\chi_{[\alpha,\beta]}$ is the operator of multiplication by the characteristic function of $[\alpha,\beta]$). This implies that $F$ has the form

$$Ff(\lambda) = \underset{\substack{c \searrow a \\ d \nearrow b}}{\text{l.i.m.}} \left( \int_c^d v_j(\lambda,x) f(x) r(x) \mathrm{d}x \right)_j$$

where the $v_j(\lambda,\cdot)$ are linearly independent solutions of $(\tau-\lambda)v = 0$ for $\rho_j$-almost every $\lambda$. Therefore the spectral representation of a Sturm-Liouville operator cannot have more than 2 components.

In the special case which we are studying here (regular at $a$, lpc at $b$, $A_\alpha$ the operator with boundary condition (3) at $a$) only the solutions are needed which satisfy the boundary condition (3) at $a$, i.e., the spectral representation has only one component:

$$F : L^2(a,b;r) \to L^2(\mathbb{R},\rho), \quad Ff(\lambda) = \underset{d\to b}{\text{l.i.m.}} \int_a^d v(\lambda,x) f(x) r(x) \mathrm{d}x,$$

where $v(\lambda,\cdot)$ is a solution of $(\tau-\lambda)v = 0$ satisfying the boundary condition (3) at $a$; for simplicity we usually choose $v(\lambda,\cdot)$ to be solution that satisfies the initial condition

$$v(\lambda,a) = \sin\alpha, \quad pv'(\lambda,a) = \cos\alpha.$$

The measure $\rho$ can be determined from Weyl's $m$-function by means of the Weyl-Titchmarsh-Kodaira formulae (cf. J. Weidmann [17], Section 9, or [19], Satz 14.5); we do not need this machinery here. [5]

For the regular approximations $A_{\alpha,d}$ (with any boundary condition at $d$) the corresponding measure $\rho_{\alpha,d}(\cdot)$ is completely determined by the fact that the measure is concentrated at the eigenvalues of $A_{\alpha,d}$ and the measure of every eigenvalue $\lambda$ is

$$\rho_{\alpha,d}(\{\lambda\}) = \|v(\lambda,\cdot)\|_{a,d}^{-2}$$

(this simply follows from the fact that $F$ should be unitary).

This determines the measure $\rho_\alpha(\cdot)$ corresponding to the operator $A_\alpha$ uniquely by means of the following:

**Theorem 10.1.** *Let $\tau$ be regular at $a$, $A_\alpha$ the self-adjoint realization with*

$$D(A_\alpha) := \Big\{ f \in D(T) : \textit{b. c. (3) at } a, [g,f]_b = 0 \textit{ if } \tau \textit{ is lcc at } b \Big\},$$

[5] For the general case (which we do not consider here) a fundamental system $\{v_1(\lambda,\cdot), v_2(\lambda,\cdot)\}$ of $(\tau-\lambda)u = 0$ can be used, for example with the fixed initial condition at some $c \in (a,b)$

$$v_1(\lambda,c) = -pv_2'(\lambda,c) = \sin\gamma, \quad v_2(\lambda,c) = pv_1'(\lambda,c) = \cos\gamma,$$

in order to get a spectral representation $F : L^2(a,b;r) \to L^2(\mathbb{R},\tilde{\rho})$, where $\tilde{\rho}(\cdot)$ is a matrix-valued measure on $\mathbb{R}$, which again can be determined by the Weyl-Titchmarsh-Kodaira formulae. But this can be explicitly evaluated only in a few very simple cases.

*$A_{\alpha,d}$ the operator in $L^2(a,d;r)$ defined by*

$$D(A_{\alpha,d}) := \left\{ f \in D(T_{a,d}) : \begin{array}{l} \textit{boundary condition (3) at } a, \textit{ any s.a.b.c. at } d \\ \textit{if } \tau \textit{ is lpc at } b, \ [g,f]_d = 0 \textit{ if } \tau \textit{ is lcc at } b \end{array} \right\}.$$

*If $\rho_\alpha$ and $\rho_{\alpha,d}$ are the spectral measures of $A_\alpha$ and $A_{\alpha,d}$, respectively, then*

$$\lim_{d\to b} \rho_{\alpha,d}((\lambda_1,\lambda_2)) = \rho_\alpha((\lambda_1,\lambda_2))$$

*for every interval $(\lambda_1,\lambda_2)$ with $\lambda_j$ not eigenvalues of $A_\alpha$.*

*Proof.* For every $f \in L^2(a,b;r)$ with compact support and every interval $(\sigma_1,\sigma_2)$ with $\sigma_j$ not eigenvalues of $A_\alpha$

$$\int_{\sigma_1}^{\sigma_2} \left| \int v(\lambda,x) f(x) r(x) \mathrm{d}x \right|^2 \mathrm{d}\rho_{\sigma,d}(\lambda) = \|E_{\alpha,d}((\sigma_1,\sigma_2))f\|^2$$
$$\to \|E_\alpha((\sigma_1,\sigma_2))f\|^2 = \int_{\sigma_1}^{\sigma_2} \left| \int v(x,\lambda) f(x) r(x) \mathrm{d}x \right|^2 \mathrm{d}\rho_\alpha(\lambda).$$

From this the result can be easily deduced (for details see J. Weidmann [19], Satz 14.13). □

The above result can be used to prove absolute continuity of the spectrum in an interval $I$ (actually we prove unitary equivalence of the part of $A_\alpha$ corresponding to the spectral interval $I$ to the multiplication with id in $L^2(I)$). This result is a very elementary analogue to the *subordinacy result* of D. Gilbert and D. B. Pearson [6] (see also the article by D. Gilbert in this volume).

In the sense of Gilbert-Pearson, a solution $u$ of $(\tau - \lambda)u = 0$ is called *subordinate* (at $b$) if for every linearly independent solution $v$ of $(\tau - \lambda)u = 0$ and $c \in (a,b)$

$$\frac{\int_c^d |u(x)|^2 r(x) \mathrm{d}x}{\int_c^d |v(x)|^2 r(x) \mathrm{d}x} \to 0 \quad \text{for } d \to b.$$

If $\tau$ is lpc at $b$ and for some $\lambda \in \mathbb{R}$ there exists no subordinate solution, then obviously there is no $L^2$-solution of $(\tau - \lambda)u = 0$ and therefore $\lambda$ is in the essential spectrum of every self-adjoint realization of $\tau$. If for no $\lambda$ in $I$ there exists a subordinate solution, then the spectrum of every self-adjoint realization is continuous in $I$. The result of Gilbert and Pearson says that the spectrum is even absolutely continuous in $I$. Actually it is also proved that the set of all $\lambda \in \mathbb{R}$ for which no subordinate solution exists is an essential support of the absolutely continuous part of the spectral measure. This implies for example that in an interval $I$ where there exists an $L^2$-solution for every $\lambda$, there is no absolutely continuous spectrum (cf. Theorem 8.1). The proof of this result is not easy; it uses detailed information about the relations between the behavior of the $m$-functions near the real axis and the type of the spectrum.

The result given below is much more elementary, with an extremely simple proof. It is not as far reaching as that of Gilbert and Pearson; under stronger assumptions it gives a somewhat sharper conclusion, but it does not give information about the absence of absolutely continuous spectrum or about existence of singular continuous spectrum if this stronger assumption of uniform nonsubordinacy is not satisfied.

We say that all solutions of $(\tau-\lambda)u=0$ for $\lambda\in I$ are *of the same size* near $b$ (or, for $\lambda\in I$ the solutions of $(\tau-\lambda)u=0$ are *uniformly nonsubordinate*), if there exists a $\vartheta\in(0,1]$ such that for every $\lambda\in I$ and all solutions $u_1,u_2$ of $(\tau-\lambda)u=0$ satisfying the *normalization condition*

$$|u(c)|^2+|pu'(c)|^2=1 \tag{5}$$

with $c\in(a,b)$ ($c\in[a,b)$ if $\tau$ is lcc at $a$) we have

$$\vartheta\int_c^d |u_1(x)|^2 r(x)\mathrm{d}x \le \int_c^d |u_2(x)|^2 r(x)\mathrm{d}x,$$

i.e., if there exists a function $k:[c,\infty)\to(0,\infty)$ such that

$$\vartheta k(d)\le \|u\|^2_{(c,d)}\le k(d)$$

for every solution $u$ of $(\tau-\lambda)u=0$ ($\lambda\in I$) which satisfies the normalization condition (5). (Since $\tau$ is lpc at $b$ this implies that $k(d)\to\infty$ for $d\to b$.)

**Theorem 10.2.** *Let $\tau$ be regular at $a$ and lpc at $b$. If for $\lambda\in I$ all solutions of $(\tau-\lambda)u=0$ are of the same size at $b$, then every self-adjoint realization $A_\alpha$ of $\tau$ has purely absolutely continuous spectrum in $I$. Actually the part of $A_\alpha$ corresponding to the spectral interval $I$, $A_\alpha|_{R(E(I))}$, is unitarily equivalent to multiplication by* id *in $L^2(I)$.*

*Proof.* By assumption the solutions $v_\alpha(\lambda,\cdot)$ used in the spectral representation, defined by

$$v_\alpha(\lambda,a)=\sin\alpha,\quad pv'_\alpha(\lambda,a)=\cos\alpha,$$

satisfy the normalization condition (5), and therefore

$$\vartheta k(d)\le\|v_\alpha(\lambda,\cdot)\|^2_{a,d}\le k(d)\ \text{ for every } \alpha\in[0,\pi).$$

By means of the formula for $\rho_{\alpha,d}$ given above Theorem 10.1 and the fact that the number of eigenvalues of $A_{\alpha,d}$ in an interval $B$ is the same up to $\pm1$ for every $\alpha$, it follows for $\alpha,\beta\in[0,\pi)$ and every interval $B\subset I$

$$\vartheta\rho_{\alpha,d}(B)\le\rho_{\beta,d}(B)\le\frac{1}{\vartheta}\rho_{\alpha,d}(B)\quad\text{for } d \text{ close to } b.$$

In the limit $d\to b$ this implies (remember that $A_\alpha$ has no eigenvalues in $I$) the equivalence of the $\rho_\alpha$ in the sense

$$\vartheta\rho_\alpha(B)\le\rho_\beta(B)\le\frac{1}{\vartheta}\rho_\alpha(B)$$

for every interval $B \subset I$. Since by a result of S. Kotani [8] (cf. also J. Weidmann [19], Satz 14.11 and Satz 14.18)

$$\int_0^\pi \rho_\alpha(B)\mathrm{d}\alpha = |B|,$$

this implies that all $\rho_\alpha$ are equivalent to the Lebesgue measure. □

Typical applications of this result are those to Sturm-Liouville operators on $[0,\infty)$ with $p = r = 1$ and

- $q = q_1 + q_2$ with $q_1 \in L^1(0,\infty)$ and $q_2$ of bounded variation with $q_2(x) \to 0$ for $x \to \infty$. In this case the spectrum of every self-adjoint realization $A_\alpha$ is absolutely continuous in $(0,\infty)$ and the part of $A_\alpha$ corresponding to the spectral interval $(0,\infty)$ is unitarily equivalent to multiplication by id in $L^2(0,\infty)$ (cf. J. Weidmann [14]).
- $q = q_{\mathrm{per}} + q_1 + q_2$ with periodic $q_{\mathrm{per}}$ and $q_1, q_2$ as above. In this case the spectrum of $A_\alpha$ is absolutely continuous in the stability intervals of $\tau_{\mathrm{per}} = -\mathrm{d}^2/\mathrm{d}x^2 + q_{\mathrm{per}}$ and the part of $A_\alpha$ corresponding to the stability intervals is unitarily equivalent to multiplication with id in $L^2(\bigcup_n I_n)$, where $I_n$ are the stability intervals.

In the proofs of these results one uses for $I$ any closed bounded interval contained in the corresponding sets.

If one considers such operators on $\mathbb{R}$, where the coefficients satisfy the corresponding assumptions also on the negative half axis, then the absolutely continuous part is unitarily equivalent to $L^2(0,\infty)^2$, resp. $L^2(\bigcup I_n)^2$, which can be shown by means of the decomposition method and a suitable trace class scattering result (cf. M. Reed-B. Simon [11], Section XI.3, or J. Weidmann [19], Section 22.4). This argument does not directly imply that the spectrum is purely absolutely continuous; but by means of an adaption of the above method this can be proved (cf. J. Weidmann [16] or [19], Satz 14.24).

On the other hand the above result implies that the spectral multiplicity of the whole line operator in $(0,\infty)$, respectively $\bigcup I_n$, is 2. Notice that in general the fact that $A, B$ have absolutely continuous spectrum $I$ does not imply that $A \oplus B$ has spectral multiplicity $\geq 2$ in $I$. This is demonstrated by the following simple example (the author does not know of a similar example in the literature).

**Example.** Define subsets $A_j$ and $B_j$ of $[0,1]$ by the following procedure:

- $A_1 := (0, 1/2)$, $B_1 := (1/2, 1)$,
- $A_{j+1}$ is constructed from $A_j$ by deleting the central $2^{-j}$th part of every interval of $A_j$ and adding the central $2^{-j}$th part of every interval of $B_j$,
- $B_{j+1}$ is constructed from $B_j$ by deleting the central $2^{-j}$th part of every interval of $B_j$ and adding the central $2^{-j}$th part of every interval of $A_j$.

(Every $A_j$ and $B_j$ has measure $1/2$. It makes no difference whether we take open or closed intervals in this procedure.)

Let now

$$r_j := \chi_{A_j}, \quad s_j := \chi_{B_j}.$$

Then the sequences $(r_j)$ and $(s_j)$ converge in $L^1(0,1)$ and almost everywhere to functions $r$ and $s$ with $r(x)s(x) = 0$ and $r(x)+s(x) = 1$ almost everywhere. Define measures $\rho$ and $\sigma$ by

$$\rho(M) := \int_M r(x)\mathrm{d}x \quad \text{and} \quad \sigma(M) := \int_M s(x)\mathrm{d}x.$$

Now take for $A$ and $B$ the operators of multiplication by id in $L^2(0,1;\rho)$ and $L^2(0,1;\sigma)$ respectively. The operators $A$ and $B$ have absolutely continuous spectrum $[0,1]$, while $A+B$ is (unitarily equivalent to) the operator of multiplication by id in $L^2(0,1)$ and therefore has multiplicity 1.

Using the Gelfand-Levitan result on the inverse problem (cf. J. M. Gelfand-B. M. Levitan [4], M. A. Naimark [9]) it follows that there exists a differential expression $\tau u = -u'' + qu$ on $(-\infty,\infty)$ such that the corresponding self-adjoint realizations in $L^2(-\infty,0)$ and $L^2(0,\infty)$ with Dirichlet boundary condition at 0 have absolutely continuous spectrum $[0,1]$, while the self-adjoint realization in $L^2(-\infty,\infty)$ has absolutely continuous spectrum $[0,1]$ with multiplicity 1.

## References


[1] P.B. Bailey, W.N. Everitt and A. Zettl, Computing eigenvalues of singular Sturm-Liouville problems, Res. Math. **20** (1991), 391–423.

[2] P.B. Bailey, W.N. Everitt, J. Weidmann and A. Zettl, Regular Approximations of singular Sturm-Liouville problems, Res. Math. **23** (1993), 3–22.

[3] N. Dunford and J.T. Schwartz, *Linear Operators Part II: Spectral Theory*, Interscience, New York, 1963.

[4] J.M. Gelfand and B.M. Levitan, On the determination of a differential equation from its spectral function (Russian), Izv. Akad. Nauk USSR **15** (1951), 309–360, Amer. Math. Soc. Translations (2) **1** (1955), 253–304.

[5] F. Gesztesy, D. Gurarie, H. Holden, M. Klaus, L. Sudan, B. Simon and P. Vogl, Trapping and Cascading of Eigenvalues in the Large Coupling Limit, Commun. Math. Phys. **118** (1988), 597–634.

[6] D. Gilbert and D.B. Pearson, On subordinacy and analysis of the spectrum of one-dimensional Schrödinger operators, J. Math. Anal. Appl. **128** (1987), 30–56.

[7] Ph. Hartman and A. Wintner, A separation theorem for continuous spectra, Amer. J. Math. **71** (1949), 650–662.

[8] S. Kotani, Lyapunov exponents and spectra for one-dimensional random Schrödinger operators, Contemp. Math. **50** (1986), 277–286.

[9] M.A. Naimark, *Linear Differential Operators Part II: Linear Differential Operators in Hilbert Space*, F. Unger Publ., New York, 1968.

[10] D.B. Pearson, *Quantum Scattering and Spectral Theory*, Academic Press, London, 1988.

[11] M. Reed and B, Simon, *Methods of Modern Mathematical Physics III: Scattering Theory*, Academic Press, New York, 1979.

[12] G. Stolz and J. Weidmann, Approximation of isolated eigenvalues of ordinary differential operators, J. Reine Angew. Math. **445** (1993), 31–44.

[13] G. Stolz and J. Weidmann, Approximation of isolated eigenvalues of general singular ordinary differential operators, Res. Math. **28** (1995), 345–358.

[14] J. Weidmann, Zur Spektraltheorie von Sturm-Liouville-Operatoren, Math. Zeitschr. **98** (1967), 268–302.

[15] J. Weidmann, Verteilung der Eigenwerte für eine Klasse von Integraloperatoren in $L^2(a,b)$, J. Reine Angew. Math. **276** (1975), 213–220.

[16] J. Weidmann, Uniform Nonsubordinacy and the Absolutely Continuous Spectrum, Analysis **16** (1996), 89–99.

[17] J. Weidmann, *Spectral Theory of Ordinary Differential Operators*, Lecture Notes in Mathematics **1258**, Springer, Berlin, 1987.

[18] J. Weidmann, *Lineare Operatoren in Hilberträumen, Teil I Grundlagen*, Teubner, Stuttgart, 2000.

[19] J. Weidmann, *Lineare Operatoren in Hilberträumen, Teil II Anwendungen*, Teubner, Stuttgart, 2003.

Joachim Weidmann
Fachbereich Mathematik
Johann Wolfgang Goethe–Universität
D–60054 Frankfurt
Germany
e-mail: weidmann@math.uni-frankfurt.de

*W.O. Amrein, A.M. Hinz, D.B. Pearson*
Sturm-Liouville Theory: Past and Present, 99–120


# Spectral Theory of Sturm-Liouville Operators on Infinite Intervals: A Review of Recent Developments

Yoram Last

**Abstract.** This review discusses some of the central developments in the spectral theory of Sturm-Liouville operators on infinite intervals over the last thirty years or so. We discuss some of the natural questions that occur in this framework and some of the main models that have been studied.

## 1. Introduction

This article discusses spectral theory of Sturm-Liouville operators on infinite intervals. More specifically, we focus on one-dimensional Schrödinger operators of the form

$$H = -\frac{d^2}{dx^2} + V(x) \tag{1.1}$$

on $L^2(\mathbb{R}, dx)$ or $L^2([0, \infty), dx)$ and on their discrete analogs defined by

$$(H\psi)(n) = \psi(n+1) + \psi(n-1) + V(n)\psi(n) \tag{1.2}$$

on $\ell^2(\mathbb{Z})$ or $\ell^2(\mathbb{N})$. We shall refer to operators on $L^2(\mathbb{R}, dx)$ or $\ell^2(\mathbb{Z})$ as *whole-line* operators and to operators on $L^2([0, \infty), dx)$ or $\ell^2(\mathbb{N})$ as *half-line* operators. We note that the half-line operators are defined by (1.1) or (1.2) along with a boundary condition at 0 which takes the form $\psi(0)\cos\theta + \psi'(0)\sin\theta = 0$ for the continuous case (1.1) and $\psi(0)\cos\theta + \psi(1)\sin\theta = 0$ for the discrete case (1.2). The boundary condition in the discrete case is equivalent to considering the operator (which acts on vectors in $\ell^2(\mathbb{N})$ with $\mathbb{N} = \{1, 2, 3, \dots\}$) as being the tridiagonal matrix defined by (1.2) for $n > 1$ and by $(H\psi)(1) = \psi(2) + (V(1) - \tan\theta)\psi(1)$ at the origin. We shall refer to $\theta$ as the *boundary phase*.

Since most of the examples we wish to discuss occur more naturally in the context of discrete operators, our primary focus will be on such operators. We note, however, that there is good correspondence between the two types of operators

Partially supported by The Israel Science Foundation (grant no. 188/02).

except for the case of continuous operators with potentials $V$ that are not bounded from below. In cases where there is "significant" unboundedness from below (e.g., potentials $V$ going to $-\infty$ at infinity), continuous operators may exhibit properties for which there is no real analog in the discrete case. Such operators are not of much interest to us here, however.

Our primary interest here is in spectral theory of the above operators, namely, in questions concerning the location and structure of their spectrum, its decomposition into various spectral types, and more generally, in properties of their spectral measures. Given a separable Hilbert space $\mathcal{H}$ and a self-adjoint operator $H$, recall [76] that for each $\psi \in \mathcal{H}$, the spectral measure $\mu_\psi$ is the unique Borel measure obeying $\langle \psi, f(H)\psi \rangle = \int f(x)\, d\mu_\psi(x)$ for any bounded Borel function $f$. By Lebesgue's decomposition theorem, every Borel measure $\mu$ decomposes uniquely as $\mu = \mu_{\rm ac} + \mu_{\rm sc} + \mu_{\rm pp}$. The absolutely-continuous part, $\mu_{\rm ac}$, gives zero weight to sets of zero Lebesgue measure. The pure-point part, $\mu_{\rm pp}$, is a countable sum of atomic measures. The singular-continuous part, $\mu_{\rm sc}$, gives zero weight to countable sets and is supported on some set of zero Lebesgue measure. Letting $\mathcal{H}_{\rm ac} \equiv \{\psi \,|\, \mu_\psi$ is purely absolutely-continuous$\}$, and similarly defining $\mathcal{H}_{\rm sc}$ and $\mathcal{H}_{\rm pp}$, one obtains a decomposition: $\mathcal{H} = \mathcal{H}_{\rm ac} \oplus \mathcal{H}_{\rm sc} \oplus \mathcal{H}_{\rm pp}$. $\mathcal{H}_{\rm ac}$, $\mathcal{H}_{\rm sc}$, and $\mathcal{H}_{\rm pp}$ are closed (in norm), mutually orthogonal subspaces, which are invariant under $H$. The absolutely-continuous spectrum, $\sigma_{\rm ac}(H)$, singular-continuous spectrum, $\sigma_{\rm sc}(H)$, and pure-point spectrum, $\sigma_{\rm pp}(H)$, are defined as the spectra of the restrictions of $H$ to the corresponding subspaces, and the spectrum $\sigma(H)$ of $H$ is their union: $\sigma(H) = \sigma_{\rm ac}(H) \cup \sigma_{\rm sc}(H) \cup \sigma_{\rm pp}(H)$.

We note that some authors, notably [76], use the term "pure point spectrum" (and the notation $\sigma_{\rm pp}$) to denote the set of eigenvalues of an operator. Our definition here coincides with defining the "pure-point spectrum" as the closure of the set of eigenvalues. Moreover, what we call here "pure-point spectrum," is often called just "point spectrum." The existence of such differing terminologies has the obvious potential of leading to confusion (e.g., some authors use the term "pure point spectrum" for what we call below "purely pure-point spectrum") and so one should note that we adhere below to the term "pure-point spectrum," as defined above.

For discrete half-line operators, the vector $\delta_1$ (where $\delta_j(n)$ is 1 if $j = n$ and 0 otherwise) is cyclic and so the spectral properties of the operator are fully determined by the spectral measure $\mu = \mu_{\delta_1}$ (which is called the spectral measure of the operator in this case). In particular, $\sigma_{\rm ac}(H)$, $\sigma_{\rm sc}(H)$, and $\sigma_{\rm pp}(H)$ coincide with the (topological) supports of the corresponding parts of the spectral measure $\mu$. For discrete whole-line operators, one needs to look at two consecutive $\delta_j$ vectors to have a cyclic family and so we get the same thing but with the spectral measure being $\mu = \mu_{\delta_0} + \mu_{\delta_1}$. Analogous natural spectral measures similarly exist for continuous operators (except that in the continuous case the corresponding spectral measures are not finite). See, e.g., [19, 30] in this volume.

By classical inverse spectral theory, one should expect the full spectral richness allowed by measure theory to find its way into such operators.

In particular, we have

**Theorem 1.1 (Gel'fand-Levitan [28]).** *Given any finite Borel measure $\nu$ on $[a,b] \subset \mathbb{R}$, there exists a continuous half-line Schrödinger operator for which the spectral measure coincides with $\nu$ on $[a,b]$.*

In the discrete case one needs to slightly broaden the class of considered operators in order to have a full inverse spectral result of this type. Explicitly, one looks at self-adjoint Jacobi matrices of the form $J(\{a_n\},\{b_n\})$, where $\{a_n\}_{n=1}^{\infty} \subset (0,\infty)$, $\{b_n\}_{n=1}^{\infty} \subset \mathbb{R}$, and

$$(J(\{a_n\},\{b_n\})\psi)(n) = a_{n+1}\psi(n+1) + a_n\psi(n-1) + b_n\psi(n). \tag{1.3}$$

We then have, by the classical theory of orthogonal polynomials (see, e.g., [91]):

**Theorem 1.2.** *Any compactly supported probability measure on $\mathbb{R}$ is the spectral measure of a unique bounded Jacobi matrix of the form $J(\{a_n\},\{b_n\})$.*

In spite of these inverse spectral results, up until the mid 1970's or so, the kind of spectrum most people had in mind in the context of spectral theory of Schrödinger operators consisted of "bands" of absolutely-continuous spectrum along with some isolated eigenvalues. These are the kind of spectra occurring for periodic potentials and for atomic and molecular Hamiltonians, and so it was generally believed that these are the kind of spectra one is likely to encounter in problems of physical interest. "Exotic" spectral phenomena such as singular-continuous spectrum, Cantor set spectrum, and thick pure-point spectrum (namely, the occurrence of eigenvalues dense in a set of positive Lebesgue measure), while obviously allowed by inverse spectral theory, were not considered as likely to occur in any problem that should be of real interest. In particular, the main role of singular-continuous spectrum in spectral theory of Schrödinger operators was that of a mathematical obscurity that needs to be excluded in many problems (in order to ensure good scattering theory). Indeed, a significant portion of [78], for example, is devoted to analytical methods for proving that singular-continuous spectrum *does not occur.*

This situation started to change, however, around the mid 1970's, as evidence started to accumulate showing that exotic spectral phenomena do actually occur in elementary mathematical models that are also of considerable interest to theoretical physics. One of the first results in this direction was the 1977 Goldsheid-Molchanov-Pastur [32] proof of Anderson localization [2] (namely, the occurrence of pure-point spectrum with eigenvalues dense in an interval) in a random Schrödinger operator. They considered a continuous one-dimensional operator of the form (1.1) on $L^2(\mathbb{R})$ with a certain type of random potential $V$. Even earlier, the Ishii-Pastur theorem (see Theorem 9.13 of [17]) indicated that some random one-dimensional Schrödinger operators have no absolutely continuous spectrum in spite of their spectrum being an interval (which says they must have either thick pure-point spectrum or singular-continuous spectrum or both). Anderson localization for discrete Schrödinger operators of the form (1.2) with potentials made of independent, identically distributed random variables, has been proven in 1980 by Kunz-Souillard [63]. Another notable result of the era is Pearson's seminal 1978 paper [72], which gave an explicit construction of a one-dimensional Schrödinger operator having purely singular-continuous spectrum.

Starting around 1980 (note, however, the 1975 paper of Dinaburg-Sinai [24]), the growing interest in spectral theory of almost periodic Schrödinger operators [82] provided many more examples of physically interesting operators exhibiting exotic spectral phenomena. In particular, studies of the *almost Mathieu operator* (namely, the operator of the form (1.2) on $\ell^2(\mathbb{Z})$ with potential $V(n) = \lambda \cos(2\pi\alpha n + \theta)$, where $\lambda, \alpha, \theta \in \mathbb{R}$) provided an example of Cantor set spectrum (namely, a case where the spectrum of the operator is nowhere dense and has no isolated points). It occurs for this operator whenever $\alpha \in \mathbb{R} \setminus \mathbb{Q}$ (something that was first conjectured in 1964 by Azbel [11] and demonstrated numerically in 1976 by Hofstadter [47]). The first mathematical result was given in 1982 by Bellissard-Simon [13] who have shown that Cantor set spectrum must occur for a generic set of parameters in this model (a proof for all $\lambda, \theta \in \mathbb{R}$ and $\alpha \in \mathbb{R} \setminus \mathbb{Q}$ has been completed only very recently; see below). The almost Mathieu operator also led to a second example of an operator with purely singular-continuous spectrum, as it was shown in 1982 by Avron-Simon [9] (using ideas of Gordon [33]) that it has such spectrum if $\alpha$ is a Liouville number and $|\lambda| > 2$ (see below).

While the 1980's interest in almost periodic Schrödinger operators seems to have started largely by itself and has been much driven by the richness of the associated spectral theory, it was also soon enhanced by strong connections with some major discoveries in physics. The 1980 discovery of the integer Quantum Hall Effect by von Klitzing [60] (for which he got the Nobel prize in 1985), led to a beautiful theory by Thouless, Kohmoto, Nightingale, and den Nijs [95], which explains the quantization of charge transport in this effect as connected with certain topological invariants. Central to their theory is the use of the almost Mathieu operator as a model for Bloch electrons in a magnetic field (in which case the frequency $\alpha$ is proportional to the magnetic flux; see below). Another strong source of interest in almost periodic problems came from the 1984 discovery of quasicrystals by Shechtman *et al.* [75], as almost periodic Schrödinger operators provide elementary models for electronic properties in such media. Yet another connection with physics arose in the context of Quantum Chaos theory, notably in works of Fishman, Grempel, and Prange [26, 36, 37], as discrete one-dimensional Schrödinger operators (and in particular, almost periodic ones) appeared in studies of dynamics of some elementary quantum models and, in particular, in studies aimed towards distinguishing quantum from classical dynamics in chaotic systems.

Another phase in these developments occurred in the mid 1990's, as it has been realized (largely by Simon and co-workers [21, 22, 20, 46, 55, 84, 85, 86, 89]) that singular-continuous spectrum is a much more common occurrence than previously believed and that it is, in fact, a "generic" phenomenon for many families of operators. At that time it has also been realized that singular-continuous spectrum is connected with rich dynamics of various quantum systems and that it is rich itself, in the sense of being naturally decomposable into many kinds of spectra associated with different dynamical behaviors [38, 39, 66].

The above discoveries, among others, led to a growing interest in spectral theory of one-dimensional Schrödinger operators which eventually turned, over the last thirty years or so, to a rich field of research in its own right. Other than extensive studies of some specific models and classes of models, much work has also gone into finding general techniques for spectral analysis of broad classes of such operators. (See Gilbert's paper [30] in this volume for a review of some analytical methods and, in particular, the Gilbert-Pearson theory of subordinacy [29, 31, 56].) In fact, the area is so vast, by now, that it would be impossible to provide a truly meaningful review of the subject in the framework of less than a thick book. We would thus not at all attempt it here, but rather take some pointwise glimpses at two specific classes of models which played important roles. One of these is the above-mentioned almost Mathieu operator and the other is the class of sparse potentials. Hopefully, this would give some idea about the nature of the field. For further reading, we recommend Simon's review paper [87], which, among other things, discusses slowly decaying potentials, and Damanik's review paper [18], which discusses some classes of potentials generated by circle maps and substitutions (these include the most natural models for one-dimensional quasicrystals). A solid, albeit somewhat outdated, introduction to the spectral theory of Schrödinger operators with random and almost periodic potentials is given in [17].

The rest of this paper is organized as follows: In Section 2, we present some natural extensions of the classical spectral types. In Section 3, we consider the almost Mathieu operator, and in Section 4, we discuss sparse potentials.

## 2. Extending the spectral types

Pre-1980 spectral theory of self-adjoint operators on separable Hilbert spaces identified five spectral types arising from natural spectral decompositions of the Hilbert space. These are the essential spectrum, $\sigma_{\text{ess}}$, discrete spectrum, $\sigma_{\text{disc}}$, absolutely-continuous spectrum, $\sigma_{\text{ac}}$, singular-continuous spectrum, $\sigma_{\text{sc}}$, and pure-point spectrum, $\sigma_{\text{pp}}$. As rich spectral properties started to appear, the desire to make more spectral distinctions arose, and more spectral types have been introduced.

The first introduction of new spectral types is due to Avron-Simon [8] in 1981. They decomposed the Hilbert space $\mathcal{H}$ into a transient subspace, $\mathcal{H}_{\text{tac}}$, and a recurrent subspace, $\mathcal{H}_{\text{rec}} = \mathcal{H}_{\text{tac}}^{\perp}$. $\mathcal{H}_{\text{tac}}$ is a subspace of $\mathcal{H}_{\text{ac}}$, which, in some sense, extracts its smoothest component. It is given by $\mathcal{H}_{\text{tac}} = P_{\text{ei}}\mathcal{H}_{\text{ac}}$, where $P_{\text{ei}}$ is the spectral projection on the essential interior of the essential support of the absolutely-continuous part of the spectral measure class of the operator $H$. The spectra $\sigma_{\text{tac}} \cup \sigma_{\text{rec}} = \sigma$ are defined by $\sigma_{\text{tac}} = \sigma(H \restriction \mathcal{H}_{\text{tac}})$ and $\sigma_{\text{rec}} = \sigma(H \restriction \mathcal{H}_{\text{rec}})$, and $\sigma_{\text{tac}} \subseteq \sigma_{\text{ac}}$. These spectral decompositions are connected with dynamics, since, as Avron-Simon [8] show, $\mathcal{H}_{\text{tac}} = \overline{\{\psi \mid \hat{\mu}_\psi(t) \in L^1\}}$, where $\hat{\mu}_\psi(t) = \langle \psi, e^{-iHt}\psi \rangle =$

$\int e^{-iEt}\, d\mu_\psi(E)$ is the Fourier transform of the spectral measure $\mu_\psi$ of the vector $\psi$. This should be compared with the classical [77] $\mathcal{H}_{\rm ac} = \overline{\{\psi \,|\, \hat{\mu}_\psi(t) \in L^2\}}$.

Additional spectral types, essentially subtypes of singular-continuous spectrum, were introduced in 1996 by Last [66]. This has been impacted by Guarneri's seminal papers on quantum dynamics [38, 39] and utilized the Rogers-Taylor [79, 80] theory of decomposing Borel measures with respect to Hausdorff measures.

Recall that for any subset $S$ of $\mathbb{R}$ and $\alpha \in [0,1]$, the $\alpha$-dimensional Hausdorff measure, $h^\alpha$, is given by

$$h^\alpha(S) \equiv \lim_{\delta\to 0} \inf_{\delta\text{-covers}} \sum_{\nu=1}^{\infty} |b_\nu|^\alpha , \tag{2.1}$$

where a $\delta$-cover is a cover of $S$ by a countable collection of intervals, $S \subset \bigcup_{\nu=1}^{\infty} b_\nu$, such that for each $\nu$ the length of $b_\nu$ is at most $\delta$. (Technically, we consider $h^\alpha$ as being defined by (2.1) also for real $\alpha$'s outside $[0,1]$, but the resulting $h^\alpha$'s are trivial in such a case.) $h^\alpha$, as defined by (2.1), is an outer measure on $\mathbb{R}$ and its restriction to Borel sets is a Borel measure. $h^1$ coincides with the Lebesgue measure and $h^0$ is the counting measure (assigning to each set the number of points in it), such that the family $\{h^\alpha \,|\, 0 \le \alpha \le 1\}$ can be viewed as a way of continuously interpolating between the counting measure and the Lebesgue measure. Given any $\emptyset \neq S \subseteq \mathbb{R}$, there exists a unique $\alpha(S) \in [0,1]$ such that $h^\alpha(S) = 0$ for any $\alpha > \alpha(S)$, and $h^\alpha(S) = \infty$ for any $\alpha < \alpha(S)$. This unique $\alpha(S)$ is called the Hausdorff dimension of $S$. A rich theory of decomposing measures with respect to Hausdorff measures and dimensions has been developed by Rogers and Taylor [79, 80]. Here we only discuss a small part of it. A much more detailed description can be found in [66].

Given $\alpha$, a measure $\mu$ is called $\alpha$-continuous ($\alpha$c) if $\mu(S) = 0$ for every set $S$ with $h^\alpha(S) = 0$. It is called $\alpha$-singular ($\alpha$s) if it is supported on some set $S$ with $h^\alpha(S) = 0$. We say that $\mu$ is one-dimensional (od) if it is $\alpha$-continuous for every $\alpha < 1$. We say that it is zero-dimensional (zd) if it is $\alpha$-singular for every $\alpha > 0$. A measure $\mu$ is said to have exact dimension $\alpha$ if, for every $\epsilon > 0$, it is both $(\alpha-\epsilon)$-continuous and $(\alpha+\epsilon)$-singular.

Given a (positive, finite) measure $\mu$ and $\alpha \in [0,1]$, we define

$$D_\mu^\alpha(x) \equiv \limsup_{\epsilon\to 0} \frac{\mu((x-\epsilon, x+\epsilon))}{(2\epsilon)^\alpha} \tag{2.2}$$

and $T_\infty \equiv \{x \,|\, D_\mu^\alpha(x) = \infty\}$. The restriction $\mu(T_\infty \cap \cdot\,) \equiv \mu_{\alpha s}$ is $\alpha$-singular, and $\mu((\mathbb{R}\setminus T_\infty) \cap \cdot\,) \equiv \mu_{\alpha c}$ is $\alpha$-continuous. Thus, each measure decomposes uniquely into an $\alpha$-continuous part and an $\alpha$-singular part: $\mu = \mu_{\alpha c} + \mu_{\alpha s}$. Moreover, an $\alpha$-singular measure must have $D_\mu^\alpha(x) = \infty$ a.e. (with respect to it) and an $\alpha$-continuous measure must have $D_\mu^\alpha(x) < \infty$ a.e. It is important to note that $D_\mu^\alpha(x)$ is defined with a limit superior. The corresponding limit need not exist.

Consider now a separable Hilbert space $\mathcal{H}$ and a self-adjoint operator $H$. We let $\mathcal{H}_{\alpha c} \equiv \{\psi \,|\, \mu_\psi \text{ is } \alpha\text{-continuous}\}$ and $\mathcal{H}_{\alpha s} \equiv \{\psi \,|\, \mu_\psi \text{ is } \alpha\text{-singular}\}$. $\mathcal{H}_{\alpha c}$

and $\mathcal{H}_{\alpha s}$ are mutually orthogonal closed subspaces which are invariant under $H$, and $\mathcal{H}$ decomposes as $\mathcal{H} = \mathcal{H}_{\alpha c} \oplus \mathcal{H}_{\alpha s}$. The $\alpha$-continuous spectrum ($\sigma_{\alpha c}$) and $\alpha$-singular spectrum ($\sigma_{\alpha s}$) are defined as the spectra of the restrictions of $H$ to the corresponding subspaces, and $\sigma = \sigma_{\alpha c} \cup \sigma_{\alpha s}$. Thus, the standard spectral theoretical scheme which uses the Lebesgue decomposition of a Borel measure into absolutely-continuous, singular-continuous, and pure-point parts can be extended to include further decompositions with respect to Hausdorff measures.

As described in [66], the full picture is somewhat richer than discussed above. For every dimension $\alpha \in (0, 1)$, there is a natural unique decomposition (of a $\sigma$-finite Borel measure $\mu$ on $\mathbb{R}$) into five parts: one below the dimension $\alpha$, one above it, and three within it – of which the middle one is absolutely continuous with respect to $h^\alpha$. Furthermore, this picture can be extended to consider more general Hausdorff measures (namely, ones that do not come from a power law) and families of such measures – as originally discussed by Rogers-Taylor [79, 80]. All of these measure decompositions lead to corresponding Hilbert space spectral decompositions. A point to note here is that continuity and singularity with respect to Hausdorff measures are completely determined by the a.e. local scaling behavior of the measure. Knowing $D_\mu^\alpha(x)$ for every $\alpha$ in $[0, 1]$ and a.e. $x$ with respect to $\mu$ completely determines the decomposition of $\mu$ with respect to dimensional Hausdorff measures. Knowing only the local dimension

$$\alpha_\mu(x) \equiv \liminf_{\epsilon \to 0} \frac{\log(\mu((x - \epsilon,\, x + \epsilon)))}{\log \epsilon} \tag{2.3}$$

(for a.e. $x$ with respect to $\mu$) determines its decomposition with respect to Hausdorff dimensions. In particular, $\mu$ is of exact dimension $\alpha$ if and only if $\alpha_\mu(x) = \alpha$ a.e. with respect to it.

It is interesting to note that the singular-continuous spectrum in Pearson's seminal example [72] turns out to be purely one-dimensional spectrum (as has been essentially shown by Simon [86]), while the singular-continuous spectrum found by Avron-Simon [9] in the almost Mathieu operator turns out to be purely zero-dimensional [54, 66]. Examples of certain sparse potentials having spectrum of exact dimension $\alpha$, for any $\alpha \in [0, 1]$, were constructed by Jitomirskaya-Last [53] using a power-law variant of the Gilbert-Pearson theory [29, 31, 56]; see below.

## 3. The almost Mathieu operator

The almost Mathieu operator (also known as the Harper operator or the Hofstadter model) is the discrete one-dimensional Schrödinger operator (acting on $\ell^2(\mathbb{Z})$) given by:

$$(H_{\alpha,\lambda,\theta}\psi)(n) = \psi(n + 1) + \psi(n - 1) + \lambda \cos(2\pi\alpha n + \theta)\psi(n), \tag{3.1}$$

where $\alpha, \lambda, \theta \in \mathbb{R}$. Its name comes from the similarity to the Mathieu equation:

$$-y''(x) + \lambda \cos(x)\, y(x) = Ey(x). \tag{3.2}$$

$H_{\alpha,\lambda,\theta}$ is a tight binding model for the Hamiltonian of an electron in a one-dimensional lattice, subject to a commensurate (if $\alpha$ is rational) or incommensurate (if $\alpha$ is irrational) potential. It is also related to the Hamiltonian of an electron in a two-dimensional lattice, subject to a perpendicular (uniform) magnetic field. There are two different ways (or limits) in which this relation can be obtained. The first, going back to Harper [42], is to start with a tight binding model of a two-dimensional rectangular lattice (which only takes into account nearest neighbor interactions), and then to consider a Landau gauge for the magnetic field. Namely, the vector potential is taken to be in one direction, parallel to one of the directions of the lattice and perpendicular to the other. This makes the Hamiltonian separable, such that the eigenfunctions are plane waves in the direction which is perpendicular to the vector potential, and one obtains $H_{\alpha,\lambda,\theta}$ for the direction of the vector potential. The number $\alpha$ is the magnetic flux per unit cell (in quantum flux units), $\theta$ is the wave-number of the plane waves in the transversal direction, and $\lambda/2$ is the ratio between the length of a unit cell in the direction of the vector potential and its length in the transversal direction. In particular, $\lambda = 2$ corresponds to a square lattice. This approach is closely related to the standard Landau gauge solution for free electrons (in the plane) in a uniform magnetic field, where one gets plane waves in one direction and the harmonic oscillator (which gives rise to Landau levels) in the other direction. In this sense, $H_{\alpha,\lambda,\theta}$ appears as a tight binding analog of the harmonic oscillator. The second way, going back at least to Rauh [74], is to start with free electrons (in the plane) in a uniform magnetic field and to consider the perturbation of a single Landau level arising from a weak periodic (sinusoidal) potential. The magnetic flux per unit cell is $1/\alpha$ in this case. In both of these ways, the relevant energy spectrum is the union over $\theta$ of the individual energy spectra of $H_{\alpha,\lambda,\theta}$. Namely, it is the set $S(\alpha,\lambda)$ defined by:

$$S(\alpha,\lambda) \equiv \bigcup_{\theta} \sigma(\alpha,\lambda,\theta), \tag{3.3}$$

where $\sigma(\alpha,\lambda,\theta)$ is the spectrum of $H_{\alpha,\lambda,\theta}$.

The almost Mathieu operator plays an important role in the study of fundamental problems related to Bloch electrons in magnetic fields. In particular, it plays a major role in the Thouless-Kohmoto-Nightingale-den Nijs [95] theory of the integer quantum Hall effect, where it gives rise to a rich set of possible integer Hall conductances. It is interesting to note that the Thouless *et al.* theory, and thus also the relevance of the almost Mathieu operator for describing electrons in magnetic fields, has been recently verified experimentally [1]. For a recent review of the quantum Hall effect and the role played by the almost Mathieu operator, see [7].

Apart from its relations to some fundamental problems in physics, $H_{\alpha,\lambda,\theta}$ is fascinating also because of the incredible spectral richness obtained by varying the parameters $\alpha, \lambda, \theta$, along with the fact that, to a large extent, this richness can be rigorously analyzed. It serves as a primary example for many spectral phenomena, and it is the most studied concrete model of a one-dimensional Schrödinger operator. The central spectral questions can be divided into two classes: questions concerned with the spectrum as a set and those concerned with the spectral types. Below we review some of the central findings for each of those.

### 3.1. The spectrum as a set

If $\alpha$ is a rational number, $\alpha = p/q$, where $p$ and $q$ are relatively prime, $H_{\alpha,\lambda,\theta}$ is a periodic Jacobi matrix, and by classical Bloch-Floquet theory, the structure of the spectrum $\sigma(\alpha,\lambda,\theta)$ is well understood. It consists of $q$ bands (closed intervals), which are usually separated by gaps. As $\theta$ is varied, these bands move, and their length may change, but this happens in such a way that different bands never overlap (other than in band edges). Namely, an energy which is inside a given band (suppose that we always label the bands by their order of occurrence on the real line) for some $\theta$, will never be in a different band for any other $\theta$ (see, e.g., [6]). Thus, the set $S(\alpha,\lambda)$ is similar to $\sigma(\alpha,\lambda,\theta)$ for any individual $\theta$ and also consists of $q$ bands. If $\alpha$ is irrational, it follows from general principles [10, 17, 82] that the spectrum is independent of $\theta$. Thus, we have in such case $S(\alpha,\lambda) = \sigma(\alpha,\lambda,\theta)$ for any $\theta$.

An important property of $H_{\alpha,\lambda,\theta}$ is the Aubry duality [3], which allows to relate eigenfunctions and spectra of the almost Mathieu operator with some given $\lambda$ to those of the almost Mathieu operator with $\lambda$ replaced by $4/\lambda$. This duality can be understood from a physical viewpoint: If we consider the magnetic field problem described above, and change the gauge of the vector potential into a Landau gauge in the transversal direction, we obtain again the almost Mathieu operator, but with $\lambda$ replaced by $4/\lambda$ and the whole operator rescaled by a factor of $\lambda/2$. Since the energy spectrum of the magnetic field Hamiltonian must be gauge independent, we should expect

$$S(\alpha,\lambda) = \frac{\lambda}{2} S(\alpha, 4/\lambda). \tag{3.4}$$

Indeed, (3.4) has been established (for any real $\alpha$) by Avron-Simon [10]. From (3.4) we see, in particular, that it is sufficient to study $S(\alpha,\lambda)$ for $0 \le |\lambda| \le 2$, since for $|\lambda| > 2$, $S(\alpha,\lambda)$ is obtained immediately from the $|\lambda| < 2$ case.

In the rational case $\alpha = p/q$, $S(\alpha,\lambda)$ (which consists of $q$ bands) can have at most $q-1$ gaps. It turns out that these gaps are always open, except for the middle gap for even $q$ (which is always closed). This fact has been proven by van Mouche [70], and also, independently, by Choi-Elliot-Yui [15]. In fact, [15] even obtains an explicit lower bound on the size of each open gap, which is:

**Theorem 3.1.** *If $\mathbb{Q} \ni \alpha = p/q$, and $|\lambda| \le 2$, then all gaps (except the middle gap for $q$ even) of $S(\alpha,\lambda)$ are open and have width larger than $(|\lambda|/16)^q$.*

Since $H_{\alpha,\lambda,\theta}$ is strongly continuous in $\alpha$, $\sigma(\alpha,\lambda,\theta)$ is also continuous, in the sense that if $E \in \sigma(\alpha,\lambda,\theta)$ and $\alpha_n \to \alpha$, then there are points $E_n \in \sigma(\alpha_n,\lambda,\theta)$ such that $E_n \to E$. The set $S(\alpha,\lambda)$ has even better continuity properties and various results on this continuity have been obtained. In particular, Avron-van Mouche-Simon [6] have shown the following uniform (in both $\alpha$ and the energy) Hölder continuity of order 1/2.

**Proposition 3.2.** *For a fixed $\lambda$ and $|\alpha - \alpha'| < C(\lambda)$, each $E \in S(\alpha, \lambda)$ has $E' \in S(\alpha', \lambda)$ with $|E - E'| < 6|\lambda(\alpha - \alpha')|^{1/2}$.*

In particular, this theorem says that for every gap in $S(\alpha, \lambda)$ of width $|g|$ larger than $12|\lambda(\alpha - \alpha')|^{1/2}$, there is a corresponding gap in $S(\alpha', \lambda)$ of width larger than $|g| - 12|\lambda(\alpha - \alpha')|^{1/2}$.

Among the notable results for the periodic case ($\alpha \in \mathbb{Q}$) are those involving the Lebesgue measure of the spectrum. While the spectrum itself clearly depends greatly on $\alpha = p/q$, as it consists of different numbers of disjoint bands for different $q$'s, the "total bandwidth" turns out to have some remarkable universal properties. This has been first noted in 1980 by Aubry-André [3] who found numerical evidence that for $p, q$ relatively prime, $\lim_{q\to\infty} |S(p/q, \lambda)| = |4 - 2|\lambda||$ (where $|\cdot|$, for subsets of $\mathbb{R}$, denotes Lebesgue measure). The first rigorous result on this issue came in a 1983 paper by Thouless [93], who showed that for every rational $p/q$, $|S(p/q, \lambda)| \geq |4 - 2|\lambda||$. In their 1990 paper [6], Avron-van Mouche-Simon have proven that for $|\lambda| \leq 2$ and $p, q$ relatively prime,

$$4 - 2|\lambda| \leq |S(p/q, \lambda)| \leq 4 - 2|\lambda| + 4\pi \left(\frac{|\lambda|}{2}\right)^{q/2}, \tag{3.5}$$

and moreover,

$$|S_-(p/q, \lambda)| = 4 - 2|\lambda|, \tag{3.6}$$

where $S_-(\alpha, \lambda) \equiv \bigcap_\theta \sigma(\alpha, \lambda, \theta)$. (3.5) is in accordance with the numerical observation of Aubry-André for any $|\lambda| \neq 2$.

For the $|\lambda| = 2$ case, it turns out that not only is $|S(p/q, \lambda)|$ vanishing as $q \to \infty$, but this occurs at a remarkably universal rate. A conjecture of Thouless [93, 94, 96, 97] states that (for $p, q$ relatively prime) $\lim_{q\to\infty} q|S(p/q, 2)| = (32/\pi)\beta = 9.32\ldots$, where $\beta$ is Catalan's constant. While heuristic analytical justifications for this conjecture have been given [43, 69, 94], it has been rigorously established only for some specific sequences [43]. However, while the above precise scaling law is generally an open problem, the following general bound has been proven by Last [65]:

$$\frac{2(\sqrt{5}+1)}{q} < |S(p/q, 2)| < \frac{8e}{q} \tag{3.7}$$

(where $e \equiv \exp(1) = 2.71\ldots$). It is interesting to note that this result relies on looking at the $\lambda \to 2$ limit of the remarkable exact equality (3.6) of Avron-van Mouche-Simon.

For the non-periodic (but almost periodic) case of $\alpha \in \mathbb{R} \setminus \mathbb{Q}$, we have the following general theorem.

**Theorem 3.3.** *For any $\alpha \in \mathbb{R} \setminus \mathbb{Q}$ and $\lambda \neq 0$, $S(\alpha, \lambda)$ is a Cantor set and has Lebesgue measure $|4 - 2|\lambda||$.*

While Theorem 3.3 is elegant and simple to formulate, its proof took a considerable effort by many individuals over a period of more than twenty years. In fact, the "last nail" that enables us to write this theorem in such a simple general form is not yet published [4].

While Theorem 3.3 holds equally well for all irrational $\alpha$'s, its proof, like some other elements of the spectral theory of $H_{\alpha,\lambda,\theta}$ discussed below, requires looking differently at different kinds of irrationals. The main issue that distinguishes irrationals here is how well they can be approximated by rationals. In particular, we say that $\alpha$ is a Liouville number if there exists a sequence $\{p_n/q_n\}$ of rationals, such that $|\alpha - p_n/q_n| < n^{-q_n}$. We say that $\alpha$ has typical Diophantine properties if for every $\epsilon > 0$ there is a $\delta > 0$, such that $|\alpha - p/q| > \delta/q^{2+\epsilon}$ for all rationals $p/q$. The numbers with typical Diophantine properties are a set of full Lebesgue measure, while the Liouville numbers are a dense $G_\delta$ set [76] of zero Lebesgue measure (and zero Hausdorff dimension). The question of how well an irrational can be approximated by rationals is strongly connected with the properties of its continued fraction expansion (by integers), which has the form:

$$\alpha = [n_1, n_2, n_3, \dots] = \cfrac{1}{n_1 + \cfrac{1}{n_2 + \cfrac{1}{n_3 + \cdots}}}\,. \tag{3.8}$$

Roughly speaking, Liouville numbers have a subsequence of $n_i$'s which grows very fast, while the $n_i$'s for numbers with typical Diophantine properties cannot grow too fast. Nevertheless, the set of irrationals for which the sequence of $n_i$'s is bounded (these are the numbers for which $\inf_{p,q\in\mathbb{Z}} q^2|\alpha - p/q| > 0$) has zero Lebesgue measure. For more information on continued fraction expansions and approximating irrationals by rationals, see [41] or [57].

Theorem 3.3 actually consists of two independent results. One is the Cantor structure of the spectrum, while the other is the precise $\alpha$-independent formula for the Lebesgue measure of the spectrum. The two are connected only at the critical point $|\lambda| = 2$, where the vanishing of the measure of the spectrum also implies that it must be a Cantor set.

The equality $|S(\alpha, \lambda)| = |4 - 2|\lambda||$ has been first conjectured by Aubry-André [3] in 1980 (due to their numerical findings for the limiting behavior for rational $\alpha$). Even before then, Hofstadter [47] conjectured the vanishing of $|S(\alpha, 2)|$. Thouless [93], showed in 1983 that $|S(\alpha, \lambda)| \geq |4 - 2|\lambda||$. In 1989, Helffer-Sjöstrand [44] proved that $S(\alpha, \lambda)$ is a Cantor set of zero Lebesgue measure for the special case $|\lambda| = 2$ and for a special class of irrationals characterized through their continued fraction expansions (3.8) as having $n_i > C$ for some large constant $C$ and all $i$. Their analysis is based on a unique spectral renormalization procedure which exploits elaborate semi-classical analysis and builds on ideas (and a heuristic analysis) of Wilkinson [98]. The class of relevant $\alpha$'s is nowhere dense and of zero Lebesgue measure. Later Last [64], in 1993, noted that the core results of Avron-van Mouche-Simon [6] (namely, (3.5) and Proposition 3.2 above) imply that $|S(\alpha,\lambda)| = |4-2|\lambda||$ holds for any $|\lambda| \neq 2$ and $\alpha$ having an unbounded continued fraction expansion (and thus for Lebesgue a.e. $\alpha$). He later extended [65] this result to also include $|\lambda| = 2$, by proving (3.7). In 1998, Jitomirskaya-Last [52] have shown that $|S(\alpha,\lambda)| = |4-2|\lambda||$ holds for any irrational $\alpha$ if $|\lambda| > 29$ (or $|\lambda| < 4/29$). Later, in 2002, Jitomirskaya-Krasovsky [51] extended this to any $|\lambda| \neq 2$. Finally, in 2003, Avila-Krikorian [5] have shown that $|S(\alpha,2)| = 0$ also for irrational $\alpha$'s with a bounded continued fraction expansion and thus completed the proof that $|S(\alpha,\lambda)| = |4-2|\lambda||$ holds for any $\lambda$ and any irrational $\alpha$.

As noted above, the Cantor structure of the spectrum has been first conjectured in 1964 by Azbel [11] and later demonstrated numerically in 1976 by Hofstadter [47]. The problem of proving it has been the most notable one in the spectral theory of the almost Mathieu operator. It has been named "the ten Martini problem" by Simon [82], due to Mark Kac offering ten Martinis to anyone who solves it (see [81, 82]). The first mathematical result for this problem has been given in 1982 by Bellissard-Simon [13] who have shown that $S(\alpha,\lambda)$ is a Cantor set for a dense $G_\delta$ of pairs $(\lambda,\alpha) \in \mathbb{R}^2$ (and thus "generically," in the commonly used topological sense). The $\alpha$'s which are relevant to their proof are essentially Liouville numbers, but are not really specified. Nor are the $\lambda$'s. Later Sinai [90], in a 1987 paper, has proven that for (Lebesgue) a.e. $\alpha$, and sufficiently large (or small) $|\lambda|$, $S(\alpha,\lambda)$ is a Cantor set. Sinai's proof is perturbative and quite complicated. The relevant $\alpha$'s are those with typical Diophantine properties and are explicitly given. The $\lambda$'s are not explicitly given and the required largeness (or smallness) of $|\lambda|$ may depend on $\alpha$. In 1989, the above-mentioned result of Helffer-Sjöstrand [44] implied the Cantor structure for $|\lambda| = 2$ and some irrational $\alpha$'s. In 1990, Choi-Elliot-Yui [15] have proven that $S(\alpha,\lambda)$ is a Cantor set for all $\lambda$'s and Liouville $\alpha$'s. In particular, their result strengthens the Bellissard-Simon result. The Choi-Elliot-Yui result is a simple consequence of their lower bound on gap sizes for rational $\alpha$'s (Theorem 3.1), along with the continuity properties of $S(\alpha,\lambda)$ (Proposition 3.2). Later, in 1994, Last's result [65] on the vanishing of $|S(\alpha,2)|$ also established the Cantor structure for $|\lambda| = 2$ and Lebesgue a.e. $\alpha$. A major development occurred recently in the work of Puig [73], who shows that $S(\alpha,\lambda)$ is a Cantor set for all $\lambda \neq 2$ and $\alpha$'s with typical Diophantine properties. While this means Lebesgue a.e. $\alpha$, there are still some irrationals which are not covered by either Puig's result or the Choi-Elliot-Yui [15] result. This last remaining issue has been resolved very recently by Avila-Jitomirskaya [4] who completed the proof of Theorem 3.3.

We note that while the Cantor structure of the spectrum is now fully established, there is also a strong (also known as dry) form of the ten Martini problem, which is to prove that all of the gaps which are allowed by the gap labelling theorem (see [82] and references therein) are really open. The above-mentioned result of Choi-Elliot-Yui [15] actually does solve this strong form for the case of Liouville numbers. Moreover, Puig [73] solves it for $\alpha$'s with typical Diophantine properties and for very large (or small) $|\lambda|$. Beyond that, however, this question is still open.

Finally, we remark that in order to fully appreciate the rich and elegant structure of the spectrum of the almost Mathieu operator, one is strongly advised to look at some of its numerical drawings. The first of those was done by Hofstadter [47] who created the famous "Hofstadter butterfly" out of it. A nice selection of many numerically computed illustrations of almost Mathieu spectra has been provided by Guillement-Helffer-Treton in [40]. For a (physically motivated) colored version of Hofstadter's butterfly, see Osadchy-Avron [71] (or [7]).

### 3.2. Spectral types and the metal-insulator transition

For the periodic case ($\alpha \in \mathbb{Q}$), $H_{\alpha,\lambda,\theta}$ (like all periodic Schrödinger operators) has purely absolutely-continuous spectrum. Once $\alpha$ is irrational, however, the nature of the spectrum of $H_{\alpha,\lambda,\theta}$ is quite rich and depends on the precise values of the parameters. Aubry-André, in their 1980 paper [3], conjectured that $H_{\alpha,\lambda,\theta}$ (for $\alpha \in \mathbb{R} \setminus \mathbb{Q}$) exhibits an elegant "metal-insulator transition" as follows: For $|\lambda| < 2$ it has purely absolutely-continuous spectrum and for $|\lambda| > 2$ it exhibits Anderson localization (namely, it has purely pure-point spectrum with exponentially decaying eigenvectors). As studies in the years that followed indicate, this conjecture is correct only in the probabilistic sense of holding for Lebesgue a.e. $\alpha, \theta$. The full picture is a lot more delicate as there are sets of parameters of zero Lebesgue measure giving rise to different spectral properties.

The following version of the Aubry-André conjecture (whose proof has been recently completed by Jitomirskaya [50]) is now known:

**Theorem 3.4.** *For Lebesgue a.e. pair $\alpha, \theta$, $H_{\alpha,\lambda,\theta}$ has spectral properties as follows:*

(i) *If $|\lambda| < 2$, purely absolutely continuous spectrum.*
(ii) *If $|\lambda| = 2$, purely singular continuous spectrum.*
(iii) *If $|\lambda| > 2$, purely pure-point spectrum with exponentially decaying eigenvectors.*

Theorem 3.4 is complemented by the following facts:

- If $|\lambda| > 2$ and $\alpha$ is a Liouville number, then $H_{\alpha,\lambda,\theta}$ has purely singular-continuous spectrum for every $\theta$. (The result for a.e. $\theta$ has been obtained by Avron-Simon [9] in 1982, following an idea of Gordon [33]. The fact that it holds for every $\theta$ is a consequence of later results [62, 66, 68]). Thus, the $\alpha$'s for which pure-point spectrum occurs are, roughly, those with typical Diophantine properties.
- If $|\lambda| > 2$ and $\alpha$ is irrational, then there exists a dense $G_\delta$ set of $\theta$'s for which $H_{\alpha,\lambda,\theta}$ has purely singular-continuous spectrum (Jitomirskaya-Simon [55], 1994).
- If $|\lambda| > 2$ and $\alpha$ is irrational, then for every $\theta$, the spectrum of $H_{\alpha,\lambda,\theta}$ is purely zero-dimensional, namely, the spectral measures are supported on a set of zero Hausdorff dimension (Jitomirskaya-Last [54], 2000).
- If $|\lambda| = 2$ and $\alpha$ is irrational, then for Lebesgue a.e. $\theta$, the spectrum of $H_{\alpha,\lambda,\theta}$ is purely singular-continuous (see below).
- If $|\lambda| < 2$, then for every $\alpha, \theta$, $H_{\alpha,\lambda,\theta}$ has no eigenvalues (Delyon [23], 1987).

- If $|\lambda| < 2$ and $\alpha$ is irrational, then for every $\theta$, $H_{\alpha,\lambda,\theta}$ has some absolutely-continuous spectrum and, moreover, the Lebesgue measure of its absolutely-continuous spectrum is equal to the Lebesgue measure of its spectrum. (This is essentially a 1993 result of Last [64], who proved it for every $\alpha$ and a.e. $\theta$ using Kotani theory [61, 83]. The fact that it holds for every $\theta$ is a consequence of later results [62, 68].)

We note that while the above listed facts provide a fairly detailed and nearly complete picture for the spectral properties of $H_{\alpha,\lambda,\theta}$, the following natural questions are still open:

- Is there ever any singular-continuous spectrum for $|\lambda| < 2$?
- Is there ever any pure-point spectrum for $|\lambda| = 2$?

Similarly to Theorem 3.3, proving Theorem 3.4 took a considerable effort over a period of almost twenty years, with many partial results along the way. Some of the main highlights (excluding results already mentioned above as complements to the theorem) are the following: Absence of absolutely-continuous spectrum for any $|\lambda| > 2$, irrational $\alpha$, and a.e. $\theta$, has been established in 1982 by Avron-Simon [10] and Figotin-Pastur [25]. They show (based on the arguments of Aubry-André [3]) positivity of the Lyapunov exponent, which has also been obtained by Herman [45], and thus conclude the absence of a.c. spectrum by the Ishii-Pastur theorem (see Theorem 9.13 of [17]). In 1983, Bellissard-Lima-Testard [12], using ideas of Dinaburg-Sinai [24], have shown (for $\alpha$ with typical Diophantine properties and a.e. $\theta$) the existence of some pure-point spectrum for very large $|\lambda|$ and some absolutely-continuous spectrum for very small $|\lambda|$. In 1987, Sinai [90] established the existence of purely pure-point spectrum for $\alpha$'s with typical Diophantine properties, very large $|\lambda|$, and a.e. $\theta$. A similar result has independently been obtained about the same time by Fröhlich-Spencer-Wittwer [27]. A little later, in 1989, Chulaevsky-Delyon [16], using Sinai's result, have shown that the spectrum is purely absolutely-continuous for $\alpha$'s with typical Diophantine properties, very small $|\lambda|$, and a.e. $\theta$. In 1997, Gordon-Jitomirskaya-Last-Simon [35] proved a version of the Aubry duality saying that for a fixed irrational $\alpha$ and a.e. $\theta$, the existence of some p.p. spectrum for $\lambda$ implies the existence of some a.c. spectrum for the dual coupling $4/\lambda$, and the occurrence of purely p.p. spectrum for $\lambda$ implies the occurrence of purely a.c. spectrum for the dual coupling $4/\lambda$. This result, along with the vanishing of the measure of the spectrum of $H_{\alpha,2,\theta}$, which says that it cannot have any a.c. spectrum, implies the part of Theorem 3.4 for $|\lambda| = 2$. Finally, in 1999, Jitomirskaya [50], strengthening her earlier results [48, 49], proved that for $\alpha$ with typical Diophantine properties, any $|\lambda| > 2$, and a.e. $\theta$, $H_{\alpha,\lambda,\theta}$ has purely pure-point spectrum with exponentially localized eigenvectors (also see Bourgain-Goldstein [14] for a slightly later proof). This establishes the part of Theorem 3.4 for $|\lambda| > 2$, and when combined with the Gordon-Jitomirskaya-Last-Simon [35] result, it also implies the part of Theorem 3.4 for $|\lambda| < 2$.

## 4. Sparse potentials

Sparse potentials are potentials which vanish outside a sequence of "bumps" which are spaced "very far" apart from each other, namely, the distances between the bumps should be fastly growing to infinity. The bumps themselves may have "sizes" which are either growing, decaying, or stay the same. The size of a bump can be made large or small by controlling either its height or its width, although in what follows, we consider only unit width bumps, such that their size is fully determined by their height.

To be more concrete, we discuss here a specific class of discrete operators on $\ell^2(\mathbb{N})$, defined by (1.2) with a potential $V$ obeying:

$$V(n) = 0 \text{ if } n \notin \{L_k\}_{k=1}^{\infty}. \tag{4.1}$$

Here $\{L_k\}_{k=1}^{\infty}$ is a rapidly growing sequence of positive integers and the $V(L_k)$'s are some non-vanishing real numbers. The growth rate of the sequence $\{L_k\}_{k=1}^{\infty}$ is at least faster than linear. A broad range of growth rates has been considered, including, as we shall see below, growth rates from quadratic to much faster than factorial.

Unlike the almost Mathieu operator discussed above, sparse Schrödinger operators of this type are generally not expected to be connected with any interesting problems in physics and so the interest in them is essentially purely mathematical. What makes them interesting is the fact that they are relatively easy to analyze, while at the same time they can exhibit a broad range of spectral properties. This makes them a very useful playground for constructing explicit examples with various prescribed (and often tightly controlled) spectral properties.

The first to realize the great usefulness of sparse potentials for constructing examples with interesting spectral properties seems to have been Pearson, who used them in his seminal 1978 paper [72] to construct explicit Schrödinger operators with purely singular-continuous spectrum. While he considered continuous operators of the form (1.1), it is also easy to prove a discrete analog of his result. In particular, the following variant of Pearson's result has been proven by Kiselev-Last-Simon [59]:

**Theorem 4.1.** *Let $H$ be a discrete sparse Schrödinger operator on $\ell^2(\mathbb{N})$, of the form defined by* (1.2) *and* (4.1)*. Suppose that as $k \to \infty$, $L_k/L_{k+1} \to 0$ and $V(L_k) \to 0$.*

(i) *If $\sum_{k=1}^{\infty} |V(L_k)|^2 < \infty$, then $H$ has purely absolutely-continuous spectrum on $(-2, 2)$.*
(ii) *If $\sum_{k=1}^{\infty} |V(L_k)|^2 = \infty$, then $H$ has purely singular-continuous spectrum on $(-2, 2)$.*

In 1990, Gordon [34] and Kirsch-Molchanov-Pastur [58] used growing sparse potentials to construct deterministic Schrödinger operators with thick pure-point spectrum. What they show, in essence, is that for a given $\{L_k\}_{k=1}^{\infty}$, it is possible to make $\{V(L_k)\}_{k=1}^{\infty}$ grow fast enough so that the operator $H$ has purely pure-point spectrum for a.e. boundary phase $\theta$. More recently, models of this type have also been studied by Last-Simon [67], who show the following explicit variant for this type of results:

**Theorem 4.2.** *Let $H$ be a discrete sparse Schrödinger operator on $\ell^2(\mathbb{N})$, of the form defined by* (1.2) *and* (4.1). *Let $L_k = k^2$ and $V(L_k) = e^{\beta k}$; then for a.e. boundary phase $\theta$, $H$ has purely pure-point spectrum on $[-2,2]$ with eigenvectors decaying like $e^{-\beta n/2}$.*

In 1996, Simon-Stolz [89] took a complementary view of unbounded sparse potentials and showed that for a given unbounded sequence of potential values $\{V(L_k)\}_{k=1}^{\infty}$, it is possible to make the sequence $\{L_k\}_{k=1}^{\infty}$ grow fast enough so that $H$ has purely singular-continuous spectrum on $(-2,2)$ (for any boundary phase $\theta$). A little later, Simon [86] showed that it is, in fact, possible to make $\{L_k\}_{k=1}^{\infty}$ grow fast enough so that the singular-continuous spectrum is purely one-dimensional, namely, the spectral measures do not give weight to sets of Hausdorff dimension less than one.

We note that for unbounded sparse potentials, absolutely-continuous spectrum is excluded by a general theorem of Simon-Spencer [88], which states that discrete half-line Schrödinger operators with unbounded potentials have no a.c. spectrum (also see [68] for another proof). Thus, the spectrum is always singular and we see that there is an interplay between the rates of growth of the two sequences $\{L_k\}_{k=1}^{\infty}$ and $\{V(L_k)\}_{k=1}^{\infty}$. That is, faster growth of $\{L_k\}_{k=1}^{\infty}$ makes the spectrum more continuous and faster growth of $\{V(L_k)\}_{k=1}^{\infty}$ makes it more singular. This general principle has been fine-tuned in 1999 by Jitomirskaya-Last [53] to construct examples of operators with singular-continuous spectrum of exact dimension $\alpha$ (for any $\alpha \in (0,1)$). That is:

**Theorem 4.3.** *Let $H$ be a discrete sparse Schrödinger operator on $\ell^2(\mathbb{N})$, of the form defined by* (1.2) *and* (4.1). *Let $\alpha \in (0,1)$, $L_k = 2^{(k^k)}$, and $V(L_k) = L_k^{(1-\alpha)/2\alpha}$; then for a.e. boundary phase $\theta$, the spectrum of $H$ in $[-2,2]$ is of exact dimension $\alpha$, namely, the restriction of the spectral measure to $[-2,2]$ is supported on a set of Hausdorff dimension $\alpha$ and does not give weight to sets of Hausdorff dimension less than $\alpha$.*

We note that while we formulated Theorem 4.3 as holding for "a.e. boundary phase $\theta$," as indeed proven in [53], Tcheremchantsev [92] has recently proven a variant of this theorem where the exact dimension holds for *all* boundary phases. Thus, "a.e." in Theorem 4.3 can actually be replaced by "every".

We further note that Zlatoš [99] recently studied discrete sparse Schrödinger operators for which $\{V(L_k)\}_{k=1}^{\infty}$ is a constant sequence and $\{L_k\}_{k=1}^{\infty}$ grows exponentially. He shows that for appropriate exponential growth rates of the $L_k$'s (depending on the common value of the $V(L_k)$'s), such operators exhibit some singular-continuous spectrum with fractional dimensionality. In fact, his model is an example of a deterministic potential leading to spectral properties which are very similar to those previously established by Kiselev-Last-Simon [59] for certain random decaying potentials.

**Acknowledgment**

The author is grateful to Werner Amrein, Andreas Hinz, David Pearson and the other organizers of the Colloquium on the occasion of the 200th Anniversary of Charles Sturm for inviting him to participate and give a talk at this very interesting conference. It has been a pleasure and an honor.

## References

[1] C. Albrecht, J.H. Smet, K. von Klitzing, D. Weiss, V. Umansky and H. Schweizer, Evidence of Hofstadters fractal energy spectrum in the quantized Hall conductance, Phys. Rev. Lett. **86** (2001), 147–150.

[2] P.W. Anderson, Absence of diffusion in certain random lattices, Phys. Rev. **109** (1958), 1492–1505.

[3] S. Aubry and G. André, Analyticity breaking and Anderson localization in incommensurate lattices, Ann. Israel Phys. Soc. **3** (1980), 133–164.

[4] A. Avila and S. Jitomirskaya, in preparation.

[5] A. Avila and R. Krikorian, *Reducibility or non-uniform hyperbolicity for quasiperiodic Schrödinger cocycles*, preprint, 2003.

[6] J. Avron, P.M.H. van Mouche and B. Simon, On the measure of the spectrum for the almost Mathieu operator, Commun. Math. Phys. **132** (1990), 103–118.

[7] J.E. Avron, D. Osadchy and R. Seiler, A topological look at the quantum Hall effect, Physics Today, August 2003, 38–42.

[8] J. Avron and B. Simon, Transient and recurrent spectrum, J. Funct. Anal. **43** (1981), 1–31.

[9] J. Avron and B. Simon, Singular continuous spectrum for a class of almost periodic Jacobi matrices, Bull. Amer. Math. Soc. **6** (1982), 81–85.

[10] J. Avron and B. Simon, Almost periodic Schrödinger operators II: The integrated density of states, Duke Math. J. **50** (1983), 369–391.

[11] M.Ya. Azbel, Energy spectrum of a conduction electron in a magnetic field, Sov. Phys. JETP **19** (1964), 634–645.

[12] J. Bellissard, R. Lima and D. Testard, A metal-insulator transition for the almost Mathieu model, Commun. Math. Phys. **88** (1983), 207–234.

[13] J. Bellissard and B. Simon, Cantor spectrum for the almost Mathieu equation, J. Funct. Anal. **48** (1982), 408–419.

[14] J. Bourgain and M. Goldstein, On nonperturbative localization with quasi-periodic potential, Ann. of Math. **152** (2000), 835–879.

[15] M.D. Choi, G.A. Elliott and N. Yui, Gauss polynomials and the rotation algebra, Invent. Math. **99** (1990), 225–246.

[16] V. Chulaevsky and F. Delyon, Purely absolutely continuous spectrum for almost Mathieu operators, J. Stat. Phys. **55** (1989), 1279–1284.

[17] H.L. Cycon, R.G. Froese, W. Kirsch and B. Simon, *Schrödinger Operators*, Springer, Berlin, 1987.

[18] D. Damanik, Gordon-type arguments in the spectral theory of one-dimensional quasicrystals, in *Directions in mathematical quasicrystals*, 277–305, CRM Monogr. Ser. **13**, Providence, R.I., Amer. Math. Soc., 2000.

[19] R. Del Rio, Boundary Conditions and Spectra of Sturm-Liouville Operators, in this volume.

[20] R. Del Rio, S. Jitomirskaya, Y. Last and B. Simon, Operators with singular continuous spectrum IV: Hausdorff dimensions, rank one perturbations, and localization, J. Analyse Math. **69** (1996), 153–200.

[21] R. Del Rio, S. Jitomirskaya, N. Makarov and B. Simon, Singular continuous spectrum is generic, Bull. Amer. Math. Soc. **31** (1994), 208–212.

[22] R. Del Rio, N. Makarov and B. Simon, Operators with singular continuous spectrum II: Rank one operators, Commun. Math. Phys. **165** (1994), 59–67.

[23] F. Delyon, Absence of localisation in the almost Mathieu equation, J. Phys. A **20** (1987), L21–L23.

[24] E. Dinaburg and Ya. Sinai, The one-dimensional Schrödinger equation with a quasiperiodic potential, Funct. Anal. Appl. **9** (1975), 279–289.

[25] A. Figotin and L. Pastur, The positivity of Lyapunov exponent and absence of the absolutely continuous spectrum for the almost-Mathieu equation, J. Math. Phys. **25** (1984), 774–777.

[26] S. Fishman, D.R. Grempel and R.E. Prange, Chaos, quantum recurrences, and Anderson localization, Phys. Rev. Lett. **49** (1982), 509–512.

[27] J. Fröhlich, T. Spencer and P. Wittwer, Localization for a class of one-dimensional quasi-periodic Schrödinger operators, Commun. Math. Phys. **132** (1990), 5–25.

[28] I.M. Gel'fand and B.M. Levitan, On the determination of a differential equation from its spectral function, Izv. Akad. Nauk SSR. Ser. Mat. **15** (1951), 309–360 (Russian); English transl. in Amer. Math. Soc. Transl. Ser. 2 **1** (1955), 253–304.

[29] D.J. Gilbert, On subordinacy and analysis of the spectrum of Schrödinger operators with two singular endpoints, Proc. Roy. Soc. Edinburgh Sect. A **112** (1989), 213–229.

[30] D.J. Gilbert, Asymptotic Methods in the Spectral Analysis of Sturm-Liouville Operators, in this volume.

[31] D.J. Gilbert and D. Pearson, On subordinacy and analysis of the spectrum of one-dimensional Schrödinger operators, J. Math. Anal. **128** (1987), 30–56.

[32] I. Goldsheid, S. Molchanov and L. Pastur, A pure point spectrum of the stochastic one-dimensional Schrödinger equation, Funct. Anal. Appl. **11** (1977), 1–10.

[33] A. Gordon, On the point spectrum of one-dimensional Schrödinger operators, Usp. Math. Nauk **31** (1976), 257–258.

[34] A.Ya. Gordon, Deterministic potential with a pure point spectrum, Math. Notes **48** (1990), 1197–1203.

[35] A. Gordon, S. Jitomirskaya, Y. Last and B. Simon, Duality and singular continuous spectrum in the almost Mathieu equation, Acta Math. **178** (1997), 169–183.

[36] D.R. Grempel, S. Fishman and R.E. Prange, Localization in an incommensurate potential: An exactly solvable model, Phys. Rev. Lett. **49** (1982), 833–836.

[37] D.R. Grempel and R.E. Prange, Quantum dynamics of a nonintegrable system, Phys. Rev. A **29** (1984), 1639–1647.

[38] I. Guarneri, Spectral properties of quantum diffusion on discrete lattices, Europhys. Lett. **10** (1989), 95–100.

[39] I. Guarneri, On an estimate concerning quantum diffusion in the presence of a fractal spectrum, Europhys. Lett. **21** (1993), 729–733.

[40] J.P. Guillement, B. Helffer and P. Treton, Walk inside Hofstadter's butterfly, J. Phys. France **50** (1989), 2019–2058.

[41] G.H. Hardy and E.M. Wright, *An Introduction to the Theory of Numbers*, Fifth ed., Oxford University Press, Oxford, 1979.

[42] P.G. Harper, Single band motion of conduction electrons in a uniform magnetic field, Proc. Phys. Soc. London A **68** (1955), 874–892.

[43] B. Helffer and P. Kerdelhue, On the total bandwidth for the rational Harper's equation, Commun. Math. Phys. **173** (1995), 335–356.

[44] B. Helffer and J. Sjöstrand, Semi-classical analysis for Harper's equation III: Cantor structure of the spectrum, Mém. Soc. Math. France (N.S.) **39** (1989), 1–139.

[45] M. Herman, Une méthode pour minorer les exposants de Lyapunov et quelques exemples montrant le caractère local d'un théorème d'Arnold et de Moser sur le tore en dimension 2, Comm. Math. Helv. **58** (1983), 453–502.

[46] A. Hof, O. Knill and B. Simon, Singular continuous spectrum for palindromic Schrödinger operators, Commun. Math. Phys. **174** (1995), 149–159.

[47] D.R. Hofstadter, Energy levels and wave functions of Bloch electrons in a rational or irrational magnetic field, Phys. Rev. B **14** (1976), 2239–2249.

[48] S. Jitomirskaya, Anderson localization for the almost Mathieu equation; A nonperturbative proof, Commun. Math. Phys. **165** (1993), 49–58.

[49] S. Jitomirskaya, Anderson localization for the almost Mathieu equation II: Point spectrum for $\lambda > 2$, Commun. Math. Phys **168** (1995), 563–570.

[50] S. Jitomirskaya, Metal-insulator transition for the almost Mathieu operator, Ann. of Math. **150** (1999), 1159–1175.

[51] S. Jitomirskaya and I.V. Krasovsky, Continuity of the measure of the spectrum for discrete quasiperiodic operators, Math. Res. Lett. **9** (2002), 413–421.

[52] S. Jitomirskaya and Y. Last, Anderson localization for the almost Mathieu operator III: Semi-uniform localization, continuity of gaps, and measure of the spectrum, Commun. Math. Phys. **195** (1998), 1–14.

[53] S. Jitomirskaya and Y. Last, Power law subordinacy and singular spectra I: Half line operators, Acta Math. **183** (1999), 171–189.

[54] S. Jitomirskaya and Y. Last, Power law subordinacy and singular spectra II: Line operators, Commun. Math. Phys. **211** (2000), 643–658.

[55] S. Jitomirskaya and B. Simon, Operators with singular continuous spectrum III: Almost periodic Schrödinger operators, Commun. Math. Phys. **165** (1994), 201–205.

[56] S. Kahn and D.B. Pearson, Subordinacy and spectral theory for infinite matrices, Helv. Phys. Acta **65** (1992), 505–527.

[57] A.Ya. Khinchin, *Continued Fractions*, Dover, Mineola, 1997.

[58] W. Kirsch, S. Molchanov and L. Pastur, The one-dimensional Schrödinger operator with unbounded potential: The pure point spectrum, Funct. Anal. Appl. **24** (1990), 176–186.

[59] A. Kiselev, Y. Last and B. Simon, Modified Prüfer and EFGP transforms and the spectral analysis of one-dimensional Schrödinger operators, Commun. Math. Phys. **194** (1998), 1–45.

[60] K. von Klitzing, G. Dorda and M. Pepper, New method for high-accuracy determination of the fine-structure constant based on quantized Hall resistance, Phys. Rev. Lett. **45** (1980), 494–497.

[61] S. Kotani, Ljaponov indices determine absolutely continuous spectra of stationary one-dimensional Schrödinger operators, in *Stochastic Analysis*, 225–248, ed. by K. Ito, North Holland, Amsterdam, 1984.

[62] S. Kotani, Generalized Floquet theory for stationary Schrödinger operators in one dimension, Chaos, Solitons & Fractals **8** (1997), 1817–1854.

[63] H. Kunz and B. Souillard, Sur le spectre des opérateurs aux différences finies aléatoires, Commun. Math. Phys. **78** (1980), 201–246.

[64] Y. Last, A relation between a.c. spectrum of ergodic Jacobi matrices and the spectra of periodic approximants, Commun. Math. Phys. **151** (1993), 183–192.

[65] Y. Last, Zero measure spectrum for the almost Mathieu operator, Commun. Math. Phys. **164** (1994), 421–432.

[66] Y. Last, Quantum dynamics and decompositions of singular continuous spectra, J. Funct. Anal. **142** (1996), 406–445.

[67] Y. Last and B. Simon, Modified Prüfer and EFGP transforms and deterministic models with dense point spectrum, J. Funct. Anal. **154** (1998), 513–530.

[68] Y. Last and B. Simon, Eigenfunctions, transfer matrices, and absolutely continuous spectrum of one-dimensional Schrödinger operators, Invent. Math. **135** (1999), 329–367.

[69] Y. Last and M. Wilkinson, A sum rule for the dispersion relations of the rational Harper's equation, J. Phys. A **25** (1992), 6123–6133.

[70] P.M.H. van Mouche, The coexistence problem for the discrete Mathieu operator, Commun. Math. Phys. **122** (1989), 23–34.

[71] D. Osadchy and J.E. Avron, Hofstadter butterfly as quantum phase diagram, J. Math. Phys. **42** (2001), 5665–5671.

[72] D.B. Pearson, Singular continuous measures in scattering theory, Commun. Math. Phys. **60** (1978), 13–36.

[73] J. Puig, Cantor Spectrum for the almost Mathieu operator, Commun. Math. Phys. **244** (2004), 297–309.

[74] A. Rauh, Degeneracy of Landau levels in crystals, Phys. Status Solidi B **65** (1974), 131–135.

[75] D. Shechtman, I. Blech, D. Gratias and J.W. Cahn, Metallic phase with long-range orientational order and no translational symmetry, Phys. Rev. Lett. **53** (1984), 1951–1953.

[76] M. Reed and B. Simon, *Methods of Modern Mathematical Physics I: Functional Analysis*, rev. ed., Academic Press, New York, 1980.

[77] M. Reed and B. Simon, *Methods of Modern Mathematical Physics III: Scattering Theory*, Academic Press, New York, 1979.

[78] M. Reed and B. Simon, *Methods of Modern Mathematical Physics IV: Analysis of Operators*, Academic Press, New York, 1978.

[79] C.A. Rogers and S.J. Taylor, The analysis of additive set functions in Euclidean space, Acta Math. **101** (1959), 273–302.

[80] C.A. Rogers and S.J. Taylor, Additive set functions in Euclidean space II, Acta Math. **109** (1963), 207–240.

[81] M.A. Shubin, Discrete magnetic Laplacian, Commun. Math. Phys. **164** (1994), 259–275.

[82] B. Simon, Almost periodic Schrödinger operators: A review, Adv. Appl. Math. **3** (1982), 463–490.

[83] B. Simon, Kotani theory for one-dimensional stochastic Jacobi matrices, Commun. Math. Phys. **89** (1983), 227–234.

[84] B. Simon, Operators with singular continuous spectrum I: General operators, Ann. Math. **141** (1995), 131–145.

[85] B. Simon, Operators with singular continuous spectrum VI: Graph Laplacians and Laplace-Beltrami operators, Proc. Amer. Math. Soc. **124** (1996), 1177–1182.

[86] B. Simon, Operators with singular continuous spectrum VII: Examples with borderline time decay, Commun. Math. Phys. **176** (1996), 713–722.

[87] B. Simon, Schrödinger operators in the twentieth century, J. Math. Phys. **41** (2000), 3523–3555.

[88] B. Simon and T. Spencer, Trace class perturbations and the absence of absolutely continuous spectra, Commun. Math. Phys. **125** (1989), 113–125.

[89] B. Simon and G. Stolz, Operators with singular continuous spectrum V: Sparse potentials, Proc. Amer. Math. Soc. **124** (1996), 2073–2080.

[90] Ya.G. Sinai, Anderson localization for one-dimensional difference Schrödinger operator with quasiperiodic potential, J. Stat. Phys. **46** (1987), 861–909.

[91] G. Szegő, *Orthogonal Polynomials*, Amer. Math. Soc. Colloq. Publ. **23**, Providence, R.I., American Mathematical Society 1939; 3rd edition, 1967.

[92] S. Tcheremchantsev, *Dynamical analysis of Schrödinger operators with growing sparse potentials*, preprint, 2003.

[93] D.J. Thouless, Bandwidth for a quasiperiodic tight binding model, Phys. Rev. B **28** (1983), 4272–4276.

[94] D.J. Thouless, Scaling for the discrete Mathieu equation, Commun. Math. Phys. **127** (1990), 187–193.

[95] D.J. Thouless, M. Kohmoto, M.P. Nightingale and M. den Nijs, Quantized Hall conductance in a two-dimensional periodic potential, Phys. Rev. Lett. **49** (1982), 405–408.

[96] D.J. Thouless and Y. Tan, Total bandwidth for the Harper equation III: Corrections to scaling, J. Phys. A **24** (1991), 4055–4066.

[97] D.J. Thouless and Y. Tan, Scaling, localization and bandwidths for equations with competing periods, Physica A **177** (1991), 567–577.

[98] M. Wilkinson, An exact renormalization group for Bloch electrons in a magnetic field, J. Phys. A **20** (1987), 4337–4354.

[99] A. Zlatoš, Sparse potentials with fractional Hausdorff dimension, J. Funct. Anal. **207** (2004), 216–252.

Yoram Last
Institute of Mathematics
The Hebrew University
91904 Jerusalem
Israel
e-mail: `ylast@math.huji.ac.il`

*W.O. Amrein, A.M. Hinz, D.B. Pearson*
Sturm-Liouville Theory: Past and Present, 121–136


# Asymptotic Methods in the Spectral Analysis of Sturm-Liouville Operators

Daphne Gilbert

**Abstract.** We consider the relationship between the asymptotic behavior of solutions of the singular Sturm-Liouville equation and spectral properties of the corresponding self-adjoint operators. In particular, we review the main features of the theory of subordinacy by considering two standard cases, the half-line operator on $L_2([0,\infty))$ and the full-line operator on $L_2(\mathbb{R})$. It is assumed that the coefficient function $q$ is locally integrable, that 0 is a regular endpoint in the half-line case, and that Weyl's limit point case holds at the infinite endpoints. We note some consequences of the theory for the well-known informal characterization of the spectrum in terms of bounded solutions. We also consider extensions of the theory to related differential and difference operators, and discuss its application, in conjunction with other asymptotic methods, to some typical problems in spectral analysis.

## 1. Introduction

In its original formulation, the Sturm-Liouville boundary value problem consists of a linear second-order ordinary differential equation expressible in the form

$$Lu := -u''(r) + q(r)u(r) = \lambda u(r), \quad r \in I \subseteq \mathbb{R}, \quad q : I \to \mathbb{R}, \quad \lambda \in \mathbb{C}, \tag{1}$$

together with suitable separated or periodic boundary conditions at the endpoints of a finite interval $I$. The spectrum of the associated self-adjoint operator consists of an increasing sequence of isolated real eigenvalues accumulating at infinity, the corresponding eigenfunctions being non-trivial solutions of (1) which satisfy the endpoint conditions [5], [12], [21], [46], [47]. Extension to the case where one of the endpoints of $I$ is singular was achieved by Weyl in 1910 [51], and if $I = [0,\infty)$ or $(-\infty,\infty)$, then (1) is often referred to as the one-dimensional time independent Schrödinger equation, following subsequent recognition of its importance in the mathematical description of quantum phenomena (see, e.g., [38]). If a singular endpoint is in Weyl's limit point case then essential spectrum, which can itself contain both discrete and continuous parts, may also or alternatively be present, and while the (generalized) eigenfunctions are still solutions of the Sturm-Liouville equation (1), their behavior is now more subtle.

Recognition of the close relationship between solutions of the differential equation and spectral properties was already evident in the 1836 paper of Sturm, where the link between the number of points of the spectrum below an eigenvalue and the number of zeros in the associated eigenfunction was noted [46]. Analogues of such properties for the singular case, as well as numerous further connections between solutions and spectra, were identified in the mid-twentieth century by Hartman, Wintner and others (see, e.g., [18], [20], [52]), while contemporary investigations in the Soviet Union contributed a number of independent results in this regard [16, Chapter V]. In more recent work linking polynomially bounded solutions to spectral properties, techniques which are applicable to both one-dimensional and higher-dimensional problems have been developed (see, e.g., [22], [39]).

Different challenges emerged in the late 1950's with the development of rigorous scattering theory and a corresponding awareness of the importance of distinguishing the absolutely continuous component from other parts of the essential spectrum, in connection with existence and completeness of the wave operators [1], [2], [25]. Subsequent efforts to identify distinguishing features of the absolutely continuous eigenfunctions include work by Carmona [6] and Weidmann [48], [50], the method of subordinacy, to be outlined in Section 3 [15], [13], [14], [34], and the use of transfer matrices by Last and Simon [30].

Prior to a seminal paper of Pearson [33], showing that apparently innocuous potentials can give rise to purely singular continuous spectrum on $\mathbb{R}^+$, significant activity was focussed on seeking conditions under which the absence of singular continuous spectrum could be assured (see, e.g., [35, Chapter XIII]). The subsequent realization that singular continuous spectrum is generically present in a variety of situations [9] has stimulated further research activity in recent years (see, e.g., [29], [40]), while the method of subordinacy and its extensions provide some insight into the hitherto obscure behavior of the associated eigenfunctions [15], [23].

The principal focus of this paper is an overview of the method of subordinacy and its extensions, together with a brief discussion of the wider historical context and some illustrative examples to demonstrate its role in applications. Details of the derivation of the theory and of related background results can be found in the cited references.

## 2. Bounded solutions and spectral properties

The usefulness in practice of methods which characterize the spectrum in terms of the asymptotic behavior of solutions is well known, and in this section we briefly discuss two such approaches, both of which have informed and are informed by the method of subordinacy. To fix ideas, we restrict attention to the half-line case where $I = [0, \infty)$, $L$ is regular at 0 and in Weyl's limit point case at infinity, and $q$ is locally integrable. In this case it is known that an initial condition of the form

$$\cos(\alpha)u(0) + \sin(\alpha)u'(0) = 0, \quad \alpha \in [0, \pi), \tag{2}$$

is needed at the origin to render the associated operator $H_\alpha$ self-adjoint and that, under the Hilbert space formulation outlined in Section 3, no endpoint condition at infinity is required. The spectrum, $\sigma(H_\alpha)$, is then the complement in $\mathbb{C}$ of the set of all $\lambda$ for which the resolvent operator $(H_\alpha - \lambda I)^{-1}$ is bounded and everywhere defined; an equivalent definition can also be formulated in terms of the corresponding spectral function, $\rho_\alpha$ (see Section 3.1).

It follows from the classical separation and comparison theorems of Sturm (see [21], [41]) that the real line may be partitioned into oscillatory and non-oscillatory regions, separated by a so-called parabolic point $\lambda^\star \in \mathbb{R} \cup \{\pm\infty\}$. For $\lambda \in (-\infty, \lambda^\star)$ any, and hence all, solutions of (1) have a finite number of zeros for $r \geq 0$, while for $\lambda > \lambda^\star$ any, and hence all, solutions of (1) have a countably infinite number of zeros accumulating only at infinity; we therefore refer to (1) as being non-oscillatory or oscillatory according as $\lambda \in (-\infty, \lambda^\star)$ or $\lambda \in (\lambda^\star, \infty)$ respectively [20].

Since (1) is oscillatory at $\lambda$ if and only if the spectrum on $(-\infty, \lambda)$ is an infinite set [18], the spectrum of $H_\alpha$ on $(-\infty, \lambda^\star)$ consists of isolated eigenvalues only, possibly accumulating at $\lambda^\star$, with every $\lambda \in (-\infty, \lambda^\star)$ being an eigenvalue for some value of $\alpha$ in (2) [19]. It follows that there exists an $L_2([0, \infty))$ solution $u_\lambda(r)$ of (1) for each $\lambda < \lambda^\star$; this is known as a *principal solution* and satisfies

$$\lim_{r \to \infty} \frac{u_\lambda(r)}{v_\lambda(r)} = 0$$

whenever $v_\lambda(r)$ is a linearly independent solution of the same equation, with $u_\lambda(r) = o(1)$ as $r \to \infty$ if $\lambda^\star < \infty$ [20], [52]. The restriction of the spectrum to $(-\infty, \lambda^\star)$ is then given by

$$\sigma(H_\alpha) \cap (-\infty, \lambda^\star) = \{\lambda < \lambda^\star : \text{there exists a principal solution of (1) satisfying the boundary condition (2)}\}.$$

A second and more pervasive approach to the spectral analysis of the singular problem, particularly among physicists, effectively bypasses the intricacies of Hilbert space theory by defining the "spectrum", $S_\alpha$, to be the set of all $\lambda$ for which a non-trivial solution $u(r)$ of (1) and (2) satisfies the further condition:

$$u(r) = O(1) \text{ as } r \to \infty. \tag{3}$$

In the case where $\alpha = 0$, $q \equiv 0$, this yields $S_\alpha = (0, \infty)$, which is not a closed set, so that $S_\alpha \neq \sigma(H_\alpha)$ in general; however, since 0 is not an eigenvalue of $H_\alpha$, and $\sigma(H_\alpha) = [0, \infty)$, we see that $S_\alpha = \sigma(H_\alpha)$ except for a set which has measure zero with respect to both Lebesgue and spectral measures.

The longstanding conjecture that $S_\alpha = \sigma(H_\alpha)$ a.e. holds in general appears to be still unresolved so far as the essential spectrum is concerned (see, e.g., [16, Chapter V], [39, Section C5]), although study of related discrete operators with almost periodic potentials suggests that for this part of the spectrum the conjecture is false [53]. In the case of isolated point spectrum, the conjecture was refuted for $\lambda^\star = \infty$ by a counterexample due to Hartman and Wintner, in which for each $\lambda \in \mathbb{R}$, the square integrable solution fails to be $O(1)$ as $r \to \infty$ (see [20, (v)(c) and p. 648]); we then have $S_\alpha = \varnothing$, while $\sigma(H_\alpha)$ consists of an infinite sequence of isolated eigenvalues accumulating at infinity. However, the conjecture is confirmed for isolated point spectrum when $\lambda^\star < \infty$, since if $\lambda$ is not in the essential spectrum in this case, the $L_2([0,\infty))$ solution of (1) is $O(r^{-N})$ as $r \to \infty$ for every fixed $N$ [52].

This informal or "working" definition of the spectrum in terms of bounded solutions seems to have originated from concern that solutions of the wave equation of quantum mechanics (i.e., the time dependent Schrödinger equation) be "physically admissible", taking into account the accepted interpretation of the square of the modulus of the value of the wave function at a point as the position probability density (see, e.g., [10, Section 38], [37, Chapter II]). Apart from its apparent agreement with physical intuition, the working definition is undoubtedly attractive in that it is conceptually simple, easy to apply and agrees with the Hilbert space definition of the spectrum (at least up to null sets) in many familiar elementary examples. We shall return to the relationship between bounded solutions and spectral properties in Section 5.1, taking into account some connections with the theory of subordinacy.

Although analysis of the spectrum in terms of principal solutions is restricted to the non-oscillatory region $(-\infty, \lambda^\star)$, the method is still of interest because it uses properties of the solution space *as a whole* to identify the spectrum. This contrasts with the informal approach in terms of bounded solutions, which aims to locate the spectrum by applying a specific criterion (3) to the particular solution of (1) which satisfies (2), without any reference to the remainder of the solution space. It will be seen that the definition of a subordinate solution given in Section 3.1 directly extends the definition of a principal solution by replacing the pointwise comparison of solutions for large $r$ with limiting ratios of Hilbert space norms, thus enabling the idea of a principal solution to be applicable in both oscillatory and non-oscillatory regions. This key definition is fundamental to the theory of subordinacy, and enables precise correlations between the *relative* asymptotic behavior of solutions of (1) and specific spectral properties of $H_\alpha$ to be established. The use of properties of the solution space as a whole is also a feature of related transfer matrix methods, which are particularly effective in connection with the absolutely continuous spectrum (see, e.g., [30], [28], [17]) and will be briefly introduced in Section 5.1.

## 3. The method of subordinacy

For self-adjoint operators associated with (1), it is the behavior of solutions at one or both endpoints of the interval $I$, not their intermediate properties, that determines the contribution to the spectrum at each fixed value of the spectral parameter $\lambda$. This situation is made precise by the theory of subordinacy, which provides rigorous criteria for locating minimal supports for the absolutely continuous and singular spectra, and also, in the case of two limit point endpoints, enables the simple and degenerate parts of the spectrum to be identified [15], [13], [14], [34]. Similar results have been shown to apply to related operators, such as the one-dimensional Dirac operator, the general Sturm-Liouville operator, infinite matrix operators and the random Schrödinger operator, and will be summarized in Section 4.

The method of subordinacy is advantageous in several respects. In the first place, the results are independent of the detailed properties of $q$; only very general requirements, as, for example, that $q$ be locally integrable and that $L$ is in the limit point case at the infinite endpoints, need to be met. Moreover, a complete analysis of the spectrum can be achieved, at least in principle, by considering the behavior of solutions at *real values only* of the spectral parameter $\lambda$. As a result, the considerable technicalities of the spectral function and Titchmarsh-Weyl $m$-function, which are key features in the derivation of the theory, can now be avoided in applications. Also, to identify the absolutely continuous component of $\sigma(H_\alpha)$, it is only necessary to consider the behavior of solutions at the limit point endpoint(s), and for certain classes of potentials, the condition of non-subordinacy can be replaced by a much simpler boundedness criterion (see Section 5.1).

### 3.1. The half-line case

We recall that in this case $L$ is assumed to be regular at $x = 0$; the differential operator $H_\alpha$ acting on $\mathcal{H} = L_2([0,\infty))$ is then defined by

$$H_\alpha f = Lf, \quad f \in \mathcal{D}(H_\alpha)$$

where

$$\mathcal{D}(H_\alpha) = \{f \in \mathcal{H} : Lf \in \mathcal{H}; f, f' \text{ locally a.c.; } \cos(\alpha)f(0) + \sin(\alpha)f'(0) = 0\}$$

for some fixed $\alpha \in [0,\pi)$, and $q$ is locally integrable. Note that since $L$ is limit point at infinity, there is at most one solution of $Lf = \lambda f$ in $\mathcal{H}$ for any $\lambda \in \mathbb{C}$, and that a boundary condition is needed only at 0 (see [15] and references therein).

Associated with $H_\alpha$ is a non-decreasing spectral function $\rho_\alpha : \mathbb{R} \to \mathbb{R}$ and it is convenient in the present context to define the spectrum, $\sigma(H_\alpha)$, to be the complement of those points of $\mathbb{R}$ in a neighborhood of which $\rho_\alpha$ is constant. The spectral function generates a corresponding Borel-Stieltjes measure, $\mu_\alpha$, on $\mathbb{R}$ in the usual way, and the minimal supports of $\mu_\alpha$ provide an indication of where $\sigma(H_\alpha)$ is concentrated. Minimal supports (sometimes also known as *essential supports*) are defined as follows.

**Definition 1.** *A subset $S$ of $\mathbb{R}$ is said to be a minimal support of a Borel-Stieltjes measure $\tau$ if*

(i) $\tau(\mathbb{R}\backslash S) = 0$,
(ii) *whenever $S_0 \subseteq S$ satisfies $\tau(S_0) = 0$, then $|S_0| = 0$, where $|\cdot|$ denotes Lebesgue measure.*

It follows from the definition that minimal supports of $\mu_\alpha$ are unique up to Lebesgue and $\mu_\alpha$-null sets; while they may differ from $\sigma(H_\alpha)$ by sets of positive Lebesgue measure, there always exists a minimal support of $\mu_\alpha$ whose closure is the spectrum [15]. Since $\mu_\alpha$ can be decomposed uniquely into absolutely continuous, singular continuous and pure point parts, Definition 1 may also be applied to $(\mu_\alpha)_{a.c.}, (\mu_\alpha)_{s.c.}$ and $(\mu_\alpha)_{p.p.}$.

The definition of a subordinate solution given below makes precise the concept of relative asymptotic smallness of a solution at a limit point endpoint and is meaningful even if all solutions are oscillatory or no solutions are in $L_2([0,\infty))$.

**Definition 2.** *If $L$ is regular at 0 and in the limit point case at infinity, then a non-trivial solution $u_s(r,\lambda)$ of $Lu = \lambda u$ is said to be subordinate at infinity if for every linearly independent solution $u(r,\lambda)$*

$$\lim_{N\to\infty} \frac{\|u_s(r,\lambda)\|_N}{\|u(r,\lambda)\|_N} = 0 \tag{4}$$

*where $\|\cdot\|_N$ denotes the $L_2([0,N])$ norm.*

Note that subordinate solutions are unique up to multiplicative constants, and that if (4) holds for one solution $u(r,\lambda)$ which is linearly independent from $u_s(r,\lambda)$, then it holds for every solution $u(r,\lambda)$ which is linearly independent from $u_s(r,\lambda)$.

The following theorem identifies precise correlations between the spectral parts of $H_\alpha$ and the asymptotics of solutions of $Lu = \lambda u$, in terms of minimal supports of $(\mu_\alpha)_{a.c.}$, $(\mu_\alpha)_{s.c.}$ and $(\mu_\alpha)_{p.p.}$. The derivation of this result is crucially dependent on the corresponding Titchmarsh-Weyl function $m_\alpha$, which is a Herglotz function on $\mathbb{C}^+$, and whose limiting behavior as the real axis is approached normally is closely related both to a generalized derivative of $\mu_\alpha$, and to the subordinacy properties of solutions of (1) [2], [15].

**Theorem 1.** *Minimal supports $\mathcal{M}_{a.c.}(H_\alpha), \mathcal{M}_{s.c.}(H_\alpha)$ and $\mathcal{M}_{p.p.}(H_\alpha)$ of $(\mu_\alpha)_{a.c.}$, $(\mu_\alpha)_{s.c.}$ and $(\mu_\alpha)_{p.p.}$ respectively are as follows:*

$$\begin{aligned}
\mathcal{M}_{a.c.}(H_\alpha) &= \{\lambda\in\mathbb{R} : \textit{no solution of } Lu=\lambda u \textit{ is subordinate at infinity}\},\\
\mathcal{M}_{s.c.}(H_\alpha) &= \{\lambda\in\mathbb{R} : \textit{a solution of } Lu=\lambda u \textit{ exists which satisfies the}\\
&\qquad \textit{boundary condition at } 0, \textit{ is subordinate at infinity, but is}\\
&\qquad \textit{not in } L_2([0,\infty))\},\\
\mathcal{M}_{p.p.}(H_\alpha) &= \{\lambda\in\mathbb{R} : \textit{a non-trivial } L_2([0,\infty)) \textit{ solution of } Lu=\lambda u\\
&\qquad \textit{exists which satisfies the boundary condition at } 0\}.
\end{aligned}$$

Theorem 1 shows that there are striking distinctions between the asymptotic behavior of solutions associated with the different parts of the spectrum. If $\lambda \in \mathcal{M}_{a.c.}(H_\alpha)$, then all solutions of $Lu = \lambda u$ are, in some sense, of comparable asymptotic size at infinity, and this implies, by the limit point property, that no (non-trivial) solutions are in $L_2([0,\infty))$. The absence of an $L_2([0,\infty))$ solution is also a feature if $\lambda \in \mathcal{M}_{s.c.}(H_\alpha)$, although not, of course, when $\lambda \in \mathcal{M}_{p.p.}(H_\alpha)$.

Generally speaking, however, the more significant distinctions are between the singular and absolutely continuous supports of $\mu_\alpha$; thus, for example, if $\beta \neq \alpha (\mathrm{mod}\pi)$ is a distinct boundary condition at 0, it is immediate from the theorem that

$$\mathcal{M}_{s.c.}(H_\alpha) \cap \mathcal{M}_{s.c.}(H_\beta) = \mathcal{M}_{p.p.}(H_\alpha) \cap \mathcal{M}_{p.p.}(H_\beta) = \varnothing,$$

whereas

$$\mathcal{M}_{a.c.}(H_\alpha) = \mathcal{M}_{a.c.}(H_\beta),$$

which confirms well-known results of Kato and others concerning the stability of the absolutely continuous spectrum under finite rank perturbations [25].

It should be noted that some care is needed in interpreting the results of Theorem 1, given the nature of the relationship between $\mathcal{M}_{a.c.}(H_\alpha)$, $\mathcal{M}_{s.c.}(H_\alpha)$ and $\mathcal{M}_{p.p.}(H_\alpha)$ and the corresponding spectra, $\sigma_{a.c.}(H_\alpha)$, $\sigma_{s.c.}(H_\alpha)$ and $\sigma_{p.p.}(H_\alpha)$, which are closed sets. For example, although it is true that $\sigma_{p.p.}(H_\alpha) = \varnothing$ if and only if $\mathcal{M}_{p.p.}(H_\alpha) = \varnothing$, analogous statements cannot be made for $\sigma_{a.c.}(H_\alpha)$ or $\sigma_{s.c.}(H_\alpha)$. Indeed we may have $\sigma_{a.c.}(H_\alpha) = \varnothing$ even if $\mathcal{M}_{a.c.}(H_\alpha) \neq \varnothing$, and similarly for $\sigma_{s.c.}(H_\alpha)$; however, the converse situation is not possible since, using well-known properties of absolutely continuous and singular continuous measures, it follows from $\sigma_{a.c.}(H_\alpha) \neq \varnothing$ that any minimal support of $(\mu_\alpha)_{a.c.}$ has positive Lebesgue measure, and from $\sigma_{s.c.}(H_\alpha) \neq \varnothing$ that any minimal support of $(\mu_\alpha)_{s.c.}$ is an uncountable set of Lebesgue measure zero.

A similar situation holds in the full-line case, which we now consider.

### 3.2. The full-line case

Let $H$ denote the one-dimensional operator associated with (1) on $\mathcal{H} = L_2(\mathbb{R})$, let $q$ be locally integrable on $(-\infty, \infty)$ and suppose that $L$ is in the limit point case at both endpoints. Then the self-adjoint operator $H$ is uniquely defined by

$$Hf = Lf, \quad f \in \mathcal{D}(H),$$

where

$$\mathcal{D}(H) = \{f \in \mathcal{H} : Lf \in \mathcal{H}; f, f' \text{ locally a.c.}\}.$$

The analogue of the spectral function is now a $2 \times 2$ positive semidefinite spectral matrix function $(\rho_{ij})$, and a suitable spectral measure, the so-called trace measure, $\mu$, is generated from the sum of its diagonal terms.

Let $H_0^-, H_0^+$ respectively denote the self-adjoint operators on $L_2((-\infty, 0])$ and $L_2([0, \infty))$, which are defined in the usual way by $L$, together with a Dirichlet boundary condition at 0. In the derivation of minimal supports for the full-line operator $H$, a delicate relationship between the trace measure $\mu$ and the Titchmarsh-Weyl $m$-functions associated with $H_0^-$ and $H_0^+$ is identified, which is then combined with application of Theorem 1 to $H_0^-$ and $H_0^+$, to give the following theorem [13]; note that the definition of a solution which is subordinate at $-\infty$ is entirely analogous to that of Definition 2, except that the $L_2([0, N])$ norm is now replaced by the $L_2([-N, 0])$ norm.

**Theorem 2.** *Minimal supports $\mathcal{M}_{a.c.}(H)$, $\mathcal{M}_{s.c.}(H)$ and $\mathcal{M}_{p.p.}(H)$ of $\mu_{a.c.}$, $\mu_{s.c.}$ and $\mu_{p.p.}$ respectively are as follows:*

$$\begin{aligned}
\mathcal{M}_{a.c.}(H) &= \{\lambda \in \mathbb{R} : \textit{no solution of } Lu = \lambda u \textit{ is subordinate at } -\infty\} \\
&\quad \cup \{\lambda \in \mathbb{R} : \textit{no solution of } Lu = \lambda u \textit{ is subordinate at } +\infty\}, \\
\mathcal{M}_{s.c.}(H) &= \{\lambda \in \mathbb{R} : \textit{a solution of } Lu = \lambda u \textit{ exists which is subordinate} \\
&\quad \textit{both at } +\infty \textit{ and at} -\infty, \textit{but is not in } L_2(\mathbb{R})\}, \\
\mathcal{M}_{p.p.}(H) &= \{\lambda \in \mathbb{R} : \textit{a non-trivial } L_2(\mathbb{R}) \textit{ solution of } Lu = \lambda u \textit{ exists}\}.
\end{aligned}$$

Using Theorems 1 and 2, the well-known result that

$$\sigma_{a.c.}(H) = \sigma_{a.c.}(H_0^-) \cup \sigma_{a.c.}(H_0^+)$$

follows easily from the fact that

$$\mathcal{M}_{a.c.}(H) = \mathcal{M}_{a.c.}(H_0^-) \cup \mathcal{M}_{a.c.}(H_0^+).$$

The analogous situation for the singular spectrum is less straightforward, since the matching of subordinate solutions at the decomposition point 0 is involved; taking this into account, we obtain, with obvious notation,

$$\mathcal{M}_{s.}(H) = \cup_\alpha (\mathcal{M}_{s.}(H_\alpha^-) \cap \mathcal{M}_{s.}(H_\alpha^+)),$$

where $\mathcal{M}_{s.}(H) = \mathcal{M}_{s.c.}(H) \cup \mathcal{M}_{p.p.}(H)$, and the union is taken over all $\alpha \in [0, \pi)$.

An important issue in the full-line case is that of spectral multiplicity, which may be 1 or 2; in the half-line case, the question effectively does not arise, since the spectrum is always simple. A significant contribution to this topic is due to I.S. Kac, who identified necessary and sufficient conditions for the existence of degenerate spectrum in terms of the boundary behavior of the $m$-functions associated with $H_\alpha^-$ and $H_\alpha^+$ [24]. By combining Kac's result with Theorem 2, we obtain the following result [14].

**Theorem 3.** *$H$ has spectral multiplicity 2 if and only if the Lebesgue measure of the set*

$$\begin{aligned}
\mathcal{M}_2(H) &= \{\lambda \in \mathbb{R} : \textit{no solution of } Lu = \lambda u \textit{ is subordinate at } -\infty\} \\
&\quad \cap \{\lambda \in \mathbb{R} : \textit{no solution of } Lu = \lambda u \textit{ is subordinate at } +\infty\}
\end{aligned}$$

*is strictly positive; otherwise the spectrum of $H$ is simple.*

Thus the degenerate spectrum of $H$, if it exists, is effectively concentrated on $\mathcal{M}_2(H)$, which is a subset of $\mathcal{M}_{a.c.}(H)$, and the simple spectrum on $\mathcal{M}_1(H) = \mathcal{M}(H) \setminus \mathcal{M}_2(H)$, where $\mathcal{M}(H) = \mathcal{M}_{a.c.}(H) \cup \mathcal{M}_{s.}(H)$ is a minimal support of $\mu$; for further details, see [14]. Note that $\mathcal{M}_2(H)$ in Theorem 3 cannot be replaced by $S = \sigma_{a.c.}(H_0^-) \cap \sigma_{a.c.}(H_0^+)$, since it is known that operators $H$ exist for which $\mid S \mid > 0$, but $\mid \mathcal{M}_2(H) \mid = \varnothing$ (see, e.g., [14, Example 6.5]).

It is interesting to observe from Theorem 3 that the simple part of $\sigma(H)$ is characterized by the existence of a relatively small solution of $Lu = \lambda u$ on $\mathbb{R}$; this smallness need only be strict at one of the endpoints. The degenerate spectrum, on the other hand, is characterized by all solutions of $Lu = \lambda u$ on $\mathbb{R}$ being indistinguishable in terms of relative asymptotic size. On $\mathbb{R} \setminus \mathcal{M}(H)$, which includes the resolvent set, the solutions of $Lu = \lambda u$ are not well ordered in the sense that, although there exists a solution, $u_{-\infty}$, which is subordinate at $-\infty$, and a solution, $u_{+\infty}$, which is subordinate at $+\infty$, these solutions are linearly independent; thus there is no non-trivial solution of $Lu = \lambda u$ which is relatively small at both endpoints.

## 4. Extensions and generalizations

The usefulness of the theory of subordinacy in a range of spectral problems has led to a number of extensions to related operators. In each of the cases which we now describe, a limit point, limit circle theory is known, which is analogous to the Weyl theory for the standard singular Sturm-Liouville case. We assume therefore, as before, that 0 is a regular endpoint in the half-line case and that the infinite endpoints are limit point.

### 4.1. Generalized Sturm-Liouville operators

Consider the generalized Sturm-Liouville operator $H_\alpha$ which is associated with the differential equation

$$Lu := -(pu')'(r) + q(r)u(r) = \lambda w(r)u(r), \qquad r \in [0, \infty),$$

and boundary condition $\cos(\alpha)u(0) + \sin(\alpha)(pu')(0) = 0$ on the weighted Hilbert space $L_2^w([0,\infty))$. Here $\alpha \in [0, \pi)$ is fixed and $p, q, w$ are real-valued functions with $p, w > 0$ and $p^{-1}, q, w$ locally integrable. Then Theorem 1 holds with $\|\cdot\|_N = (\int_0^N |\cdot|^2\, w(r)dr)^{1/2}$ [8].

### 4.2. Separated Dirac operators

For spherically symmetric potentials, the separated Dirac equation may be written in system form

$$\begin{pmatrix} 0 & -1 \\ 1 & 0 \end{pmatrix} \underline{u}'(r) + \begin{pmatrix} q_1(r) & q_2(r) \\ q_2(r) & q_3(r) \end{pmatrix} \underline{u}(r) = \lambda \underline{u}(r), \qquad r \in [0, \infty),$$

where $q_1$, $q_2$, $q_3$ are real-valued, locally integrable functions, and $\underline{u}(r) = (u_1, u_2)^t$. To construct the corresponding self-adjoint operator $H_\alpha$, it is necessary to impose an initial condition of the form $\cos(\alpha)u_1(0) + \sin(\alpha)u_2(0) = 0$, for some $\alpha \in [0, \pi)$. Then Theorem 1 holds for $H_\alpha$ under the usual $L_2([0,\infty)) \otimes \mathbb{C}^2$ norm, and Theorem 2 is also valid for the corresponding operator $H$ on $L_2(\mathbb{R}) \otimes \mathbb{C}^2$ [3].

### 4.3. Infinite matrix operators

Let $H_\alpha$ be the self-adjoint operator associated with the semi-infinite matrix equation

$$\begin{pmatrix} a_0 & b_0 & & & & \\ b_0 & a_1 & b_1 & & 0 & \\ & b_1 & a_2 & b_2 & & \\ & & b_2 & a_3 & b_3 & \\ & 0 & & \dots & \dots & \dots \\ & & & & \dots & \dots \end{pmatrix} \begin{pmatrix} u_0 \\ u_1 \\ u_2 \\ u_3 \\ \dots \\ \dots \end{pmatrix} = \lambda \begin{pmatrix} w_0u_0 \\ w_1u_1 \\ w_2u_2 \\ w_3u_3 \\ \dots \\ \dots \end{pmatrix}$$

and boundary condition $\cos(\alpha)u_1 + \sin(\alpha)b_0(u_1 - u_0) = 0$, where $\alpha \in [0, \pi)$ is fixed, and $a_i, b_i, w_i \in \mathbb{R}, b_i \neq 0, w_i > 0$, for $i = 0, 1, 2, \dots$. This case includes the Jacobi matrix operator for which $b_i = w_i = 1$, and Theorem 1 holds for $H_\alpha$ under the usual $\ell_2^w$ norm with $\|\underline{u}\|_N = (\sum_{n=0}^N w_n \mid u_n \mid^2)^{1/2}$ [7], [26], [44].

An interesting variant is the matrix operator associated with orthogonal polynomials on the unit circle and the Szegő recurrence relations,

$$\overrightarrow{X}(z,n) = \frac{1}{(1- \mid a_n \mid^2)^{1/2}} \begin{pmatrix} z & a_n \\ \overline{a_n} z & 1 \end{pmatrix} \overrightarrow{X}(z, n-1), \; n \in \mathbb{N}, \mid a_n \mid < 1, \mid z \mid \in [0,1],$$

with initial condition $\overrightarrow{X}(z,0) = (1,1)^t$. Here the interior of the unit circle takes the place of $\mathbb{C}^+$ in the construction of an analogue, $F(z)$, of the Titchmarsh-Weyl $m$-function, and $F(z)$ in turn is associated with an orthogonality measure $\mu$ on the unit circle which has a similar role to the spectral measure $\mu_\alpha$ on $\mathbb{R}$ in the half-line case. Theorem 1 now holds for $\mu$ with $\|\overrightarrow{X}(z)\|_N = (\sum_{n=0}^N \|\overrightarrow{X}(z,n)\|^2)^{1/2}$, where one has $\overrightarrow{X}(z) = (\overrightarrow{X}(z,0), \dots, \overrightarrow{X}(z,n), \dots)$, $\overrightarrow{X}(z,n) = (x_1(z,n), x_2(z,n))^t$ and $\|\overrightarrow{X}(z,n)\| = (\mid x_1(z,n) \mid^2 + \mid x_2(z,n) \mid^2)^{1/2}$ [17].

### 4.4. Random Schrödinger operators

Here the general form of the differential operator considered is

$$L_w u := -u''(r) + q_w(r)u(r) = \lambda u(r), \qquad r \in [0, \infty),$$

where $\{q_w(r) : r \in [0, \infty)\}$ is a random function on a probability space $(\Omega, \mathcal{F}, P)$, or a random generalized function. In this context, we assume that $L_w$ is in the limit point case for $P$-almost all $w$, so that $H_{w,\alpha}$, defined from $L_w$ together with the boundary condition $\cos(\alpha)u(0) + \sin(\alpha)u'(0) = 0$, is self-adjoint with probability 1; the unique self-adjoint operator $H_w$ is defined by $L_w$ on $\mathbb{R}$ in a similar way. Then under the assumption that a technical condition related to the well-known Kotani trick holds, analogous results to Theorems 1 and 2 can be established for $\lambda$-intervals of the real line [31].

**Remark 1.** In addition to extensions of the theory to a number of related operators, some refinements to the concept of a subordinate solution have been proposed and used in applications; for example, *power law subordinacy* [23], *strong non-subordinacy* [8], *uniform non-subordinacy* [49] and *sequential subordinacy* [34], [43]. Power law subordinacy is a significant generalization of the original concept, which enables dimensional Hausdorff properties of the singular continuous spectrum to be investigated, and has provided detailed results for the almost Mathieu operator and Fibonacci Hamiltonian. Strong and uniform non-subordinacy are associated with the relatively "well-behaved" situation where no solution of (1) is subordinate for any $\lambda$ in a compact interval $I$; in such cases, various uniformity properties on $I$ can be inferred for the ratios of norms of solutions and for the spectral density functions $\rho'_\alpha$. From Definition 2 it is evident that, to establish non-subordinacy of solutions, it is only necessary (and usually much more convenient) to demonstrate non-sequential subordinacy for a single sequence.

## 5. Applications

In principle, Theorems 1, 2, 3 and their extensions provide distinguishing criteria which enable a complete and detailed analysis of the spectrum to be carried out for a range of differential and difference operators. However, in practice it is rare for explicit expressions for the solutions of the governing equations to be available, so that in applications the theory is often used indirectly or in conjunction with suitable asymptotic estimates of the behavior of the solutions at the endpoints. Some typical strategies include use of the Liouville-Green approximation, application of Levinson's theorem and its extensions, various results on boundedness and non-subordinacy, and related transfer matrix methods.

Before considering specific examples, we briefly return to the relationship between bounded solutions and spectral properties.

### 5.1. Boundedness revisited

The following result, linking boundedness of solutions to non-subordinacy, can greatly simplify the analysis when absolutely continuous spectrum is present [43].

**Lemma 1.** *Let $L$ be as in (1) with $I = [0, \infty)$, and suppose that*

$$\sup_{x\geq 0} \int_x^{x+1} q_-(r)dr < \infty, \tag{5}$$

*where $q_-$ denotes the negative part of $q$. Then if all solutions of $Lu = \lambda u$ are bounded for some fixed $\lambda \in \mathbb{R}$, $L$ is in the limit point case at infinity and, for the same fixed $\lambda$,*

(i) *$u'$ is bounded for all solutions $u$ of $Lu = \lambda u$,*
(ii) *no solution of $Lu = \lambda u$ is subordinate at infinity.*

Note that Lemma 1 shows that, provided (5) is satisfied, the boundedness of solutions of $Lu = \lambda u$ is a sufficient condition for $\lambda$ to be in $\mathcal{M}_{a.c.}(H_\alpha)$; however, it is not a necessary condition, as may be seen by considering a fundamental set of solutions, $\{(r+1)^{1/2}\cos(\frac{1}{2}\ln(r+1)), (r+1)^{1/2}\sin(\frac{1}{2}\ln(r+1))\}$, of $Lu = \lambda u$ when $q(r) = -2^{-1}(r+1)^{-2}$ and $\lambda = 0$. Variants of Lemma 1 for the Dirac and Jacobi matrix operators may be found in [4], [39], [44], and an alternative proof is contained in [40].

It turns out that Lemma 1 may be recast in terms of transfer matrices. To see this, let $\lambda \in \mathbb{R}$ be fixed and suppose that for $x \in I = [0, \infty)$, $\{u_\lambda(x), v_\lambda(x)\}$ is a fundamental set of solutions of $Lu = \lambda u$ satisfying $u_\lambda(0) = v'_\lambda(0) = 0$, $u'_\lambda(0) = v_\lambda(0) = 1$. Then setting

$$T_\lambda(x) := \begin{pmatrix} u'_\lambda(x) & v'_\lambda(x) \\ u_\lambda(x) & v_\lambda(x) \end{pmatrix},$$

it is easy to check that for any solution $y_\lambda(x) = au_\lambda(x) + bv_\lambda(x)$ of $Lu = \lambda u$, with $a, b \in \mathbb{C}$,

$$T_\lambda(x) \begin{pmatrix} y'_\lambda(0) \\ y_\lambda(0) \end{pmatrix} = \begin{pmatrix} y'_\lambda(x) \\ y_\lambda(x) \end{pmatrix}.$$

If we now define

$$\|T_\lambda(x)\| := \sup_{\|y\|=1} \|T_\lambda(x)y\| = \sup_{|a|^2+|b|^2=1} (| \, y'_\lambda(x) \, |^2 + | \, y_\lambda(x) \, |^2)^{1/2},$$

where $y = (y'_\lambda(0), y_\lambda(0))^t$, then it may be seen from Lemma 1 that, subject to (5), all solutions of $Lu = \lambda u$ are bounded if and only if $\limsup_{x\to\infty} \|T_\lambda(x)\| < \infty$. It follows that issues relating to boundedness of solutions and spectral properties can be investigated using results from subordinacy and from the study of transfer matrices. Interest in the latter approach has already led to a number of promising new developments (see, e.g., [17], [30], [28]).

### 5.2. Examples

In each of the following examples, it is readily verified that, as appropriate, 0 is a regular endpoint and the infinite endpoints are limit point. The operators $H_\alpha$ and $H$ are defined as in Sections 2 and 3 unless otherwise stated.

**Example 1.** Let $q = p + s + w$ on $[0, \infty)$, where $p \in L_1([0, \infty))$, $s$ is smooth and long range with $s' \in L_1([0, \infty))$, and the von Neumann-Wigner part, $w$, satisfies $w(r) \to 0$ as $r \to \infty$, and is conditionally integrable, i.e., $\lim_{r\to\infty} \int_0^r w(r)dr$ exists. Then for all $\alpha \in [0, \pi)$, $\sigma(H_\alpha)$ is purely absolutely continuous on $(0, \infty)$ apart from at most a countable set of embedded eigenvalues, known as resonances. This result is obtained by using asymptotic integration to obtain suitable estimates of solutions of $Lu = \lambda u$ for $\lambda > 0$ and large $r$, and applying Theorem 1 [3]; some generalizations are obtained in [4], where Lemma 1 and a limiting absorption principle are also used.

**Example 2.** Let $L_0 := -d^2/dr^2 + \cos(r), \;\; r \in (-\infty, \infty)$,

$$L := -\frac{d^2}{dr^2} + \cos(r) + \delta \cos(|r|^\gamma), \;\; 0 < \gamma < 1, \;\; r \in (-\infty, \infty),$$

and denote by $(a_n, b_n), n = 1, 2, \ldots$, the stability intervals of the self-adjoint operator $H_0$ associated with $L_0$. Then it is well known that $\sigma(H_0) = \cup_n [a_n, b_n]$, and is purely absolutely continuous, with $b_n < a_{n+1}$ for all $n \in \mathbb{N}$; moreover, as $n \to \infty$, the length of the stability intervals is $O(n)$ and the length of the gaps becomes arbitrarily small (see, e.g., [11], [35, Chapter XIII, Section 16, Example 1]). However, the essential spectrum of the perturbed operator $H$ associated with $L$ is a countable union of closed intervals, each consisting of a central band of absolutely continuous spectrum and two outer bands of singular spectrum, these banded intervals being disjoint for suitably chosen $\delta$ and sufficiently small $n$. In addition, every interval of absolutely continuous spectrum, respectively gap in the essential spectrum, of the perturbed operator $H$ is a subset of a stability interval, respectively subset of a spectral gap, of the unperturbed operator $H_0$. To achieve these results, trace class methods are used to show that if $(b, a)$ is a spectral gap of $H_0$, then $(b - \delta, a + \delta) \cap \sigma_{a.c.}(H) = \varnothing$, and adaptations of arguments used in [44] are combined with Lemma 1 and Theorems 1 and 2 to show that if $(a, b)$ is a stability interval of $H_0$, then $(a + \delta, b - \delta)$ is a stability interval of $H$ [45]. The degeneracy of $\sigma_{a.c.}(H)$ can be inferred from Theorem 3.

**Example 3.** Using the notation of Section 4, let $H_w$ denote the Goldsheid-Molchanov-Pastur model on $\mathbb{R}$ and $H_{w,\alpha}$ a self-adjoint operator arising from $L_w := -d^2/dr^2 + kW(r), r \in [0, \infty)$, where $k > 0$ and $W(r)$ is Gaussian white noise. If $F \geq 0$ is the intensity of a constant electric field, then with probability 1 the perturbed GMP operator $H_w^F = H_w - Fr$ has purely absolutely continuous spectrum with $\sigma(H_w^F) = \mathbb{R}$, and the perturbed operator $H_{w,\alpha}^F = H_{w,\alpha} - Fr$ has dense pure point spectrum on $\mathbb{R}$ for $F < k^2/2$ and purely singular continuous spectrum with $\sigma(H_{w,\alpha}^F) = \mathbb{R}$ for $F \geq k^2/2$. These results are established by using the asymptotic expansions of Airy functions in a transformation of the eigenvalue equations and applying analogues of Theorems 1 and 2 for random operators [31].

**Example 4.** Let $L$ be as in (1), with $I = [0, \infty)$ and $q : I \to \mathbb{C}$ satisfying $\operatorname{Im} q \geq 0$ and $\operatorname{Im} q(r) \to 0$ as $r \to \infty$. Let $H_\alpha$ denote the non-self-adjoint operator associated with $L$ and a self-adjoint boundary condition at the origin (cf. (2)). Then if $\operatorname{Im} q \notin L_1([0, \infty))$, the absolutely continuous spectrum of $H_\alpha$ is empty. This result depends on a characterization in terms of Hardy spaces of the absolutely continuous subspace, $N_e$, of a maximally dissipative, completely non-self-adjoint operator with essential spectrum on the real axis (see, e.g., [32]). Also involved are an analogue of the Titchmarsh-Weyl function and an associated spectral measure, which together enable a theory of subordinacy to be established for the class of operators considered, through which the proof is completed [36].

**Acknowledgements**

I should like to express my gratitude to Professor Amrein and the organizing committee for inviting me to contribute to the Sturm Bicentennial Conference. I also wish to thank the referees of this paper for their careful reading of the manuscript and constructive suggestions.

## References


[1] W.O. Amrein and V. Georgescu, Strong asymptotic completeness of wave operators for highly singular potentials, Helv. Phys. Acta **47** (1974), 517–533.

[2] N. Aronszajn, On a problem of Weyl in the theory of singular Sturm-Liouville equations, Amer. J. Math. **79** (1957), 597–610.

[3] H. Behncke, Absolute continuity of Hamiltonians with von Neumann-Wigner potentials, Proc. Amer. Math. Soc. **111** (1991), 373–384.

[4] H. Behncke, Absolute continuity of Hamiltonians with von Neumann-Wigner potentials II, Manuscripta Math. **71** (1991), 163–181.

[5] M. Bôcher, *Leçons sur les Méthodes de Sturm*, Gauthier-Villars, Paris, 1917.

[6] R. Carmona, One-dimensional Schrödinger operators with random or deterministic potentials: New spectral types, J. Funct. Anal. **51** (1983), 229–258.

[7] S.L. Clark, A spectral analysis for selfadjoint operators generated by a class of second order difference equations, J. Math. Anal. Appl. **197** (1996), 267–285.

[8] S.L. Clark and D.B. Hinton, Strong nonsubordinacy and absolutely continuous spectra for Sturm-Liouville equations, Differential Integral Equations **6** (1993), 573–586.

[9] R. del Rio, S. Jitomirskaya, N. Makarov and B. Simon, Singular continuous spectrum is generic, Bull. Amer. Math. Soc. **31** (1994), 208–212.

[10] P.A.M. Dirac, *Principles of Quantum Mechanics*, 4th Edition, Oxford University Press, 1957.

[11] P. Djakov and B. Mityagin, Spectral gaps of the periodic Schrödinger operator when its potential is an entire function, Adv. in Appl. Math. **31** (2003), 562–596.

[12] W.N. Everitt, Charles Sturm and the Development of Sturm-Liouville Theory in the Years 1900 to 1950, in this volume.

[13] D.J. Gilbert, On subordinacy and analysis of the spectrum of Schrödinger operators with two singular endpoints, Proc. Royal Soc. Edinburgh Sect. A **112** (1989), 213–229.

[14] D.J. Gilbert, On subordinacy and spectral multiplicity for a class of singular differential operators, Proc. Royal Soc. Edinburgh Sect. A **128** (1998), 549–584.

[15] D.J. Gilbert and D.B. Pearson, On subordinacy and analysis of the spectrum of one-dimensional Schrödinger operators, J. Math. Anal. Appl. **128** (1987), 30–56.

[16] I.M. Glazman, *Direct Methods of Qualitative Spectral Analysis of Singular Differential Operators*, Israel Program for Scientific Trans., Jerusalem, 1965.

[17] L. Golinskii and P. Nevai, Szegő difference equations, transfer matrices and orthogonal polynomials on the unit circle, Comm. Math. Phys. **223** (2001), 223–259.

[18] P. Hartman, A characterisation of the spectra of one-dimensional wave equations, Amer. J. Math. **71** (1949), 915–920.

[19] P. Hartman and A. Wintner, An oscillation theorem for continuous spectra, Proc. Nat. Acad. Sci. U.S.A. **33** (1947), 376–379.

[20] P. Hartman and A. Wintner, Oscillatory and non-oscillatory linear differential equations, Amer. J. Math. **71** (1949), 627–649.

[21] Don Hinton, Sturm's 1836 Oscillation Results. Evolution of the Theory, in this volume.

[22] A.M. Hinz and G. Stolz, Polynomial boundedness of eigensolutions and the spectrum of Schrödinger operators, Math. Ann. **294** (1992), 195–211.

[23] S. Jitomirskaya and Y. Last, Power-law subordinacy and singular spectra I: Half-line operators, Acta Math. **183** (1999), 171–179; II: Line operators, Comm. Math. Phys. **211** (2000), 643–658.

[24] I.S. Kac, On the multiplicity of the spectrum of a second order differential operator, Soviet Math. Doklady **3** (1962), 1035–1039; On the multiplicity of the spectrum of a second order differential operator and the associated eigenfunction expansion (Russian), Izv. Akad. Nauk. SSSR, Ser. Mat. **27** (1963), 1081–1112.

[25] T. Kato, On finite-dimensional perturbations of selfadjoint operators, J. Math. Soc. Japan **9** (1957), 239–249.

[26] S. Khan and D.B. Pearson, Subordinacy and spectral theory for infinite matrices, Helv. Phys. Acta **65** (1992), 505–527.

[27] A. Kiselev, Absolutely continuous spectrum of one-dimensional Schrödinger operators and Jacobi matrices with slowly decreasing potentials, Comm. Math. Phys. **179** (1996), 377–400.

[28] A. Kiselev, Y. Last and B. Simon, Modified Prüfer transforms and absolutely continuous spectrum of one-dimensional Schrödinger operators, Comm. Math. Phys. **194** (1998), 1–45.

[29] Y. Last, Quantum dynamics and decompositions of singular continuous spectra, J. Funct. Anal. **142** (1996), 406–445.

[30] Y. Last and B. Simon, Eigenfunctions, transfer matrices and absolutely continuous spectrum of one-dimensional Schrödinger operators, Invent. Math. **135** (1999), 329–367.

[31] N. Minami, Random Schrödinger operators with a constant electric field, Ann. Inst. H. Poincaré Phys. Théor. **56** (1992), 307–344.

[32] S.N. Naboko, A functional model of perturbation theory and its applications to scattering theory (Russian), Trudy Mat. Inst. Steklov **147** (1980), 86–114; English transl. in Proc. Steklov Inst. Math. **2** (1981), 85–116.

[33] D.B. Pearson, Singular continuous measures in scattering theory, Comm. Math. Phys. **60** (1978), 13–36.

[34] D.B. Pearson, *Quantum Scattering and Spectral Theory*, Academic Press, London, 1988.

[35] M. Reed and B. Simon, *Methods of Modern Mathematical Physics, Vol. IV: Analysis of Operators*, Academic Press, London, 1978.

[36] R. Romanov, On unstableness of absolutely continuous spectrum of dissipative Schrödinger operators and Jacobi matrices, Algebra i Analiz **17** (2005), 145–169 (Russian); English translation in St. Petersburg Math. J.

[37] L.I. Schiff, *Quantum Mechanics*, McGraw-Hill, New York, 1949.

[38] E. Schrödinger, Quantisierung als Eigenwertproblem, Ann. Physik **80** (1926), 437–490.

[39] B. Simon, Schrödinger semigroups, Bull. Amer. Math. Soc. **7** (1982), 447–526.

[40] B. Simon, Bounded eigenfunctions and absolutely continuous spectrum for one-dimensional Schrödinger operators, Proc. Amer. Math. Soc. **124** (1996), 3361–3369.

[41] B. Simon, Sturm Oscillation and Comparison Theorems, in this volume.

[42] B. Simon and G. Stolz, Operators with singular continuous spectrum V. Sparse potentials, Proc. Amer. Math. Soc. **124** (1996), 2073–2080.

[43] G. Stolz, Bounded solutions and absolute continuity of Sturm-Liouville operators, J. Math. Anal. Appl. **169** (1992), 210–228.

[44] G. Stolz, Spectral theory for slowly oscillating potentials, I. Jacobi matrices, Manuscripta Math. **84** (1994), 245–260.

[45] G. Stolz, Spectral theory for slowly oscillating potentials, II. Schrödinger operators, Math. Nachr. **183** (1997), 275–294.

[46] C. Sturm, Mémoire sur les Équations différentielles linéaires du second ordre, J. Math. Pures Appl. **1** (1836), 106–186.

[47] C. Sturm and J. Liouville, Extrait d'un mémoire sur le développement des fonctions en séries dont les différents termes sont assujettis à satisfaire à une même équation différentielle linéaire, contenant un paramètre variable, J. Math. Pures Appl. **2** (1837), 220–223.

[48] J. Weidmann, *Spectral Theory of Ordinary Differential Operators*, Lecture Notes in Mathematics **1258**, Springer, Berlin, 1987.

[49] J. Weidmann, Uniform nonsubordinacy and the absolutely continuous spectrum, Analysis **16** (1996), 89–99.

[50] J. Weidmann, Spectral Theory of Sturm-Liouville Operators. Approximation by Regular Problems, in this volume.

[51] H. Weyl, Über gewöhnliche Differentialgleichungen mit Singularitäten und die zugehörigen Entwicklungen willkürlicher Funktionen, Math. Ann. **68** (1910), 220–269.

[52] A. Wintner, On the smallness of isolated eigenfunctions, Amer. J. Math. **71** (1949), 603–611.

[53] S.Ya. Zhitomirskaya, Singular spectral properties of a one-dimensional Schrödinger operator with almost periodic potential, Adv. Soviet Math. **3** (1991), 215–254.

Daphne Gilbert
School of Mathematical Sciences
Dublin Institute of Technology
Kevin Street
Dublin 8
Ireland
e-mail: `daphne.gilbert@dit.ie`

*W.O. Amrein, A.M. Hinz, D.B. Pearson*
Sturm-Liouville Theory: Past and Present, 137–171


# The Titchmarsh-Weyl Eigenfunction Expansion Theorem for Sturm-Liouville Differential Equations

Christer Bennewitz and W. Norrie Everitt

*Dedicated to the achievements and memory of*
*Hermann Weyl 1885–1955*
*and*
*Edward Charles Titchmarsh 1899–1963*

**Abstract.** This paper involves a revisit to the original works of Hermann Weyl in 1910 and of Edward Charles Titchmarsh in 1941, concerning Sturm-Liouville theory and the corresponding eigenfunction expansions.

For this account the essential results of Weyl concern the regular, limit-circle and limit-point classifications of Sturm-Liouville differential equations; the eigenfunction expansion theory from Titchmarsh is based on classical function theory methods, in particular complex function theory.

The eigenfunction expansion theory presented in this paper is based on the Titchmarsh-Weyl $m$-coefficient; the proofs are essentially in classical function theory but are related to operator theoretic methods in Hilbert space.

One important innovation here is that for the Sturm-Liouville problem considered on the open interval $(a,b)$, the endpoint $a$ can be classified as either regular or limit-circle, whilst the endpoint $b$ can be regular, limit-circle or limit-point; nevertheless it is shown that these conditions lead to the definition of a single Titchmarsh-Weyl $m$-coefficient. From this coefficient the complex function theory methods of Titchmarsh provide a guide to a new proof of the general eigenfunction expansion theorem.

**Mathematics Subject Classification (2000).** Primary; 34B24, 34L10, 47E05: Secondary; 34B20, 34B27, 47A05.

**Keywords.** Sturm-Liouville theory; Titchmarsh-Weyl $m$-coefficient; eigenfunction expansion.

## 1. Introduction

This paper presents a new approach to the now classical theory of Sturm-Liouville eigenfunction expansions, based essentially on the existence of the Titchmarsh-Weyl $m$-coefficient. The main method follows the Titchmarsh use of the theory of functions of a complex variable and contour integration.

The Sturm-Liouville differential equation is

$$-(py')' + qy = \lambda wy \text{ on } (a,b); \tag{1.1}$$

the standard minimal conditions on the coefficients $p, q, w$, defined on the open interval $(a,b)$, where $-\infty \leq a < b \leq +\infty$, of the real line $\mathbb{R}$, are:

$$\begin{cases} \text{(i)} & p, q, w : (a,b) \to \mathbb{R} \\ \text{(ii)} & p^{-1}, q, w \in L^1_{\text{loc}}(a,b) \\ \text{(iii)} & w(x) > 0 \text{ for almost all (Lebesgue) } x \in (a,b). \end{cases} \tag{1.2}$$

The spectral parameter $\lambda$ belongs to $\mathbb{C}$, the complex field.

The only additional restriction of the coefficients $p, q, w$ is that the left-hand endpoint $a$ of the open interval $(a,b)$ is either in the regular or limit-circle case, in the framework provided by the Hilbert function space $L^2((a,b);w)$. If endpoint $a$ is limit-point then the methods work if the endpoint $b$ is regular or limit-circle in $L^2((a,b);w)$. Thus the only case that is not considered is when both endpoints $a$ and $b$ are in the limit-point case in $L^2((a,b);w)$; there are technical reasons, discussed below, for this exclusion.

The Sturm-Liouville boundary value problems which give rise to the eigenfunction expansions here considered, are determined by imposing separated boundary conditions on the solutions of the differential equation (1.1). If both endpoints $a$ and $b$ are, independently, in the regular or limit-circle case, coupled boundary conditions can be introduced, but then it is not possible to work with a single $m$-coefficient and this case has been excluded from consideration. There are technical reasons, discussed below, for this exclusion.

The methods used in this paper involve two innovations:

(a) The initial value problem for the differential equation (1.1) can be solved not only at a regular endpoint but also at a singular limit-circle endpoint.

(b) The application to the $m$-coefficient of the Herglotz-Nevanlinna-Pick-Riesz integral representation of Cauchy analytic functions that are holomorphic in an open half-plane of $\mathbb{C}$, and for which the imaginary part of the function values are of one sign.

The complete statements of the Sturm-Liouville expansion results are given in the theorem and corollaries of Section 11.

## 2. Notations

The real and complex fields are represented by $\mathbb{R}$ and $\mathbb{C}$ respectively; a general interval of $\mathbb{R}$ is represented by $I$; compact and open intervals of $\mathbb{R}$ are represented by $[\alpha,\beta]$ and $(a,b)$ respectively. The prime symbol $'$ denotes classical differentiation on the real line $\mathbb{R}$.

Lebesgue integration on $\mathbb{R}$ is denoted by $L$, and $L^1(I)$ denotes the Lebesgue integration space of complex-valued functions defined on an interval $I$. The local integration space $L^1_{\text{loc}}(a,b)$ is the set of all complex-valued functions on $I$ which are Lebesgue integrable on all compact sub-intervals $[\alpha,\beta] \subset (a,b)$.

Absolute continuity, with respect to Lebesgue measure, is denoted by $AC$; the space of all complex-valued functions defined on $(a,b)$ which are absolutely continuous on all compact sub-intervals $[\alpha,\beta]$ of $(a,b)$, is denoted by $AC_{\text{loc}}(a,b)$.

The weight function $w$ on $(a,b)$ is a Lebesgue measurable function $w : (a,b) \to \mathbb{R}$ satisfying $w(x) > 0$ for almost all $x \in (a,b)$.

Given an interval $(a,b)$ and a weight function $w$ the space $L^2((a,b);w)$ is defined as the set of all complex-valued, Lebesgue measurable functions $f : (a,b) \to \mathbb{C}$ such that

$$\int_a^b |f(x)|^2 \, w(x) \, dx < +\infty.$$

Taking equivalence classes into account, $L^2((a,b);w)$ is a Hilbert function space with inner product and norm

$$(f,g)_w := \int_a^b f(x)\overline{g}(x)w(x) \, dx \text{ for all } f,g \in L^2((a,b);w),$$

$$\|f\|_w := \left\{ \int_a^b |f(x)|^2 \, w(x) \, dx \right\}^{1/2} \quad \text{for all } f \in L^2((a,b);w).$$

The class of Cauchy analytic functions that are holomorphic on the open set $U \subseteq \mathbb{C}$ is denoted by $\mathbf{H}(U)$; if $U = \mathbb{C}$ then $\mathbf{H}$ denotes the class of entire or integral functions on $\mathbb{C}$.

## 3. History

We refer to the paper by Everitt, [11] in this volume, on the development of Sturm-Liouville theory up to the year 1950. This paper covers the first structured account, by Weyl, of singular Sturm-Liouville boundary value problems, [23]; the first operator theoretic account, by Stone, of the boundary value problems under the minimal conditions on the coefficients, [19]; the first structured account, by Titchmarsh, of the use of complex variable techniques in singular Sturm-Liouville boundary value problems, [22].

The development of classical, rather than the operator theoretic, Sturm-Liouville theory in the years after 1950 can be found in various sources; in particular in the texts of Atkinson [4], Coddington and Levinson [8], Levitan and Sargsjan [17]. The operator theoretic development is given in the texts by Naimark [20] and Akhiezer and Glazman [2].

This paper is based on results in an unpublished manuscript of Bennewitz [6], and the paper by Everitt [10].

## 4. Sturm-Liouville differential expressions and equations

Given the interval $(a,b)$, a set of Sturm-Liouville coefficients $\{p,q,w\}$ has to satisfy the minimal conditions given in (1.2). Note that in general there is no sign restriction on the leading coefficient $p$.

The associated Sturm-Liouville differential expression $M(p,q) \equiv M[\cdot]$ is the linear operator defined by

$$\begin{cases} \text{(i)} & \text{domain } D(M) := \{f : (a,b) \to \mathbb{C} : f, pf' \in AC_{\text{loc}}(a,b)\} \\ \text{(ii)} & \begin{cases} M[f](x) := -(p(x)f(x)')' + q(x)f(x) \text{ for all } f \in D(M) \\ \text{and almost all } x \in (a,b). \end{cases} \end{cases} \tag{4.1}$$

We note that $M[f] \in L^1_{\text{loc}}(I)$ for all $f \in D(M)$; it is shown in [20, Chapter V, Section 17] that $D(M)$ has a dense intersection with the Hilbert space $L^2((a,b);w)$; also $D(M)$ is dense in the locally convex linear topological space $L^1_{\text{loc}}(a,b)$, see [3].

Given the interval $(a,b)$ and the set of Sturm-Liouville coefficients $\{p,q,w\}$, the associated Sturm-Liouville differential equation is the second-order linear ordinary differential equation

$$M[y](x) \equiv -(p(x)y'(x))' + q(x)y(x) = \lambda w(x)y(x) \text{ for all } x \in (a,b), \tag{4.2}$$

where $\lambda \in \mathbb{C}$ is a complex-valued spectral parameter.

The above minimal conditions on the set of coefficients $\{p,q,w\}$ imply that the Sturm-Liouville differential equation has a solution to any initial value problem at a point $c \in (a,b)$. See the existence theorem in [20, Chapter V, Section 15]; *i.e.*, given two complex numbers $\xi, \eta \in \mathbb{C}$ and any value of the parameter $\lambda \in \mathbb{C}$, there exists a unique solution of the differential equation, say $y(\cdot,\lambda) : (a,b) \to \mathbb{C}$, with the properties:

$$\begin{cases} \text{(i)} & y(\cdot,\lambda) \text{ and } (py')(\cdot,\lambda) \in AC_{\text{loc}}(a,b) \\ \text{(ii)} & y(c,\lambda) = \xi \text{ and } (py')(c,\lambda) = \eta \\ \text{(iii)} & y(x,\cdot) \text{ and } (py')(x,\cdot) \text{ are holomorphic on } \mathbb{C} \\ \text{(iv)} & \text{if } \xi,\eta \in \mathbb{R} \text{ then } \overline{y}(\cdot,\lambda) = y(\cdot,\overline{\lambda}) \text{ and} \\ & (p\overline{y}')(\cdot,\lambda) = (py')(\cdot,\overline{\lambda}) \text{ for all } \lambda \in \mathbb{C}. \end{cases} \tag{4.3}$$

The Green's formula for the differential expression $M$ is, for any compact interval $[\alpha,\beta] \subset (a,b)$,

$$\int_\alpha^\beta \left\{\overline{g}(x)M[f](x) - f(x)\overline{M[g]}(x)\right\} dx = [f,g](\beta) - [f,g](\alpha) \text{ for all } f,g \in D(M), \tag{4.4}$$

where the symplectic form $[\cdot,\cdot](\cdot) : D(M) \times D(M) \times (a,b) \to \mathbb{C}$ is defined by

$$[f,g](x) := f(x)(p\overline{g}')(x) - (pf')(x)\overline{g}(x). \tag{4.5}$$

For the operator theoretic results given below see [20, Chapter V].

Incorporating now the weight function $w$ and the Hilbert function space $L^2((a,b);w)$, the maximal operator $T_1$ generated from $M$ is defined by

$$\begin{cases} \text{(i)} & T_1 : D(T_1) \subset L^2((a,b);w) \to L^2((a,b);w) \\ \text{(ii)} & D(T_1) := \{f \in D(M) : f, w^{-1}M[f] \in L^2((a,b);w)\} \\ \text{(iii)} & T_1 f := w^{-1}M[f] \text{ for all } f \in D(T_1). \end{cases} \tag{4.6}$$

We note, from the Green's formula, that the symplectic form of $M$ has the property that the following limits

$$[f,g](a) := \lim_{x\to a}[f,g](x) \text{ and } [f,g](b) := \lim_{x\to b}[f,g](x) \tag{4.7}$$

both exist and are finite in $\mathbb{C}$.

The minimal operator $T_0$ generated by $M$ is defined by

$$\begin{cases} \text{(i)} & T_0 : D(T_0) \subset L^2((a,b);w) \to L^2((a,b);w) \\ \text{(ii)} & D(T_0) := \{f \in D(T_1) : [f,g](b) = [f,g](a) = 0 \text{ for all } g \in D(T_1) \\ \text{(iii)} & T_0 f := w^{-1}M[f] \text{ for all } f \in D(T_0). \end{cases} \tag{4.8}$$

With these definitions the following properties hold for $T_0$, $T_1$ and their adjoint operators,

$$\begin{cases} \text{(i)} & T_0 \subseteq T_1 \\ \text{(ii)} & T_0 \text{ is closed and symmetric in } L^2((a,b);w) \\ \text{(iii)} & T_0^* = T_1 \text{ and } T_1^* = T_0 \\ \text{(iv)} & T_1 \text{ is closed in } L^2((a,b);w) \\ \text{(v)} & T_0 \text{ has equal deficiency indices } (d,d) \text{ with } 0 \le d \le 2. \end{cases} \tag{4.9}$$

Self-adjoint extensions $T$ of $T_0$ exist and satisfy

$$T_0 \subseteq T \subseteq T_1 \tag{4.10}$$

where the domain $D(T)$ is determined, as a restriction of the domain $D(T_1)$, by applying symmetric boundary conditions to the elements of the maximal domain $D(T_1)$.

## 5. The generalized initial value problem

We require the following lemma:

**Lemma 5.1.** *Let the pair* $(M,w)$ *be given on the interval* $(a,b)$*, satisfying the minimal conditions* (1.2)*;*

1. *If the endpoint* $a$ *is in the limit-point case in* $L^2((a,b);w)$ *then, see* (4.7),

$$[f,g](a) \equiv \lim_{x\to a}[f,g](x) = 0 \text{ for all } f,g \in D(T_1). \tag{5.1}$$

2. *If the endpoint $a$ is regular or limit-circle in $L^2((a,b);w)$ then there exists a continuum of pairs $\{\gamma,\delta\}$ such that*

$$\begin{cases} \text{(i)} & \gamma,\delta : (a,b) \to \mathbb{R} \\ \text{(ii)} & \gamma,\delta \in D(M) \\ \text{(iii)} & \gamma,\delta \in L^2((a,\beta];w) \text{ for all } \beta \in (a,b) \\ \text{(iv)} & w^{-1}M[\gamma], w^{-1}M[\delta] \in L^2((a,\beta];w) \text{ for all } \beta \in (a,b) \\ \text{(v)} & [\gamma,\delta](x) = 1 \text{ for all } x \in (a,b). \end{cases} \tag{5.2}$$

*There are similar results to both* (1) *and* (2) *for the endpoint $b$.*

*Proof.* These results are known: see [10, Section 5] and [20, Chapter V]. □

If $a$ is regular or limit-circle then, given the pair $\{\gamma,\delta\}$ as in Lemma 5.1, all separated boundary conditions at the endpoint $a$ can be determined in the form

$$[f, A\gamma + B\delta](a) = 0 \text{ with } f \in D(T_1), \tag{5.3}$$

where $A, B \in \mathbb{R}$ with $A^2 + B^2 > 0$. If $a$ is regular then this separated condition is equivalent to the classical form

$$f(a)\cos(\boldsymbol{\alpha}) + (pf')(a)\sin(\boldsymbol{\alpha}) = 0. \tag{5.4}$$

for some $\boldsymbol{\alpha} \in [0,\pi)$, see [22, Chapter V, Section 5.3]. In both cases (5.3) and (5.4) there is a continuum of separated boundary conditions at the endpoint $a$, each determined by the choice of the pair $\{A,B\}$ or the parameter $\boldsymbol{\alpha}$, respectively.

**Remark 5.1.** We use bold face $\boldsymbol{\alpha}$ for the boundary condition parameter in (5.4) to distinguish from $\alpha$ for the endpoint of the compact interval $[\alpha,\beta] \subset (a,b)$; similarly for $\boldsymbol{\beta}$ and $\beta$.

We can now state

**Theorem 5.1.** *Let the pair $(M,w)$ be given on the interval $(a,b)$, satisfying the minimal conditions* (1.2)*; let the endpoint $a$ be either regular or limit-circle in $L^2((a,b);w)$; let the pair $\{\gamma,\delta\}$ satisfy the conditions* (5.2)*; let $\xi(\cdot)$ and $\eta(\cdot) : \mathbb{C} \to \mathbb{C}$ and satisfy $\xi,\eta \in \mathbf{H}$ (see Section* 2 *above); then there exists a unique mapping*

$$\psi : (a,b) \times \mathbb{C} \to \mathbb{C} \tag{5.5}$$

*with the properties*

$$\begin{cases} \text{(i)} & \psi(\cdot,\lambda) \text{ and } (p\psi')(\cdot,\lambda) \in AC_{loc}(a,b) \text{ for all } \lambda \in \mathbb{C} \\ \text{(ii)} & \psi(x,\cdot) \text{ and } (p\psi')(x,\cdot) \in \mathbf{H} \text{ for all } x \in (a,b) \\ \text{(iii)} & [\psi(\cdot,\lambda),\gamma(\cdot)](a) = \xi(\lambda) \text{ and } [\psi(\cdot,\lambda),\delta(\cdot)](a) = \eta(\lambda) \text{ for all } \lambda \in \mathbb{C} \\ \text{(iv)} & \psi(\cdot,\lambda), \text{ for each } \lambda \in \mathbb{C}, \text{ satisfies the differential equation (1.1)} \\ & \text{almost everywhere on } (a,b) \\ \text{(v)} & \psi(\cdot,\lambda) \in L^2((a,\beta];w) \text{ for all } \beta \in (a,b) \text{ and all } \lambda \in \mathbb{C}. \end{cases} \tag{5.6}$$

*There is a similar result for the endpoint $b$.*

*Proof.* For the proof of this result see [15, Section 5, Theorem 2]. □

**Corollary 5.1.** *If in the conditions for Theorem* 5.1 *the initial value functions* $\xi(\cdot)$ *and* $\eta(\cdot)$ *are replaced by real numbers* $\xi$ *and* $\eta$, *then*

$$\overline{\psi}(\cdot,\lambda) = \psi(\cdot,\overline{\lambda}) \text{ and } (p\overline{\psi}')(\cdot,\lambda) = (p\psi')(\cdot,\overline{\lambda}) \text{ for all } \lambda \in \mathbb{C}. \tag{5.7}$$

*Proof.* This result follows on examination of the proof of Theorem 5.1. □

**Remark 5.2.** We make the following remarks concerning the results of Theorem 5.1:

1. When the endpoint $a$ is regular the generalized initial value result in (5.6) reduces to the classical initial value result in (4.3).
2. The result (iii) contains the generalized initial conditions at the endpoint $a$.
3. If the endpoint $b$ is also, independently, regular or limit-circle then the result (v) becomes $\psi(\cdot,\lambda) \in D(T_1)$ for all $\lambda \in \mathbb{C}$.
4. In general it is impossible to solve the initial value problem, either in classical or generalized form, at a limit-point endpoint.

## 6. The basic solutions $\theta$ and $\varphi$

To create the framework for the extended Titchmarsh eigenfunction expansion theorem to be stated and proved we first use the Titchmarsh-Weyl theory to give the $m$-coefficient, and the basic solution of the Sturm-Liouville differential equation (1.1) in the Hilbert function space $L^2((a,b);w)$.

Suppose given the open interval $(a,b) \subseteq \mathbb{R}$, and the pair $\{M,w\}$ of the differential expression $M$ and the weight function $w$, as in Sections 4 and 5 above.

The only additional restriction required in the endpoint classification of $M$ in $L^2((a,b);w)$ concerns the endpoint $a$; we suppose that:

$$\begin{cases} \text{(i)} & \text{the endpoint } a \text{ is regular or limit-circle} \\ \text{(ii)} & \text{the endpoint } b \text{ may be regular or limit-circle or limit-point.} \end{cases} \tag{6.1}$$

**Remark 6.1.** We make the following comments:

1. The only restriction required is that the endpoint $a$ is not in the limit-point case.
2. The conditions on the endpoints $a$ and $b$ can be reversed.
3. Taken together the main restriction is that the differential expression $M$ is not to be in the limit-point case at both endpoints $a$ and $b$; see item 4 of Remark 5.2 above.

In the results of Weyl [23] and Titchmarsh [22, Chapters II and III], the conditions on $M$ with the interval $(a,b)$ require that the endpoint $a$ is regular in $L^2((a,b);w)$. In the Titchmarsh notation, see [22, Chapter II, Section 1] but with some minor alterations here used, the two solutions $\theta$ and $\varphi$ of the differential equation (1.1) are defined, using the classical initial value result (4.3), by

$$\begin{cases} \theta(a,\lambda) = \cos(\boldsymbol{\alpha}) & (p\theta')(a,\lambda) = \sin(\boldsymbol{\alpha}) \\ \varphi(a,\lambda) = -\sin(\boldsymbol{\alpha}) & (p\varphi')(a,\lambda) = \cos(\boldsymbol{\alpha}); \end{cases} \tag{6.2}$$

here the parameter $\boldsymbol{\alpha}$ is chosen to satisfy $\boldsymbol{\alpha} \in [0, \pi)$. Note that for all $x \in (a, b)$ and $\lambda \in \mathbb{C}$

$$\overline{\theta}(x, \lambda) = \theta(x, \overline{\lambda}) \text{ and } \overline{\varphi}(x, \lambda) = \varphi(x, \overline{\lambda}). \tag{6.3}$$

For the Wronskian $W(\theta, \varphi) \equiv \theta(p\varphi') - (p\theta')\varphi$ we have

$$1 = W(\theta, \varphi)(a, \lambda) = W(\theta, \varphi)(x, \lambda) \text{ for all } x \in (a, b) \text{ and all } \lambda \in \mathbb{C}, \tag{6.4}$$

and for the symplectic form $[\theta, \varphi]$, see (4.5) and (6.3),

$$[\theta(\cdot, \lambda), \varphi(\cdot, \overline{\lambda})](x) = W(\theta, \varphi)(x, \lambda) = 1 \text{ for all } x \in (a, b) \text{ and all } \lambda \in \mathbb{C}. \tag{6.5}$$

To define the basic solutions $\theta$ and $\varphi$ of the Sturm-Liouville differential equation (1.1) in the generalized endpoint case, under the conditions (6.1), we use the results of Theorem 5.1, having chosen the endpoint $a$, and the pair $\{\gamma, \delta\}$ of elements as given in (5.2). The solutions $\theta$ and $\varphi$ are determined by the generalized initial conditions, given $\boldsymbol{\alpha} \in [0, \pi)$,

$$[\theta(\cdot, \lambda), \gamma(\cdot)](a) = \cos(\boldsymbol{\alpha}) \text{ and } [\theta(\cdot, \lambda), \delta(\cdot)](a) = \sin(\boldsymbol{\alpha}) \tag{6.6}$$

$$[\varphi(\cdot, \lambda), \gamma(\cdot)](a) = -\sin(\boldsymbol{\alpha}) \text{ and } [\varphi(\cdot, \lambda), \delta(\cdot)](a) = \cos(\boldsymbol{\alpha}). \tag{6.7}$$

These two solutions $\theta$ and $\varphi$ satisfy the property (6.3), see Corollary 5.1. Note also that, see item (ii) of (5.6),

$$\theta(x, \cdot), (p\theta')(x, \cdot), \varphi(x, \cdot), (p\varphi')(x, \cdot) \in \mathbf{H} \text{ for all } x \in (a, b), \tag{6.8}$$

and from item (v) of (5.6)

$$\theta(\cdot, \lambda) \text{ and } \varphi(\cdot, \lambda) \in L^2((a, \beta]; w) \text{ for all } \beta \in (a, b) \text{ and all } \lambda \in \mathbb{C}. \tag{6.9}$$

In the case when the endpoint $a$ is regular, Weyl [23] proved that $\theta$ and $\varphi$, determined by (6.2), satisfy either

(i) in the limit-point case at the endpoint $b$

$$\theta(\cdot, \lambda) \notin L^2((a, b); w) \text{ and } \varphi(\cdot, \lambda) \notin L^2((a, b); w) \text{ for all } \lambda \in \mathbb{C} \setminus \mathbb{R} \tag{6.10}$$

or

(ii) in the regular or limit-circle case at the endpoint $b$

$$\theta(\cdot, \lambda) \in L^2((a, b); w) \text{ and } \varphi(\cdot, \lambda) \in L^2((a, b); w) \text{ for all } \lambda \in \mathbb{C}. \tag{6.11}$$

These results also hold when the endpoint $a$ is regular or limit-circle and $\theta$ and $\varphi$ are determined by the generalized initial conditions (6.6) and (6.7).

To prove the Wronskian result (6.4) for this pair $\theta, \varphi$ of solutions, we have to use the Plücker identity, see [15, Section 5, Remark 1], for the symplectic form $[\cdot, \cdot]$ of (4.5). Let $f_r, g_r \in D(M)$ for $r = 1, 2, 3$ be any six functions from the domain $D(M)$, where $D(M)$ is given by (4.1), and define the complex-valued $3 \times 3$ matrix $A(\cdot)$, for all $x \in (a, b)$, by

$$A(x) := \begin{bmatrix} [f_1, g_1](x) & [f_1, g_2](x) & [f_1, g_3](x) \\ [f_2, g_1](x) & [f_2, g_2](x) & [f_2, g_3](x) \\ [f_3, g_1](x) & [f_3, g_2](x) & [f_3, g_3](x) \end{bmatrix}.$$

Then

$$\det A(x) = 0 \text{ for all } x \in (a,b). \tag{6.12}$$

In this identity substitute as follows, for any $\lambda \in \mathbb{C}$,

$$f_1, f_2, f_3 : \theta(\cdot,\lambda), \gamma(\cdot), \delta(\cdot) \text{ and } g_1, g_2, g_3 : \varphi(\cdot,\overline{\lambda}), \gamma(\cdot), \delta(\cdot);$$

on evaluating the determinant in (6.12) and recalling that $[\gamma(\cdot), \delta(\cdot)](x) = 1$ for all $x \in (a,b)$, we obtain

$$[\theta(\cdot,\lambda), \varphi(\cdot,\overline{\lambda})](x) = 1 \text{ for all } x \in (a,b) \text{ and all } \lambda \in \mathbb{C}. \tag{6.13}$$

In turn this gives the result, using (6.3),

$$W(\theta,\varphi)(x,\lambda) = 1 \text{ for all } x \in (a,b) \text{ and all } \lambda \in \mathbb{C}. \tag{6.14}$$

We also use the Plücker identity to prove the result

$$[\varphi(\cdot,\lambda_1), \varphi(\cdot,\lambda_2)](a) = 0 \text{ for all } \lambda_1, \lambda_2 \in \mathbb{C}; \tag{6.15}$$

for this result substitute in (6.12) as follows

$$f_1, f_2, f_3 : \varphi(\cdot,\lambda_1), \gamma(\cdot), \delta(\cdot) \text{ and } g_1, g_2, g_3 : \varphi(\cdot,\lambda_2), \gamma(\cdot), \delta(\cdot)$$

and take into account the conditions (6.6) and (6.7).

## 7. Boundary value problems

Although not part of the Titchmarsh-Weyl theory it is essential to introduce the GKN (Glazman-Krein-Naimark) method of constructing boundary conditions for linear ordinary boundary value problems, see [20, Chapter V, Section 18] and [13].

This method determines the domains $D(T)$ of self-adjoint restrictions $T$ of the maximal operator $T_1$; see (4.8), (4.9) and (4.10) of Section 4 above.

In the Sturm-Liouville case with the differential equation (1.1) we start with the endpoint conditions (6.1) for the interval $(a,b)$. Restricting the determination of the boundary value problem by separated boundary conditions, the domain $D(T) \subset L^2((a,b);w)$ of any self-adjoint restriction $T$ of $T_1$ is defined by selecting two non-null (modulo $D(T_0)$), real-valued elements $\chi, \kappa \in D(T_1)$ such that

$$[\chi,\chi](a) = [\kappa,\kappa](b) = 0. \tag{7.1}$$

The domain $D(T)$ and operator $T$ are then defined by

$$\begin{cases} \text{(i)} \quad D(T) := \{f \in D(T_1) : [f,\chi](a) = [f,\kappa](b) = 0\} \\ \text{(ii)} \quad Tf := w^{-1}M[f] \text{ for all } f \in D(T). \end{cases} \tag{7.2}$$

To link this boundary condition at the endpoint $a$ with the Titchmarsh type boundary condition [22, Chapter V, Section 5.3] when $a$ is regular, we take the element $\chi$ to be defined, for any fixed $\lambda \in \mathbb{C}$ and for some $c \in (a,b)$, by

$$\chi(x) := \varphi(x,\lambda) \text{ for all } x \in (a,c]$$

and then patch $\chi$ to zero in some interval $[d,b)$ for $d \in (c,b)$, using the Naimark patching lemma [20, Chapter V, Section 17.3, Lemma 2]. Thus the separated boundary conditions at the endpoints of the interval to determine the domain $D(T)$ are, for any fixed $\lambda \in \mathbb{C}$, and any non-null $\kappa \in D(T_1)$ satisfying $[\kappa, \kappa](b) = 0$,

$$D(T) := \{f \in D(T_1) : [f, \varphi(\cdot, \lambda)](a) = [f, \kappa](b) = 0\} \tag{7.3}$$

Note that this definition of the domain $D(T)$ implies that

$$[f, g](a) = [f, g](b) = 0 \text{ for all } f, g \in D(T), \tag{7.4}$$

and for the self-adjoint operator $T$, in $L^2((a, b); w)$,

$$(Tf, g)_w = (f, Tg)_w \text{ for all } \ f, g \in D(T). \tag{7.5}$$

For the proof of the results in this section see [20, Chapter V] and [13].

## 8. The Titchmarsh-Weyl $m$-coefficient

With the information and results of the above sections now available it is possible to state a theorem giving the existence of the Titchmarsh-Weyl $m$-coefficient, but now with the generalized basic solutions $\theta$ and $\varphi$ as defined in Section 6 above. Since the analysis required for this theorem is similar to the Titchmarsh analysis as given in [22, Chapter II, Sections 2.1 and 2.2], we state the following results without proof. The reader may consult the same reference for a discussion of Weyl circles. See also the paper [10] for some additional details related to this section.

**Theorem 8.1.** (*Titchmarsh-Weyl*)

1. *Given the interval $(a, b)$ let the coefficients $p, q, w$ satisfy the minimal conditions*

$$\begin{cases} \text{(i)} & p, q, w : (a, b) \to \mathbb{R} \\ \text{(ii)} & p^{-1}, q, w \in L^1_{loc}(a, b) \\ \text{(iii)} & w(x) > 0 \textit{ for almost all (Lebesgue) } x \in (a, b), \end{cases} \tag{8.1}$$

*to give the Sturm-Liouville differential equation*

$$-(py')' + qy = \lambda wy \textit{ on } (a, b). \tag{8.2}$$

2. *Let the endpoints of the interval $(a, b)$ satisfy the classification conditions*

$$\begin{cases} \text{(i)} & \textit{the endpoint } a \textit{ is regular or limit-circle} \\ \text{(ii)} & \textit{the endpoint } b \textit{ may be regular or limit-circle or limit-point.} \end{cases} \tag{8.3}$$

*in the weighted Hilbert function space $L^2((a, b); w)$.*

3. *Let the solutions $\theta$ and $\varphi$ of the differential equation (8.2) be determined by the generalized initial value conditions, for some $\boldsymbol{\alpha} \in [0, \pi)$,*

$$[\theta(\cdot, \lambda), \gamma(\cdot)](a) = \cos(\boldsymbol{\alpha}) \textit{ and } [\theta(\cdot, \lambda), \delta(\cdot)](a) = \sin(\boldsymbol{\alpha}) \tag{8.4}$$

*and*

$$[\varphi(\cdot, \lambda), \gamma(\cdot)](a) = -\sin(\boldsymbol{\alpha}) \textit{ and } [\varphi(\cdot, \lambda), \delta(\cdot)](a) = \cos(\boldsymbol{\alpha}), \tag{8.5}$$

*for some given pair* $\{\gamma, \delta\}$ *satisfying the conditions, see* (5.2),

$$\begin{cases} \text{(i)} & \gamma, \delta : (a,b) \to \mathbb{R} \\ \text{(ii)} & \gamma, \delta \in D(M) \\ \text{(iii)} & \gamma, \delta \in L^2((a,\beta];w) \text{ for all } \beta \in (a,b) \\ \text{(iv)} & w^{-1}M[\gamma], w^{-1}M[\delta] \in L^2((a,\beta];w) \text{ for all } \beta \in (a,b) \\ \text{(v)} & [\gamma,\delta](x) = 1 \text{ for all } x \in (a,b). \end{cases} \tag{8.6}$$

*Then there exists at least one Cauchy analytic function* $m : \mathbb{C} \setminus \mathbb{R} \to \mathbb{C}$ *with the properties:*

(i) $m$ *is regular on the open set* $\mathbb{C} \setminus \mathbb{R}$

(ii) $\overline{m}(\lambda) = m(\overline{\lambda})$ *for all* $\lambda \in \mathbb{C} \setminus \mathbb{R}$

(iii) *the solution* $\psi$ *of the differential equation* (8.2) *defined by*

$$\psi(x,\lambda) := \theta(x,\lambda) + m(\lambda)\varphi(x,\lambda) \text{ for all } x \in (a,b) \text{ and all } \lambda \in \mathbb{C} \setminus \mathbb{R} \tag{8.7}$$

*satisfies*

$$\psi(\cdot,\lambda) \in L^2((a,b);w) \text{ for all } \lambda \in \mathbb{C} \setminus \mathbb{R}. \tag{8.8}$$

*Proof.* See [22, Chapter II, Sections 2.1 and 2.2], and the account in [10, Sections 5, 7 and 11] which depends upon the prior definition of the self-adjoint operator $T$ in $L^2((a,b);w)$, as determined in (7.2).

In both these accounts of the Titchmarsh-Weyl $m$-coefficient a critical role is played by the Vitali convergence theorem in complex analysis, see [21, Chapter V, Sections 5.21 and 5.22], rather than involving points on the Weyl circles. □

**Remark 8.1.** For the $m$-coefficient the following cases occur, see [22, Chapter II, Section 2.1],

1. If the differential equation (8.2) is in the limit-point case at the singular endpoint $b$ then for each choice of the boundary condition parameter $\boldsymbol{\alpha} \in [0,\pi)$ there is a unique $m$-coefficient, which depends upon $\boldsymbol{\alpha}$, with the above properties; for all $\lambda \in \mathbb{C} \setminus \mathbb{R}$ the unique value $m(\lambda)$ is the limit-point of the Weyl circles for that value of $\lambda$.
2. If the differential equation (8.2) is in the regular or limit-circle case at the singular endpoint $b$, then for each choice of the boundary condition parameter $\boldsymbol{\alpha} \in [0,\pi)$ there is a continuum of $m$-coefficients. In that case, it is possible, for each fixed $\boldsymbol{\alpha}$, to use a separated boundary condition function $\kappa \in D(T_1)$, see (7.1) and (7.2), to distinguish an $m$-coefficient from those satisfying the conditions (i)–(iii) of Theorem 8.1. For this particular $m$-coefficient the corresponding solution $\psi$ from (8.7) should satisfy the separated boundary condition at $b$, *i.e.*,

$$[\psi(\cdot,\lambda),\kappa](b) = 0 \text{ for all } \lambda \in \mathbb{C} \setminus \mathbb{R}. \tag{8.9}$$

One then has a unique $m$-coefficient for each choice of the pair $\boldsymbol{\alpha}$ and $\kappa$. The determination of this $m$-coefficient can be made by use of the limit-circle process, but see the application of the Vitali convergence theorem in [22, Chapter II, Section 2.2].

**Corollary 8.1.** *We have, for any $m$-coefficient determined as in Remark* 8.1, *see Theorem* 8.1:

1. *For all* $\lambda_1, \lambda_2 \in \mathbb{C} \setminus \mathbb{R}$
$$[\psi(\cdot, \lambda_1), \psi(\cdot, \lambda_2)](b) = \lim_{\beta \to b} [\psi(\cdot, \lambda_1), \psi(\cdot, \lambda_2)](\beta) = 0. \tag{8.10}$$
*and*
$$\int_a^b w(x)\psi(x, \lambda_1)\overline{\psi}(x, \lambda_2)\, dx = \frac{m(\lambda_1) - \overline{m}(\lambda_2)}{\lambda_1 - \overline{\lambda_2}}. \tag{8.11}$$
2. *For all* $\lambda \in \mathbb{C} \setminus \mathbb{R}$
$$0 < \int_a^b w(x)\, |\psi(x, \lambda)|^2\, dx = \frac{\operatorname{Im}(m(\lambda))}{\operatorname{Im}(\lambda)}. \tag{8.12}$$

*Proof.* See [22, Chapter II, Sections 2.3, 2.4 and 2.5], and [10]. □

## 9. Analytic properties

We require the following results for the basic solutions $\theta$ and $\varphi$, and for the Titchmarsh-Weyl solution $\psi$.

In all the results here given $\beta$ is any number satisfying $\beta \in (a, b)$.

**Lemma 9.1.** *The two integral limits*
$$\lim_{\alpha \to a} \int_\alpha^\beta w(x)\, |\theta(x, \lambda)|^2\, dx \text{ and } \lim_{\alpha \to a} \int_\alpha^\beta w(x)\, |\varphi(x, \lambda)|^2\, dx$$
*converge uniformly for* $\lambda \in K$, *where* $K$ *is any compact subset of* $\mathbb{C}$.

*Proof.* The proof of this result is to be found in the paper [14, Section 3, (3.6)] and [8, Chapter 9, Theorem 2.1]. □

**Lemma 9.2.** *We have:*

(i) *The mappings*
$$\lambda \longmapsto \int_a^\beta w(x)\, |\theta(x, \lambda)|^2\, dx \text{ and } \int_a^\beta w(x)\, |\varphi(x, \lambda)|^2\, dx$$
*are locally bounded on* $\mathbb{C}$, *i.e., bounded on any compact subset* $K$ *of* $\mathbb{C}$.

(ii) *The mapping*
$$\lambda \longmapsto \int_\beta^b w(x)\, |\psi(x, \lambda)|^2\, dx$$
*is locally bounded on the open set* $\mathbb{C} \setminus \mathbb{R}$.

*Proof.* (i) The uniform convergence of the integrals as given in Lemma 9.1 implies that these mappings are continuous on $K$, and hence bounded on this set.
(ii) This boundedness property follows from the result given in (8.12). □

**Lemma 9.3.** *We have:*

(i) *For any* $f \in L^2((a,b);w)$

$$\int_a^\beta w(x)\theta(x,\cdot)f(x)\ dx \in \mathbf{H} \ \ \textit{and} \ \ \int_a^\beta w(x)\varphi(x,\cdot)f(x)\ dx \in \mathbf{H}.$$

(ii) *For any* $f \in L^2((a,b);w)$

$$\int_\beta^b w(x)\psi(x,\cdot)f(x)\ dx \in \mathbf{H}(\mathbb{C}\setminus\mathbb{R}).$$

*Proof.* These results follow from the boundedness properties in Lemma 9.2, and the general theorem given in the paper [12, Theorem 1]. □

## 10. Nevanlinna functions

The properties (i), (ii) given before (8.7) in the statement of Theorem 8.1, and the result (8.12) above imply that the analytic coefficient $m(\cdot)$ is a Nevanlinna (Herglotz, Pick, Riesz) function, *i.e.*, if the open half planes $\mathbb{C}_+$ and $\mathbb{C}_-$ are defined by

$$\mathbb{C}_\pm := \{\lambda \in \mathbb{C} : \operatorname{Im}(\lambda) \gtrless 0\},$$

then

$$\begin{array}{ll} \text{(i)} & m(\cdot) : \mathbb{C}_\pm \to \mathbb{C}_\pm \\ \text{(ii)} & m(\cdot) \text{ is regular on } \mathbb{C}_\pm \\ \text{(iii)} & \overline{m}(\lambda) = m(\overline{\lambda}) \text{ for all } \lambda \in \mathbb{C}_\pm. \end{array}$$

Such an analytic function has an integral representation of the following form, see Akhiezer and Glazman [2, Chapter 6, Section 69, Theorem 2], where $A, B \in \mathbb{R}$ with $B \geq 0$,

$$m(\lambda) = A + B\lambda + \int_{(-\infty,+\infty)} \left\{\frac{1}{t-\lambda} - \frac{t}{t^2+1}\right\} d\rho(t) \text{ for all } \lambda \in \mathbb{C}\setminus\mathbb{R}. \tag{10.1}$$

Here the function $\rho : \mathbb{R} \to \mathbb{R}$ is monotonic non-decreasing on $\mathbb{R}$ and satisfies the growth restriction

$$\int_{-\infty}^{+\infty} \frac{1}{1+t^2}\ d\rho(t) < +\infty; \tag{10.2}$$

this function $\rho$ is the spectral function for the $m$-coefficient. In [2, Chapter 6, Section 69, Theorem 2] the integrals in (10.1) and (10.2) are given as generalized Riemann-Stieltjes integrals; however these integrals are best interpreted as Lebesgue-Stieltjes integrals with the symbol $\rho$ representing a non-negative Baire measure on the Borel $\sigma$-algebra of the real line $\mathbb{R}$; see the text of Royden [18, Chapter 12, Section 3]. For this measure $\rho$ every bounded Borel set has finite measure but in general $\rho(\mathbb{R})$ may be $+\infty$.

In the text [2, Chapter 6, Section 69, Theorem 2] it is shown that there are two integral representations for the Nevanlinna function $m(\cdot)$; in addition to (10.1) there is a representation in the form

$$m(\lambda) = C + D\lambda + \int_{(-\infty,+\infty)} \frac{1+sz}{s-\lambda}\, d\sigma(s) \text{ for all } \lambda \in \mathbb{C} \setminus \mathbb{R}; \tag{10.3}$$

here $\sigma(\cdot)$ is monotonic non-decreasing on $\mathbb{R}$ and $C, D \in \mathbb{R}$ with $D \geq 0$. As before $\sigma(\cdot)$ generates a non-negative Baire measure $\sigma$ on the Borel $\sigma$-algebra of the real line $\mathbb{R}$ but now this measure is bounded on all Borel sets, *i.e.*, $\sigma(\mathbb{R}) < +\infty$.

However in the analysis which follows it is more convenient to use the representation (10.2) and to call $\rho$ the spectral measure of the $m$-coefficient. This non-negative Baire measure $\rho$ determines a Hilbert function space $L^2(\mathbb{R};\rho)$ of equivalence classes of functions $F : \mathbb{R} \to \mathbb{C}$ such that

$$\int_{(-\infty,+\infty)} |F(t)|^2\, d\rho(t) < +\infty$$

with inner-product and norm determined by, for all $F, G \in L^2(\mathbb{R};\rho)$,

$$(F,G)_\rho = \int_{(-\infty,+\infty)} F(t)\overline{G}(t)\, d\rho(t) \text{ and } \|F\|_\rho = \left\{ \int_{(-\infty,+\infty)} |F(t)|^2\, d\rho(t) \right\}^{1/2}. \tag{10.4}$$

**Remark 10.1.** There is a connection between the two measures $\rho$ and $\sigma$ best expressed using the Radon-Nikodym theorem, see [18, Chapter 11, Section 6]. The two measures are absolutely continuous with respect to each other; in particular

$$\sigma(E) = \int_E \frac{1}{1+t^2}\, d\rho(t) \text{ and } \rho(E) = \int_E (1+s^2)\, d\sigma(s) \tag{10.5}$$

for all Borel sets $E \subseteq \mathbb{R}$. In terms of the Radon-Nikodym derivatives, see [18, Chapter 11, Section 6, Problem 34], we have

$$\left[\frac{d\sigma}{d\rho}\right](t) = \frac{1}{1+t^2} \text{ for all } t \in \mathbb{R}, \text{ and } \left[\frac{d\rho}{d\sigma}\right](s) = 1+s^2 \text{ for all } s \in \mathbb{R}.$$

**Remark 10.2.** All Lebesgue-Stieltjes integrals are written in the form (10.5) even in the case when the Borel set $E$ is an interval.

## 11. Expansion theorem

We can now state the generalized form of the Titchmarsh-Weyl eigenfunction expansion theorem.

**Theorem 11.1.** *Let the open interval $(a,b) \subseteq \mathbb{R}$ be given; let the coefficients $p, q, w$ of the Sturm-Liouville differential equation, with spectral parameter $\lambda \in \mathbb{C}$,*

$$M[y] \equiv -(py')' + qy = \lambda wy \text{ on } (a,b) \tag{11.1}$$

*satisfy the minimal conditions*

$$\begin{cases} \text{(i)} & p, q, w : (a,b) \to \mathbb{R} \\ \text{(ii)} & p^{-1}, q, w \in L^1_{loc}(a,b) \\ \text{(iii)} & w(x) > 0 \textit{ for almost all (Lebesgue) } x \in (a,b). \end{cases} \tag{11.2}$$

*Let the endpoint classification of the equation* (11.1) *in the Hilbert space* $L^2((a,b);w)$ *satisfy*

$$\begin{cases} \text{(i)} & \textit{the endpoint } a \textit{ is regular or limit-circle} \\ \text{(ii)} & \textit{the endpoint } b \textit{ may be regular or limit-circle or limit-point;} \end{cases} \tag{11.3}$$

*let the pair* $\{\gamma, \delta\}$ *of boundary functions for the endpoint* $a$ *be chosen so that, see* (5.2),

$$\begin{cases} \text{(i)} & \gamma, \delta : (a,b) \to \mathbb{R} \\ \text{(ii)} & \gamma, \delta \in D(M) \\ \text{(iii)} & \gamma, \delta \in L^2((a,\beta];w) \textit{ for all } \beta \in (a,b) \\ \text{(iv)} & w^{-1}M[\gamma], w^{-1}M[\delta] \in L^2((a,\beta];w) \textit{ for all } \beta \in (a,b) \\ \text{(v)} & [\gamma,\delta](x) = 1 \textit{ for all } x \in (a,b). \end{cases} \tag{11.4}$$

*Let the basic solutions* $\theta$ *and* $\varphi$ *of the equation* (11.1) *be determined, for the parameter* $\boldsymbol{\alpha} \in [0,\pi)$, *by the generalized initial conditions at the endpoint* $a$

$$[\theta(\cdot,\lambda),\gamma(\cdot)](a) = \cos(\boldsymbol{\alpha}) \textit{ and } [\theta(\cdot,\lambda),\delta(\cdot)](a) = \sin(\boldsymbol{\alpha}) \tag{11.5}$$

*and*

$$[\varphi(\cdot,\lambda),\gamma(\cdot)](a) = -\sin(\boldsymbol{\alpha}) \textit{ and } [\varphi(\cdot,\lambda),\delta(\cdot)](a) = \cos(\boldsymbol{\alpha}). \tag{11.6}$$

*Let the Sturm-Liouville boundary value problem be determined, in operator theoretic form, by*

$$\begin{cases} \text{(i)} & D(T) := \{f \in D(T_1) : [f,\varphi](a) = [f,\kappa](b) = 0\} \\ \text{(ii)} & Tf := w^{-1}M[f] \textit{ for all } f \in D(T), \end{cases} \tag{11.7}$$

*where* $\varphi$ *is the basic solution of the differential equation subject to* (11.6), *and* $\kappa \in D(T_1)$ *defines the non-null boundary condition for the endpoint* $b$*; note that*

$$[\varphi(\cdot,\lambda),\varphi(\cdot,\lambda)](a) = 0 \textit{ for all } \lambda \in \mathbb{C}, \textit{ and } [\kappa,\kappa](b) = 0. \tag{11.8}$$

*Let* $m(\cdot)$ *be the* m*-coefficient for this boundary value problem, defined in Remark* 8.1, *see Theorem* 8.1. *Let* $\rho$ *be the spectral measure of* $m(\cdot)$ *and* $L^2(\mathbb{R};\rho)$ *be its generated Hilbert function space.*

1. *For any function* $f \in L^2((a,b);w)$ *define the family*

$$\{F_{\alpha,\beta}(\cdot) : \mathbb{R} \to \mathbb{C} \textit{ for all } \alpha,\beta \textit{ with } a < \alpha < \beta < b\}$$

*by*

$$F_{\alpha,\beta}(t) := \int_\alpha^\beta \varphi(x,t)f(x)w(x)\ dx \textit{ for all } t \in \mathbb{R} \textit{ and all } \alpha,\beta \in (a,b); \tag{11.9}$$

*then:*

(i) $F_{\alpha,\beta}(\cdot) \in L^2(\mathbb{R};\rho)$ *for all* $\alpha,\beta \in (a,b)$

(ii) *the family* $\{F_{\alpha,\beta} : \alpha,\beta \in (a,b)\}$ *converges in mean, as* $\alpha \to a$ *and* $\beta \to b$, *in* $L^2(\mathbb{R};\rho)$ *to, say,* $F$

(iii) *Let* $f,g \in L^2((a,b);w)$ *generate* $F,G \in L^2(\mathbb{R};\rho)$, *respectively, as above; then*

$$(F,G)_\rho = (f,g)_w;$$

*in particular* $\|F\|_\rho = \|f\|_w$.

2. *For any function* $G \in L^2(\mathbb{R};\rho)$ *define the family* $\{g_\tau(\cdot) : (a,b) \to \mathbb{C}$ *for all* $\tau \in (0,\infty)\}$ *by*

$$g_\tau(x) := \int_{[-\tau,\tau]} \varphi(x,t)G(t)\ d\rho(t) \text{ for all } x \in (a,b) \text{ and all } \tau \in (0,\infty); \tag{11.10}$$

*then:*

(i) $g_\tau(\cdot) \in L^2((a,b;w)$ *for all* $\tau \in (0,\infty)$

(ii) *the family* $\{g_\tau : \tau \in (0,\infty)\}$ *converges in mean, as* $\tau \to +\infty$, *in* $L^2((a,b);w)$ *to, say,* $g \in L^2((a,b);w)$

(iii) *Let* $G,F \in L^2(\mathbb{R};\rho)$ *generate* $g,f \in L^2((a,b);w)$, *respectively, as above; then*

$$(f,g)_w = (F,G)_\rho;$$

*in particular* $\|g\|_w = \|G\|_\rho$.

3. *Given* $f \in L^2((a,b);w)$ *and any* $G \in L^2(\mathbb{R};\rho)$ *let* $F$ *be defined as in item* 1, (ii) *and* $g$ *defined as in item* 2, (ii) *above; then* $f = g$ *in* $L^2((a,b);w)$ *if and only if* $G = F$ *in* $L^2(\mathbb{R};\rho)$.

4. *For all* $f,F$ *and* $G,g$ *defined as above we have the generalized Parseval equality*

$$\int_a^b f(x)\overline{g}(x)w(x)\ dx = \int_{(-\infty,+\infty)} F(t)\overline{G}(t)\ d\rho(t). \tag{11.11}$$

*Proof.* See the lemmas stated and proved in the following sections; in particular Lemmas 15.2 and 15.3 of Section 15. □

**Remark 11.1.** Terminology

(i) $F \in L^2(\mathbb{R};\rho)$ is termed the **generalized transform** of $f \in L^2((a,b);w)$

(ii) $g \in L^2((a,b);w)$ is termed the **generalized inverse transform** of $G \in L^2(\mathbb{R};\rho)$.

**Corollary 11.1.** (*The eigenfunction expansion theorem*)

*Given* $f \in L^2((a,b);w)$ *define* $F \in L^2(\mathbb{R};\rho)$, *using convergence in mean in* $L^2(\mathbb{R};\rho)$, *as in item* 1 *of Theorem* 11.1 *above, i.e.,*

$$\lim_{\alpha\to a,\beta\to b} \int_{(-\infty,+\infty)} \left| F(t) - \int_\alpha^\beta f(x)\varphi(x,t)w(x)\ dx \right|^2 d\rho(t) = 0; \tag{11.12}$$

*then, with convergence in mean in* $L^2((a,b);w)$, *as in item* 2 *of Theorem* 11.1 *above,*

$$\lim_{\tau\to+\infty}\int_a^b \left| f(x) - \int_{[-\tau,\tau]} F(t)\varphi(x,t)\ d\rho(t)\right|^2 w(x)\ dx = 0. \tag{11.13}$$

*Proof.* See the lemmas stated and proved in the following sections; in particular Lemma 15.3 of Section 15. □

**Corollary 11.2.** *We have*

1. *Given* $f \in L^2((a,b);w)$ *define* $F \in L^2(\mathbb{R};\rho)$ *as in* (11.12) *above, and define the mapping* $U$ *by*
$$U : f \in L^2((a,b);w) \to L^2(\mathbb{R};\rho) \text{ by } Uf := F \text{ for all } f \in L^2((a,b);w); \tag{11.14}$$
*then* $U$ *is an isomorphic, isometric mapping on* $L^2((a,b);w)$ *onto* $L^2(\mathbb{R};\rho)$.
2. *Given* $F \in L^2(\mathbb{R};\rho)$ *define* $f \in L^2((a,b);w)$ *as in* (11.13) *above, and define the mapping* $V$ *by*
$$V : L^2(\mathbb{R};\rho) \to L^2((a,b);w) \text{ by } VF := f \text{ for all } F \in L^2(\mathbb{R};\rho); \tag{11.15}$$
*then* $V$ *is an isomorphic, isometric mapping on* $L^2(\mathbb{R};\rho)$ *onto* $L^2((a,b);w)$.
3. *The mappings* $U$ *and* $V$ *are inverse to each other, i.e.,* $V = U^{-1}$ *and* $U = V^{-1}$.

*Proof.* See the lemmas stated and proved in the following sections; in particular Lemma 16.2 of Section 16. □

**Remark 11.2.** Given $f \in L^2((a,b);w)$ with $F := Uf$ in (11.14), then $F$ represents the generalized "Fourier coefficients" of $f$; similarly with $f := VF$ in (11.15) this result represents the generalized "Fourier expansion" of the original function $f$.

## 12. The resolvent function $\Phi$

Following [22, Chapter II, Section 2.6] we define the mapping

$$\Phi : (a,b) \times \mathbb{C}\setminus\mathbb{R} \times L^2((a,b);w) \to \mathbb{C}$$

by

$$\Phi(x,\lambda;f) := \psi(x,\lambda)\int_a^x \varphi(t,\lambda)f(t)w(t)\ dt + \varphi(x,\lambda)\int_x^b \psi(t,\lambda)f(t)w(t)\ dt \tag{12.1}$$

for all $x \in (a,b)$, all $\lambda \in \mathbb{C}\setminus\mathbb{R}$ and all $f \in L^2((a,b);w)$; this mapping is well defined using (8.8) and Remark 8.1.

**Lemma 12.1.** *For all* $x \in (a,b)$, *all* $\lambda,\mu \in \mathbb{C}\setminus\mathbb{R}$ *and all* $f \in L^2((a,b);w)$ *the following properties of the resolvent function* $\Phi$ *hold, where* $I$ *is the identity operator in* $L^2((a,b);w)$:

1. $\Phi(\cdot,\lambda;f) \in L^2((a,b);w)$, *in particular*

$$\|\Phi(\cdot,\lambda;f)\|_w^2 = \int_a^b w(x)\,|\Phi(x,\lambda;f)|^2\,dx \le \big(\operatorname{Im}(\lambda)\big)^{-2}\int_a^b |f(x)|^2\,w(x)\,dx \quad (12.2)$$

2. $-(p(x)\Phi'(x,\lambda;f))' + q(x)\Phi(x,\lambda;f) = \lambda w(x)\Phi(x,\lambda;f) + w(x)f(x)$
3. $[\Phi(\cdot,\lambda;f),\varphi(\cdot,\lambda)](a) = [\Phi(\cdot,\lambda;f),\kappa(\cdot)](b) = 0$
4. $\Phi(\cdot,\lambda;f) \in D(T)$, *with* $D(T)$ *given by* (7.2)
5. $\Phi(x,\lambda;\cdot)$ *is a bounded linear operator in* $L^2((a,b);w)$
6. $(T-\lambda I)\,(\Phi(\cdot,\lambda;f)) = f$ *and* $\Phi(\cdot,\lambda;(T-\lambda I)f) = f$ *for* $f \in L^2((a,b);w)$ *and* $f \in D(T)$ *respectively*
7. $(\Phi(\cdot,\lambda;f),g)_w = (f,\Phi(\cdot,\overline{\lambda};g))_w$
8. $\Phi(x,\lambda;f) - \Phi(x,\mu;f) = (\lambda-\mu)\Phi(x,\lambda;\Phi(\cdot,\mu;f))$
9. $\Phi(x,\cdot;f) \in \mathbf{H}(\mathbb{C}\setminus\mathbb{R})$ *for all* $x \in (a,b)$ *and all* $f \in L^2((a,b);w)$.

*Proof.* The proof of these results proceeds along the same lines as given by Titchmarsh [22, Chapter II, Sections 2.6, 2.7 and 2.8] but with some additional analysis due to the possible limit-circle case at the endpoint $a$.

The proof of item 8 is best given in the method of Kato [16, Chapter 1, Section 5.2] and the operator results given in item 6.

The proof of item 9 follows from the analytical results given in Section 9, Lemma 9.3. □

**Remark 12.1.** Essentially, the Titchmarsh resolvent function $\Phi$ is the resolvent operator for the self-adjoint operator $T$ in the Hilbert space $L^2((a,b);w)$.

## 13. Representation of the resolvent function $\Phi$

We have the following lemma

**Lemma 13.1.** *Given the resolvent function* $\Phi$ *for all* $f,g \in L^2((a,b);w)$ *there exists a unique mapping*

$$\sigma(f,g;\cdot) : \mathbb{R} \to \mathbb{C} \quad (13.1)$$

*with the properties:*

(i) $\sigma(f,g;\cdot)$ *is right-continuous on* $\mathbb{R}$

(ii) $\sigma(f,g;\cdot)$ *is of total bounded variation on* $\mathbb{R}$ *with* $\lim_{t\to-\infty}\sigma(f,g;t) = 0$

(iii) $\sigma(f,f;\cdot)$ *is monotone non-decreasing on* $\mathbb{R}$

(iv) $\sigma(f,g;\cdot)$ *generates a totally bounded complex-valued Baire measure on the Borel* $\sigma$*-algebra of* $\mathbb{R}$ *such that the Lebesgue-Stieltjes integral has the property*

$$\int_{(-\infty,+\infty)} |d\sigma(f,g;\cdot)| \le \|f\|_w\,\|g\|_w \quad (13.2)$$

(v) *for all* $\lambda \in \mathbb{C}\setminus\mathbb{R}$ *there is the Lebesgue-Stieltjes representation*

$$(\Phi(\cdot,\lambda;f),g)_w = \int_{(-\infty,+\infty)} \frac{1}{t-\lambda}\,d\sigma(f,g;t) \text{ for all } f,g \in L^2((a,b);w). \quad (13.3)$$

*Proof.* To show uniqueness assume $\sigma(t) := \sigma(f, g; t)$, for all $t \in \mathbb{R}$, to be of total bounded variation and

$$\int_{(-\infty,+\infty)} \frac{1}{t-\lambda} d\sigma(t) = 0 \text{ for all } \lambda \in \mathbb{C} \setminus \mathbb{R}.$$

It follows that

$$\int_{(-\infty,+\infty)} \left\{ \frac{1}{t-\lambda} - \frac{1}{t-\overline{\lambda}} \right\} d\sigma(t) = 0$$

so that, for $\lambda = \mu + i\nu$,

$$\int_{(-\infty,+\infty)} \frac{\nu}{(t-\mu)^2+\nu^2} d\sigma(t) = 0 \text{ for all } \mu \in \mathbb{R} \text{ and all } \nu > 0.$$

Let $A, B$ be points of continuity of $\sigma(\cdot)$; by absolute convergence of the integrals we have

$$\begin{aligned} 0 &= \int_{[A,B]} d\mu \int_{(-\infty,+\infty)} \frac{\nu}{(t-\mu)^2+\nu^2} d\sigma(t) \\ &= \int_{(-\infty,+\infty)} d\sigma(t) \int_A^B \frac{\nu}{(t-\mu)^2+\nu^2} d\mu \\ &= \int_{(-\infty,+\infty)} \left( \arctan\left(\frac{B-t}{\nu}\right) + \arctan\left(\frac{t-A}{\nu}\right)\right) d\sigma(t). \end{aligned}$$

Since the points $A, B$ do not carry $\sigma$-mass this last integral converges as $\nu \to 0^+$, by bounded convergence, to

$$\pi \int_{[A,B]} d\sigma(t)$$

and so $\sigma(\cdot)$ is constant on $\mathbb{R}$.

Since the form $(\Phi(\cdot,\lambda; f), g)_w$ is sesquilinear in $f, g$ then by item 7 of Lemma 12.1 it follows that $\sigma(f, g; \cdot)$ is hermitian, supposing it to exist.

However, by items 7 and 8 of Lemma 12.1

$$\frac{\operatorname{Im}\left((\Phi(\cdot,\lambda; f), f)_w\right)}{\operatorname{Im}(\lambda)} = \|\Phi(\cdot,\lambda; f)\|_w^2 \geq 0 \text{ for all } \lambda \in \mathbb{C} \setminus \mathbb{R} \tag{13.4}$$

so that we may apply the Nevanlinna theory of Section 10 to the analytic function

$$(\Phi(\cdot,\lambda; f), f)_w$$

defined on $\mathbb{C}\setminus\mathbb{R}$; this application gives the existence of a monotonic non-decreasing function, denoted by $\sigma(f, f; \cdot)$ to indicate its dependence on the choice of $f \in L^2((a,b); w)$, such that

$$\int_{(-\infty,+\infty)} \frac{1}{1+t^2} d\sigma(f, f; t) < +\infty,$$

and for real numbers $A, B$ with $B \geq 0$

$$(\Phi(\cdot,\lambda; f), f)_w = A + B\lambda + \int_{(-\infty,+\infty)} \left\{ \frac{1}{t-\lambda} - \frac{t}{t^2+1} \right\} d\sigma(f, f; t) \text{ for all } \lambda \in \mathbb{C}\setminus\mathbb{R}.$$

From item 1 of Lemma 12.1 the norm $\|f\|_w^2$ is a bound for the mapping

$$\nu \to \nu \left(\Phi(\cdot, i\nu; f), f\right)_w \, as \, |\nu| \to +\infty;$$

the imaginary part of this mapping is

$$B\nu^2 + \int_{(-\infty,+\infty)} \frac{\nu^2}{t^2+\nu^2} d\sigma(f,f;t).$$

It follows that $B = 0$ and for all $\nu, \varepsilon > 0$ we have

$$\frac{1}{1+\varepsilon^2} \int_{[-\nu\varepsilon,\nu\varepsilon]} d\sigma(f,f;t) \leq \|f\|_w^2 ;$$

thus first letting $\nu \to +\infty$ and then letting $\varepsilon \to 0^+$, we obtain

$$\int_{(-\infty,+\infty)} d\sigma(f,f;t) \leq \|f\|_w^2 .$$

We may now assume, without loss of generality, that the normalized condition

$$\lim_{t \to -\infty} \sigma(f,f;t) = 0$$

is satisfied.

Clearly

$$\int_{(-\infty,+\infty)} \frac{|t|}{1+t^2} d\sigma(f,f;t) < +\infty$$

and so we may write, with absolute convergence,

$$\left(\Phi(\cdot,\lambda;f),f\right)_w = A' + \int_{(-\infty,+\infty)} \frac{1}{t-\lambda} d\sigma(f,f;t) \text{ for all } \lambda \in \mathbb{C} \setminus \mathbb{R}.$$

For $\lambda = i\nu$ this last integral tends to zero as $\nu \to +\infty$, by bounded convergence; similarly for the term on the left-hand side on using (12.2); thus $A' = 0$. Thus (13.3) is now established in the case when $f = g$.

In the general case we have, from the polarization formula,

$$\left(\Phi(\cdot,\lambda;f),g\right)_w = \tfrac{1}{4} \sum_{k=0}^{3} i^k \left(\Phi(\cdot,\lambda;f+i^k g), f+i^k g\right)_w ;$$

we now obtain the required result (13.3)

$$\left(\Phi(\cdot,\lambda;f),g\right)_w = \int_{(-\infty,+\infty)} \frac{1}{t-\lambda} d\sigma(f,g;t)$$

on making the definition

$$\sigma(f,g;t) := \tfrac{1}{4} \sum_{k=0}^{3} i^k \sigma(f+i^k g, f+i^k g;t) \text{ for all } t \in \mathbb{R}.$$

The mapping $\sigma(f,g;\cdot) : \mathbb{R} \to \mathbb{C}$ has the required normalization

$$\lim_{t \to -\infty} \sigma(f,g;t) = 0$$

for all $f, g \in L^2((a,b);w)$.

For $f, g \in L^2((a,b);w)$ the mapping

$$f, g \to \int_\Delta d\sigma(f,g;t),$$

where $\Delta$ is any interval of the real line, is a hermitian, semi-positive and bounded (since it is bounded for $f = g$) form on the product of the Hilbert function spaces. For any $\Delta$ the Cauchy-Schwarz inequality yields

$$\left| \int_\Delta d\sigma(f,g;t) \right| \le \left\{ \int_\Delta d\sigma(f,f;t) \int_\Delta d\sigma(g,g;t) \right\}^{1/2}.$$

If $\{\Delta_j : j = 1, 2, \ldots, N\}$ is any finite partition of $\mathbb{R}$ into disjoint intervals it follows that

$$\begin{aligned}
\sum_{j=1}^N \left| \int_{\Delta_j} d\sigma(f,g;t) \right| &\le \sum_{j=1}^N \left\{ \int_{\Delta_j} d\sigma(f,f;t) \int_{\Delta_j} d\sigma(g,g;t) \right\}^{1/2} \\
&\le \left\{ \sum_{j=1}^N \int_{\Delta_j} d\sigma(f,f;t) \sum_{j=1}^N \int_{\Delta_j} d\sigma(g,g;t) \right\}^{1/2} \\
&\le \|f\|_w \|g\|_w \text{ for all } f, g \in L^2((a,b);w),
\end{aligned}$$

and so the total variation of $\sigma(f,g;\cdot)$ on $\mathbb{R}$ does not exceed $\|f\|_w \|g\|_w$ to give the required result (13.2).

Note that we have invoked the use of complex signed Baire measures in defining the integrals from the function $\sigma(f,g;\cdot)$ of total bounded variation on $\mathbb{R}$; see the definitions for such integration methods in [18, Chapter 11, Section 5].

This completes the proof of the lemma. □

**Lemma 13.2.** *We have*

$$\int_{(-\infty,+\infty)} d\sigma(f,g;t) = (f,g)_w \text{ for all } f, g \in L^2((a,b);w) \tag{13.5}$$

*Proof.* Let $f \in D(T)$ as defined in (7.2). Then from item 6 of Lemma 12.1 we can write

$$f = \Phi(\cdot, \lambda; (T - \lambda I)f) \text{ for all } \lambda \in \mathbb{C} \setminus \mathbb{R},$$

*i.e.*,

$$(f,g)_w = (\Phi(\cdot,\lambda;Tf),g)_w - \lambda\,(\Phi(\cdot,\lambda;f),g)_w \text{ for all } g \in L^2((a,b);w).$$

Now take $\lambda = i\nu$ and let $\nu$ tend to $+\infty$ in this last result; using the representation (13.3) for $\Phi$ it follows that the first term on the right-hand side tends to zero, whilst the second term gives

$$\lim_{\nu \to +\infty} \left\{ -i\nu \int_{(-\infty,+\infty)} \frac{1}{t - i\nu} d\sigma(f,g;t) \right\} = \int_{(-\infty,+\infty)} d\sigma(f,g;t)$$

on using bounded convergence.

Thus the required result (13.5) holds for all $f \in D(T)$ and all $g \in L^2((a,b);w)$; since

$$f, g \to \int_{(-\infty,+\infty)} d\sigma(f,g;t)$$

is a bounded, Hermitian form, and since the domain $D(T)$ of the self-adjoint operator $T$ is dense in $L^2((a,b);w)$, the general result (13.5) now follows by continuity arguments. □

## 14. Further properties of the resolvent $\Phi$

**Definition 14.1.** *For all $f \in L^2((a,b);w)$ with compact support in the open interval $(a,b)$ define $F : \mathbb{C} \to \mathbb{C}$ by*

$$F(\lambda) := \left(f, \varphi(\cdot, \overline{\lambda})\right)_w \text{ for all } \lambda \in \mathbb{C}, \tag{14.1}$$

*where $\varphi$ is the solution of the differential equation (1.1) defined by the generalized initial condition (6.7).*

**Lemma 14.1.** *With the function $F$ defined by (14.1) it follows that*

$$F(\cdot) \in \mathbf{H} \text{ for all } f \in L^2((a,b);w). \tag{14.2}$$

*Proof.* This result follows from item (i) of Lemma 9.3. □

**Lemma 14.2.** *Let $f, g \in L^2((a,b);w)$ both with compact support in $(a,b)$; let the $m$-coefficient be determined as in the statement of Theorem (11.1); then the function*

$$(\Phi(\cdot, \lambda; f), g)_w - F(\lambda)\overline{G(\overline{\lambda})}m(\lambda),$$

*defined for all $\lambda \in \mathbb{C}$, belongs to the family of entire (integral) functions $\mathbf{H}$.*

*Proof.* This result follows from the definitions of the functions $\Phi, F, G$ and a straightforward calculation. □

**Lemma 14.3.** *Let the function $\rho : \mathbb{R} \to \mathbb{R}$ be monotonic non-decreasing and right-continuous on $\mathbb{R}$; assume that $\rho$ is differentiable at the point 0; then the integral*

$$\int_{-1}^{1} \left\{ \int_{(-1,1]} \frac{1}{\sqrt{t^2+s^2}} d\rho(t) \right\} ds \tag{14.3}$$

*is convergent.*

*Proof.* We have on integration by parts

$$\int_{(-1,1]} \frac{1}{\sqrt{t^2+s^2}} d\rho(t) = \frac{\rho(1)-\rho(-1)}{\sqrt{1+s^2}} - \int_{-1}^{1} \frac{\rho(t)-\rho(0)}{t} t \frac{d}{dt} \frac{1}{\sqrt{t^2+s^2}} dt.$$

The first factor in the last integral is bounded since $\rho'(0)$ exists, and the second factor is negative since $(t^2+s^2)^{-1/2}$ decreases with $|t|$. Furthermore

$$\begin{aligned}-\int_{-1}^{1} t\frac{d}{dt}\frac{1}{\sqrt{t^2+s^2}}dt &= \frac{-2}{\sqrt{1+s^2}} + \int_{-1}^{1}\frac{1}{\sqrt{t^2+s^2}}dt \\ &= \frac{-2}{\sqrt{1+s^2}} - \frac{1}{2}\ln\left(1+\sqrt{1+s^2}\right) + \frac{1}{2}\ln(|s|)\end{aligned}$$

which is locally integrable. This gives the required result. □

**Lemma 14.4.** *Let $f,g \in L^2((a,b);w)$ have compact supports in $(a,b)$, and let $A<B$ be points of differentiability of the spectral function $\rho$ of the defined $m$-coefficient; see Section 8 and (10.2) and (10.3). Then*

$$\frac{1}{2\pi i}\int_\Gamma (\Phi(\cdot,\lambda;f),g)_w\ d\lambda = -\int_{[A,B]} F(t)\overline{G}(t)\ d\rho(t), \tag{14.4}$$

*where $\Gamma$ is the positively orientated boundary of the rectangle with corners at $A\pm i$ and $B\pm i$, and the integrals are absolutely convergent.*

*Proof.* According to Lemma 14.2 and the Cauchy theory of analytic functions

$$\frac{1}{2\pi i}\int_\Gamma (\Phi(\cdot,\lambda;f),g)_w\ d\lambda = \frac{1}{2\pi i}\int_\Gamma F(\lambda)\overline{G(\overline{\lambda})}m(\lambda)\ d\lambda, \tag{14.5}$$

if either of these two integrals exists. However, by (10.1),

$$\begin{aligned}&\frac{1}{2\pi i}\int_\Gamma F(\lambda)\overline{G(\overline{\lambda})}m(\lambda)\ d\lambda \\ &= \frac{1}{2\pi i}\int_\Gamma\left\{F(\lambda)\overline{G(\overline{\lambda})}\int_{(-\infty,+\infty)}\left(\frac{1}{t-\lambda}-\frac{t}{1+t^2}\right)d\rho(t)\right\}d\lambda.\end{aligned} \tag{14.6}$$

This double integral converges except possibly where $t=\lambda$; thus it is enough to verify existence of the integral

$$\frac{1}{2\pi}\int_{-1}^{1}\left\{\int_{[\mu-1,\mu+1]} F(\mu+is)\overline{G}(\mu-is)\frac{1}{t-\mu-is}d\rho(t)\right\}ds$$

for $\mu=A,B$. However, Lemma 14.3 assures the absolute convergence of this last integral.

Changing then the order of integration in (14.6) we obtain

$$\begin{aligned}&\frac{1}{2\pi i}\int_\Gamma F(\lambda)\overline{G(\overline{\lambda})}m(\lambda)\ d\lambda \\ &= \int_{(-\infty,+\infty)}\left\{\frac{1}{2\pi i}\int_\Gamma F(\lambda)\overline{G(\overline{\lambda})}\left(\frac{1}{t-\lambda}-\frac{t}{1+t^2}\right)d\lambda\right\}d\rho(t) \\ &= -\int_{[A,B]} F(t)\overline{G}(t)\ d\rho(t)\,.\end{aligned}$$

This last result follows since for any $t \in (A, B)$ the residue of the inner integral is $-F(t)\overline{G}(t)$, whereas the points $A, B$ do not carry any mass; also the inner integrand is regular for $t < A$ and $t > B$.

Thus the integral on the left-hand side of (14.5) exists and the required result (14.4) follows. □

## 15. Transform theory

We start now on the proof of the transform theory as given in Theorem 11.1.

**Lemma 15.1.** *Let $f, g \in L^2((a, b); w)$ be given with compact support in $(a, b)$; let $F, G$ be defined, respectively, as in Definition* 14.1. *Then*

(i) $F, G \in L^2(\mathbb{R}; \rho)$

(ii) $\sigma(f, g; t) = \int_{(-\infty,t]} F(s)\overline{G}(s)\ d\rho(s)$ *for a dense set of values of $t$ in $\mathbb{R}$*

(iii) $(f, g)_w = (F, G)_\rho$.

*Proof.* Let $A, B \in \mathbb{R}$, with $A < B$, be points of differentiability of $\rho(\cdot)$ and of continuity of $\sigma(f, g; \cdot)$; these points exist almost everywhere (Lebesgue) on $\mathbb{R}$. From (13.3) of Lemma 13.1 we have

$$(\Phi(\cdot, \lambda; f), g)_w = \int_{(-\infty,+\infty)} \frac{1}{t - \lambda} d\sigma(f, g; t) \text{ for all } \lambda \in \mathbb{C} \setminus \mathbb{R}.$$

Repeating the method used in Lemma 14.4 with the rectangular contour $\Gamma$, and inverting the integrals,

$$\begin{aligned} \frac{1}{2\pi i} \int_\Gamma (\Phi(\cdot, \lambda; f), g)_w \ d\lambda &= \frac{1}{2\pi i} \int_\Gamma \left\{ \int_{(-\infty,+\infty)} \frac{1}{t - \lambda} d\sigma(f, g; t) \right\} d\lambda \\ &= \int_{(-\infty,+\infty)} \left\{ \frac{1}{2\pi i} \int_\Gamma \frac{1}{t - \lambda} d\lambda \right\} d\sigma(f, g; t) \\ &= -\int_{[A,B]} d\sigma(f, g; t). \end{aligned}$$

From (14.4) we now obtain

$$\int_{[A,B]} d\sigma(f, g; t) = \int_{[A,B]} F(t)\overline{G}(t)\ d\rho(t)$$

for this dense set of points $A, B$. Letting $A \to -\infty$ we obtain, changing notations,

$$\int_{(-\infty,t]} d\sigma(f, g; s) = \lim_{\tau \to -\infty} \int_{[\tau,t]} F(s)\overline{G}(s)\ d\rho(s) \text{ for a dense set of } t \text{ in } \mathbb{R}. \quad (15.1)$$

Now take $g = f$ to give

$$\int_{(-\infty,t]} d\sigma(f, f; s) = \int_{(-\infty,t]} |F(s)|^2\ d\rho(s)$$

and

$$\int_{(-\infty,+\infty)} d\sigma(f,f;s) = \int_{(-\infty,+\infty)} |F(s)|^2 \ d\rho(s).$$

Thus $F \in L^2(\mathbb{R};\rho)$, and similarly we have $G \in L^2(\mathbb{R};\rho)$; this establishes item (i) of this theorem. Hence

$$\int_{(-\infty,t]} d\sigma(f,g;s) = \int_{(-\infty,t]} F(s)\overline{G}(s) \ d\rho(s) \text{ for a dense set of } t \text{ in } \mathbb{R}. \tag{15.2}$$

From (15.1) it now follows that item (ii) is established, *i.e.*, for a dense set of values of $t$ in $\mathbb{R}$

$$\sigma(f,g;t) = \int_{(-\infty,t]} d\sigma(f,g;s) = \int_{(-\infty,t]} F(s)\overline{G}(s) \ d\rho(s).$$

Finally, from (13.5), we have

$$(f,g)_w = \int_{(-\infty,+\infty)} d\sigma(f,g;s) = \int_{(-\infty,+\infty)} F(s)\overline{G}(s) \ d\rho(s) = (F,G)_\rho \tag{15.3}$$

to give item (iii) and to complete the proof of the lemma. □

**Remark 15.1.** Given $f \in L^2((a,b);w)$ with compact support the function $F \in L^2(\mathbb{R};\rho)$, as defined above, is called the generalized Fourier transform of $f$; in the next lemma this definition is extended to the case of any $f \in L^2((a,b);w)$ as required for Theorem 11.1.

**Lemma 15.2.** *Given $f,g \in L^2((a,b);w)$ the generalized Fourier transforms $F,G \in L^2(\mathbb{R};\rho)$ can be defined with the properties:*
(i) *$F,G \in L^2(\mathbb{R};\rho)$*
(ii) *$\sigma(f,g;t) = \int_{(-\infty,t]} F(s)\overline{G}(s) \ d\rho(s)$ for a dense set of values of $t$ in $\mathbb{R}$*
(iii) *$(f,g)_w = (F,G)_\rho$.*

*Proof.* Given any $f \in L^2((a,b);w)$ and any compact interval $[\alpha,\beta] \subset (a,b)$ define $f_{\alpha\beta} : (a,b) \to \mathbb{C}$ by

$$\begin{aligned} f_{\alpha\beta}(x) &:= f(x) \text{ for all } x \in [\alpha,\beta] \\ &:= 0 \text{ for all } x \in (a,b) \setminus [\alpha,\beta]; \end{aligned}$$

then $f_{\alpha\beta}$ has compact support and its transform $F_{\alpha\beta} : \mathbb{R} \to \mathbb{C}$ is defined as above by

$$F_{\alpha\beta}(t) := \int_a^b f_{\alpha\beta}(x)\varphi(x,t)w(x) \ dx = \int_\alpha^\beta f_{\alpha\beta}(x)\varphi(x,t)w(x) \ dx;$$

then $F_{\alpha\beta} \in L^2(\mathbb{R};\rho)$ and

$$\|F_{\alpha\beta}\|_\rho = \|f_{\alpha\beta}\|_w .$$

Given the interval $[\alpha',\beta']$ such that $[\alpha,\beta] \subset [\alpha',\beta'] \subset (a,b)$ then

$$\|F_{\alpha'\beta'} - F_{\alpha\beta}\|_\rho = \|f_{\alpha'\beta'} - f_{\alpha\beta}\|_w ;$$

if now we choose a family $\{[\alpha,\beta]\subset[\alpha',\beta']\}$ such that $\lim[\alpha,\beta]=(a,b)$ then, since $f\in L^2((a,b);w)$,

$$\lim\left\|F_{\alpha'\beta'}-F_{\alpha\beta}\right\|_\rho=\lim\left\|f_{\alpha'\beta'}-f_{\alpha\beta}\right\|_w=0$$

and there exists a unique element $F\in L^2(\mathbb{R};\rho)$ such that

$$\lim_{[\alpha,\beta]\to(a,b)}\left\|F-F_{\alpha\beta}\right\|_\rho=0.$$

This element $F\in L^2(\mathbb{R};\rho)$ is by definition the generalized transform of $f\in L^2((a,b);w)$; similarly for the pair $g,G$; item (i) of this lemma now follows.

Since $\sigma(f,g)(\cdot)$ is a bounded hermitian form on $L^2((a,b);w)$ item (ii) of this lemma follows now by continuity.

To prove item (iii) we have, from above and the result (15.3) for compact support functions,

$$(f,g)_w=\lim_{[\alpha,\beta]\to(a,b)}(f_{\alpha\beta},g_{\alpha\beta})_w=\lim_{[\alpha,\beta]\to(a,b)}(F_{\alpha\beta},G_{\alpha\beta})_\rho=(F,G)_\rho. \tag{15.4}$$

□

**Remark 15.2.** We note now that item 1 of Theorem 11.1 has been proved.

**Lemma 15.3.** *Let $G\in L^2(\mathbb{R};\rho)$; then the generalized inverse Fourier transform of $G$ is $g\in L^2((a,b);w)$ defined by*

$$\lim_{\tau\to+\infty}\int_a^b\left|g(x)-\int_{[-\tau,\tau]}G(t)\varphi(x,t)\ d\rho(t)\right|^2w(x)\ dx=0;$$

*moreover $\|g\|_w=\|G\|_\rho$.*

*If $G$ is the transform of $g_0\in L^2((a,b);w)$ then $g=g_0$ in $L^2((a,b);w)$.*

*The inverse transform $g$ is the null element of $L^2((a,b);w)$ if and only if $G$ is orthogonal in $L^2(\mathbb{R};\rho)$ to all generalized transforms from $L^2((a,b);w)$.*

*Proof.* Let $G\in L^2(\mathbb{R};\rho)$ have compact support in $(-\infty,+\infty)$; if $g:(a,b)\to\mathbb{C}$ is defined by

$$g(x):=\int_{(-\infty,+\infty)}G(t)\varphi(x,t)\ d\rho(t)\ \text{for all}\ x\in(a,b),$$

then $g$ is continuous on $(a,b)$.

Given $[\alpha,\beta]\subset(a,b)$ define $g_{\alpha\beta}:(a,b)\to\mathbb{C}$ by

$$\begin{aligned}g_{\alpha\beta}(x)&:=g(x)\ \text{for all}\ x\in[\alpha,\beta]\\&:=0\ \text{for all}\ x\in(a,b)\setminus[\alpha,\beta];\end{aligned}$$

then $g_{\alpha\beta}\in L^2((a,b);w)$ and has a transform $G_{\alpha\beta}\in L^2(\mathbb{R};\rho)$, where

$$G_{\alpha\beta}(t)=\int_\alpha^\beta g_{\alpha\beta}(x)\varphi(x,t)w(x)\ dx\ \text{for all}\ t\in(-\infty,+\infty).$$

We have, using the results in Lemma 15.2 and inverting the double integral by absolute convergence,

$$\begin{aligned}\int_{\alpha}^{\beta} |g(x)|^2 w(x)\ dx &= \int_{\alpha}^{\beta} \left\{ \int_{(-\infty,+\infty)} G(t)\varphi(x,t)\ d\rho(t) \right\} \overline{g}(x)w(x)\ dx \\ &= \int_{(-\infty,+\infty)} G(t) \left\{ \int_{\alpha}^{\beta} \overline{g}(x)\varphi(x,t)w(x)\ dx \right\} d\rho(t) \qquad (15.5) \\ &= \int_{(-\infty,+\infty)} G(t) \left\{ \int_{\alpha}^{\beta} \overline{g_{\alpha\beta}}(x)\varphi(x,t)w(x)\ dx \right\} d\rho(t) \\ &\le \|G\|_{\rho} \|G_{\alpha\beta}\|_{\rho} = \|G\|_{\rho} \|g_{\alpha\beta}\|_{w}\,.\end{aligned}$$

Now let $[\alpha,\beta] \to (a.b)$ to give $\|g\|_w \le \|G\|_\rho$ and so $g \in L^2((a,b);w)$.

If now $G \in L^2(\mathbb{R};\rho)$ is arbitrary, then for any $\tau \in (1,\infty)$ define $g_\tau : (a,b) \to \mathbb{C}$ by

$$g_\tau(x) := \int_{[-\tau,\tau]} G(t)\varphi(x,t)\ d\rho(t) \text{ for all } x \in (a,b). \qquad (15.6)$$

This integral exists since for each $x \in (a,b)$ the solution $\varphi(x,\cdot)$ is bounded on $[-\tau,\tau]$.

Regarding, in this definition (15.6), $G$ as having compact support in $[-\tau,\tau] \subset (-\infty,+\infty)$ it follows, from (15.5), that

$$\int_a^b |g_\tau(x)|^2 w(x)\ dx \le \int_{[-\tau,\tau]} |G(t)|^2\, d\rho(t) \le \|G\|_\rho^2 < +\infty;$$

thus $g_\tau \in L^2((a,b);w)$.

If now $\tau' \in (\tau,\infty)$ then a computation shows that

$$\frac{1}{2} \|g_{\tau'} - g_\tau\|_w^2 \le \int_{[-\tau',\tau]} |G(t)|^2\, d\rho(t) + \int_{[\tau,\tau']} |G(t)|^2\, d\rho(t).$$

Letting $\tau,\tau' \to +\infty$ shows that $\{g_\tau : \tau \in (0,+\infty)\}$ is a Cauchy family in $L^2((a,b);w)$; let $g \in L^2((a,b);w)$ be the limit in this space of this Cauchy family.

In passing we note that, from the definitions (15.6) and of $g$,

$$\lim_{\tau\to+\infty} \int_a^b \left| g(x) - \int_{[-\tau,\tau]} G(t)\varphi(x,t)\ d\rho(t) \right|^2 w(x)\ dx = 0$$

as required for (11.13) of Corollary 11.1.

If now $f$ is any element of $L^2((a,b);w)$ let $F$ be the generalized Fourier transform of $f$, as in Lemma 15.2. Then

$$\int_{[-\tau,\tau]} G(t)\left\{\int_\alpha^\beta \overline{f}(x)\varphi(x,t)w(x)\ dx\right\} d\rho(t)$$
$$= \int_\alpha^\beta \overline{f}(x)\left\{\int_{[-\tau,\tau]} G(t)\varphi(x,t)\ d\rho(t)\right\} w(x)\ dx,$$

since the integrals may be inverted by absolute convergence. In this last result let $\tau \to +\infty$ to give

$$\int_{(-\infty,+\infty)} G(t)\left\{\int_\alpha^\beta \overline{f}(x)\varphi(x,t)w(x)\ dx\right\} d\rho(t) = \int_\alpha^\beta \overline{f}(x)g(x)w(x)\ dx,$$

and then let $[\alpha,\beta] \to (a,b)$ to obtain

$$(G,F)_\rho = (g,f)_w \text{ for all } f \in L^2((a,b);w). \tag{15.7}$$

Here we have called upon dominated convergence in the two limit processes.

If, additionally, we suppose that $G$ is the transform of $g_0 \in L^2((a,b);w)$ then from (15.4) we have

$$(G,F)_\rho = (g_0,f)_w \text{ for all } \ f \in L^2((a,b);w). \tag{15.8}$$

Thus $(g-g_0,f)_w = 0$ for all $f \in L^2((a,b);w)$ and so $g_0 = g$ in $L^2((a,b);w)$. Similarly $G$ is orthogonal in $L^2(\mathbb{R};\rho)$ to all transforms $F$ if and only if $g = 0$. □

## 16. Unitary mappings

This section and succeeding sections are concerned with the proof of Corollaries 11.1 and 11.2.

Let $G \in L^2(\mathbb{R};\rho)$ and $f \in L^2((a,b);w)$; then in the notation of these corollaries we can write, from (15.7),

$$(f,VG)_w = (Uf,G)_\rho. \tag{16.1}$$

This shows that the generalized inverse Fourier transform

$$V : L^2(\mathbb{R};\rho) \to L^2((a,b);w)$$

is the adjoint operator of the generalized Fourier transform $U : L^2((a,b);w) \to L^2(\mathbb{R};\rho)$.

The basic difficulty that remains is to prove that these mappings $U$ and $V$ are one-to-one and inverses of each other.

**Lemma 16.1.** *Let $f \in L^2((a,b);w)$ and let $\lambda \in \mathbb{C}\setminus\mathbb{R}$; then the generalized Fourier transform $U(\Phi(\cdot,\lambda;f))$ of the resolvent function $\Phi(\cdot,\lambda;f)$ is given by*

$$U(\Phi(\cdot,\lambda;f))(t) = \frac{F(t)}{t-\lambda} \text{ for all } t \in \mathbb{R}, \tag{16.2}$$

*where $F = Uf$.*

*Proof.* From item (ii) of Lemma 15.2

$$\sigma(f, g; t) = \int_{(-\infty,t]} F(s)\overline{G}(s)\, d\rho(s) \tag{16.3}$$

for a dense set of values of $t$ in $\mathbb{R}$. From item (v) of Lemma 13.1, see (13.3),

$$(\Phi(\cdot, \lambda; f), g)_w = \int_{(-\infty,+\infty)} \frac{1}{t-\lambda} d\sigma(f, g; t) \text{ for all } f, g \in L^2((a,b); w). \tag{16.4}$$

From (16.3) and (16.4) (using Radon-Nikodym theory, see [18, Chapter 11, Section 6]) we obtain by setting $F_\lambda(t) := F(t)/(t-\lambda)$

$$(\Phi(\cdot, \lambda; f), g)_w = (F_\lambda, G)_\rho. \tag{16.5}$$

Using item (iii) of Lemma 15.2, (13.4), (16.4) and finally (16.3) in the case $g = f$, we have

$$\begin{aligned} \|U(\Phi(\cdot, \lambda; f))\|_\rho^2 = \|\Phi(\cdot, \lambda; f)\|_w^2 &= \frac{\operatorname{Im}((\Phi(\cdot, \lambda; f), f)_w)}{\operatorname{Im}(\lambda)} \\ &= \int_{(-\infty,+\infty)} \frac{1}{|t-\lambda|^2} d\sigma(f, f; t) = \|F_\lambda\|_\rho^2. \end{aligned}$$

It follows, using (16.5) with $g = \Phi(\cdot, \lambda; f)$, that

$$\|F_\lambda - U(\Phi(\cdot, \lambda; f))\|_\rho^2 = 0$$

and (16.2) is established. □

**Lemma 16.2.** *Let the mappings $U$ and $V$ be defined as in Corollary* 11.2*; then*
(i) *$U$ is defined on $L^2((a,b); w)$ and is onto $L^2(\mathbb{R}; \rho)$*
(ii) *$V$ is defined on $L^2(\mathbb{R}; \rho)$ and is onto $L^2((a,b); w)$*
(iii) *$V^{-1} = U$ and $U^{-1} = V$*
*(iv) $U$ and $V$ are unitary mappings.*

*Proof.* We need to show, following Lemma 15.3, that if $F \in L^2(\mathbb{R}; \rho)$ has a null inverse map in $L^2((a,b); w)$ then $F = 0$.

Now, according to Lemma 16.1, if $G$ is a transform then so is the function $(t-\lambda)^{-1}G(\cdot)$ for all $\lambda \in \mathbb{C}\backslash\mathbb{R}$. Thus if $F$ is orthogonal in $L^2(\mathbb{R}; \rho)$ to all transforms then so is $(t-\lambda)^{-1}F(\cdot)$ and so is $F(\cdot)\left((t-\lambda)^{-1} - (t-\overline{\lambda})^{-1}\right)$, both for all $\lambda \in \mathbb{C}\backslash\mathbb{R}$. Hence for all $\mu \in \mathbb{R}$ and all $\nu > 0$ the function, for all $t \in \mathbb{R}$,

$$t \longmapsto \frac{\nu F(t)}{(t-\mu)^2 + \nu^2}$$

is orthogonal to all transforms. Thus even the function

$$t \longmapsto F(t) \int_A^B \frac{\nu}{(t-\mu)^2 + \nu^2} d\mu \tag{16.6}$$

is orthogonal to all transforms for any compact interval $[A, B]$ and all $\nu > 0$, since the corresponding Riemann sums in forming the integral converge in $L^2(\mathbb{R}; \rho)$, being bounded by $F(\cdot)$. If now we choose $A$ and $B$ to be points of continuity of the monotonic non-decreasing function $\rho$ and let $\nu \to 0^+$ in (16.6) then we obtain in this limit, in the space $L^2(\mathbb{R};\rho)$, the function $F_{A,B}$ where

$$F_{A,B}(t) = \left\{ \begin{array}{ll} F(t) & \text{for all } t \in [A,B] \\ 0 & \text{for all } t \in \mathbb{R} \setminus [A,B]. \end{array} \right\}$$

The inverse transform of $F_{A,B}$ is thus the null function in $L^2((a,b);w)$ for almost all $A, B$ with respect to Lebesgue measure.

Since, by the existence theorem for solutions of the differential equation (1.1), the functions $\varphi(\cdot,t)$ and $(p\varphi')(\cdot,t)$ are both locally absolutely continuous on $(a,b)$, and locally uniformly so with respect to $t \in \mathbb{R}$, the two integrals

$$\int_A^B F(t)\varphi(\cdot,t)\ d\rho(t) \text{ and } \int_A^B F(t)(p\varphi')(\cdot,t)\ d\rho(t) \tag{16.7}$$

are both locally absolutely continuous on $(a,b)$, and identically zero for almost all $A, B \in \mathbb{R}$. Thus, using the symplectic form $[\cdot,\cdot]$ of (4.4), and $\{\gamma,\delta\}$ of Lemma 5.1, we have

$$\begin{aligned} 0 &= \left[\int_A^B F(t)\varphi(\cdot,t)\ d\rho(t), \cos(\boldsymbol{\alpha})\delta(\cdot) - \sin(\boldsymbol{\alpha})\gamma(\cdot)\right] \\ &= \int_A^B F(t)\left[\varphi(\cdot,t), \cos(\boldsymbol{\alpha})\delta(\cdot) - \sin(\boldsymbol{\alpha})\gamma(\cdot)\right] d\rho(t) = \int_A^B F(t)\ d\rho(t) \end{aligned}$$

for almost all $A, B \in \mathbb{R}$, thus showing that $F$ is the null element of $L^2(\mathbb{R};\rho)$.

This completes the proof of the lemma. □

## 17. Self-adjoint operator $T$

The self-adjoint operator $T$ in $L^2((a,b);w)$ is defined in Section 7 above, see (7.2) and (7.3); if $f \in D(T)$ then $f$ satisfies the separated boundary conditions

$$[f,\chi](a) \equiv [f,\varphi(\cdot,\lambda)](a) = 0 \text{ and } [f,\kappa](b) = 0. \tag{17.1}$$

**Lemma 17.1.** *Let $f \in D(T)$ with transform $F$. Then for some $\tilde{f} \in L^2((a,b);w)$ it follows that $M[f] = w\tilde{f}$; this implies that the transform of $\tilde{f}$ is $G \in L^2(\mathbb{R};\rho)$ where*

$$G(t) = tF(t) \text{ for } \rho\text{-almost all } t \in \mathbb{R}. \tag{17.2}$$

*Conversely, if $F$ and $G$ are both in $L^2(\mathbb{R};\rho)$ then $F$ is the transform of some $f \in D(T)$.*

*Proof.* The element $f$ satisfies the condition $f \in D(T)$ if and only if

$$f(x) = \Phi(x,\lambda;\tilde{f} - \lambda f) \text{ for all } x \in (a,b), \tag{17.3}$$

see (6) of Lemma 12.1; according to Lemmas 15.3 and 16.1 this holds if and only if

$$F(t) = \frac{\tilde{F}(t) - \lambda F(t)}{t - \lambda}, \text{ i.e., } \tilde{F}(t) = tF(t) \text{ for } \rho\text{-almost all } t \in \mathbb{R}.$$ □

## 18. The Green's function

We define the Green's function $g$ for the differential operator $T$ in $L^2((a,b);w)$, where

$$g : (a,b) \times (a,b) \times \mathbb{C} \setminus \mathbb{R} \to \mathbb{C}, \tag{18.1}$$

by

$$g(x,\xi;\lambda) := \left\{ \begin{array}{ll} \psi(x,\lambda)\varphi(\xi,\lambda) & a < \xi < x < b \\ \varphi(x,\lambda)\psi(\xi,\lambda) & a < x \leq \xi < b \end{array} \right\}. \tag{18.2}$$

From this definition it follows that, for all $x, \xi \in (a,b)$, all $\lambda, \mu \in \mathbb{C} \setminus \mathbb{R}$ and all $f \in L^2((a.b);w)$

(i) $$\Phi(x,\lambda;f) = \int_a^b f(\xi)g(x,\xi;\lambda)w(\xi)\, d\xi = (f, \overline{g(x,\cdot;\lambda)})_w = (f, g(x,\cdot;\overline{\lambda}))_w \tag{18.3}$$

(ii) $$g(x,\cdot;\lambda) \in L^2((a,b);w) \tag{18.4}$$

(iii) $$g(x,\xi;\lambda) = \overline{g(\xi,x;\overline{\lambda})} \tag{18.5}$$

(iv) from items 7 and 8 of Lemma 12.1

$$\begin{aligned} g(x,\xi;\lambda) - g(x,\xi;\mu) &= (\lambda - \mu) \int_a^b g(x,\zeta;\lambda)g(\zeta,\xi;\mu)w(\zeta)\, d\zeta \\ &= (\lambda - \mu)(g(x,\cdot;\lambda), \overline{g(\cdot,\xi;\mu)})_w. \end{aligned} \tag{18.6}$$

**Lemma 18.1.** *We have*

(i) *The transform of the Green's function $g(x,\cdot;\lambda)$ is given, for all $x \in (a,b)$ and all $\lambda \in \mathbb{C} \setminus \mathbb{R}$, by*

$$G(x,t;\lambda) = \frac{\varphi(x,t)}{t - \overline{\lambda}} \text{ for almost all } t \in \mathbb{R}. \tag{18.7}$$

(ii) *The transform of the solution $\psi(\cdot,\lambda)$ is given, for all $\lambda \in \mathbb{C} \setminus \mathbb{R}$, by*

$$\Psi(t,\lambda) = \frac{1}{t - \lambda} \text{ for all } t \in \mathbb{R}. \tag{18.8}$$

*Proof.* We have

$$\Phi(x,\lambda;f) = (f, g(x,\cdot;\overline{\lambda}))_w = (F(\cdot), G(x,\cdot;\overline{\lambda}))_\rho \tag{18.9}$$

by the formula of Parseval. On the other hand the Fourier transform of $\Phi(\cdot,\lambda;f)$ is $F(t)/(t-\lambda)$ so that

$$\lim_{\tau \to +\infty} \int_{-\tau}^{+\tau} \frac{F(t)}{t - \lambda} \varphi(\cdot,t)\, d\rho(t) = \Phi(\cdot,\lambda;f) \tag{18.10}$$

with mean convergence; thus with ordinary convergence almost everywhere on $(a,b)$ as $\tau \to +\infty$, for some infinite sequence. It follows that

$$G(x,t;\overline{\lambda}) = \frac{\varphi(x,t)}{t - \lambda}, \tag{18.11}$$

and incidentally that $\varphi(x,\cdot)/(\cdot - \lambda) \in L^2(\mathbb{R};\rho)$, both for all $x \in (a,b)$.

Since, for all $x \in (a,b)$,

$$p(x)\frac{\partial \Phi}{\partial x}(x,\lambda;f) = \int_a^b f(\xi)p(x)\frac{\partial g}{\partial x}(x,\xi;\lambda)w(\xi)\ d\xi \tag{18.12}$$

we obtain similarly

$$p(x)\frac{\partial G}{\partial x}(x,t;\lambda) = \frac{(p\varphi')(x,t)}{t-\overline{\lambda}}. \tag{18.13}$$

Thus, from the definition (18.2) of $g$ and the representation (18.7) of its generalized transform $G$, the right-hand side of (18.13) is also in the space $L^2(\mathbb{R};\rho)$ for all $x \in (a,b)$.

In particular, the generalized Fourier transform of

$$\psi(x,\lambda) = \lim_{\xi \to a}[g(\xi,x;\lambda), \cos(\boldsymbol{\alpha})\delta(\xi) - \sin(\boldsymbol{\alpha})\gamma(\xi)] \tag{18.14}$$

is given by, for all $t \in \mathbb{R}$,

$$\begin{aligned} \Psi(t,\lambda) &= \lim_{\xi \to a}[G(\xi,t;\lambda), \cos(\boldsymbol{\alpha})\delta(\xi) - \sin(\boldsymbol{\alpha})\gamma(\xi)] \\ &= \lim_{\xi \to a}\frac{1}{t-\lambda}[\varphi(\xi,\lambda), \cos(\boldsymbol{\alpha})\delta(\xi) - \sin(\boldsymbol{\alpha})\gamma(\xi)] \\ &= \frac{1}{t-\lambda}. \end{aligned} \tag{18.15}$$

□

To prove Corollary 18.1, as given below, we require

**Lemma 18.2.** *If the sequence* $\{f_j \in L^2((a,b);w) : j \in \mathbb{N}\}$ *converges to* $f$ *in* $L^2((a,b);w)$ *then the two following sequences converge locally uniformly in* $(a,b)$ *and* $\mathbb{C}$ *to the limits shown*

$$\lim_{j\to\infty} \Phi(x,\lambda;f_j) = \Phi(x,\lambda;f) \ \text{and} \ \lim_{j\to\infty} (p\Phi')(x,\lambda;f_j) = (p\Phi')(x,\lambda;f). \tag{18.16}$$

*Proof.* A direct calculation shows that $\|g(x,\cdot;\lambda)\|_w$ is locally uniformly bounded in $(a,b)$ and $\mathbb{C}$. Thus

$$|\Phi(x,\lambda;f_j) - \Phi(x,\lambda;f)| = |\Phi(x,\lambda;f_j - f)| \le \|f_j - f\|_w \|g(x,\cdot;\lambda)\|_w$$

from which result the first sequential convergence in (18.16) follows.

The second sequential convergence in (18.16) follows from a similar inequality. □

**Corollary 18.1.** *If, see Section* 7, $f \in D(T)$ *and so satisfies the boundary conditions* (7.3), *and if* $M[f] = w\tilde{f}$ *with* $\tilde{f} \in L^2(\mathbb{R};\rho)$, *then the generalized inverse transform of* $F$, *see Remark* 11.1, *converges absolutely and locally uniformly in* $(a,b)$ *to* $f$, *i.e.,*

$$\begin{aligned} f(x) &= \lim_{\tau\to+\infty}\int_{[-\tau,\tau]} F(t)\varphi(x,t)\ d\rho(t) \ \text{for all } x \in (a,b) \\ &= \int_{(-\infty,+\infty)} F(t)\varphi(x,t)\ d\rho(t) \ \text{for all } x \in (a,b). \end{aligned} \tag{18.17}$$

*Proof.* From Corollary 11.1 and the theory of mean convergence in $L^2((a,b);w)$, we obtain

$$f(x) = \lim_{\tau \to +\infty} \int_{[-\tau,\tau]} F(t)\varphi(x,t)\ d\rho(t)$$

with convergence in $\mathbb{C}$ for almost all (Lebesgue) $x \in (a,b)$.

Thus, since $f$ is continuous on $(a,b)$, it is sufficient to prove that $F(\cdot)\varphi(x,\cdot) \in L^1(\mathbb{R};\rho)$ for all $x \in (a,b)$; this last result follows from, for all $\lambda \in \mathbb{C} \setminus \mathbb{R}$,

$$\begin{aligned}\lim_{\tau \to +\infty} \int_{[-\tau,\tau]} F(t)\varphi(x,t)\ d\rho(t) &= \lim_{\tau \to +\infty} \int_{[-\tau,\tau]} [(t-\lambda)F(t)] \left[\frac{\varphi(x,t)}{t-\lambda}\right]\ d\rho(t) \\ &= \int_{(-\infty,+\infty)} \left[\tilde{F}(t) - \lambda F(t)\right] G(x,t;\lambda)\ d\rho(t),\end{aligned}$$

and the last integral is absolutely convergent.

Now, since $f \in D(T)$ it follows that, for all $\lambda \in \mathbb{C} \setminus \mathbb{R}$,

$$f(x) = \Phi(x,\lambda; Tf - \lambda f) \text{ for all } x \in (a,b).$$

Setting

$$G = U(Tf - \lambda f),$$

so that $G \in L^2(\mathbb{R};\rho)$, and defining the sequence $\{G_j \in L^2(\mathbb{R};\rho) : j \in \mathbb{N}\}$ by

$$G_j(t) = \left\{ \begin{array}{ll} G(t) & \text{for all } t \in [-j,j] \\ 0 & \text{for all } t \in \mathbb{R} \setminus [-j,j] \end{array} \right\}$$

we have $\lim_{j\to\infty} G_j = G$ in $L^2(\mathbb{R};\rho)$; hence, defining $g_j := V(G_j)$ for all $j \in \mathbb{N}$, we have $\lim_{j\to\infty} g_j = Tf - \lambda f$ in the space $L^2((a,b);w)$.

From Lemma 18.2 it now follows that

$$\lim_{j\to\infty} \int_{[-j,j]} F(t)\varphi(x,t)\ d\rho(t) = f(x)$$

with local uniform convergence on $(a,b)$, as required.

This result completes the proof of the corollary. □

## 19. A property of the $m$-coefficient

A final result in this account is, from the results in Section 18,

**Corollary 19.1.** *In the Nevanlinna representation of the m-coefficient, see* (10.1) *of Section* 10, *the coefficient* $B = 0$, *i.e.*,

$$m(\lambda) = A + \int_{(-\infty,+\infty)} \left\{\frac{1}{t-\lambda} - \frac{t}{t^2+1}\right\} d\rho(t) \text{ for all } \lambda \in \mathbb{C} \setminus \mathbb{R}. \tag{19.1}$$

*Proof.* We have

$$B + \int_{(-\infty,+\infty)} \frac{1}{|t-\lambda|^2} d\rho(t) = \frac{\operatorname{Im}(m(\lambda))}{\operatorname{Im}(\lambda)} = \|\psi(\cdot,\lambda)\|_w^2 = \left\|\frac{1}{t-\lambda}\right\|_\rho^2 \tag{19.2}$$

to give $B = 0$. □

**Acknowledgements**

Both authors extend personal thanks to Werner Amrein and to David Pearson for advice and help in the preparation of this manuscript.

Norrie Everitt extends his gratitude to his co-author Christer Bennewitz for collaboration down the years since 1974.

## References

[1] N.I. Akhiezer, *The classical moment problem*, Oliver and Boyd, Edinburgh, 1965. Translated from the Russian edition of 1961.

[2] N.I. Akhiezer and I.M. Glazman, *Theory of linear operators in Hilbert space*: I & II, Pitman, London and Scottish Academic Press, Edinburgh, 1981.

[3] R.R. Ashurov and W.N. Everitt, Linear quasi-differential operators in locally integrable spaces on the real line, Proc. Roy. Soc. Edinburgh A **130** (2000), 671–698.

[4] F.V. Atkinson, *Discrete and continuous boundary problems*, Academic Press, New York, 1964.

[5] F.V. Atkinson, On bounds for the Titchmarsh-Weyl $m$-coefficients and for spectral functions for second-order differential equations, Proc. Royal Soc. Edinburgh A **97** (1984), 1–7.

[6] C. Bennewitz, The Titchmarsh eigenfunction expansion theory and the $m$-coefficient, unpublished manuscript, University of Birmingham, 1983.

[7] C. Bennewitz and W.N. Everitt, Some remarks on the Titchmarsh-Weyl $m$-coefficient, in *Proceedings of the Pleijel Conference, University of Uppsala, 1979*, 49–108, published by the Department of Mathematics, University of Uppsala, Sweden.

[8] E.A. Coddington and N. Levinson, *Theory of ordinary differential equations*, McGraw-Hill, New York, 1955.

[9] W.N. Everitt, On the transformation theory of ordinary second-order linear symmetric differential equations, Czechoslovak Mathematical Journal **32** (107) (1982), 275–306.

[10] W.N. Everitt, Some remarks on the Titchmarsh-Weyl $m$-coefficient and associated differential operators, in *Differential Equations, Dynamical Systems and Control Science: A festschrift in Honor of Lawrence Markus*, 33–53, edited by K.D. Elworthy, W.N. Everitt, E.B. Lee, Lecture Notes in Pure and Applied Mathematics **152**, Marcel Dekker, New York, 1994.

[11] W.N. Everitt, Charles Sturm and the Development of Sturm-Liouville Theory in the Years 1900 to 1950, in this volume.

[12] W.N. Everitt, W.K. Hayman and G. Nasri-Roudsari, On the representation of holomorphic functions by integrals, Applicable Analysis **65** (1997), 95–102.

[13] W.N. Everitt and L. Markus, The Glazman-Krein-Naimark theorem for ordinary differential operators, Operator Theory: Advances and Applications **98** (1997), 118–130.

[14] W.N. Everitt, G. Schöttler and P.L. Butzer, Sturm-Liouville boundary value problems and Lagrange interpolation series, Rendiconti di Matematica VII **14** (1994), 87–126.

[15] W.N. Everitt, C. Shubin, G. Stolz and A. Zettl, Sturm-Liouville problems with an infinite number of interior singularities, in *Spectral theory and computational methods of Sturm-Liouville problems*, 211–249, Lecture Notes in Pure and Applied Mathematics **191**, Marcel Dekker, New York, 1997.

[16] T. Kato, *Perturbation theory for linear operators*, Springer-Verlag, Heidelberg, 1980.

[17] B.M. Levitan and I.S. Sargsjan, *Sturm-Liouville and Dirac operators*, Nauka, Moscow, 1988. Translated from the Russian *Mathematics and its Applications*, Soviet Series **59**, Kluwer Academic Publishers Group, Dordrecht, 1991.

[18] H.L. Royden, *Real analysis*, Prentice Hall, Englewood Cliffs, New Jersey, Third edition, 1987.

[19] M.H. Stone, *Linear transformations in Hilbert space*, American Mathematical Society Colloquium Publications **15**, American Mathematical Society, Providence, Rhode Island, 1932.

[20] M.A. Naimark, *Linear differential operators II*, translated from the second Russian edition, Ungar, New York, 1968.

[21] E.C. Titchmarsh, *The theory of functions*, Oxford University Press, second edition, 1952.

[22] E.C. Titchmarsh, *Eigenfunction expansions I*, Oxford University Press, second edition, 1962.

[23] H. Weyl, Über gewöhnliche Differentialgleichungen mit Singularitäten und die zugehörigen Entwicklungen willkürlicher Funktionen, Math. Ann. **68** (1910), 220–269.

Christer Bennewitz
Department of Mathematics
University of Lund
P.O. Box 118
SE-221 00, Lund, Sweden
e-mail: `christer.bennewitz@math.lu.se`

W. Norrie Everitt
School of Mathematics and Statistics
University of Birmingham
Edgbaston
Birmingham B15 2TT, England, UK
e-mail: `w.n.everitt@bham.ac.uk`

*W.O. Amrein, A.M. Hinz, D.B. Pearson*
Sturm-Liouville Theory: Past and Present, 173–199


# Sturm's Theorems on Zero Sets in Nonlinear Parabolic Equations

Victor A. Galaktionov and Petra J. Harwin

**Abstract.** We present a survey on applications of Sturm's theorems on zero sets for linear parabolic equations, established in 1836, to various problems including reaction-diffusion theory, curve shortening and mean curvature flows, symplectic geometry, etc. The first Sturm theorem, on nonincrease in time of the number of zeros of solutions to one-dimensional heat equations, is shown to play a crucial part in a variety of existence, uniqueness and asymptotic problems for a wide class of quasilinear and fully nonlinear equations of parabolic type. The survey covers a number of the results obtained in the last twenty-five years and establishes links with earlier ones and those in the ODE area.

**Mathematics Subject Classification (2000).** 35K55.

**Keywords.** Sturm theorems, multiple zeros, nonlinear parabolic equations, intersection comparison, blow-up, asymptotic behavior.

## 1. Introduction: Sturm's theorems for parabolic equations

In 1836 C. Sturm published two celebrated papers in the first volume of J. Liouville's Journal de Mathématique Pures et Appliquées. The first paper [125] on zeros of solutions $u(x)$ of second-order ordinary differential equations such as

$$u'' + q(x)u = 0, \quad x \in \mathbb{R}, \tag{1.1}$$

very quickly exerted a great influence on the general theory of ODEs. Then and nowadays Sturm's oscillation, comparison and separation theorems can be found in most textbooks on ODEs with various generalizations to other equations and systems of equations. In general, such theorems classify and compare zeros and zero sets $\{x \in \mathbb{R} : u(x) = 0\}$ of different solutions $u_1(x)$ and $u_2(x)$ of (1.1) or solutions of equations with different continuous ordered potentials $q_1(x) \geq q_2(x)$. We refer to other papers of the present volume containing a detailed survey of this classical theory.

---

Research supported by RTN network HPRN-CT-2002-00274.

The second paper [126] was devoted to the *evolution* analysis of zeros and zero sets $\{x : u(x,t) = 0\}$ for solutions $u(x,t)$ of partial differential equations of parabolic type, for instance,

$$u_t = u_{xx} + q(x)u, \quad x \in [0, 2\pi], \; t > 0, \tag{1.2}$$

with the same ordinary differential operator as in (1.1) and the Dirichlet boundary condition $u = 0$ at $x = 0$ and $x = 2\pi$ and given smooth initial data at $t = 0$. Two of Sturm's results on PDEs like (1.2) can be stated as follows:

**First Sturm Theorem**: nonincrease with time of the number of zeros (or sign changes) of solutions;

**Second Sturm Theorem**: a classification of blow-up self-focusing formations and collapses of multiple zeros.

We will refer to both of Sturm's Theorems together as the *Sturmian argument* on zero set analysis. Most of Sturm's PDE paper [126] was devoted to the second Theorem on striking evolution "dissipativity" properties of zeros of solutions of linear parabolic equations, where a detailed backward-forward continuation analysis of the collapse of multiple zeros of solutions was performed. The first Theorem was formulated as a consequence of the second one (it is a form of the strong Maximum Principle (MP) for parabolic equations). As a by-product of the first Theorem, Sturm presented an evolution proof of bounds on the number on zeros of eigenfunction expansions. For finite Fourier series

$$f(x) = \sum_{L \le k \le M} (a_k \cos kx + b_k \sin kx), \quad x \in [0, 2\pi], \tag{1.3}$$

by using the PDE (1.2), $q \equiv 0$ (with periodic boundary conditions), it was proved that $f(x)$ has at least $2L$ and at most $2M$ zeros.[1] Sometimes the lower bound on zeros is referred to as the Hurwitz Theorem, which was better known than the first Sturm PDE Theorem. This Sturm-Hurwitz Theorem is the origin of many striking results, ideas and conjectures in topology of curves and symplectic geometry.

Unlike the classical Sturm theorems on zeros of solutions of second-order ODEs, Sturm's evolution zero set analysis for parabolic PDEs did not attract much attention in the nineteenth century and, in fact, was forgotten for almost a century. It seems that G. Pólya (1933) [112] was the first person in the twentieth century to revive interest in the first Sturm Theorem for the heat equation. (The earlier extension by A. Hurwitz (1903) [71] of Sturm's result on zeros of (1.3) to infinite Fourier series with $M = \infty$ did not use PDEs.) Since the 1930s the Sturmian argument has been rediscovered in part several times. For instance, a key idea of the Lyapunov monotonicity analysis in the famous KPP-problem, by A.N. Kolmogorov, I.G. Petrovskii and N.S. Piskunov (1937) [82] on the stability of travelling waves (TWs) in reaction-diffusion equations, was based on the first Sturm Theorem in a simple geometric configuration with a single intersection between solutions. This was separately proved there by the Maximum Principle.

[1] Sturm also presented an ODE proof.

From the 1980s the Sturmian argument for PDEs began to penetrate more and more into the theory of linear and nonlinear parabolic equations and was found to have several fundamental applications. These include asymptotic stability theory for various nonlinear parabolic equations, orbital connections and transversality of stable-unstable manifolds for semilinear parabolic equations such as Morse-Smale systems, unique continuation theory, Floquet bundles and a Poincaré-Bendixson theorem for parabolic equations and problems of symplectic geometry and curve shortening flows. A survey on Sturm's ideas in PDEs will be continued in Section 2, where we present the statements of both of Sturm's Theorems, and in Section 3, where we describe further related results and generalizations achieved in the twentieth century.

## 2. Sturm's theorems for linear parabolic equations

### 2.1. First Sturm Theorem: nonincrease of the number of sign changes

Let $D$ and $J$ be open bounded intervals in $\mathbb{R}$. Consider in $S = D \times J$ the linear parabolic equation

$$u_t = a(x,t)u_{xx} + b(x,t)u_x + c(x,t)u. \tag{2.1}$$

Given a constant $\tau \in J$, we denote the parabolic boundary of the domain $S_\tau = S \cap \{t < \tau\}$, i.e., the lateral sides and the bottom of the boundary of $S_\tau$, by $\partial S_\tau$. Given a solution $u$ defined on $S_\tau$, the positive and negative sets of $u$ are defined as follows:

$$U^+ = \{(x,t) \in S_\tau : \ u(x,t) > 0\}, \quad U^- = \{(x,t) \in S_\tau : \ u(x,t) < 0\}. \tag{2.2}$$

A *component* of $U^+$ (or $U^-$) is a maximal open connected subset of $U^+$ (or $U^-$).

Given a $t \in \overline{J}$, the number (finite or infinite) of components of $\{x \in D : u(x,t) \neq 0\}$ minus one is called the *number of sign changes* of $u(x,t)$ and is denoted by $Z(t,u)$. Alternatively, let $K$ be the supremum over all natural numbers $k$ such that there exist $k$ points from $D$, $x_1 < x_2 < \cdots < x_k$, satisfying

$$u(x_j,t) \cdot u(x_{j+1},t) < 0 \quad \text{for all } j = 1,2,\ldots,k-1,$$

then $Z(t,u) = K - 1$.

**Theorem 2.1 (First Sturm Theorem: sign changes).** *Let $a$, $b$, $c$ be continuous, bounded and $a \geq \mu > 0$ in $S$ for some constant $\mu$. Let $u(x,t)$ be a solution of* (2.1) *in $S$ that is continuous on $\overline{S}$.*

(i) *Suppose that on $\partial S_\tau$ there are precisely $n$ (respectively $m$) disjoint intervals where $u$ is positive (respectively negative). Then $U^+$ (resp. $U^-$) has at most $n$ (resp. $m$) components in $S_\tau$ and the closure of each component must intersect $\partial S_\tau$ in at least one interval.*

(ii) *The number of sign changes $Z(\tau,u)$ of $u(x,\tau)$ on $D$ is not greater than the number of sign changes of $u$ on $\partial S_\tau$.*

The first Sturm Theorem is formulated on p. 431 in [126]. The present proof of Theorem 2.1 is taken from [118] (similar to that in [105]).

*Proof.* The proof is based on the strong Maximum Principle.
(i) Let $I \subset \partial S_\tau$ be a maximal interval where $u > 0$. Suppose that two open connected subsets $F_1$, $F_2 \subset U^+$ intersect $\partial S_\tau$ in disjoint open intervals $I_1$, $I_2 \subset I$. Since $u$ is continuous in $\overline{S}_\tau$, there exists an open set $G \subset U^+$ whose closure in $\overline{S}_\tau$ contains $I$. Then $G$ must contain points of both $F_1$ and $F_2$ so that these must belong to the same open component of $U^+$. Thus, at most one component of $U^+$ intersects each of the $n$ open intervals on $\partial S_\tau$ where $u > 0$. The same result holds for the components of $U^-$. Therefore, it suffices to show that every component of $U^+$ (or $U^-$) intersects $\partial S_\tau$ in one or more intervals.

We can assume that $c \leq 0$ in $S_\tau$. Otherwise, we set $u = e^{\lambda t} v$ ($U^\pm$ stay the same for $v$), where $v$ then solves equation (2.1) with the last coefficient $c$ on the right-hand side replaced by $c - \lambda$ and we can choose the constant $\lambda \geq \sup c$.

Let $F \subset U^+$ be a component in $S_\tau$. Since $u$ is continuous, it must attain a positive maximum on $\overline{F}$. Then $c \leq 0$ implies $u_t \leq a u_{xx} + b u_x$ in $F$, and, by continuity, $u = 0$ at any boundary point of $F$ which is interior to $S_\tau$. By the MP, $u$ cannot attain its maximum at an interior point of $F$ or on the line $\{t = \tau\}$. Hence, $F$ must have a boundary point $Q \in \partial S_\tau$ such that $u(Q) > 0$ and by continuity $u$ is positive in an interval of $\partial S_\tau$ about $Q$.
(ii) is a straightforward consequence of (i). □

The first Sturm Theorem is true for wider classes of linear parabolic equations that are sufficiently regular (so the strong MP can be applied). An important example is the radial parabolic equation in $\mathbb{R}^N$ with continuous coefficients and $a \geq \mu > 0$,

$$u_t = a(r,t)\Delta u + b(r,t)u_r + c(r,t)u, \tag{2.3}$$

where $r = |x| \geq 0$ denotes the radial variable and $\Delta = \frac{\mathrm{d}^2}{\mathrm{d}r^2} + \frac{N-1}{r}\frac{\mathrm{d}}{\mathrm{d}r}$ is the radial Laplace operator. Bearing in mind that we consider smooth bounded solutions satisfying the symmetry condition at the origin, $u_r(0,t) = 0$ for $t \in J$, the MP applies to equation (2.3) in $S = D \times J$, where $D = \{r < R\}$ is a ball in $\mathbb{R}^N$, and the first Sturm Theorem holds.

### 2.2. Second Sturm Theorem: formation and collapse of multiple zeros

*Results in the class of analytic functions.* We consider parabolic equations with analytic coefficients admitting analytic solutions. Then any zero of $u(x,t)$ has finite multiplicity. Under this assumption, the following result is true:

**Theorem 2.2 (Second Sturm Theorem: multiple zeros).** *Let $O = (0,0) \in S$ and $u \in C^\infty(S) \cap C(\overline{S})$ be a solution of equation (2.1) with $C^\infty$-coefficients $a, b, c$, where $a \geq \mu > 0$ in $S$. Assume that $u(x,t)$ does not change sign on the lateral boundary of $S$, and $u(x,0)$ has a zero of order $m \geq 2$ at the origin $x = 0$, i.e.,*

$$D_x^k u(0,0) = 0 \quad \textit{for } k = 0, 1, \ldots, m-1 \;\; \textit{and } D_x^m u(0,0) = m!A \neq 0. \tag{2.4}$$

*Then $Z(t,u)$ decreases at $t = 0$, and for any $t_1 < 0 < t_2$ near $t = 0$, there holds*

$$Z(t_1,u) - Z(t_2,u) \geq \{m, \text{ if } m \text{ is even}; \quad m-1, \text{ if } m \text{ is odd}\}. \tag{2.5}$$

In the proof of Theorem 2.2 we will follow Sturm's original computations and analysis in [126], pp. 417–427, which was done for the following semilinear parabolic equation on a bounded interval

$$gu_t = (ku_x)_x - lu, \quad x \in (\mathrm{x}, \mathrm{X}),\ t > 0, \tag{2.6}$$

with smooth functions $g, k$ and $l$ depending on $x$ and $t$. The main calculations were performed for $g, k, l$ depending on $x$ only. A comment on p. 431 extends the results to allow dependence on $t$. Third type (Robin) boundary conditions were incorporated:

$$ku_x - hu = 0 \quad \text{at } x = \mathrm{x}, \quad ku_x + Hu = 0 \quad \text{at } x = \mathrm{X}, \tag{2.7}$$

where $h, H$ are constants but also can depend on $t$, see p. 431. (Zero Dirichlet boundary conditions are also mentioned there.) Sturm's analysis on pp. 428–430 includes the case of multiple zeros occurring at boundary points x or X.

*Proof.* By Taylor's formula near the origin we have

$$u(x,0) = Ax^m + O(x^{m+1}). \tag{2.8}$$

Using a Taylor expansion in $t$, we have

$$u(x,t) = u(x,0) + u_t(x,0)t + \tfrac{1}{2!}\, u_{tt}(x,0)t^2 + \cdots + \tfrac{1}{n!}\, D_t^n u(x,0)t^n + O(t^{n+1}), \tag{2.9}$$

where $n = m/2$ if $m$ is even and $n = (m-1)/2$ if $m$ is odd. Let us estimate the coefficients. Let $d_j = m!/(m-2j)!$ for $j = 0, 1, \ldots, n$. It follows from the parabolic equation (2.1) and (2.8) that $u_t(x,0) = a(x,0)u_{xx}(x,0) + b(x,0)u_x(x,0) + c(x,0)u(x,0) = a_0 A d_1 x^{m-2} + O(x^{m-1})$, where $a_0 = a(0,0)$ and $a(x,0) = a_0 + O(x)$. Differentiating the equation and using expansion (2.8) again, we obtain, keeping the leading terms only,

$$u_{tt}(x,0) = au_{txx} + \cdots = a_0^2 A d_2 x^{m-4} + O(x^{m-3}),$$

and finally $D_t^n u(x,0) = D_t^{n-1} a_0 u_{xx}(0,0) + \cdots = a_0^n A d_n x^{m-2n} + O(x^{m-2n+1})$. The Taylor expansion in both independent variables, $x$ and $t$, takes the form

$$u(x,t) = A(x^m + a_0 d_1 x^{m-2} t + \tfrac{1}{2!}\, a_0^2 d_2 x^{m-4} t^2 + \cdots + \tfrac{1}{n!}\, a_0^n d_n x^{m-2n} t^n) + O(\cdot) \tag{2.10}$$

with the remainder $O(\cdot) = O(|x|^{m+1} + |x|^{m-1}|t| + \cdots + |x|^{m-2n+1}|t|^n + |t|^{n+1})$.

**(i) Backward continuation.** Consider the behavior for $t \approx 0^-$. The dimensional structure of the right-hand side of (2.10) suggests rewriting this expansion in terms of the rescaled *Sturm backward continuation variable*

$$z = x/\sqrt{a_0(-t)} \quad \text{for } t < 0. \tag{2.11}$$

Substituting $x = z\sqrt{a_0(-t)}$, we obtain that

$$A^{-1} a_0^{-m/2} (-t)^{-m/2} u(x,t) = P_m(z) + O((-t)^{1/2}(1 + |z|^{m+1})), \tag{2.12}$$

where $P_m(z) = \sum_{j=0}^{n}(-1)^j \frac{d_j}{j!} z^{m-2j}$. The $m$th order polynomial $P_m(z)$ is the Hermite polynomial $H_m(z)$ (up to a constant multiplier which we omit in what follows). Each orthogonal polynomial $H_m(z)$ has exactly $m$ simple zeros $\{z_i,\ i = 1, \ldots, m\}$ with $H_m'(z_i) \neq 0$. Sturm proved this separately on p. 426. This is the classical theory of orthogonal polynomials, see G. Szegö's book [128], Chapter 6.

A similar expansion for the derivative $u_x(x,t)$ shows that (2.12) can be differentiated in $x$ giving the derivative $P_m'(z)$ in the right-hand side. It follows from the expansions of $u(x,t)$ and $u_x(x,t)$ near the multiple zero that for any $t \approx 0^-$, the solution $u(x,t)$ has $m$ simple zeros $\{x_i(t),\ i = 1, \ldots, m\}$, $u_x(x_i(t),t) \neq 0$, with the following asymptotic behavior: $x_i(t) = z_i(-t)^{1/2} + O(-t) \to 0$ as $t \to 0^-$, so that exactly $m$ smooth zero curves intersect each other at the origin $(0,0)$.

**(ii) Forward continuation.** Following Sturm's analysis, we consider the behavior of the solution $u(x,t)$ as $t \to 0^+$. Introducing the heat kernel rescaled variable of the *forward continuation*

$$z = x/\sqrt{a_0 t} \quad \text{for } t > 0, \tag{2.13}$$

instead of (2.12) we obtain another polynomial on the right-hand side

$$A^{-1} a_0^{-m/2} t^{-m/2} u(x,t) = Q_m(z) + O(t^{1/2}(1 + |z|^{m+1})), \tag{2.14}$$

where $Q_m(z) = \sum_{j=0}^{n} \frac{d_j}{j!} z^{m-2j}$. The $m$th order polynomial $Q_m(z)$ has positive coefficients. If $m$ is odd, then it is strictly increasing with $Q_m(0) = 0$. If $m$ is even, then it has a single positive minimum at $z = 0$. Therefore, (2.14) implies that for small $t > 0$ on compact subsets $\{|x| \leq ct^{1/2}\}$ with any $c > 0$, the solution $u(x,t)$ has a unique simple zero $\tilde{x}_1(t) = O(t)$ if $m$ is odd, and no zeros if $m$ is even. This is Sturm's analysis on p. 423.

In order to complete the proof, it suffices to observe that if $m$ is even and, say, $A > 0$, by continuity and the strong MP, there exists a small interval $(-\varepsilon, \varepsilon)$ such that $u(x,t)$ becomes strictly positive on $(-\varepsilon, \varepsilon)$ for all small $t > 0$. This means that at least $m$ zero curves disappear at $(0,0)$. If $m$ is odd and $A > 0$, then applying Theorem 2.1 to the domain $S = (-\varepsilon, \varepsilon) \times (0, \varepsilon)$ we have that on $(-\varepsilon, \varepsilon)$ for $t > 0$ there exists a unique continuous curve of simple zeros $\tilde{x}_1(t)$ starting from $(0,0)$. In this case at least $m - 1$ zero curves disappear at the origin as $t \to 0^-$. □

Such a complete analysis of the evolution of multiple zeros in 1D applies to more general parabolic equations. In particular, in $N$-dimensional geometry similar results are true for radial solutions $u = u(r,t)$ of parabolic equations (2.3) with analytic coefficients; see the next section.

Sturm's proof, consisting of two parts (i) and (ii), exhibits typical features of the asymptotic evolution analysis for general linear uniformly parabolic equations:

(i) A finite-time formation of a multiple zero as $t \to 0^-$ as a *singularity formation* (single point blow-up self-focusing of zero curves);
(ii) Disappearance of multiple zeros at $t = 0^+$, i.e., instantaneous *collapse of a singularity* and a unique continuation of the solution beyond the singularity.

Regarding this part of Sturm's analysis, we present the result separately as follows.

**Corollary 2.1.** *Under the assumptions of Theorem 2.2, the following results hold:*

(i) *As $t \to 0^-$, the rescaled solution converges uniformly on any compact subset $\{|z| \le \text{const.}\}$ to the mth order Hermite polynomial with finite oscillations:*

$$A^{-1} a_0^{-m/2} (-t)^{-m/2} u(x,t) \to H_m(z). \tag{2.15}$$

(ii) *As $t \to 0^+$, the rescaled solution converges uniformly on compact subsets to the non-oscillating mth order polynomial:*

$$A^{-1} a_0^{-m/2} t^{-m/2} u(x,t) \to Q_m(z). \tag{2.16}$$

Phenomena of singularity blow-up formation, collapse and proper solution extensions beyond singularities are important subjects of general PDE theory. In applications to semilinear and quasilinear parabolic equations of reaction-diffusion type, the perturbation techniques for infinite-dimensional dynamical systems plays a key role; see various examples in [58]. We briefly comment on Sturm's analysis using the perturbation theory of linear operators.

**(i) Formation of multiple zeros: backward continuation.** Using Sturm's backward rescaled variable (2.11), we introduce the rescaled solution

$$u(x,t) = \theta(z,\tau), \quad z = x/\sqrt{a_0(-t)}, \tag{2.17}$$

where $\tau = -\ln(-t) \to +\infty$ as $t \to 0^-$ is the new time variable. Substituting (2.17) into equation (2.1) yields the rescaled equation

$$\theta_\tau = \mathbf{B}\,\theta + \mathbf{C}(\tau)\theta, \tag{2.18}$$

where $\mathbf{B}$ is the linear operator

$$\mathbf{B} = \frac{\mathrm{d}^2}{\mathrm{d}z^2} - \frac{1}{2} z \frac{\mathrm{d}}{\mathrm{d}z} \equiv \frac{1}{\rho}\frac{\mathrm{d}}{\mathrm{d}z}\left(\rho \frac{\mathrm{d}}{\mathrm{d}z}\right), \quad \text{where } \rho(z) = e^{-z^2/4}, \tag{2.19}$$

which is symmetric in $L^2_\rho(\mathbb{R}^N)$ (see below). The non-autonomous perturbation in (2.18) has the form

$$\mathbf{C}(\tau)\theta = \left(\frac{a - a_0}{a_0}\right)\theta_{zz} + e^{-\tau/2} \frac{b}{\sqrt{a_0}}\,\theta_z + e^{-\tau} c\,\theta,$$

where for the regular coefficient $a$, $(a(x,t) - a_0)/a_0 \equiv (a(z[a_0(-t)]^{1/2}, t) - a_0)/a_0 = O(e^{-\tau/2})$. This means that for smooth solutions, the perturbation

$$\mathbf{C}(\tau)\theta = e^{-\tau/2}[\theta_{zz}\, O(1) + b a_0^{-1/2} \theta_z + e^{-\tau/2} c\,\theta]$$

is exponentially small as $\tau \to \infty$. Equation (2.18) is an exponentially small perturbation of the autonomous equation

$$\theta_\tau = \mathbf{B}\,\theta. \tag{2.20}$$

The operator $\mathbf{B}$ is known to be self-adjoint in the weighted space $L^2_\rho(\mathbb{R})$ with the inner product $(v,w)_\rho = \int_{-\infty}^{\infty} \rho(z) v(z) w(z) dz$. Its domain $\mathcal{D}(\mathbf{B}) = H^2_\rho(\mathbb{R})$ is a Hilbert space of functions $v$ satisfying $v, v', v'' \in L^2_{\text{loc}}(\mathbb{R})$ with the inner product

$\langle v, w \rangle_\rho = (v, w)_\rho + (v', w')_\rho + (v'', w'')_\rho$ and the induced norm $\|v\|_\rho^2 = \langle v, v \rangle_\rho$. Moreover, $\mathbf{B}$ has compact resolvent and its spectrum only consists of eigenvalues:

$$\sigma(\mathbf{B}) = \{\lambda_k = -\tfrac{k}{2}, \ k = 0, 1, \dots\}.$$

The eigenfunctions are orthonormal Hermite polynomials $\tilde{H}_k(z) = c_k H_k(z)$, $c_k$ being normalization constants. These are classical results of the theory of linear self-adjoint operators in Hilbert spaces. We refer to the first chapters of the book [22] (see p. 48 on Hermite polynomials in $\mathbb{R}^N$). Using eigenfunction expansions and semigroup estimates (see Section 3) yields that the exponentially perturbed dynamical system (2.18) on $L^2_\rho(\mathbb{R}^N)$ admits a discrete subset of asymptotic patterns. These coincide with those for the unperturbed equation (2.20) exhibiting the asymptotic behavior on tangent stable ($\lambda_m < 0$) eigenspaces of $\mathbf{B}$. Hence (2.15) holds. As $\tau \to \infty$, uniformly on compact subsets we have

$$\theta(z, \tau) = C e^{\lambda_m \tau} H_m(z) + O(e^{\lambda_{m+1}\tau}) \quad \text{with a constant } C \neq 0. \tag{2.21}$$

**(ii) Collapse of multiple zero on the spatial structure of adjoint polynomials: forward continuation.** For $t > 0$, we use the forward rescaled variable (2.13). Similarly, we deduce that the rescaled function $u(x, t) = g(z, s)$, where the time variable is $s = \ln t \to -\infty$ as $t \to 0^+$, solves the exponentially perturbed equation as $s \to -\infty$

$$g_s = (\mathbf{B}^* - \tfrac{1}{2} I) g + \mathbf{C}(s) g, \tag{2.22}$$

where $I$ denotes identity and $\mathbf{B}^*$ is the adjoint differential operator

$$\mathbf{B}^* = \frac{\mathrm{d}^2}{\mathrm{d}z^2} + \frac{1}{2} z \frac{\mathrm{d}}{\mathrm{d}z} + \frac{1}{2} I \equiv \frac{1}{\nu} \frac{\mathrm{d}}{\mathrm{d}z} \left( \nu \frac{\mathrm{d}}{\mathrm{d}z} \right) + \frac{1}{2} I \quad \text{with weight } \nu(z) = e^{z^2/4}.$$

As in the backward analysis, the perturbation term $\mathbf{C}(s) g = O(e^{s/2}) \to 0$ as $s \to -\infty$ and is exponentially small for smooth solutions on compact subsets. $\mathbf{B}^*$ is self-adjoint in $L^2_\nu(\mathbb{R})$, $\mathcal{D}(\mathbf{B}^*) = H^2_\nu(\mathbb{R})$, with the point spectrum $\sigma(\mathbf{B}^*) = \sigma(\mathbf{B})$ and a complete set of orthonormal eigenfunctions.

Unlike the phenomenon of the *evolution* blow-up formation of multiple zeros, in the asymptotic analysis as $s \to -\infty$ spectral properties and eigenfunctions of $\mathbf{B}^*$ play no role. The limit $t \to 0^+$ corresponds to the collapse of the *initial singularity* created by the preceding singularity formation as $t \to 0^-$. The behavior of $u(x, t)$ as $t \to 0^+$ is uniquely determined by the initial data $u(x, 0)$. Consider (2.21) for $|z| \gg 1$. Since $P_m(z) \equiv H_m(z) = z^m + \cdots$ as $z \to \infty$, it can be shown (a compactness argument is necessary at this step to extend the behavior from compact subsets $\{|z| \le c\}$ to $\{0 < |x| \ll 1\}$) that passing to the limit $t \to 0^-$ gives $u(x, 0)$ as follows:

$$u(x, t) = C(-t)^{-\lambda_m} x^m a_0^{-m/2} (-t)^{-m/2} + \cdots \to C a_0^{-m/2} x^m + \cdots. \tag{2.23}$$

The solution $g(z, s)$ of the rescaled equation (2.22) with initial data calculated in (2.23) has the expansion

$$g(z, s) = \tilde{C} e^{-\lambda_m s} Q_m(z) + \cdots, \quad \tilde{C} \neq 0, \tag{2.24}$$

where $Q_m$ is the polynomial solution of the linear equation $(\mathbf{B}^* - \frac{1}{2}I)Q_m = \frac{m}{2}Q_m$. We thus arrive at the linear problem for the "adjoint" polynomials $\{Q_m\}$. Notice that these have nothing to do with the orthogonal subset of eigenfunctions $\{\exp(-z^2/4)\, H_m(z)\}$ of the adjoint operator $\mathbf{B}^*$. Moreover $Q_m \notin L^2_\nu(\mathbb{R})$. In order to match (2.24) and the initial condition (2.23), by a similar local extension to $\{0 < |x| \ll 1\}$ we have that

$$g(z,s) = \tilde{C} t^{-\lambda_m} x^m a_0^{-m/2} t^{-m/2} + \cdots \to \tilde{C} a_0^{-m/2} x^m + \cdots \quad \text{as } t \to 0^+.$$

By matching with (2.23), this uniquely determines the constant $\tilde{C} = C$ in (2.24) and completes the asymptotic analysis of both the backward and forward evolution of multiple zeros.

*Results in classes of finite regularity.* Fix finite $T > 0$ and let $J = (0,T)$. If $u(x,t) \not\equiv 0$ is a solution, analytic in $x$, of the linear parabolic equation (2.1) with analytic coefficients $a, b, c$, then for any $t \in (0,T)$, all the zeros of $u(x,t)$ are isolated and hence the number of sign changes $Z(t,u)$ is finite even if $Z(0,u) = \infty$. A similar result holds in classes of solutions and equations of finite regularity. We present without proofs two results by S. Angenent [7]; more references are given in Section 3. We begin with initial-boundary value problems.

**Theorem 2.3.** *Let $u$ be a bounded solution of* (2.1) *in $S = D \times (0,T)$ which does not change sign on the lateral boundary of $S$. Assume that the coefficients $a, b$ and $c$ of the equation are such that*

$$a, a^{-1}, a_x, a_{xx}, b, b_t, b_x, c \in L^\infty(S).$$

*Then the number of sign changes of $u(\cdot,t)$ satisfies:*

(i) *$Z(t,u)$ is finite and nonincreasing on $(0,T)$;*
(ii) *If $x = x_0 \in D$ is a multiple zero of $u(x,t_0)$ for some $t_0 \in (0,T)$, then for all $0 < t_1 < t_0 < t_2 < T$ the strict inequality $Z(t_1,u) > Z(t_2,u)$ holds, so that $Z(t,u)$ is strictly decreasing at $t = t_0$.*

As a consequence, any global solution $u(x,t)$ defined in $S = D \times \mathbb{R}_+$ has only simple zeros for all $t \gg 1$. A similar result is valid for parabolic equations in unbounded domains if we restrict the analysis to classes of functions with a fixed growth at infinity, similar to Tikhonov's classes of uniqueness. Let $D = \mathbb{R}$, and consider the following linear parabolic equation:

$$u_t = u_{xx} + q(x,t)u \quad \text{in } S = \mathbb{R} \times (0,T). \tag{2.25}$$

**Theorem 2.4.** *Let $q \in L^\infty(S)$, and let $u(x,t)$ be a solution of* (2.25) *in the class $\{|u(x,t)| \le Ae^{Bx^2}$ in $S\}$ for some positive constants $A$ and $B$. Then for each $t \in (0,T)$, the zero set of the solution $\{x \in \mathbb{R} : u(x,t) = 0\}$ is a discrete subset of $\mathbb{R}$.*

As a direct consequence of this we have that if $x = \pm\infty$ are not accumulation points of zeros of $u(x,0)$, then statements (i) and (ii) of Theorem 2.3 hold. Theorem 2.4 is true for more general equations like (2.1) in unbounded domains in suitable classes of uniqueness. Equation (2.1) can be reduced to (2.25) by the Liouville transformation. Using the new spatial coordinate $y = \int_0^x (a(s,t))^{-1/2}\,ds$, we have that $u = u(y,t)$ satisfies the equation

$$u_t = u_{yy} + \tilde{b}(y,t)u_y + \tilde{c}(y,t)u.$$

Substituting $v(y,t) = \exp\{\frac{1}{2}\int_0^y \tilde{b}(s,t)\,ds\}u(y,t)$, yields equation (2.25) for $v(y,t)$ with a potential $\tilde{q}(y,t)$. Checking necessary properties of $\tilde{q}(y,t)$ one deduces that Sturm's results are valid in the corresponding uniqueness classes.

## 3. Survey on Sturm's theorems and ideas in parabolic PDEs

We begin our survey with those ODE results that fall into the scope of the PDE theory or can admit a PDE treatment or proof. The rest is devoted to applications of Sturm's Theorems in areas where parabolic PDEs occur.

### 3.1. On some ODE results

Classical Sturm results on zeros for a single second-order ODE like

$$y'' + q(t)y = 0, \quad t \in (0, 2\pi), \tag{3.1}$$

can be stated in a topological form describing rotations in the phase space of equations (this form is convenient for extensions to higher-order equations). Let

$$Y(t) = \begin{pmatrix} y_1(t) & y_2(t) \\ y_1'(t) & y_2'(t) \end{pmatrix} \quad \text{satisfying } Y(0) = E_2 = \begin{pmatrix} 1 & 0 \\ 0 & 1 \end{pmatrix}$$

be a matrix solution of (3.1), where $y_1(t)$ and $y_2(t)$ are linearly independent solutions. Then the vector $z(t) = y_1(t) + iy_2(t)$ moves counterclockwise in the complex plane. Indeed, since by construction the Wronskian $W(y_1, y_2)(t) = \det Y(t) \equiv 1$, we have that $\arg z(t) = \tan^{-1}(y_2(t)/y_1(t))$ satisfies $\frac{d}{dt}\arg z = W(y_1, y_2)/(y_1^2 + y_2^2) = 1/(y_1^2 + y_2^2) > 0$. Sturm's theorems follow from this monotonicity property.

The first generalizations of Sturm's theorems to the case of vector-valued operators and to systems (3.1) with symmetric matrices $q(t)$ are due to M. Morse (1930) [101], [102], where variational methods are applied. Oscillatory theorems for general canonical systems of $2k$th order were first established by V.B. Lidskii (1955) [88] for the equation

$$y' = IH(t)y, \quad I = \begin{pmatrix} 0 & E_k \\ -E_k & 0 \end{pmatrix},$$

where $E_k$ is the $k \times k$ identity matrix and $H(t)$ is a $2k \times 2k$ real continuous symmetric matrix (the Hamiltonian). We present brief comments on these results. Let $Y(t)$ with $Y(0) = E_{2k}$ be a matrix solution. Then $Y(t)$ is symplectic: $Y^* I Y \equiv I$. Denote

$$H(t) = \begin{pmatrix} h_{11}(t) & h_{12}(t) \\ h_{21}(t) & h_{22}(t) \end{pmatrix} \quad \text{and} \quad Y(t) = \begin{pmatrix} y_{11}(t) & y_{12}(t) \\ y_{21}(t) & y_{22}(t) \end{pmatrix},$$

where $h_{ij}(t)$ and $y_{ij}(t)$ are $k \times k$ blocks. Consider the non-singular matrix $z(t) = y_{11}(t) + iy_{12}(t)$ (cf. the case $k = 1$ above), and set $u(t) = (\overline{z}(t))^{-1}z(t)$. Then $u(t)$

is unitary and symplectic. The alternation theorem of Lidskii is as follows. Let $h_{22}(t) > 0$ (for (3.1) with $k = 1$, $h_{22} \equiv 1$). Then the eigenvalues $\rho_1(t), \dots, \rho_k(t)$ of $u(t)$ move counterclockwise around the unit circle: $\frac{d}{dt} \arg \rho_s(t) > 0$ for $s = 1, \dots, k$. For $\rho_s(t) = -1$ (resp., $\rho_s(t) = +1$) the matrix $u(t)$ has the same zero subspace as $y_{11}(t)$ (resp., $y_{12}(t)$), i.e., the "zeros" of the matrices $y_{11}(t)$ and $y_{12}(t)$ alternate. Lidskii also proved an analogue of the Sturm comparison theorem. Consider two canonical systems

$$Y_1' = IH_1(t)Y_1 \quad \text{and} \quad Y_2' = IH_2(t)Y_2, \quad \text{where } H_1(t) > H_2(t).$$

Then specially enumerated eigenvalues $\rho_s^{(1)}(t)$ and $\rho_s^{(2)}(t)$ of the unitary matrices $u^{(1)}(t)$ and $u^{(2)}(t)$ satisfy $\arg \rho_s^{(1)}(t) > \arg \rho_s^{(2)}(t)$, $s = 1, \dots, k$, i.e., $\rho_s^{(1)}(t)$ moves "ahead" of $\rho_s^{(2)}(t)$.

Variational approaches to Sturm's theorems for self-adjoint linear $2k$th order systems were also developed by R. Bott (1959) [24] and by H.H. Edwards (1964) [39]. (See the books [115] and [20] for a detailed presentation.) These results were related to the Maslov index [95]. In 1985 V.I. Arnold [14] characterized this as follows: "... numerous authors writing on the Maslov index, symplectic geometry, geometric quantization, Lagrangian analysis, etc., starting with [13], have not noticed the earlier works by Lidskii [88], as well as the earlier works of Bott [24] and Edwards [39], in which a Hermitian version of the theory of the Maslov index and Sturm intersections were constructed."

A survey of earlier results concerning distribution and alternation of zeros for $n$th order linear ODEs can also be found in [87], where, as well as in the books mentioned above, various links to other related subjects are described in detail. These include S.A. Chaplygin's comparison theorem (1932) [30] closely connected with the theory of positive operators, W.A. Markov's theorem (1916) [94] on the conservation of the alternation of zeros of polynomials under differentiation, C. de la Vallée-Poussin's theorem (1929) [38] and G. Pólya's (1924) [111] criterion on non-oscillation (the first non-oscillation test of best-possible character is due to N.E. Zhukovskii (1892) [136]), F.R. Gantmakher (1936) [59] and M.G. Krein's (1939) [83] theory of oscillating kernels [60] (a direction originated with O.D. Kellogg's work (1922) [78] on symmetric kernels), S.N. Bernstein results (1938) [21] on connections between Chebyshev and Cartesian systems, etc. See also Hinton's survey [69].

Sturmian methods for ODEs can be applied to investigations in the complex plane, see [68], Chapter 8. The classical Sturm comparison theorem for ODEs admits special extensions to linear and quasilinear elliptic and parabolic PDEs, see first results in [108], the book [127] and [2], as well as to ODEs in Hilbert spaces [75]. More recent extensions of Sturm's comparison theorems to quasilinear elliptic equations can be found in [3], [4], where extra references are available.

Sturm's Theorem on the number of distinct real roots of polynomials by computing the number of sign changes in Sturm sequences (1835) [124] is well known in algebra, see, e.g., [86] and [23]. In constructing Sturm sequences the first step is differentiation, establishing a link to ODEs (Sturm's comparison or oscillation theorems).

As with respect to ODEs, Sturm's ideas have applications in the classical problem on zeros of complete Abelian integrals defined by means of a planar Hamiltonian flow, which is closely related to Hilbert's 16th problem (the so-called weakened, infinitesimal or tangential Hilbert problem). Abelian integrals were known to satisfy a system of Picard-Fuchs ODEs [61], see also [72] for further references. This is a part of a general problem on zeros of Pfaffian functions and the fewnomials theory, [79], [80], where the eventual reduction to polynomial structures is used. In particular, algorithmic consistency problems for systems of Pfaffian equations and inequalities occur (with applications to computer sciences); see [50] and references therein.

Let us return to the Sturm-Hurwitz theorem establishing that the finite Fourier series (1.3) has at least $2L$ and at most $2M$ zeros. On pp. 436–444 of the PDE paper [126], Sturm presented an ODE proof of the result. Sturm's ODE proof, as well as Liouville's one in [89] published in the same volume, exhibit certain features of a discrete evolution analysis (to be compared with Sturm's PDE proof via parabolic evolution equation with continuous time variable). A. Hurwitz (1903) [71] extended this result to Fourier series with $M = \infty$.

Further extension is due to O.D. Kellog (1916) [77] who proved oscillation theorems for linear combinations of real continuous functions $\phi_0(x), \phi_1(x), \dots, \phi_n(x)$ that are orthonormal in $L^2((0,1))$. These are not eigenfunctions of a Sturm-Liouville problem. The main assumption is as follows (we keep the original notation). For any $n \geq 1$, let the determinants

$$D(x_0, x_1, \dots, x_n) = \begin{vmatrix} \phi_0(x_0) & \phi_1(x_0) & \dots & \phi_n(x_0) \\ \phi_0(x_1) & \phi_1(x_1) & \dots & \phi_n(x_1) \\ \dots & \dots & \dots & \dots \\ \phi_0(x_n) & \phi_1(x_n) & \dots & \phi_n(x_n) \end{vmatrix}$$

be positive for any $0 < x_0 < x_1 < \dots < x_n < 1$ ($D_0(x_0)$ being understood as $\phi_0(x_0)$). Let

$$\Phi_{m,n}(x) = c_m \phi_m(x) + \dots + c_n \phi_n(x).$$

Then, among other results, it is established that:

(i) $\Phi_{0,n}(x)$ cannot vanish at $n+1$ distinct points in $(0,1)$ without vanishing identically;

(ii) $\phi_n(x)$ vanishes exactly $n$ times and changes sign at each zero;

(iii) every continuous function $\psi(x)$ orthogonal to $\phi_0(x), \dots, \phi_n(x)$ changes sign at least $n+1$ times;

(iv) $\Phi_{m,n}(x)$ changes sign at least $m$ times and at most $n$ times.

The infinitesimal version of the discriminants with $x_{k+1} - x_k \to 0$, $k = 0, 1, \dots, n-1$, defines the Wronskians of the given functions. Hence some of the assumptions are valid for eigenfunctions of regular Sturm-Liouville problems. On the other hand, Kellogg's results do not cover those of Sturm, see p. 5 in [77].

The Sturm-Hurwitz Theorem plays a fundamental role in topological problems in wave propagation theory (topology of caustics and wave fronts), the geometry of plane and spherical curves and in general symplectic geometry and topology, see [14], [16], [17], [19] and references therein. Alternating, oscillating and non-oscillating Sturm theorems have multi-dimensional symplectic analogues and describe rotation of a Lagrangian subspace of the phase space [14]. For instance, the Sturm-Hurwitz theorem proves a generalization [129] of the classical four vertex theorem by S. Mukhopadyaya [103] and A. Kneser [81] asserting that a plane closed non-self-intersecting curve has at least four vertices (critical points of the curvature). It is pointed out in [17] that the same minimal number occurs in:

(i) theorems on four cusps of general caustics on every convex surface of positive curvature (the related conjecture goes back to C.G.J. Jacobi (1884) [74]),
(ii) four cusps of the envelope of the family of perturbed Larmor orbits of given energy,
(iii) the tennis-ball theorem (a closed curve on the sphere without self-intersections, a smooth embedding $S^1 \to S^2$, dividing the sphere into two parts of equal area, has at least four points of spherical inflection with zero curvature),
(iv) the four equilibrium points theorem,
(v) the four flattening points theorem for perturbed convex curves of positive curvature on a plane lying in three-dimensional space, etc.

Infinitesimal versions of such topological theorems (for infinitely small perturbations of curves) follow from the Sturm-Hurwitz theorem. For finite perturbations, some of these results can be proved by means of evolution Sturm theorems on zeros for parabolic PDEs to be discussed later on.

Half of Arnold's third lecture in the Fields Institute (1997) [18] was devoted to Sturm's theory on Fourier series, which "provides one of the manifestations of the general principle of economy in algebraic geometry" (related to Arnold's conjecture (1965) and the symplectification of topology). In particular, the Morse inequality (in the simplest version it says that the number of critical points of functions on the circle is at least 2) is the Sturm-Hurwitz theorem with $L = 1$.

The Sturm-Hurwitz theorem was first proved by the PDE method [126], pp. 431–436, in the general form including any (finite) series composed from eigenfunctions of a Sturm-Liouville problem. These extensions of Sturm's ideas have many other applications to be discussed below.

Extensions of Sturm's results on zeros (nodal sets) of linear combinations of eigenfunctions to standard self-adjoint elliptic operators (e.g., the Laplacian $\Delta$) in bounded smooth domains $\Omega \subset \mathbb{R}^N$, $N \geq 2$, are unknown; see [17] and [18]. In particular, the so-called Herrmann theorem announced in [37], p. 454: a linear combination of the first $n$ eigenfunctions divides the domain, by means of its nodes (piecewise smooth nodal surfaces), into not more than $n$ subdomains, fails to hold for the spherical Laplacian [18]. Courant's Theorem on p. 452 asserts that the nodes of the $n$th eigenfunction divide the domain into no more than $n$ subdomains.

In dimensions $N \geq 2$, given a linear combination $f(x)$ of eigenfunctions of $\Delta$, the structure of the nodal set itself $\mathcal{N}(f) = \{x \in \Omega : f(x) = 0\}$ is not sufficient to define a kind of a Sturmian "index" of the surface $z = f(x)$, similar to the number of zeros in 1D, which can inherit a certain numerical property (say, a lower bound) from the lowest harmonic of the series. Such an index should depend on global properties of $f(x)$ at all points $x \in \Omega$ including those far away from $\mathcal{N}(f)$. It seems reasonable that for a proper definition of a Sturmian index, it is necessary to control the intersections of the graph of the function $f$ with the graphs of the functions in the finite-dimensional set $B = \{V_\nu(x)\}$ containing functions associated with the operator $\Delta$. Roughly speaking, this would mean that such a "local" characteristic as the number of zeros of $f(x)$ on an interval from $\mathbb{R}$ cannot work in $\mathbb{R}^N$, where any possible nonincreasing property of, say, the number of maximal connected subdomains of the positivity subset $\{f(x) > 0\}$ should include some global properties of the function formulated in an unknown way. In any case, a proper definition of Sturmian index of surfaces governed by parabolic equations in $\mathbb{R}^N$ is not expected to admit a simple formulation or such easy and effective applications as it has in the 1D case.

### 3.2. Parabolic PDEs and Sturm's theorems

The Sturmian argument for 1D parabolic equations turns out to be an extremely effective technique in the study of different aspects of the theory of nonlinear parabolic equations. In the twentieth century the argument was partially and independently rediscovered several times. We will mention some of the papers published at least twenty years ago, but of course there are many other interesting and important papers published more recently, which are not referred to here.

G. Pólya (1933) [112] paid special attention to Sturm's zero set properties of periodic solutions to the heat equation. He studied the number of "Nullstellen" of $u(x,t)$, i.e., the number of $x \in [0, 2\pi]$ such that $u(x,t) = 0$, on the basis of Sturm's approach with a reference to [126]. Radial and cylindrical solutions were considered and zero properties of convolution integrals were also studied.

The celebrated KPP-paper (1937) [82] was devoted to the stability analysis of the minimal travelling wave (TW) for a semilinear heat equation

$$u_t = u_{xx} + f(u) \quad \text{in } \mathbb{R} \times \mathbb{R}_+,$$

with the typical nonlinearity $f(u) = u(1-u)$. There the construction of a geometric Lyapunov function in Theorem 11 was based on the following intersection comparison argument: the initial 1-step function $u_0(x) = 1$ for $x > 0$ and 0 for $x \leq 0$ intersects any smooth travelling wave profile exactly at a single point and there exists a unique intersection curve for $t > 0$. In our notation this means that the number of intersections $\text{Int}(t, V) \equiv 1$ for any TW $V(x,t) = g(x - \lambda_0 t + a)$ and any $t > 0$, where $\lambda_0 > 0$ is the minimal speed. In general, the number of intersections can be treated as a discrete nonincreasing Lyapunov function. On the other hand, it gives a standard monotone Lyapunov function: on any fixed level $\{u(x,t) = c \in (0,1)\}$ the derivative $u_x(x,t) < 0$ is monotone increasing in

$t$ and bounded above. Then passage to the limit $t \to \infty$ establishes the convergence to the minimal TW profile in the hodograph plane $\{u, u_x\}$ or in the moving coordinate system in the $\{x, u\}$-plane.

K. Nickel's paper (1962) [105] (see also [106]) established nonincrease of the number of sign changes of solutions of parabolic equations (more precisely, of the number of relative maxima of a solution profile, i.e., the number of zeros of the derivative $u_x(x,t)$). Nickel's results are explained in detail relative to general fully nonlinear parabolic equations in W. Walter's books [134] and [133], Section 27. R.M. Redheffer and W. Walter (1974) [114] extended such results to more general classes of equations. For particular linear parabolic equations in $\mathbb{R}$, these results were proved by S. Karlin (1964) [76], whose analysis was based on ideas of total positivity of Green's functions and applied to Brownian motion processes. Related questions and techniques were discussed by I.K. Ivanov (1965) [73] (the number of changes of sign was considered), by E.K. Godunova and V.I. Levin (1966) [62] (a proof of existence of a single maximum was based on the theory of probabilistic distributions; eventual single maximum distribution and eventual concavity of solutions were also established) and by E.M. Landis (1966) [85] (properties of evolution of level sets for (2.1) were investigated). D.H. Sattinger's results (1969) [118] on sign changes for linear parabolic equations are similar to those obtained by Nickel and Walter. Observe that in the proof of Theorem 7 on exponential decay of total variation, Sattinger uses a reflection technique and studies zeros of the differences $u(x,t)$ and the reflected solution $u(2l - x, t)$, see p. 88 in [118]. Such a combination of Sturm's theorems and A.D. Aleksandrov's Reflection Principle and ideas (1960) [1] later became a powerful tool in the asymptotic theory for nonlinear singular parabolic equations. Papers by A.N. Stokes (1977) [122] and [123] used the nonincrease of zero number with application to stability analysis of travelling waves. Here the basic idea of proving a Lyapunov monotonicity property in the hodograph plane is essentially the same as in the KPP-analysis [82]. A general stability analysis of TWs in analytic semilinear parabolic equations via zero set properties was performed in [12].

H. Matano (1978) [96] proved the first Sturm Theorem and applied it to establishing that the $\omega$-limit set of any bounded solution to a semilinear parabolic equation $u_t = (a(x)u_x)_x + f(x,u)$ on $(0, L) \times \mathbb{R}_+$, $a \geq a_0 > 0$, with smooth coefficients and Robin boundary conditions contains at most one stationary point. At that time such a result was already known [135] for smooth uniformly parabolic equations $u_t = a(x, u, u_x)u_{xx} + b(x, u, u_x)$ with general nonlinear boundary conditions. It was proved by constructing a standard (integral) Lyapunov functional by the method of characteristics, a fruitful idea which applies to 1D quasilinear parabolic equations. The geometric proof by Matano is more general and can be applied to fully nonlinear parabolic equations

$$u_t = F(x, u, u_x, u_{xx}). \tag{3.2}$$

More detailed results related to the first Sturm Theorem were published in [97]. A finite difference approach to some of these Sturmian properties was developed earlier by M. Tabata (1980) [130]. An application of intersection comparison to blow-up solutions of quasilinear parabolic equations $u_t = (k(u)u_x)_x + Q(u)$ was given in [52].

Computations similar to those of Sturm in the proof of Theorem 2.2 in Section 2 can be found in [12], Section 5. For radial equations (2.3) with $N > 1$ such computations for $t < 0$ lead to Laguerre polynomials $L_m^\gamma(z)$ of order $\gamma = N/2$, see Section 3 in [8]. Perturbation techniques for the operator (2.19) were developed in [65], [7], [32]. Sturm's backward parabolic rescaling with $z = x/(-t)^{1/2}$ plays an important role in continuation theorems and topology of nodal sets for linear parabolic equations in $\mathbb{R}^N$ [32]. A weak form of the continuation analysis [113] based on a monotonicity formula and weighted inequalities (this idea goes back to T. Carleman (1939) [29] with applications to elliptic equations), which are convolutions with the backward heat kernel, uses the same Sturm backward variable.

The evolution proof of the Sturm-Hurwitz Theorem on zeros of (finite) linear combinations of eigenfunctions $\{V_k(x),\ k = 1, 2, \ldots\}$, where each $V_k$ has exactly $k - 1$ simple transversal zeros, of a Sturm-Liouville operator given by (2.6), (2.7),

$$Y(x) = C_i V_i(x) + C_{i+1} V_{i+1}(x) + \cdots + C_p V_p(x)$$

is given on pp. 431-444 in [126] and is as follows (we keep the original notation). Consider the solution

$$u(x,t) = C_i V_i(x) e^{-\rho_i t} + C_{i+1} V_{i+1(x)} e^{-\rho_{i+1} t} + \cdots + C_p V_p(x) e^{-\rho_p t} \tag{3.3}$$

of the parabolic equation (2.6) with $u(x, 0) \equiv Y(x)$, where the sequence of eigenvalues $\{-\rho_k\}$ is strictly decreasing. Then for $t \gg 1$, the first harmonic is dominant and hence $u(x,t)$ has exactly $i-1$ zeros. Since the number of zeros of $u(x,t)$ does not increase, $u(x,t)$ has at least $i-1$ zeros for all $t \in \mathbb{R}$, and hence at $t = 0$. On the other hand, for $t \ll -1$ the last harmonic in (3.3) is dominant, $u(x,t)$ has exactly $p-1$ zeros, so that by Sturm's Theorem, $u(x,t)$ has at most $p-1$ zeros for all $t \in \mathbb{R}$.

On p. 436 Sturm compares his proof with that by J. Liouville [89] "... without using consideration of the auxiliary variable $t$ ..." (by means of an ODE argument). In Section XXVI Sturm presents his own ODE proof. Corollary 2.1 is a paraphrase of Sturm's calculations. The proof of Theorems 2.3 and 2.4 are given in [7]. Finiteness of $Z(t,u)$ on $(0,1)$ for $t > 0$ was also established in [84] for coefficients $a \in H^1$, $b \in W^{1,\infty}$ and $c \in L^\infty$ depending on $x$ only. The second Sturm Theorem on formation of multiple zeros remains valid for $W^{2,1}_{p,\mathrm{loc}}$ solutions ($p > 1$) from Tikhonov's uniqueness class for linear uniformly parabolic equations in $\mathbb{R}^N$ with bounded coefficients [32] (the proof uses Sturmian backward rescaling). The analytic case was treated in [12]. Eventual simplicity of zeros was first observed in [26].

An evolution approach to connections of equilibria for semilinear parabolic equations was introduced by D. Henry [65], where such a time-dependent Sturm-Liouville theory was rigorously established (including completeness of asymptotic limits in Theorem 4 proved by Agmon's estimates). This theory was used in completing the proof that, under some hypotheses, a general semilinear parabolic equation $u_t = u_{xx} + f(x, u, u_x)$ in $(0, 1) \times \mathbb{R}_+$, with Dirichlet or nonlinear boundary conditions, represents a Morse-Smale system. It is established that given a heteroclinic connection $\bar{u}(x, t)$ of two hyperbolic (linearly nondegenerate) equilibria $\phi_\pm$, $u(x, -\infty) = \phi_-(x)$ and $u(x, +\infty) = \phi_+(x)$, the stable manifold $W^s(\phi_+)$ and the unstable one $W^u(\phi_-)$ meet transversally at $\bar{u}(\cdot, t)$ for each $t$. See also [6] for the case $f = f(x, u) \in C^2$. This transversality result was used in [65] to describe all connecting orbits between equilibria for the Chafee-Infante problem with $f = f(u)$, $f(0) = 0$. For earlier results on connections for parabolic equations see [64] and [25]. For more general $f \in C^2$ such connections were established in [27]. See also the survey [44].

A spectrum of Hermite polynomials occurred in the zero set analysis by D. Henry [65] and S.B. Angenent [7]. Zero set results played a role in the analyticity study of solutions of the porous medium equation (PME) [8]. A few years after papers [65], [6] and [7] on parabolic Morse-Smale systems, the same linearized operators, with eigenfunctions composed from Hermite polynomials, were obtained in the center and stable manifold behavior in the study of blow-up solutions of the semilinear parabolic equations from combustion theory $u_t = \Delta u + u^p$, $p > 1$ and $u_t = \Delta u + e^u$ (the nonstationary Frank-Kamenetskii equation), see [132], [47], [66], [131] and [99].

Sturm's Theorems play a key role in the analysis of other aspects of behaviour in infinite-dimensional dynamical systems associated with nonlinear parabolic equations. These are convergence to periodic solutions and related questions for periodic equations [33], [28] (results apply to general 1D fully nonlinear equations), [43], [31] (transversality properties), [109], [67] and [34] (applied to $N$-dimensional semilinear parabolic equations by means of symmetrization and moving plane techniques), [120] (almost periodicity). Zero set analysis is a leading ingredient of a Poincaré-Bendixson theorem for semilinear heat equations, [12], [98], [45], and in the construction of G. Floquet bundles (see [48] and results by A.M. Lyapunov [91]) for linear parabolic equations in periodic and nonperiodic cases (solutions $u_n(x, t)$ having exactly $n$ zeros for all $t \in \mathbb{R}$) [35], [36] (a generalization of Sturm-Liouville theory to the time-dependent case, results include exponential dichotomies and other estimates). Such Floquet-type solutions $\{u_n(x, t), t > 0\}$ exist for the semilinear heat equation $u_t = u_{xx} - |u|^{p-1}u$ in $\mathbb{R} \times \mathbb{R}_+$ with exponential decay as $t \to \infty$ depending on $n$ [100]. The nonincreasing number of zeros plays a key role in the problems of Morse decomposition [92] and connections of Morse sets [46] for the monotone feedback differential delay equation

$$\dot{u}(t) = f(u(t), u(t-1)), \quad u \in \mathbb{R}.$$

Nonincrease of the number of zeros per unit interval for such linear equations was first established by A.D. Myschkis (1955), see Theorem 32 in [104]. It is also true for monotone cyclic feedback systems [93] $\dot{u}_i = f_i(u_i, u_{i-1})$, $u_i \in \mathbb{R}$, $i \bmod n$.

Sturm's intersection ideas play a fundamental role in *curve shortening* or *flows by mean curvature* problems for curves on surfaces. For curves on a surface $M$ with a Riemannian metric $g$, such a motion is described by the curve shortening equation

$$v^{\perp} = V(t, k), \tag{3.4}$$

where $v^{\perp}$ is the normal velocity of the curve, $k$ is the curvature and $V$ is a $C^{1,1}$ function satisfying $\partial V / \partial k > 0$. The reason that Sturm's results apply to such evolution problems (though some of the properties are intuitively obvious for intersections of curves) is that (3.4) reduces to a nonlinear parabolic equation for the curvature $k$ or, after a suitable parametrization, for a function $u(x,t)$ satisfying a fully nonlinear parabolic equation (3.2), where $F$ depends on $V$. [See the first results in [70], [119] and [51] (a parabolic PDE for curvature $k_\tau = k^2(k_{\theta\theta} + k)$ was derived for the flow $v^{\perp} = k$), and [41], [63].] A general approach to curve shortening flows via 1D parabolic equations was developed in [9], [10] (where Sturm's intersection theory is described), see also [117]. The mean curvature flows can generate different types of singularities.

Parabolic properties of a curve shortening evolution can be used in a number of well-known problems concerning plane curves. As a first example, a Birkhoff curve shortening evolution was a basic idea in proving the theorem of the three geodesics (any Riemannian 2-sphere has at least three simple closed geodesics) by L.A. Lusternik and L.G. Schnirelman (1929) [90]. A smooth evolution via curvature was used in [63] based on Uhlenbeck's suggestion of using the curvature flow.

Sturm's evolution PDEs approach on zero sets can give a new insight to a number of topological problems of plane and spherical curves, caustics, and related topics of symplectic geometry briefly outlined above. For instance, three of Arnold's theorems [15] on the number of inflection points (at least four for any embedded curve in $S^2$, the "tennis ball theorem"; and at least three for any noncontractible embedded curve in $\mathbb{RP}^2$) and extatic points (at least six for any plane convex curve) can be proved by using a suitable parabolic mean curvature evolution (the affine one for extatic points), see [11] and comments in [18]. Namely, the asymptotic expansion of the solution $u(x,t)$ as $t \to \infty$ describing the convergence to limiting geodesics via a 1D parabolic equation determines a minimally possible number of critical points. Then the result follows from Sturm's result on the nonincrease with time of the number of such points (e.g., inflections which are zeros of the curvature). While the Sturm-Hurwitz theorem can deal with infinitesimal perturbations of curves (see above), Sturm's evolution analysis extends the results to any finite perturbation. It follows that the statements from [17], p. 14, " The tennis ball theorem asserts that the result remains true for finite perturbations, even very large ones," and "... the tennis-ball theorem may be considered as a generalization of Hurwitz' theorem to the case of multi-valued functions" are covered by the first Sturm Theorem on zeros of single-valued functions (solutions of the PDE) since a suitable parabolic 1D evolution is available. The case of finite perturbations reduces via parabolic evolution to the infinitesimal one, and then Sturm's Theorem establishes that the number of critical points (zeros, inflections, extatic points, etc.) cannot be less than the eventual, infinitesimal one for arbitrarily small perturbations where a standard linearization applies. If a suitable parabolic evolution exists, the Sturm-Hurwitz theorem guarantees that the "infinitesimal geometric characteristic" of convergence (the number of critical points) is the optimal lower bound for any finite, arbitrarily large perturbation.

After a suitable surface parametrization, the quasilinear parabolic equation

$$u_t = u_{xx}/[1 + (u_x)^2] - (N-2)/u$$

describes the evolution of cylindrically symmetric hypersurfaces moving by mean curvature in $\mathbb{R}^N$, $N \geq 3$, [42], [121], [5]. A similar singular lower-order term occurs in the Prandtl boundary layer equations, which by von Mises non-local transformation reduce to the PME with an extra term $u_t = (uu_x)_x + g(t)/u$ where $g$ depends on the velocity of the potential flow (though in the original setting no singularities occur); see Section 30 in [134].

It is known that the first Sturm Theorem cannot be generalized to parabolic equations in $\mathbb{R}^N$ in the sense that such a general "order structure" does not exist; see [49] and the detailed survey [110].

The Sturmian classification of multiple zeros holds for a system of parabolic inequalities. Rescaling by Sturm's backward variable shows that Sturm's Theorems are true for $W^{2,1}_{p,\mathrm{loc}}$ solutions (from Tikhonov's class) of a system of parabolic inequalities

$$|u_t - u_{xx}| \leq M_1|u_x| + M_0|u|, \quad x \in \mathbb{R}, \ t \in J.$$

See [32], where such rescaling detailed analyses of nodal sets were carried out for equations in $\mathbb{R}^N$, namely the heat equation:

$$u_t = \Delta u \quad \text{in } \mathbb{R}^N \times (-\infty, 0).$$

In terms of Sturm's backward variable $z = x/(-t)^{1/2}$ this reduces to the rescaled equation

$$u_\tau = \mathbf{B}u \quad \text{in } \mathbb{R}^N \times \mathbb{R}_+, \quad \text{where } \tau = -\ln(-t) \to \infty \ \text{as } t \to 0^-, \tag{3.5}$$

with the symmetric second-order operator

$$\mathbf{B}u = \Delta u - \tfrac{1}{2}\, z \cdot \nabla u \equiv \tfrac{1}{\rho}\, \nabla \cdot (\rho \nabla u), \ \ \rho(z) = e^{-|z|^2/4}. \tag{3.6}$$

It is self-adjoint in $L^2_\rho(\mathbb{R}^N)$ with the domain $H^2_\rho(\mathbb{R}^N)$ and a point spectrum $\sigma(\mathbf{B}) = \{\lambda_\beta = -|\beta|/2, |\beta| = 0, 1, \ldots\}$ ($\beta = (\beta_1, \ldots, \beta_N)$ is a multiindex, $|\beta| = \beta_1 + \cdots + \beta_N$) and the eigenfunctions $\Phi = \{H_\beta(z) = \rho^{-1}(z) D^\beta \rho(z)\}$ are Hermite polynomials in $\mathbb{R}^N$; see [22], p. 48. The asymptotic structures $Ce^{\lambda_\beta \tau} H_\beta(z)$ with any eigenvalue $\lambda_\beta < 0$ describe for $\tau \to \infty$ all possible types of multiple zeros of the heat equation in $\mathbb{R}^N$. This makes it possible to study general properties (e.g., Hausdorff dimension) of nodal sets of general solutions [32].

The main principles of Sturm's evolution analysis of multiple zeros also remain valid for $2m$th order linear parabolic equations. Since the analysis is essentially local in a shrinking neighborhood of zero (according to Sturm's variable $z = x/(-t)^{1/2}$), without loss of generality, we consider the canonical $2m$th order parabolic equation with constant coefficients

$$u_t = -(-\Delta)^m u \quad \text{in } \mathbb{R}^N \times (-\infty, 0).$$

Sturm's backward variable takes the form

$$z = x/(-t)^{1/2m}$$

and we arrive at the equation (cf. (3.5))

$$u_\tau = \mathbf{B}u, \quad \text{where } \mathbf{B} = -(-\Delta)^m - \tfrac{1}{2m}\, z \cdot \nabla, \quad \tau = -\ln(-t). \tag{3.7}$$

For any $m > 1$, this operator is not self-adjoint in any weighted space $L^2_\rho(\mathbb{R}^N)$ unlike the second-order case $m = 1$. We introduce the space $L^2_\rho(\mathbb{R}^N)$ with the exponential weight $\rho(z) = e^{-a|z|^\alpha} > 0$ in $\mathbb{R}^N$, where $\alpha = 2m/(2m-1) \in (1, 2)$ and $a = a(m, N) > 0$ is a sufficiently small constant. For $m = 1$ we have $\alpha = 2$, $a = 1/4$ and $\rho(z) = e^{-|z|^2/4}$ is the rescaled Gaussian kernel as in (3.6). In $L^2_\rho(\mathbb{R}^N)$ the operator $\mathbf{B}$, with domain $H^{2m}_\rho(\mathbb{R}^N)$ being a weighted Sobolev space, admits the point spectrum $\sigma(\mathbf{B}) = \{\lambda_\beta = -|\beta|/2m \le 0,\ |\beta| = 0, 1, \dots\}$. The subset of eigenfunctions $\{\psi_\beta(z)\}$ (Kummer's polynomials in $\mathbb{R}^N$ of order $|\beta|$) is complete in $L^2_\rho(\mathbb{R}^N)$ [40], [54]. For $m = 1$, these are the Hermite polynomials. In view of completeness of polynomials, in the existence class $\{|u(x,t)| \le Ae^{a|x|^\alpha}\}$, $a, A > 0$, any solution of (3.5), (3.7) has the eigenfunction expansion $u(z,\tau) = \sum C_\beta e^{\lambda_\beta \tau} \psi_\beta(z)$. As a consequence, the complete subset of polynomials $\{\psi_\beta(z)\}$ describes in the rescaled form possible types of formation of multiple zeros occurring for this higher-order parabolic equation and describing local properties of nodal sets, [55]. Of course, the first Sturm Theorem in 1D (nonincrease of the number of zeros) is no longer available for $2m$th order equations, where new zeros can occur with evolution.

Finally, we notice that Sturm's zero-set ideas often play a crucial role in the asymptotic analysis of nonlinear parabolic PDEs admitting finite-time singularities or free boundaries of different types. A large amount of mathematical literature was devoted to these subjects during the last twenty years. An extensive list of references on geometric Sturmian approaches to nonlinear parabolic equations with applications to singularity formation phenomena (like blow-up, extinction or focusing) and regularity analysis of free-boundary problems are available in books [58] and [56] and in the survey papers [57] and [53].

## References


[1] A.D. Aleksandrov, Certain estimates for the Dirichlet problem, Soviet Math. Dokl. **1** (1960), 1151–1154.

[2] W. Allegretto, A comparison theorem for nonlinear operators, Ann. Scuola Norm. Sup. Pisa, Cl. Sci (4), **25** (1971), 41–46.

[3] W. Allegretto, Sturm type theorems for solutions of elliptic nonlinear problems, Nonl. Differ. Equat. Appl. **7** (2000), 309–321.

[4] W. Allegretto, Sturm Theorems for degenerate elliptic equations, Proc. Amer. Math. Soc. **129** (2001), 3031–3035.

[5] S. Altschuler, S. Angenent and Y. Giga, Mean curvature flow through singularities for surfaces of rotation, J. Geom. Anal. **5** (1995), 293–358.

[6] S.B. Angenent, The Morse-Smale property for a semi-linear parabolic equation, J. Differ. Equat. **62** (1986), 427–442.

[7] S. Angenent, The zero set of a solution of a parabolic equation, J. reine angew. Math. **390** (1988), 79–96.

[8] S. Angenent, Solutions of the one-dimensional porous medium equation are determined by their free boundary, J. London Math. Soc. (2) **42** (1990), 339–353.

[9] S. Angenent, Parabolic equations for curves on surfaces. Part I. Curves with $p$-integrable curvature, Ann. Math. **132** (1990), 451–483.

[10] S. Angenent, Parabolic equations for curves on surfaces. Part II. Intersections, blow-up and generalized solutions, Ann. Math. **133** (1991), 171–215.

[11] S. Angenent, Inflection points, extatic points and curve shortening, In: *Proceedings of the Conference on Hamiltonian Systems with* 3 *or More Degrees of Freedom*, S'Agarro, Catalunia, Spain, 1995.

[12] S.B. Angenent and B. Fiedler, The dynamics of rotating waves in scalar reaction diffusion equations, Trans. Amer. Math. Soc. **307** (1988), 545–568.

[13] V.I. Arnold, On the characteristic class entering in quantization condition, Funct. Anal. Appl. **1** (1967), 1–14.

[14] V.I. Arnold, The Sturm theorems and symplectic geometry, Funct. Anal. Appl. **19** (1985), 251–259.

[15] V.I. Arnold, *Topological Invariants of Plane Curves and Caustics*, A.M.S. University Lecture Series **5**, 1994.

[16] V.I. Arnol'd, The geometry of spherical curves and the algebra of quaternions, Russian Math. Surveys **50** (1995), 3–68.

[17] V.I. Arnold, Topological problems of the theory of wave propagation, Russian Math. Surveys **51** (1996), 1–47.

[18] V.I. Arnold, Topological problems in wave propagation theory and topological economy principle in algebraic geometry, Third Lecture by V. Arnold at the Meeting in the Fields Institute Dedicated to His 60th Birthday, Fields Inst. Commun., 1997.

[19] V.I. Arnold, Symplectic geometry and topology, J. Math. Phys. **41** (2000), 3307–3343.

[20] F.V. Atkinson, *Discrete and Continuous Boundary Problems*, Acad. Press, New York/London, 1964.

[21] S.N. Bernstein, The basis of a Chebyshev system, Izv. Akad. Nauk SSSR, Ser. Mat. **2** (1938), 499–504.

[22] M.S. Birman and M.Z. Solomjak, *Spectral Theory of Self-Adjoint Operators in Hilbert Space*, D. Reidel Publ. Comp., Dordrecht/Tokyo, 1987.

[23] J. Bochnak, M. Coste and M.-F. Roy, *Real Algebraic Geometry*, Springer-Verlag, Berlin/New York, 1998.

[24] R. Bott, On the iteration of closed geodesics and the Sturm intersection theory, Comm. Pure Appl. Math. **70** (1959), 313–337.

[25] P. Brunovský and B. Fiedler, Number of zeros on invariant manifolds in reaction-diffusion equations, Nonlinear Anal. TMA **10** (1986), 179–193.

[26] P. Brunovský and B. Fiedler, Simplicity of zeros in scalar parabolic equations, J. Differ. Equat. **62** (1986), 237–241.

[27] P. Brunovský and B. Fiedler, Connecting orbits in scalar reaction diffusion equations. II: The complete solution, J. Differ. Equat. **81** (1989), 106–135.

[28] P. Brunovský, P. Polácik and B. Sandstede, Convergence in general periodic parabolic equations in one space dimension, Nonlinear Anal. TMA **18** (1992), 209–215.

[29] T. Carleman, Sur un problème d'unicité pour le système d'équations aux dérivées partielles à deux variables indépendantes, Ark. Mat. Astr. Fys. **26B** (1939), 1–9.

[30] S.A. Chaplygin, *A New Method of Approximate Integration of Differential Equations*, GTTI, Moscow/Leningrad, 1932.

[31] M. Chen, X.-Y. Chen and J.K. Hale, Structural stability for time-periodic one-dimensional parabolic equations, J. Differ. Equat. **96** (1992), 355–418.

[32] X.-Y. Chen, A strong unique continuation theorem for parabolic equations, Math. Ann. **311** (1998), 603–630.

[33] X.-Y. Chen and H. Matano, Convergence, asymptotic periodicity, and finite-point blow-up in one-dimensional semilinear parabolic equations, J. Differ. Equat. **78** (1989), 160–190.

[34] X.-Y. Chen and P. Polácik, Asymptotic periodicity of positive solutions of reaction diffusion equations on a ball, J. reine angew. Math. **472** (1996), 17–51.

[35] S.-N. Chow, K. Lu and J. Mallet-Pare, Floquet theory for parabolic differential equations, J. Differ. Equat. **109** (1993), 147–200.

[36] S.-N. Chow, K. Lu and J. Mallet-Pare, Floquet bundles for scalar parabolic equations, Arch. Rational Mech. Anal. **129** (1995), 245–304.

[37] R. Courant and D. Hilbert, *Methods of Mathematical Physics*, Vol. **1**, Intersci. Publ., Inc., New York, 1953.

[38] C. de la Vallée-Poussin, Sur l'équation différentielle linéaire du second ordre. Détermination d'une intégrale par deux valeurs assignées. Extension aux équations d'ordre $n$, J. Math. Pures Appl. **8** (1929), 125–144.

[39] H.M. Edwards, A generalized Sturm theorem, Ann. Math. **80** (1964), 22–57.

[40] Yu.V. Egorov, V.A. Galaktionov, V.A. Kondratiev and S.I. Pohozaev, Global solutions of higher-order semilinear parabolic equations in the supercritical range, Comptes Rendus Acad. Sci. Paris, Série I, **335** (2002), 805–810.

[41] C.L. Epstein and M.I. Weinstein, A stable manifold theorem for the curve shortening equations, Comm. Pure Appl. Math. **40** (1987), 119–139.

[42] L.C. Evans and J. Spruck, Motion of level sets by mean curvature. I, J. Differ. Geom. **33** (1991), 635–681.

[43] E. Feireisl and P. Poláčik, Structure of periodic solutions and asymptotic behavior for time-periodic reaction-diffusion equations on $\mathbb{R}$, Adv. Differ. Equat. **5** (2000), 583–622.

[44] B. Fiedler, Discrete Ljapunov functionals and $\omega$-limit sets, Math. Model. Numer. Anal. **23** (1989), 415–431.

[45] B. Fiedler and J. Mallet-Paret, A Poincaré-Bendixson theorem for scalar reaction diffusion equations, Arch. Rational Mech. Anal. **107** (1989), 325–345.

[46] B. Fiedler and J. Mallet-Paret, Connections between Morse sets for delay-differential equations, J. reine angew. Math. **397** (1989), 23–41.

[47] S. Filippas and R.V. Kohn, Refined asymptotics for the blow-up of $u_t - \Delta u = u^p$, Comm. Pure Appl. Math. **45** (1992), 821–869.

[48] G. Floquet, Sur les équations différentielles linéaires à coefficients périodiques, Ann. Sci. École Norm. Sup. **12** (1883), 47–89.

[49] G. Fusco and S.M. Verduyn Lunel, Order structures and the heat equation, J. Differ. Equat. **139** (1997), 104–145.

[50] A. Gabriaelov and N. Vorobjov, Complexity of stratification of semi-Pfaffian sets, Discr. Comput. Geom. **14** (1995), 71–91.

[51] M. Gage and R.S. Hamilton, The heat equation shrinking convex plane curves, J. Differ. Geom. **23** (1986), 69–96.

[52] V.A. Galaktionov, On localization conditions for unbounded solutions of quasilinear parabolic equations, Soviet Math. Dokl. **25** (1982), 775–780.

[53] V.A. Galaktionov, Geometric theory of one-dimensional nonlinear parabolic equations I. Singular interfaces, Adv. Differ. Equat. **7** (2002), 513–580.

[54] V.A. Galaktionov, On a spectrum of blow-up patterns for a higher-order semilinear parabolic equation, Proc. Royal Soc. London A **457** (2001), 1–21.

[55] V.A. Galaktionov, Sturmian nodal set analysis for higher-order parabolic equations and applications, Trans. Amer. Math. Soc., submitted.

[56] V.A. Galaktionov, *Geometric Sturmian Theory of Nonlinear Parabolic Equations*, Chapman and Hall/CRC, Boca Raton, Florida, 2004.

[57] V.A. Galaktionov and J.L. Vazquez, The problem of blow-up in nonlinear parabolic equations, Discr. Cont. Dyn. Syst. **8** (2002), 399–433.

[58] V.A. Galaktionov and J.L. Vazquez, *A Stability Technique for Evolution Partial Differential Equations. A Dynamical Systems Approach*, Birkhäuser, Boston/Berlin, 2003.

[59] F.R. Gantmakher, Non-symmetric Kellogg kernels, Doklady Akad. Nauk SSSR **1** (1936), 3–5.

[60] F.R. Gantmakher and M.G. Krein, *Oscillation Matrices and Kernels and Small Vibrations of Mechanical Systems*, Gostekhizdat (Izdat. Tekhn. Teor. Lit.), Moscow/ Leningrad, 1950; English translation from the 1950 2nd Russian ed.: AEC-tr-4481, U.S. Atomic Energy Commission, Oak Ridge, Tenn., 1961.

[61] A. Givental, Sturm theorem for hyperelliptic integrals, Leningrad J. of Math. **1** (1990), 1157–1163.

[62] E.K. Godunova and V.I. Levin, Certain qualitative questions of heat conduction, USSR Comp. Math. Math. Phys. **6** (1966), 212–220.

[63] M.A. Grayson, Shortening embedded curves, Ann. Math. **129** (1989), 71–111.

[64] J.K. Hale, Dynamics in parabolic equations – an example, In: *Systems of Nonlinear Partial Differential Equations*, J.M. Ball, ed., pp. 461–472, Reidel, Dordrecht, 1983.

[65] D.B. Henry, Some infinite-dimensional Morse-Smale systems defined by parabolic partial differential equations, J. Differ. Equat. **59** (1985), 165–205.

[66] M.A. Herrero and J.J.L. Velázquez, Blow-up behavior of one-dimensional semilinear parabolic equations, Ann. Inst. Henri Poincaré, Analyse non linéaire **10** (1993), 131–189.

[67] P. Hess and P. Polácik, Symmetry and convergence properties for non-negative solutions of nonautonomous reaction-diffusion problems, Proc. Royal Soc. Edinburgh **124A** (1994), 573–587.

[68] E. Hille, *Ordinary Differential Equations in the Complex Domain*, Dover Publications, Mineola, New York, 1997.

[69] D. Hinton, Sturm's 1836 Oscillation Results. Evolution of the Theory, this volume.

[70] G. Huisken, Flow by mean curvature of convex surfaces into spheres, J. Differ. Geom. **20** (1984), 237–266.

[71] A. Hurwitz, Über die Fourierschen Konstanten integrierbarer Funktionen, Math. Ann. **57** (1903), 425–446.

[72] Y. Ilýashenko and S. Yakovenko, Counting real zeros of analytic functions satisfying linear ordinary differential equations, J. Differ. Equat. **126** (1996), 87–105.

[73] I.K. Ivanov, A relation between the number of changes of sign of the solution of the equations $\partial[a(t,x)\partial u/\partial x]/\partial x - \partial u/\partial t = 0$ and the nature of its diminution, Godisnik Viss. Tehn. Ucebn. Zaved. Mat. **1** (1964), kn. 1, 107–116 (1965).

[74] C.G.J. Jacobi, *Vorlesungen über Dynamik*, G. Reiner, Berlin, 1884.

[75] J. Jones, Jr. and T. Mazumdar, On zeros of solutions of certain Emden-Fowler like differential equations in Hilbert space, Nonlinear Anal. TMA **12** (1988), 365–373.

[76] S. Karlin, Total positivity, absorption probabilities and applications, Trans. Amer. Math. Soc. **111** (1964), 33–107.

[77] O.D. Kellogg, The oscillation of functions of an orthogonal set, Amer. J. Math. **38** (1916), 1–5.

[78] O.D. Kellogg, On the existence and closure of sets of characteristic functions, Math. Ann. **86** (1922), 14–17.

[79] A.G. Khovanskii (A.G. Hovanskii), On a class of systems of transcendental equations, Soviet Math. Dokl. **22** (1980), 762–765.

[80] A.G. Khovanskii, *Fewnomials*, AMS Transl. Math. Monogr. **88**, Amer. Math. Soc., Providence, Rhode Island, 1991.

[81] A. Kneser, Festschrift zum 70. Geburtstag von H. Weber, Leipzig, 1912, pp. 170–192.

[82] A.N. Kolmogorov, I.G. Petrovskii and N.S. Piskunov, Study of the diffusion equation with growth of the quantity of matter and its application to a biological problem, Byull. Moskov. Gos. Univ., Sect. A, **1** (1937), 1–26. See [107], pp. 105–130 for an English translation.

[83] M.G. Krein, Sur les fonctions de Green non-symétriques oscillatoires des opérateurs différentiels ordinaires, Doklady Akad. Nauk SSSR **25** (1939), 643–646.

[84] K. Kunisch and G. Peichl, On the shape of the solutions of second-order parabolic differential equations, J. Differ. Equat. **75** (1988), 329–353.

[85] E.M. Landis, A property of solutions of a parabolic equation, Soviet Math. Dokl. **7** (1966), 900–903.

[86] S. Lang, *Algebra*, Addison-Wesley, Reading/Tokyo, 1984.

[87] A.Yu. Levin, Non-oscillation of solutions of the equation $x^{(n)} + p_1(t)x^{(n-1)} + \cdots + p_n(t)x = 0$, Russian Math. Surveys **24** (1969), 43–99.

[88] V.B. Lidskii, Oscillatory theorems for a canonical system of differential equations, Doklady Acad. Nauk SSSR **102** (1955), 877–880.

[89] J. Liouville, Démonstration d'un théorème dû à M. Sturm et relatif à une classe de fonctions transcendantes, J. Math. Pures Appl. **1** (1836), 269–277.

[90] L. Lusternik and L. Schnirelman, Sur le problème de trois géodésiques fermées sur les surfaces de genre O, Comptes Rendus Acad. Sci. Paris **189** (1929), 269–271.

[91] A.M. Lyapunov, *The General Problem of the Stability of Motion*, Kharkov, 1892 (in Russian); Taylor & Francis, London, 1992. A. Liapunoff, Problème général de la stabilité du mouvement, Ann. Fac. Sciences de Toulouse, Second series **9** (1907), 203–469; Reprinted as Ann. of Math. Studies, No. **17**, Princeton Univ. Press, 1947.

[92] J. Mallet-Paret, Morse decomposition for delay-differential equations, J. Differ. Equat. **72** (1988), 270–315.

[93] J. Mallet-Paret and H.L. Smith, The Poincaré-Bendixson theorem for monotone cyclic feedback systems, J. Dyn. Differ. Equat. **2** (1990), 367–421.

[94] W.A. Markov, Über Polynome, die in einem gegebenen Intervalle möglichst wenig von Null abweichen, Math. Ann. **77** (1916), 213–258.

[95] V.P. Maslov, *Théorie des Perturbations et Méthodes Asymptotiques*, Thesis, Moscow State Univ., 1965; Dunod, Paris, 1972.

[96] H. Matano, Convergence of solutions of one-dimensional semilinear parabolic equations, J. Math. Kyoto Univ. (JMKYAZ) **18** (1978), 221–227.

[97] H. Matano, Nonincrease of the lap-number of a solution for a one-dimensional semilinear parabolic equation, J. Fac. Sci. Univ. Tokyo, Sect. IA, Math. **29** (1982), 401–441.

[98] H. Matano, Asymptotic behavior of solutions of semilinear heat equations on $S^1$, In: *Nonl. Diff. Equat. Equil. States*, Vol. **II**, J. Serrin, W.-M. Ni and L.A. Peletier, Eds, Springer, New York, 1988, pp. 139–162.

[99] F. Merle and H. Zaag, Optimal estimates for blowup rate and behavior for nonlinear heat equations, Comm. Pure Appl. Math. **51** (1998), 139–196.

[100] N. Mizoguchi and E. Yanagida, Critical exponents for the decay rate of solutions in a semilinear parabolic equation, Arch. Rat. Mech. Anal. **145** (1998), 331–342.

[101] M. Morse, A generalization of the Sturm theorems in $n$ space, Math. Ann. **103** (1930), 52–69.

[102] M. Morse, *The Calculus of Variation in the Large*, AMS Colloquium Publications Vol. **18**, New York, 1934.

[103] S. Mukhopadyaya, New methods in the geometry of a plane arc I, Bull. Calcutta Math. Soc. **1** (1909), 31–37.

[104] A.D. Myschkis, *Lineare Differentialgleichungen mit nacheilendem Argument*, Deutscher Verlag Wiss., Berlin, 1955.

[105] K. Nickel, Gestaltaussagen über Lösungen parabolischer Differentialgleichungen, J. Reine Angew. Math. **211** (1962), 78–94.

[106] K. Nickel, Einige Eigenschaften von Lösungen der Prandtlschen Grenzschicht-Differentialgleichungen, Arch. Rational Mech. Anal. **2** (1958), 1–31.

[107] P. Pelcé, *Dynamics of Curved Fronts*, Acad. Press, New York, 1988.

[108] M. Picone, Un teorema sulle soluzioni delle equazioni ellittiche autoaggiunte alle derivate parziali del secondo ordine, Atti Accad. Naz. Lincei Rend. **20** (1911), 213–219.

[109] P. Poláčik, Transversal and nontransversal intersections of stable and unstable manifolds in reaction diffusion equations on symmetric domains, Differ. Integr. Equat. **7** (1994), 1527–1545.

[110] P. Poláčik, Parabolic equations: asymptotic behavior and dynamics on invariant manifolds, *Handbook of Dynamical Systems*, Vol. **2**, pp. 835–883, North-Holland, Amsterdam, 2002.

[111] G. Pólya, On the mean-value theorem corresponding to a given linear homogeneous differential equation, Trans. Amer. Math. Soc. **24** (1924), 312–324.

[112] G. Pólya, Qualitatives über Wärmeausgleich, Z. Angew. Math. Mech. **13** (1933), 125–128.

[113] C.-C. Poon, Unique continuation for parabolic equations, Comm. Part. Differ. Equat. **21** (1996), 521–539.

[114] R.M. Redheffer and W. Walter, The total variation of solutions of parabolic differential equations and a maximum principle in unbounded domains, Math. Ann. **209** (1974), 57–67.

[115] W.T. Reid, *Sturmian Theory for Ordinary Differential Equations*, Springer-Verlag, Berlin/New York, 1980.

[116] A.A. Samarskii, V.A. Galaktionov, S.P. Kurdyumov and A.P. Mikhailov, *Blow-up in Quasilinear Parabolic Equations,* Nauka, Moscow, 1987; English transl., rev.: Walter de Gruyter, Berlin/New York, 1995.

[117] G. Sapiro and A. Tannenbaum, An affine curve evolution, J. Funct. Anal. **119** (1994), 79–120.

[118] D.H. Sattinger, On the total variation of solutions of parabolic equations, Math. Ann. **183** (1969), 78–92.

[119] J.A. Sethian, Curvature and the evolution of fronts, Comm. Math. Phys. **101** (1985), 487–499.

[120] W. Shen and Y. Yi, Asymptotic almost periodicity of scalar parabolic equations with almost periodic time dependence, J. Differ. Equat. **122** (1995), 373–397.

[121] H.M. Soner and P.E. Souganidis, Singularities and uniqueness of cylindrically symmetric surfaces moving by mean curvature, Comm. Partial Differ. Equat. **18** (1993), 859–894.

[122] A.N. Stokes, Intersections of solutions of nonlinear parabolic equations, J. Math. Anal. Appl. **60** (1977), 721–727.

[123] A.N. Stokes, Nonlinear diffusion waveshapes generated by possibly finite initial disturbances, J. Math. Anal. Appl. **61** (1977), 370–381.

[124] C. Sturm, Mémoire sur la résolution des équations numériques, Mém. Savants Étrangers, Acad. Sci. Paris **6** (1835), 271–318.

[125] C. Sturm, Mémoire sur les Équations différentielles linéaires du second ordre, J. Math. Pures Appl. **1** (1836), 106–186.

[126] C. Sturm, Mémoire sur une classe d'Équations à différences partielles, J. Math. Pures Appl. **1** (1836), 373–444.

[127] C.A. Swanson, *Comparison and Oscillation Theory of Linear Differential Equations*, Acad. Press, New York/London, 1968.

[128] G. Szegö, *Orthogonal Polynomials*, Amer. Math. Soc., Providence, Rhode Island, 1975.

[129] S.L. Tabachnikov, Around four vertices, Russian Math. Surveys **45** (1990), 229–230.

[130] M. Tabata, A finite difference approach to the number of peaks of solutions for semilinear parabolic problems, J. Math. Soc. Japan **32** (1980), 171–191.

[131] J.J.L. Velazquez, Estimates on $(N-1)$-dimensional Hausdorff measure of the blow-up set for a semilinear heat equation, Indiana Univ. Math. J. **42** (1993), 445–476.

[132] J.J. Velazquez, V.A. Galaktionov and M.A. Herrero, The space structure near a blow–up point for semilinear heat equations: a formal approach, Comput. Math. and Math. Phys. **31** (1991), 46–55.

[133] W. Walter, *Differential- und Integral-Ungleichungen*, Springer Tracts in Natural Philosophy, Vol. **2**, Springer, Berlin/New York, 1964.

[134] W. Walter, *Differential and Integral Inequalities*, Springer, Berlin/New York, 1970.

[135] T.I. Zelenyak, Stabilization of solutions of boundary value problems for a second order parabolic equation with one space variable, Differ. Equat. **4** (1968), 17–22.

[136] N.F. Zhukovskii, Conditions of finiteness of integrals of the equation $d^2y/dx^2+py=0$, Mat. Sbornik **3** (1892), 582–591.



Victor A. Galaktionov
Department of Mathematical Sciences
University of Bath
Bath, BA2 7AY
UK

*and*

Keldysh Institute of Applied Mathematics
Miusskaya Sq. 4
125047 Moscow, Russia
e-mail: `vag@maths.bath.ac.uk`

Petra J. Harwin
Department of Mathematical Sciences
University of Bath
Bath, BA2 7AY, UK
e-mail: `P.J.Harwin@bath.ac.uk`

W.O. Amrein, A.M. Hinz, D.B. Pearson
Sturm-Liouville Theory: Past and Present, 201–216


# A Survey of Nonlinear Sturm-Liouville Equations

Chao-Nien Chen

**Abstract.** This note gives a brief survey of existence, uniqueness and bifurcation results for nonlinear Sturm-Liouville equations. Early in 1960, Nehari made an interesting proposal to study solutions with a prescribed number of nodes. His ideas have had a great influence on critical point theory as a branch of the calculus of variations. Rabinowitz established a global bifurcation theorem based on the nodal properties of solutions. Some results on bifurcation from the lowest point of the continuous spectrum will also be discussed.

## 1. Introduction

Let $\mathcal{L}u = -(p(x)u')' + q(x)u$, where $p$ is positive and continuously differentiable and $q$ is continuous. Consider the Sturm-Liouville boundary value problem

$$\begin{gathered}\mathcal{L}u = \lambda r(x)u + h(x, u, u', \lambda), \quad a < x < b, \\ \alpha_1 u(a) - \beta_1 u'(a) = 0, \quad \alpha_2 u(b) - \beta_2 u'(b) = 0,\end{gathered} \tag{1.1}$$

where $\lambda$ is a parameter, $r$ is positive and continuous, and $(\alpha_1^2 + \beta_1^2)(\alpha_2^2 + \beta_2^2) \neq 0$. A well-known example of a nonlinear Sturm-Liouville equation is the pendulum equation

$$-u'' = \lambda \sin u.$$

Here the position of the pendulum is described by $u$, which is the angle between the rod and the downward vertical direction; $\lambda$ is a constant depending on the gravitational acceleration and on properties of the pendulum.

Early in 1960, Nehari [23] started to investigate nonlinear Sturm-Liouville equations of the form

$$\begin{gathered}-u'' = F(x, u^2)u, \qquad a < x < b, \\ u(a) = u(b) = 0,\end{gathered} \tag{1.2}$$

This work is supported in part by the National Science Council of the Republic of China.

where the function $F(x,\xi)$ satisfies the following conditions:

(F1) $F(x,\xi)$ is continuous in $x$ and $\xi$ for $a \le x \le b$ and $0 \le \xi < \infty$ respectively, and $F(x,\xi) > 0$ if $\xi > 0$.

(F2) There exists a positive number $\sigma$ such that, for any $x$ in $[a,b]$, $\xi^{-\sigma}F(x,\xi)$ is a non-decreasing function of $\xi$ for $\xi \in [0,\infty)$.

The solutions of (1.2) obtained in [23] are continuously differentiable functions. If $F(x,\xi)$ is jointly continuous, i.e., $F \in C([a,b]\times[0,\infty),\mathbb{R})$, then such solutions are twice continuously differentiable. In order to study the oscillation properties of second-order nonlinear differential equations, Nehari made an interesting proposal to study solutions with a prescribed number of nodes. Using constrained minimization arguments, he established the following existence result for (1.2).

**Theorem 1.1.** *Under the above hypotheses on $F$, for each $j \in \mathbb{N}$ there exist at least a pair of solutions which possess precisely $j-1$ zeros in $(a,b)$.*

A simple example for (1.2) is

$$-u'' = w(x)|u|^{\delta}u,$$

where $\delta > 0$ and $w$ is a positive continuous function on $[a,b]$. Recall that in case $h \equiv 0$, associated with (1.1) is the linear Sturm-Liouville eigenvalue problem

$$\begin{aligned} &\mathcal{L}v = \lambda r(x)v, \quad a < x < b, \\ &\alpha_1 v(a) - \beta_1 v'(a) = 0, \quad \alpha_2 v(b) - \beta_2 v'(b) = 0. \end{aligned} \tag{1.3}$$

Let $\lambda_1 < \lambda_2 < \cdots < \lambda_j < \cdots$ be the eigenvalues of (1.3). It is known that $\lambda_j$ is a simple eigenvalue, and that any eigenfunction $v_j$ corresponding to $\lambda_j$ has exactly $j-1$ zeros in $(a,b)$ where all zeros of $v_j$ in $[a,b]$ are simple. (A simple zero of $v_j$ is a point $x$ at which $v_j(x) = 0$ and $v_j'(x) \neq 0$.) It is convenient to introduce for each $j \in \mathbb{N}$ the set $S_j^+$ of $\psi \in C^1([a,b],\mathbb{R})$ such that $\psi$ satisfies the boundary conditions of (1.1), $\psi > 0$ in a deleted neighborhood of $x = a$, $\psi$ has exactly $j-1$ zeros in $(a,b)$, and all zeros of $\psi$ in $[a,b]$ are simple. Set $S_j^- = -S_j^+$ and $S_j = S_j^+ \cup S_j^-$. Then $S_j^+$, $S_j^-$ and $S_j$ are open subsets of

$$E = \{\psi \in C^1[a,b],\mathbb{R}) \mid \psi \text{ satisfies the boundary conditions of (1.1)}\}.$$

Moreover $v_j$ defined above belongs to $S_j$ and can be made unique by requiring that $v_j \in S_j^+$ and $\|v_j\| = 1$. Here the $C^1$-norm is taken as the norm of $E$:

$$\|\psi\| = \max_{x\in[a,b]} |\psi(x)| + \max_{x\in[a,b]} |\psi'(x)|.$$

The above remarks show that (1.3) possesses the family of trivial solutions $\{(\lambda,0) \mid \lambda \in \mathbb{R}\}$ together with, for each $j \in \mathbb{N}$, a line of nontrivial solutions

$$\{(\lambda_j, \alpha v_j) \mid \alpha \in \mathbb{R}\} \subset (\{\lambda_j\} \times S_j) \cup \{(\lambda_j, 0)\}.$$

Rabinowitz [25] showed that a nonlinear analogue of this situations holds for (1.1) if $h$ satisfies the following condition:

(h1) $h$ is a continuous function and $h(x, \xi, \eta, \lambda) = o((\xi^2 + \eta^2)^{\frac{1}{2}})$ as $(\xi, \eta) \to (0, 0)$ uniformly on bounded $\lambda$ intervals.

**Theorem 1.2.** *If* (h1) *is satisfied, then, for each* $j \in \mathbb{N}$, (1.1) *possesses a continuum of solutions* $C_j$ *in* $\mathbb{R} \times E$ *with* $C_j \subset (\mathbb{R} \times S_j) \cup \{(\lambda_j, 0)\}$ *and* $C_j$ *unbounded.*

In fact, by a strong version of Theorem 1.2, (1.1) possesses two unbounded continua of solutions, $C_j^+$ and $C_j^-$, with $C_j^+ \subset (\mathbb{R} \times S_j^+) \cup \{(\lambda_j, 0)\}$ and $C_j^- \subset (\mathbb{R} \times S_j^-) \cup \{(\lambda_j, 0)\}$.

A familiar physical example of bifurcation is the buckling of a column [1]. This example was first studied by Euler in 1744. Consider a thin straight column subjected to an axial compression. Suppose the column lies on the x-axis. If the magnitude of the compression is small, the column remains linear; however if the force exceeds a certain critical level, the column buckles, i.e., it deflects out of its linear state. If the length of the column is $\pi$, $u(x)$ denotes the angle between the tangent to the column at point $x$ and the x-axis, and $\lambda$ is the applied thrust (a measure of the magnitude of the compressive force), then $u$ satisfies the equation

$$\begin{aligned} -u'' &= \lambda \sin u, \qquad 0 < x < \pi, \\ u'(0) &= u'(\pi) = 0. \end{aligned}$$

Thus $u \equiv 0$ corresponds to the unbuckled state and is a solution of (1.2) for all values of $\lambda$. As will be seen, a buckled state appears when $\lambda$ exceeds 1.

There is a sizable literature [2, 4–10, 14–25, 27–29, 32–37] on the study of nonlinear Sturm-Liouville equations. Therefore it is far beyond the scope of this report to cover all related results of this subject. Nehari's works initiated new developments in the study of nonlinear differential equations. These ideas have had a great influence on critical point theory as a branch of the calculus of variations, and will be considered in Section 2.

The global bifurcation theorem established by Rabinowitz covers a large variety of nonlinearities. As will be seen in Section 3, by carefully studying the qualitative behavior of the sets $C_j^+$ and $C_j^-$, many existence results can be unified by skillful applications of the global bifurcation theorem. A useful tool for this purpose is the Sturm Comparison Theorem. The interested reader may consult [29] for more complete references and related results.

Section 4 deals with nonlinear Sturm-Liouville problems on unbounded intervals, where the linearization at zero solution could have no point spectrum. Some examples of bifurcation from the lowest point of the continuous spectrum will be discussed [6, 8, 14–16, 18, 20, 21, 32–34]. Unlike Theorem 1.2, two continua with different nodal properties could merge together, in which case the phenomenon of "losing nodes at infinity" takes place.

## 2. Nehari's variational method

Consider the functional

$$I(u) = \int_a^b \left[(u'(x))^2 - \int_0^{(u(x))^2} F(x,\xi)d\xi\right]dx \tag{2.1}$$

within the class $\Gamma_0$ of continuous functions $u(x)$ which have a piecewise continuous derivative in $[a,b]$ and satisfy $u(a) = u(b) = 0$. Although (1.2) is the Euler-Lagrange equation corresponding to (2.1), it is easy to check that $I(u)$ is neither bounded from above nor from below as $u(x)$ ranges over the class of functions in question.

To obtain an extremum of $I$, Nehari imposed a constraint on $u$:

$$\int_a^b (u'(x))^2 dx = \int_a^b u^2 F(x,u^2)dx. \tag{2.2}$$

This constraint is satisfied by solutions of (1.2), which can easily be confirmed by multiplying both sides of (1.2) by $u(x)$ and integrating by parts. For any $y \not\equiv 0$, it is always possible to find a positive number $\alpha$ such that $u(x) = \alpha y(x)$ satisfies (2.2). This is equivalent to finding an $\alpha$ such that

$$\int_a^b (y'(x))^2 dx = \int_a^b y^2 F(x,\alpha^2 y^2)dx, \tag{2.3}$$

and the truth of the assertion follows from the observation that the right-hand side of (2.3) is a continuous function of $\alpha$ which, in accordance with (F2), tends to 0 as $\alpha \to 0$ and to $\infty$ as $\alpha \to \infty$.

For given $b > a$, define $\Gamma = \{u \mid u \in \Gamma_0 \text{ and } u \text{ satisfies (2.2)}\}$ and $\nu(a,b) = \inf_{u\in\Gamma} I(u)$.

**Theorem 2.1.** *For any fixed interval $(a,b)$, one has $\nu(a,b) > 0$ and there exists a function $u \in \Gamma$ such that $I(u) = \nu(a,b)$. Moreover, if $u \in \Gamma$ and $I(u) = \nu(a,b)$, then $|u| > 0$ in $(a,b)$ and $|u|$ is a positive solution of* (1.2).

Nehari showed that (1.2) has an infinite number of other solutions in addition to the positive one. These can be obtained by minimizing $I$ under increasingly restrictive side conditions. Pick $j+1$ distinct points $x_k$ such that $a = x_0 < x_1 < x_2 < \cdots < x_{j-1} < x_j = b$. In the interval $[x_{k-1}, x_k]$, consider functions $y$ which are piecewise continuously differentiable, vanish at $x = x_{k-1}$ and $x = x_k$ (but not identically) and are normalized by

$$\int_{x_{k-1}}^{x_k} (y'(x))^2 dx = \int_{x_{k-1}}^{x_k} y^2 F(x,y^2)dx.$$

Observe that $-|u|$ is a negative solution of (1.2) if $I(u) = \nu(a,b)$. Theorem 2.1 shows that it is sufficient to consider the following minimum problem (2.4) for functions $y(x)$ which in the intervals $[x_{k-1}, x_k]$ coincide, respectively, with the solutions $u_k(x)$ of (1.2) which vanish at $x = x_{k-1}$ and $x = x_k$, and whose existence has already been established. For $x \in [x_{k-1}, x_k]$, assign $y(x) = (-1)^{k+1}|u_k(x)|$ and set $I(y) = \nu_{j-1}(x_1, x_2, \ldots, x_{j-1})$ with

$$\nu_{j-1}(x_1, x_2, \ldots, x_{j-1}) = \sum_{k=1}^{j} \nu(x_{k-1}, x_k), \tag{2.4}$$

where $y$ ranges over the class of all functions with the indicated properties. Nehari proved that the set of numbers $x_1, \ldots, x_{j-1}$ for which the right-hand side of (2.4) attains its minimum is such that the corresponding solutions $(-1)^{k+1}|u_k(x)|$ of (1.2) combine to a single solution $y(x)$ of (1.2) in the interval $[a, b]$. This solution $y(x)$ vanishes for $x = a$ and $x = b$ and has precisely $j - 1$ zeros in $(a, b)$.

The existence of a solution $y \in S_j$ of (1.2) will be a consequence of the following properties of $\nu(a, b)$ as a function of a and b.

**Lemma 2.1.** *The function $\nu(a, b)$ is continuous with respect to both $a$ and $b$. If $a \leq \bar{a} < \bar{b} \leq b$, then $\nu(a, b) \leq \nu(\bar{a}, \bar{b})$. Moreover, $\nu(a, b) \to \infty$ as $b - a \to 0$.*

Lemma 2.1 implies that $\nu_{j-1}$ is bounded from below and that there exists a set of $j - 1$ distinct points $x_1, x_2, \ldots, x_{j-1} \in (a, b)$ for which $\nu_{j-1}$ attains its minimum. It remains to show that

$$\lim_{x \to x_k^-} y'(x) = \lim_{x \to x_k^+} y'(x). \tag{2.5}$$

Observe that (2.5) is equivalent to

$$|u_k'(x_k)| = |u_{k+1}'(x_k)|. \tag{2.6}$$

If (2.6) fails to hold at some $x_k$, Nehari proved that $\nu_{j-1}(x_1, x_2, \ldots, x_{j-1})$ is not a minimum of (2.4). The details, as well as the proof of Lemma 2.1, can be found in [23].

## 3. A global bifurcation theorem and its applications

Let $E$ be a real Banach space and $G : \mathbb{R} \times E \to E$. Consider the bifurcation problem for solutions of the operator equation

$$u = G(\lambda, u). \tag{3.1}$$

Suppose $G(\lambda, 0) = 0$ for all $\lambda \in \mathbb{R}$ so $G$ possesses the family of trivial solutions $\mathcal{T} = \{(\lambda, 0) \mid \lambda \in \mathbb{R}\}$. A point $(\hat{\lambda}, 0)$ is said to be a bifurcation point for solutions of (3.1) if every neighborhood of $(\hat{\lambda}, 0)$ contains a solution of (3.1) not in $\mathcal{T}$. Let $S$ denote the closure in $\mathbb{R} \times E$ of the set of nontrivial solutions of (3.1). The following bifurcation result was proved by Rabinowitz [26].

**Theorem 3.1.** *Suppose $G(\lambda, u) = \lambda L u + H(\lambda, u)$ where $L$ is a linear compact operator, $H$ is compact and $H(\lambda, u) = o(\|u\|)$ as $u \to 0$ uniformly on bounded $\lambda$ intervals. Let $\sigma(L)$ be the spectrum of $L$. If $\mu^{-1} \in \sigma(L)$ is of odd multiplicity, then $S$ contains a component $C$ containing $(\mu, 0)$. Moreover $C$ is either unbounded in $\mathbb{R} \times E$ or $C$ contains $(\gamma, 0)$, where $\gamma^{-1} \in \sigma(L)$ and $\gamma \neq \mu$.*

**Remark 3.1.** The multiplicity of the spectral point $\mu^{-1}$ is defined to be the dimension of $\bigcup_{j\in\mathbb{N}} N\left((id-\mu L)^j\right)$, where $N(A)$ denotes the null space of $A$.

Theorem 1.2 is a global bifurcation theorem for nonlinear Sturm-Liouville equations and can be deduced from Theorem 3.1. The proof of Theorem 1.2 may be divided into two steps. The case that zero is not an eigenvalue of (1.3) is treated first and then this restriction is removed by using an approximation argument. Two preliminary results are needed.

**Lemma 3.1.** *If* $(\lambda, u)$ *is a solution of* (1.1) *and* $u$ *has a double zero, i.e., there is a* $\xi \in [a,b]$ *such that* $u(\xi)=0$ *and* $u'(\xi)=0$*, then* $u(\xi)\equiv 0$.

*Proof.* Consider the following linear equation:

$$\mathcal{L}v = \lambda r v + \frac{h(x,u(x),u'(x),\lambda)}{u(x)^2+u'(x)^2}(u(x)v(x)+u'(x)v'(x)). \tag{3.2}$$

If $u(x)$ is a solution of (1.1), assumption (h1) implies that (3.2) has continuous coefficients. With the initial conditions $v(\xi)=v'(\xi)=0$ we see that $v(x)\equiv 0$ is a solution. But $v(x)=u(x)$ is also a solution. The basic existence and uniqueness theorem for the initial value problem implies that $u(x)\equiv 0$. □

**Lemma 3.2.** *There exists a neighborhood* $N_j$ *of* $(\lambda_j,0)$ *such that if* $(\lambda,u)$ *is a nontrivial solution of* (1.1) *and* $(\lambda,u)\in N_j$*, then* $u\in S_j$.

*Proof.* Suppose the assertion of the lemma is false. Then there is a sequence of nontrivial solutions $(\gamma_n,u_n)$ such that $(\gamma_n,u_n)\to(\lambda_j,0)$ and $u_n\notin S_j$. Since $u_n/\|u_n\|$ is bounded in $E$, the $C^1$ bound for $\{u_n\}$ together with (1.1) shows $\{u_n\}$ is bounded in $C^2$. The Arzelà-Ascoli Theorem implies that, for some subsequence, $(\gamma_n,u_n/\|u_n\|)$ converges to $(\lambda_j,v)$, where $\|v\|=1$. Furthermore, condition (h1) implies that $v$ satisfies (1.3). In other words, $v=\beta v_j$ for some $\beta\neq 0$. Since $S_j$ is an open set, it follows that for this subsequence $u_n/\|u_n\|\in S_j$ for large $n$. This leads to a contradiction, in view of the fact that $S_j$ is invariant under multiplication by nonzero scalars. □

*Proof of Theorem 1.2.* In case zero is not an eigenvalue of (1.3), there exists a Green's function $g$ with the aid of which (1.1) can be converted into an equivalent operator equation of the form

$$u = \lambda L u + H(\lambda,u).$$

Furthermore, $L$ and $H$ satisfy the hypotheses of Theorem 3.1 and one has $\sigma(L)=\{\lambda_j^{-1}\mid j\in\mathbb{N}\}$. Hence for each $j\in\mathbb{N}$, there is a component $C_j$ satisfying the alternative of Theorem 3.1. By Lemma 3.2, $C_j\cap N_j\subset(\mathbb{R}\times S_j)\cup\{(\lambda_j,0)\}$.

Suppose there is a point $(\bar\lambda,\bar u)\in\partial(\mathbb{R}\times S_j)\cap C_j$ and $(\bar\lambda,\bar u)\notin N_j$, then $\partial(\mathbb{R}\times S_j)=\mathbb{R}\times\partial S_j$ implies that $\bar u$ has a double zero. Lemma 3.1 shows that $\bar u\equiv 0$, consequently $(\bar\lambda,\bar u)=(\lambda_k,0)$ for some $k\neq j$. Since $S_j\cap S_k=\phi$ if $j\neq k$, Lemma 3.2 shows $C_j$ cannot contain $(\lambda_k,0)$ for $k\neq j$ and hence must be unbounded in $(\mathbb{R}\times S_j)\cup\{(\lambda_j,0)\}$.

Next, consider the case where zero is an eigenvalue of (1.3). In this case, an approximation argument will be employed, variants of which are useful in other situations in which the Global Bifurcation Theorem is not directly applicable. Let $\mathcal{L}_\varepsilon = \mathcal{L} + \varepsilon r$. The eigenvalues of $\mathcal{L}_\varepsilon$ are $\lambda_j + \varepsilon$ and the corresponding eigenfunctions are $v_j$. Hence for $\varepsilon \neq 0$ and small, there is a component $C_j(\varepsilon)$ of solutions of

$$\begin{aligned} \mathcal{L}_\varepsilon u = \lambda r(x)u + h(x, u, u', \lambda),\ a < x < b,\\ \alpha_1 u(a) - \beta_1 u'(a) = 0,\ \alpha_2 u(b) - \beta_2 u'(b) = 0, \end{aligned} \tag{3.3}$$

with $C_j(\varepsilon)$ unbounded in $(\mathbb{R} \times S_j) \cup \{(\lambda_j + \varepsilon, 0)\}$. Let $O$ be any bounded open neighborhood of $(\lambda_j, 0)$. Then for $\varepsilon \neq 0$ and small, $(\lambda_j + \varepsilon, 0) \in O$ and $C_j(\varepsilon) \cap \partial O \neq \phi$. Choose a sequence $\varepsilon_n \to 0$ and $(\gamma_n, u_n) \in C_j(\varepsilon_n) \cap \partial O$. Then $(\gamma_n, u_n)$ are bounded in $\mathbb{R} \times E$, and the $C^1$ bound for $\{u_n\}$ together with (3.3) shows that $\{u_n\}$ is bounded in $C^2$. The Arzelà-Ascoli Theorem implies that a subsequence of $(\gamma_n, u_n)$ converges in $\mathbb{R} \times E$, and (3.3) shows that it converges in $\mathbb{R} \times C^2$ to a solution $(\bar{\beta}, \bar{u})$ of (1.1). Moreover, $(\bar{\beta}, \bar{u}) \in \partial O \cap (\mathbb{R} \times \bar{S}_j)$. If $\bar{u} \in \partial S_j$, Lemma 3.1 implies $\bar{u} \equiv 0$. Using an argument analogous to the proof of Lemma 3.2 shows that $\bar{\beta} = \lambda_j$. But $(\lambda_j, 0) \in O$ and $(\bar{\beta}, \bar{u}) \in \partial O$, so this is impossible. Hence $(\bar{\beta}, \bar{u}) \in \partial O \cap (\mathbb{R} \times S_j)$.

Since $O$ can be any bounded open neighborhood of $(\lambda_j, 0)$, this implies that there exists $C_j$ as in the statement of Theorem 1.2. □

Next, the qualitative behavior of the sets $C_j$ will be explored under further assumptions on $h(x, u, u', \lambda)$. As an instructive example of what the sets $C_j$ can look like, consider

$$\begin{aligned} -u'' = \lambda(1 + \phi(u^2 + (u')^2, \lambda))u,\ 0 < x < \pi,\\ u(0) = 0 = u(\pi), \end{aligned} \tag{3.4}$$

where $\phi(0, \lambda) = 0$. Trying for a solution of (3.4) of the form $(\lambda, \delta \sin x)$ yields $\lambda^{-1} = \phi(\delta^2, \lambda) + 1$ and the variety of possible choices of $\phi$ indicates the wealth of possibilities for $C_1$.

For ease of exposition, most examples displayed in the remainder of this section will be of a simpler form such as

$$\begin{aligned} -u'' = \lambda(r(x) + \psi(x, u, u'))u\\ u(0) = 0 = u(\pi), \end{aligned} \tag{3.5}$$

where $\psi$ is a continuous function, $\psi(x, 0, 0) = 0$ and $r$ is a positive function. All of the results are true for some general $\mathcal{L}'s$ and boundary conditions, although possibly with some qualification in their statements.

**Proposition 3.1.** *If $(\lambda, u)$ is a solution of (3.5), $\psi \geq 0$ (resp. $\psi \leq 0$), and $u \in S_j$, then $\lambda \leq \lambda_j$ (resp. $\lambda \geq \lambda_j$) with strict inequality unless $\psi(x, u, u') \equiv 0$.*

*Proof.* It is known that $\lambda_j > 0$ for every $j \in \mathbb{N}$. Consider the case $\psi \geq 0$. If $\lambda > \lambda_j$ or $\lambda = \lambda_j$ and $\psi(x, u, u') \not\equiv 0$, $\lambda(a + \psi) \geq \lambda_j a$ with strict inequality for at least one point. By the Sturm Comparison Theorem, $u$ must have a zero between any pair of adjacent zeros of $v_j$; that is, $u$ has at least $j$ zeros in $(0, \pi)$, contrary to $u \in S_j$.

The case $\psi \leq 0$ is treated as above, reversing the roles of $v_j$ and $u$. □

In the next two propositions, $S$ denotes the closure in $\mathbb{R} \times E$ of the set of nontrivial solutions of (3.5).

**Proposition 3.2.** *Suppose for* (3.5) *there is a function* $M_j \in C([0, \infty), \mathbb{R}^+)$ *such that* $(\lambda, u) \in S \cap (\mathbb{R} \times S_j)$ *implies* $\|u\| \leq M_j(\lambda)$. *Then the projection of* $C_j$ *on* $\mathbb{R}^+$ *contains the interval* $(\lambda_j, \infty)$. *Moreover if* $M_j$ *is independent of* $j$, *then for each* $\lambda \in (\lambda_j, \infty)$, (3.5) *possesses at least* $j$ *distinct solutions.*

*Proof.* It is easy to see that $u \equiv 0$ is the only solution of (3.5) if $\lambda = 0$. Since $\lambda_j > 0$, it follows that $C_j$ is unbounded in $\{(\lambda, u) \mid \lambda > 0, \ \|u\| \leq M_j(\lambda)\}$. Therefore its projection on $\mathbb{R}$ contains $(\lambda_j, \infty)$. If furthermore $M_j$ does not depend on $j$, the projections of each of $C_1, \ldots, C_j$ contain $\lambda$ if $\lambda > \lambda_j$. □

**Example 1.** Consider

$$\begin{aligned} -u'' = \lambda \sin u, \ 0 < x < \pi \\ u(0) = 0 = u(\pi). \end{aligned} \tag{3.6}$$

This is of the form (3.5) with $\psi(u) = (\sin u - u)u^{-1}$ and $r(x) \equiv 1$. Moreover, $\psi(u) \leq 0$ for small $u$, so $C_j$ bifurcates to the right of $(\lambda_j, 0)$ for each $j \in \mathbb{N}$. If $(\lambda, u)$ is a solution of (3.6), $u'$ has at least one zero, say at $z \in (0, \pi)$. Thus

$$u(x) = \lambda \int_0^x ds \int_s^z \sin u(t) dt,$$

from which it follows that $\|u\|_{L^\infty} \leq \lambda \pi^2$. Similarly, since

$$u'(x) = \lambda \int_x^z \sin u(t) dt, \tag{3.7}$$

we have $\|u'\|_{L^\infty} \leq \lambda \pi$. Hence $\|u\| \leq \lambda(\pi + \pi^2)$ and Proposition 3.2 applies here with $M_j$ independent of $j$.

Similar reasoning can be used to obtain information on the solutions of Euler's elasticity equation, where different boundary conditions need to be dealt with.

**Example 2.** Consider

$$\begin{aligned} -u'' = \lambda(a(x) - \psi(x, u))u, \ 0 < x < \pi, \\ u(0) = 0 = u(\pi), \end{aligned} \tag{3.8}$$

where $\psi \geq 0$ and $\psi(x, \pm A) > a(x)$ for some $A > 0$ and all $x \in [0, \pi]$. Set $\bar{\psi}(x, \xi) = \psi(x, \xi)$ if $|\xi| \leq A$, $\bar{\psi}(x, \xi) = \psi(x, A)$ if $\xi > A$, and $\bar{\psi}(x, \xi) = \psi(x, -A)$ if $\xi < -A$. Suppose $u$ is a solution of

$$\begin{aligned} -u'' = \lambda(a(x) - \bar{\psi}(x, u))u, \ 0 < x < \pi, \\ u(0) = 0 = u(\pi). \end{aligned} \tag{3.9}$$

At a positive maximum of $u$, $-u''(x) \geq 0$ while if $u(x) > A$, the right-hand side of (3.9) is less than zero. Hence $u(x) \leq A$. Using a similar argument at a negative minimum of $u$ yields $u(x) \geq -A$. Thus any solution of (3.9) is a solution of (3.8). Moreover the $L^\infty$ bound for solutions of (3.9) and an argument as in (3.7) show there is an $M(\lambda)$ such that $\|u\| \leq M(\lambda)$ for all solutions $(\lambda, u)$ of (3.9). It then follows from Proposition 3.2 that for any $\lambda > \lambda_k$, (3.8) has at least $k$ distinct solutions.

**Proposition 3.3.** *Suppose for* (3.5), $\psi \geq 0$ *and there exists a function* $M_j \in C((0, \lambda_j], \mathbb{R})$ *such that* $(\lambda, u) \in S \cap (\mathbb{R} \times S_j)$ *implies* $\|u\| \leq M_j(\lambda)$. *Then the projection of* $C_j$ *on* $\mathbb{R}$ *contains the interval* $(0, \lambda_j)$.

*Proof.* If $\lambda = 0$, $u \equiv 0$ is the only solution of (3.5). Moreover Proposition 3.1 shows $(\lambda, u) \in S \cap (\mathbb{R} \times S_j)$ implies $\lambda \leq \lambda_j$. Thus $C_j$ must be in the region $(0, \lambda_j] \times E \cap \{(\lambda, u) \mid \|u\| \leq M_j(\lambda)\}$. □

**Remark 3.2.** Proposition 3.3 can be applied to (3.5) where the nonlinear term $\psi(x, u)$ has the property $\psi(x, \xi)\xi^{-1} \to \infty$ as $|\xi| \to \infty$.

There are some applications of Theorem 1.2 to situations in which bifurcation does not occur. For example, consider

$$\begin{gathered} -u'' = f(x, u)u, \ 0 < x < \pi, \\ u(0) = u(\pi) = 0. \end{gathered} \tag{3.10}$$

**Theorem 3.2.** *Suppose* $f \in C([0, \pi] \times \mathbb{R}, \mathbb{R})$, $f \geq 0$, $f(x, 0) = 0$, *and* $f(x, \xi) \to \infty$ *as* $|\xi| \to \infty$ *uniformly for* $x \in [0, \pi]$. *Then for each* $j \in \mathbb{N}$, (3.10) *possesses a solution* $u_j \in S_j$.

A strategy used here is to apply Theorem 1.2 to another related equation:

$$\begin{gathered} -u'' + u = \lambda(1 + f(x, u))u, \ 0 < x < \pi, \\ u(0) = 0 = u(\pi). \end{gathered} \tag{3.11}$$

Thus (3.11) possesses a component of solutions $C_j$ which is unbounded in $\mathbb{R} \times S_j$. Moreover, by Proposition 3.1, $C_j$ is in $(0, j^2 + 1] \times S_j$. Since for $\lambda = 1$, any solution of (3.11) is a solution of (3.10), it suffices to find a function $M_j \in C([1, j^2 + 1], \mathbb{R})$ such that $(\lambda, u) \in [1, j^2 + 1] \times S_j$ implies $\|u\| \leq M_j(\lambda)$. We refer to [29] for detailed analysis of such estimates.

The next application concerns a rather different kind of situation:

$$\begin{gathered} -u'' = \lambda f(x, u)u, \qquad 0 < x < \pi, \\ u(0) = u(\pi) = 0, \end{gathered} \tag{3.12}$$

where $f$ is continuous in its arguments, $f(x, 0) = 0$ and $f$ satisfies

(f1) there is a $\rho > 0$ such that $f(x, \xi) > 0$ for $0 < |\xi| < \rho$,

(f2) there is a $z > 0$ such that $f(x, \xi) < 0$ if $|\xi| = z$.

A simple example is $f(x, \xi) = \xi^2 - \xi^4$.

**Theorem 3.3.** *Under the above hypotheses on* $f$, *for each* $j \in \mathbb{N}$, *there exists* $d_j > 0$ *such that for each* $\lambda > d_j$, (3.12) *possesses at least two distinct solutions* $(\lambda, \bar{u}_j), (\lambda, \underline{u}_j)$ *with* $\bar{u}_j, \underline{u}_j \in S_j$.

The main idea used in the proof of Theorem 3.3 is as follows: For $\varepsilon > 0$, consider

$$\begin{aligned} -u'' = \lambda\big(\varepsilon + f(x,u)\big)u, \qquad 0 < x < \pi, \\ u(0) = u(\pi) = 0. \end{aligned} \tag{3.13}$$

As a consequence of Theorem 1.2, (3.13) possesses a component, $C_j(\varepsilon)$, of solutions unbounded in $(\mathbb{R} \times S_j) \cup \{(j^2\varepsilon^{-1}, 0)\}$. By carefully studying the resulting behavior of $C_j(\varepsilon)$ as stated in the next lemma, the desired solutions of (3.12) can be obtained through the approximation as $\varepsilon \to 0$. The details can be found in [29].

**Lemma 3.3.**

(i) *There exist $\bar{\varepsilon} > 0$ and $K > 0$ such that, if $(\lambda, u)$ is a solution of (3.13) with $\varepsilon \in [0, \bar{\varepsilon}]$, then $\|u\|_{L^\infty} \leq K$.*
(ii) *There exists $\underline{\lambda} > 0$ such that if $(\lambda, u)$ is a solution of (3.13) with $\lambda \in [0, \underline{\lambda}]$, then $u \equiv 0$.*
(iii) *If $(\lambda, u)$ is a solution of (3.13) with $u \in S_j$ and $\|u\|_{L^\infty} < \rho$, then $\lambda \leq j^2\varepsilon^{-1}$.*
(iv) *Let $\delta \in (0, \rho)$. For each $j \in \mathbb{N}$, there exists $d_j = d_j(\delta)$ such that if $(\lambda, u)$ is a solution of (3.13) with $\varepsilon \in [0, \bar{\varepsilon}]$, $u \in S_j$, and $\|u\|_{L^\infty} = \delta$, then $\lambda < d_j(\delta)$.*

## 4. Bifurcation from continuous spectrum

In this section we turn to the Sturm-Liouville problem on the half-line:

$$\begin{aligned} -u'' = \lambda r(x)u - F(x,u)u, \quad 0 < x < +\infty, \\ u(a)\cos\theta - u'(a)\sin\theta = 0, \quad u \in L^2[a, \infty), \end{aligned} \tag{4.1}$$

where $a \geq 0$, $\theta \in [0, \frac{\pi}{2}]$, $r \in C([0,\infty),(0,\infty))$ and $0 < r_1 \leq r(x) \leq r_2 < +\infty$ for $x \in [0, \infty)$. It is assumed that $F$ satisfies the following conditions:

(F3) $F : [0, \infty) \times \mathbb{R} \to [0, \infty)$ is continuous, and $\lim_{|\xi| \to 0} F(x, \xi) = 0$ uniformly on compact subsets of $[0, \infty)$.
(F4) There exist positive numbers $\sigma_i$ and continuous functions $\omega_i : [0, \infty) \to (0, \infty)$ which satisfy $\int_0^\infty \omega_i^{-2/\sigma_i} dx < +\infty$, $i = 1, 2$, such that $F(x, \xi) \geq \omega_1(x)|\xi|^{\sigma_1}$ for $x \in [0, \infty)$, $\xi \geq 0$ and $F(x, \xi) \geq \omega_2(x)|\xi|^{\sigma_2}$ for $x \in [0, \infty)$, $\xi < 0$.
(F5) For fixed $x \in [0, \infty)$, $F(x, \xi)$ is a strictly increasing function of $\xi$ if $\xi > 0$ and a strictly decreasing function of $\xi$ if $\xi < 0$.

For $j \in \mathbb{N}$, let $\Omega_j^+(\lambda)$ (resp. $\Omega_j^-(\lambda)$) denote the set of $u \in C^2[0, \infty) \cap H^1(0, \infty)$ such that $u$ satisfies (4.1), $u > 0$ (resp. $< 0$) in a deleted neighborhood of $x = 0$ and $u$ has exactly $j - 1$ simple zeros in $(0, \infty)$.

It is known that the linearization of (4.1) at $u \equiv 0$ has a purely continuous spectrum equal to $[0, \infty)$. Although Theorem 3.1 is not directly applicable, variational methods provide a way to obtain the following existence result.

**Theorem 4.1.** *Let $\lambda > 0$ and $\theta \in [0, \pi/2]$ be given. Then one of the following alternatives must occur:*

(i) $\Omega_j^+(\lambda) \neq \phi$ *and* $\Omega_j^-(\lambda) \neq \phi$ *for all* $j \in \mathbb{N}$.

(ii) $\Omega_{2k-1}^-(\lambda) \neq \phi$ *and* $\Omega_{2k}^+(\lambda) \neq \phi$ *for all* $k \in \mathbb{N}$. *Moreover, there exist an even number* $m \in \mathbb{N}$ *and an odd number* $\ell \in \mathbb{N}$, *with* $|m - \ell| = 1$, *such that* $\Omega_{2k}^-(\lambda) \neq \phi$ *for* $2k < m$, $\Omega_{2k}^-(\lambda) = \phi$ *for* $2k \geq m$ *and* $\Omega_{2k-1}^+(\lambda) \neq \phi$ *for* $2k - 1 \leq \ell$, $\Omega_{2k-1}^+(\lambda) = \phi$ *for* $2k - 1 > \ell$.

(iii) *Interchange "+" and "−" in statement* (ii).

An example of solutions having infinitely many zeros was given by Heinz [16], where he also gave a sufficient condition which prohibits the existence of such solutions.

In the proof of Theorem 4.1, since the case of $\theta \in (0, \frac{\pi}{2}]$ requires more work, only the case $u(a) = 0$ will be treated. The existence of positive and negative solutions of (4.1) can be obtained by using an approximation method. The assumption (F5) implies that positive and negative solutions are unique. To obtain solutions with a prescribed number of nodes, Nehari's idea that pieces together alternately positive and negative solutions on adjacent intervals is used. Nevertheless, different techniques are needed for using variational arguments to find solutions of (4.1). It is known [17] that there exist a unique positive solution $V_+(\lambda, a, b, x)$ and a unique negative solution $V_-(\lambda, a, b, x)$ of

$$\begin{gathered} -u'' = \lambda r(x)u - F(x, u)u, \qquad a < x < b, \\ u(a) = 0, \qquad u(b) = 0, \end{gathered} \tag{4.2}$$

provided that $\lambda > \lambda_1(a, b)$, where $\lambda$ is an eigenvalue parameter and $\lambda_j(a, b)$ denotes the $j$th eigenvalue of

$$\begin{gathered} -v'' = \lambda r(x)v, \qquad a < x < b, \\ v(a) = 0, \qquad v(b) = 0. \end{gathered}$$

On the other hand, if $\lambda \leq \lambda_1(a, b)$ the only solution of (4.2) is $u \equiv 0$. For given $\lambda > 0$ and $0 \leq a < b \leq \infty$, define the number $\Lambda^+[a, b]$ (resp. $\Lambda^-[a, b]$) by

$$\begin{aligned} &\Lambda^+[a, b] \text{ (resp. } \Lambda^-[a, b]) \\ &= \int_a^b \left[\lambda r(x)u^2(x) - (u'(x))^2 - 2\int_0^{u(x)} F(x, \xi)\xi d\xi\right] dx, \end{aligned}$$

where $\lambda_1(a, \infty)$ is set to be zero and

$$u = \begin{cases} 0 & \text{if} \quad \lambda \leq \lambda_1(a, b) \\ V_+(\lambda, a, b, \cdot) \text{(resp. } V_-) & \text{if} \quad \lambda > \lambda_1(a, b). \end{cases}$$

Also, it is convenient to adopt the notation $V_+(\lambda, a, b, x) \equiv 0$ and $V_-(\lambda, a, b, x) \equiv 0$ whenever $\lambda \leq \lambda_1(a, b)$. Let $V_+'(\lambda, a, b, x) = d/dx\, V_+(\lambda, a, b, x)$ and $V_-'(\lambda, a, b, x) = d/dx\, V_-(\lambda, a, b, x)$. The following proposition plays an important role in finding solutions with a prescribed number of nodes.

**Proposition 4.1.** *For $a < b \le \infty$, $\Lambda^+[a,b]$ (resp. $\Lambda^-[a,b]$) is a differentiable function of $a$ and $b$, with derivatives given by*

$$\frac{\partial \Lambda^+}{\partial a} = -(V'_+(\lambda,a,b,a))^2 \text{ (resp. } \frac{\partial \Lambda^-}{\partial a} = -(V'_-(\lambda,a,b,a))^2\text{)}$$

*and*

$$\frac{\partial \Lambda^+}{\partial b} = (V'_+(\lambda,a,b,b))^2 \text{ (resp. } \frac{\partial \Lambda^-}{\partial b} = (V'_-(\lambda,a,b,b))^2\text{).}$$

The proposition was proved by Hempel [17] in the case of bounded intervals. The generalization to the unbounded case was obtained in [5].

If $F$ is an even function of $\xi$, $u \in \Omega_j^+(\lambda)$ if and only if $-u \in \Omega_j^-(\lambda)$; in other words, only (i) of Theorem 4.1 can occur. In this situation, $\Lambda^+[a,b] = \Lambda^-[a,b]$ and will be simply denoted by $\Lambda[a,b]$. For $j \in \mathbb{N}$, set $x_0 = a$, $x_{j+1} = \infty$ and

$$A_j = \{(x_1,x_2,\ldots,x_j) \mid a \le x_1 \le x_2 \cdots \le x_j < +\infty\}.$$

Then, for fixed $\lambda > 0$, define a function $G_j$ on $A_j$ by

$$G_j(x_1,x_2,\ldots,x_j) = \sum_{i=1}^{j+1} \Lambda[x_{i-1},x_i].$$

Since $\Lambda[x_{i-1},x_i] \ge 0$ and is positive for at least one value of $i$, $G_j(x_1,x_2,\ldots,x_j) > 0$ for all $(x_1,x_2,\ldots,x_j) \in A_j$ and hence $\mathrm{Inf}_{A_j}\, G_j(x_1,x_2,\ldots,x_j)$ exists. Suppose $G_j$ attains its global infimum at an interior point $(z_1,z_2,\ldots,z_j)$ of $A_j$; that is, $0 < z_1 < z_2 < \cdots < z_j < \infty$. Then for $i = 1,2,\ldots,j$,

$$\begin{aligned} &\frac{\partial G_j}{\partial x_i}(z_1,z_2,\ldots,z_j) \\ &= [V'_+(\lambda,z_{i-1},z_i,z_i)]^2 - [V'_+(\lambda,z_i,z_{i+1},z_i)]^2 = 0, \end{aligned} \tag{4.3}$$

where $z_0 = 0$ and $z_{j+1} = \infty$. Also, if $\lambda \le \lambda_1(z_{k-1},z_k)$ for some $k$, let $\ell \ge k$ be the largest value such that $\lambda < \lambda_1(z_{\ell-1},z_\ell)$ and $\lambda > \lambda_1(z_\ell,z_{\ell+1})$. This implies $V'_+(\lambda,z_{\ell-1},z_\ell,z_\ell) = 0$ and $V'_+(\lambda,z_\ell,z_{\ell+1},z_\ell) \ne 0$, which contradicts (4.3). Thus for $i = 1,2,\ldots,j+1$, $\lambda > \lambda_1(z_{i-1},z_i)$ and if

$$u(x) = (-1)^i V_+(\lambda,z_i,z_{i+1},x) \qquad \text{for} \quad x \in [z_i,z_{i+1}),$$

then $u \in C^1[a,\infty)$ is the desired $j$-node solution with nodes $z_1,z_2,\ldots,z_j$.

It now remains to show that $G_j$ attains its infimum at an interior point $(z_1,z_2,\ldots,z_j)$ of $A_j$. To achieve this goal it is sufficient to prove, by induction, the following statement:

If, for $1 \le k \le j-1$,

$$G_k \text{ attains its global minimum at an interior point of } A_k$$

and

$$\mathrm{Min}_{A_k} G_k(x_1,x_2,\ldots,x_k) > \mathrm{Inf}_{A_{k+1}} G_{k+1}(x_1,x_2,\ldots,x_{k+1}),$$

then these statements also hold for $k = j$.

In case of $j = 1$, it is easy to check that $G_1(a) = \Lambda[a,\infty]$ and $\lim_{x\to\infty} G_1(x) = \Lambda[a,\infty]$. Since $\lim_{b\to a^+} \lambda_1(a,b) = \infty$, there is an $\varepsilon > 0$ such that $\lambda < \lambda_1(a,x)$ for $x \in (a, a+\varepsilon)$. Hence, $G_1'(x) < 0$ if $x \in (a, a+\varepsilon)$. Consequently $G_1$ must attain its infimum at some point $z \in (a+\varepsilon, \infty)$.

The remainder of the proof is technically quite involved. Detailed arguments can be found in [5, 7].

The second part of this section will deal with uniqueness theorems [6, 9, 15] for (4.1) under the assumptions that $r(x) \equiv 1$ and $F$ has a special form as follows:

(F6) There are $\psi_1, \psi_2 \in C^1([0,\infty),[0,\infty))$, with $\psi_1(0) = \psi_2(0) = 0$, $\psi_1' > 0$, $\psi_2' > 0$ in $(0,\infty)$, and a positive number $\sigma$ such that

$$F(x,\xi) = \begin{cases} \psi_1(w(x)|\xi|^\sigma), & \xi \geq 0, x \in [0,\infty) \\ \psi_2(w(x)|\xi|^\sigma), & \xi < 0, x \in [0,\infty), \end{cases}$$

where $w \in C^1([0,\infty),[0,\infty))$ and $w'w^{-1}$ is nondecreasing on $[0,\infty)$.

**Theorem 4.2.** *Under the above hypotheses on $r$ and $F$, for each $j \in \mathbb{N}$, (4.1) possesses at most one solution in $\Omega_j^+(\lambda)$, and also in $\Omega_j^-(\lambda)$.*

Now with the aid of Theorem 4.2, it can be shown that the lowest point $\lambda = 0$ of the continuous spectrum is a bifurcation point.

**Theorem 4.3.** *Let $E$ be the Banach space $H^1[0,\infty) \cap L^\infty[0,\infty)$. If $F(x,\xi) = F(x,-\xi)$ then for each $j \in \mathbb{N}$, (4.1) possesses two curves of solutions $C_j^\pm$ in $\mathbb{R}\times E$, with $C_j^\pm = \{(\lambda, u_j^\pm(\lambda)) \mid \lambda > 0\} \cup \{(0,0)\}$ and $u_j^\pm(\lambda) \in \Omega_j^\pm(\lambda)$.*

In case $F$ is not an even function of $\xi$, $C_j^+$ could merge with $C_{j-1}^+$ as illustrated in the next example. Let

$$F(x,\xi) = \begin{cases} \delta e^x|\xi| & \text{if} \quad \xi \geq 0 \\ e^x|\xi| & \text{if} \quad \xi < 0. \end{cases}$$

Here $\delta$ can be chosen small enough so that for $\lambda = 1$ alternative (ii) of Theorem 4.1 holds with $m = 2$ and $\ell = 1$; that is

$$\begin{aligned} \Omega_j^-(1) &\neq \phi \quad \text{for} \quad j \in I_1 = \{2k-1 \mid k \in \mathbb{N}\}, \\ \Omega_j^+(1) &\neq \phi \quad \text{for} \quad j \in I_2 = \{1\} \cup \{2k \mid k \in \mathbb{N}\}, \\ \Omega_j^-(1) &= \phi \quad \text{for} \quad j \in I_3 = \{2k \mid k \in \mathbb{N}\} \text{ and} \\ \Omega_j^+(1) &= \phi \quad \text{for} \quad j \in I_4 = \{2k+1 \mid k \in \mathbb{N}\}. \end{aligned}$$

Moreover, the same is true for all $\lambda \geq 1$. By Theorem 4.2, if $\lambda > 0$ is fixed, there is at most one solution in each nodal class. Indeed, there are infinitely many curves of solutions $C_j^+$ and $C_j^-$ in $\mathbb{R}\times E$, emanating from $(0,0)$, such that if $(\lambda,u) \in C_j^\pm$ and $\lambda \in (0,\delta^2)$ then $u \in \Omega_j^\pm(\lambda)$. Also, there exists a continuous function $M : (0,\infty) \to (0,\infty)$ such that if $u \in \Omega_j^\pm(\lambda)$ then $\|u\|_E \leq M(\lambda)$. Hence, for all $\lambda > 0$, $C_j^+ \cap (\{\lambda\}\times E) \neq \phi$ if $j \in I_2$ and $C_j^- \cap (\{\lambda\}\times E) \neq \phi$ if $j \in I_1$.

However, for those $C_j^+$, $j \in I_4$, and $C_j^-$, $j \in I_3$, they merge with $C_{j-1}^+$ and $C_{j-1}^-$ respectively at some $\lambda^* \in (\delta^2, 1]$. We refer to [8] for a detailed proof.

Theorem 4.2 can be proved using a shooting argument. Let $\lambda$ be fixed and let $U(\zeta, x)$ denote the unique solution of

$$-u'' = \lambda u - F(x,u)u, \quad u(a) = 0, u'(a) = \zeta,$$

which is understood to be extended to its maximal interval of definition. For $j \in \mathbb{N}$, let $D_j$ be the set of $\zeta > 0$ such that $U(\zeta, \cdot)$ has at least $j$ zeros in $(a, \infty)$. Ordering the zeros as an increasing sequence, $a < z_1(\zeta) < z_2(\zeta) < \cdots < z_j(\zeta) < \cdots$, the function $z_j(\zeta)$ satisfies the equation $U(\zeta, z_j(\zeta)) = 0$. By the Implicit Function Theorem there is a maximal open neighborhood of $\zeta$ on which $z_j$ is of class $C^1$ in its argument. The goal here is to show that there exists at most one $\xi_j \in (0, \infty)$ such that $U(\xi_j, \cdot) \in \Omega_j^+(\lambda)$. With the aid of the Sturm Comparison Theorem, it was proved [6, 15] that for every $j \in \mathbb{N}$, $D_j = (0, \tau_j)$ and $\partial z_j / \partial \zeta > 0$ on $D_j$, where $\{\tau_j\}$ is a non-increasing sequence. This together with other comparison arguments completes the proof of Theorem 4.2. Details can be found in [6]. An alternative proof of Theorem 4.2 was given in [9].

**Acknowledgements**

The author would like to thank Professors Amrein, Hinz and Pearson for their warm hospitality.

## References

[1] S.S. Antman, Bifurcation problems for nonlinearly elastic structures, in *Applications of Bifurcation Theory*, 73–125, P.H. Rabinowitz (ed.), Academic Press, New York, 1977.

[2] H. Berestycki and M. Esteban, Existence and bifurcation of solutions for an elliptic degenerate problem, J. Differential Equations **134** (1997), 1–25.

[3] M.S. Berger and P.C. Fife, On von Karman's equation and the buckling of a thin elastic plate II, Comm. Pure Appl. Math. **21** (1968), 227–247.

[4] C.-C. Chen and C.-S. Lin, Uniqueness of the ground state solutions of $\triangle u + f(u) = 0$ in $R^n$, $n \geq 3$, Commun. Partial Differential Equations **16** (1991), 1549–1572.

[5] C.-N. Chen, Multiple solutions for a class of nonlinear Sturm-Liouville problems on the half-line, J. Differential Equations **85** (1990), 236–275.

[6] C.-N. Chen, Uniqueness and bifurcation for solutions of nonlinear Sturm-Liouville eigenvalue problems, Arch. Rational Mech. Anal. **111** (1990), 51–85.

[7] C.-N. Chen, Multiple solutions for a class of nonlinear Sturm-Liouville problems when nonlinearities are not odd, J. Differential Equations **89** (1991), 138–153.

[8] C.-N. Chen, Some existence and bifurcation results for solutions of nonlinear Sturm-Liouville eigenvalue problems, Math. Zeitschrift **208** (1991), 177–192.

[9] C.-N. Chen, Uniqueness of solutions of some second-order differential equations, Differential Integral Equations **6** (1993), 825–834.

[10] M.G. Crandall and P.H. Rabinowitz, Nonlinear Sturm-Liouville eigenvalue problems and topological degree, J. Math. Mech. **29** (1970), 1083–1102.

[11] E.N. Dancer, Global solution branches for positive mappings, Arch. Rational Mech. Anal. **52** (1973), 181–192.

[12] P.C. Fife, Branching phenomena in fluid dynamics and chemical reaction diffusion theory, in *Eigenvalues of Nonlinear Problems*, 23–83, G. Prodi (ed.), Edizioni Cremonese, Roma, 1974.

[13] K.O. Friedrichs and J. Stoker, The nonlinear boundary value problem of the buckled plate, Amer J. Math. **63** (1941), 839–888.

[14] H.-P. Heinz, Nodal properties and variational characterizations of solutions to nonlinear Sturm-Liouville problems, J. Differential Equations **62** (1986), 299–333.

[15] H.-P. Heinz, Nodal properties and bifurcation from the essential spectrum for a class of nonlinear Sturm-Liouville problems, J. Differential Equations **64** (1986), 79–108.

[16] H.-P. Heinz, Free Ljusternik-Schnirelman theory and the bifurcation diagrams of certain singular nonlinear problems, J. Differential Equations **66** (1987), 263–300.

[17] J.A. Hempel, Multiple solutions for a class of nonlinear boundary value problems, Indiana Univ. Math. J. **20** (1971), 983–996.

[18] C. Jones and T. Kupper, Characterization of bifurcation from the continuous spectrum by nodal properties, J. Differential Equations **54** (1984), 196–220.

[19] C. Jones and T. Kupper, On the infinitely many solutions of a semilinear elliptic equation, SIAM J. Math. Anal. **17** (1986), 803–835.

[20] T. Kupper, The lowest point of the continuous spectrum as a bifurcation point, J. Differential Equations **34** (1979), 212–217.

[21] T. Kupper, On minimal nonlinearities which permit bifurcation from the continuous spectrum, Math. Methods Appl. Sci. **1** (1979), 572–580.

[22] M.K. Kwong, Uniqueness of positive solutions of $\triangle u - u + u^p = 0$ in $R^n$, Arch. Rational Mech. Anal. **105** (1989), 234–266.

[23] Z. Nehari, Characteristic values associated with a class of nonlinear second-order differential equations, Acta Math. **105** (1961), 141–175.

[24] W.M. Ni and R. Nussbaum, Uniqueness and nonuniqueness for positive radial solutions of $\triangle u + f(u, r) = 0$, Comm. Pure Appl. Math. **38** (1985), 67–108.

[25] P.H. Rabinowitz, Nonlinear Sturm-Liouville problems for second-order ordinary differential equations, Comm. Pure Appl. Math. **23** (1970), 939–961.

[26] P.H. Rabinowitz, Some global results for nonlinear eigenvalue problems, J. Funct. Anal. **7** (1971), 487–513.

[27] P.H. Rabinowitz, A global theorem for nonlinear eigenvalue problems and applications, in *Nonlinear Functional Analysis*, 11–36, E.H. Zaratonello (ed.), Academic Press, New York, 1971.

[28] P.H. Rabinowitz, Some aspects of nonlinear eigenvalue problems, Rocky Mountain J. Math. **3** (1973), 161–202.

[29] P.H. Rabinowitz, Global aspects of bifurcation, in *Topological Methods in Bifurcation Theory*, 63–112, Séminaire de Mathématiques Supérieures **91**, Presses Univ. Montréal, 1985.

[30] D.H. Sattinger, *Topics in stability and bifurcation theory*, Lecture Notes in Mathematics **309**, Springer, Berlin, 1973.

[31] C.A. Stuart, Some bifurcation theory for $k$-sets contractions, Proc. London Math. Soc. (3) **27** (1973), 531–550.

[32] C.A. Stuart, Bifurcation for Neumann problems without eigenvalues, J. Differential Equations **36** (1980), 391–407.

[33] C.A. Stuart, Global properties of components of solutions of non-linear second-order ordinary differential equations on the half-line, Ann. Scuola Norm. Sup. Pisa Cl. Sci. (4) **2** (1975), 265–286.

[34] J.F. Toland, Global bifurcation for Neumann problems without eigenvalues, J. Differential Equations **44** (1982), 82–110.

[35] R.E.L. Turner, Superlinear Sturm-Liouville problems, J. Differential Equations **13** (1973), 157–171.

[36] E. Yanagida, Uniqueness of positive radial solutions of $\triangle u + g(r)u + h(r)u^p$ in $R^n$, Arch. Rational Mech. Anal. **115** (1991), 257–274.

[37] E. Yanagida and S. Yotsutani, Existence of positive radial solutions to $\triangle u + K(|x|)u^p = 0$ in $R^n$, J. Differential Equations **115** (1995), 477–502.

Chao-Nien Chen
Department of Mathematics
National Changhua University of Education
Taiwan, R.O.C.
e-mail: chenc@math.ncue.edu.tw

*W.O. Amrein, A.M. Hinz, D.B. Pearson*
Sturm-Liouville Theory: Past and Present, 217–235


# Boundary Conditions and Spectra of Sturm-Liouville Operators

Rafael del Río

**Abstract.** This is a discussion of some aspects of the relation between boundary conditions and spectra of Sturm-Liouville operators. It is intended to review results which show how the spectrum behaves when the boundary condition changes. The absolutely continuous part will normally be stable and the more interesting problems concern the behavior of the singular part and coexistence of different spectral types.

## 1. Introduction

This paper is a survey of some aspects of the relation between boundary conditions and spectrum of Sturm-Liouville operators. The main problem is to understand how the spectrum behaves when the boundary condition varies. In particular it is of interest to study the behavior of the different parts of the spectrum (for example singular and absolutely continuous). These operators were introduced in [38, 39].

A key tool of some developments I intend to describe is the so-called Weyl $m$-function. This function is analytic in the upper half-plane and closely connected to the resolvent of the operator. Its study will allow us to clarify what happens with the spectrum when the boundary condition changes.

In 1910 H. Weyl proved that the essential spectrum, which in this case is just the set of accumulation points of the spectrum, is stable when the boundary condition is modified. What changes in fact are the isolated points. In 1957 several remarkable papers were published. M. Rosenblum [35] and T. Kato [23] proved stability of absolutely continuous spectra for self-adjoint operators under trace class perturbations and N. Aronszajn [1] showed that the absolutely continuous parts of spectral measures of Sturm-Liouville problems corresponding to different boundary conditions are equivalent[1], whereas their singular parts are mutually

Partially supported by Project 37444E CONACyT.
[1] *i.e.*, they lead to the same null sets.

singular measures[2]. To the best of my knowledge this is the first study of singular continuous spectra of differential operators.

Since the absolutely continuous part is stable, it is of particular interest to understand the behavior of the singular part, especially that part which is embedded in the essential spectrum. As was already noticed by Aronszajn, this part is very unstable. In fact it is not possible for all boundary conditions to allow eigenvalues embedded in the essential spectrum. This kind of spectra will have to disappear when the boundary condition changes. We can nevertheless have singular spectrum which is embedded in the absolutely continuous spectrum, for all boundary conditions. In this review the above-mentioned result of Aronszajn will be described and several other theorems which clarify to some extent the behavior of the embedded singular part will be sketched. I will concentrate mainly on results that are more familiar to me.

After a brief description of the result on stability proven by Aronszajn, a theorem due to Hartman and Wintner on the behavior of isolated eigenvalues is stated. This result mainly says that isolated eigenvalues, that is eigenvalues that are in gaps of the essential spectrum, behave smoothly under the considered perturbations and that the complement of the essential spectrum is contained in the interior of the set of points $\lambda \in \mathbb{R}$ for which there are $L^2$ solutions of $-u'' + v(x)u = \lambda u$.

Following this, the problem of embedded singular spectrum is considered. The basic tools used are properties of the Weyl $m$-function and some results of [1]. It is shown that coexistence of singular and absolutely continuous spectrum is possible for large sets of boundary conditions. Here is explained in more detail a result that gives conditions on the length of an interval for the parameter of boundary conditions which imply that this interval contains a set of full measure where singular and absolutely continuous spectra coexist.

Thereafter, an example of a very explicit spectral function is given which generates a situation with mixed spectra for all boundary conditions with the exception of one. Using the inverse spectral theorem of Gelfand-Levitan it is known that Sturm-Liouville operators with this kind of spectral function exist, and therefore that mixed situations even for large sets of boundary conditions are possible. It remains to carry out an explicit construction of such operators.

Finally some results on inverse spectral theory of regular problems are considered. If the spectra are known for two boundary conditions, the Sturm-Liouville operator can be uniquely reconstructed [2] (see also the article by M. Malamud in this volume). It happens that if we know something about the potential then we need less information about the spectra. This kind of problem is very different from those considered above and illustrates the role of the relation between boundary conditions and spectra in other settings.

I hope this text will be useful to those wishing to understand the important relations between boundary conditions and spectra of Sturm-Liouville operators.

---

[2] *i.e.*, they are concentrated on mutually disjoint sets of Lebesgue measure zero.

## 2. Some classical results

We consider one-dimensional Schrödinger equations

$$\mathcal{L}y = -y''(x) + v(x)\,y(x) = \lambda y(x), \quad 0 \le x < \infty, \tag{1}$$

and the associated self-adjoint operators

$$H_\alpha = -\frac{d^2}{dx^2} + v(x) \quad \text{in } L^2(0,\infty)$$

generated by the boundary condition

$$y(0)\cos\alpha - y'(0)\sin\alpha = 0, \quad \alpha \in [0,\pi). \tag{2}$$

Here we assume that the real function $v$ is locally of class $L^1$ on $[0,\infty)$ and that the limit point case holds at $\infty$.

Let $u_1(x,z)$ and $u_2(x,z)$ be solutions of

$$\mathcal{L}u = zu \tag{3}$$

which satisfy

$$\begin{aligned} u_1(0,z) &= \sin\alpha, & u_1'(0,z) &= \cos\alpha, \\ u_2(0,z) &= -\cos\alpha, & u_2'(0,z) &= \sin\alpha. \end{aligned}$$

For every non-real $z$ there exists a function

$$\varphi_\alpha(x,z) = u_2(x,z) + m_\alpha(z)u_1(x,z)$$

which is a solution of (3) and belongs to $L^2(0,\infty)$. Note that $u_1$ satisfies the boundary condition (2). In the limit point case at $\infty$ (see [41]), for each $z$ the complex number $m_\alpha(z)$ is defined uniquely. This is called the Weyl $m$-function for the boundary condition (2) given by $\alpha$ and has an integral representation of the form

$$m_\alpha(z) = c + \int_{\mathbb{R}} \left( \frac{1}{\mu - z} - \frac{\mu}{\mu^2+1} \right) d\rho_\alpha(\mu), \tag{4}$$

where $\rho_\alpha$ is a Lebesgue-Stieltjes measure uniquely determined by $m_\alpha$. The measure $d\rho_\alpha$ is called the spectral measure, and $\rho_\alpha$ is called the spectral function, of the operator $H_\alpha$. We shall denote by $\rho_\alpha(S)$ the spectral measure of a set $S$.

The spectral density $d\rho_\alpha/d\lambda$ is given almost everywhere by

$$\frac{d\rho_\alpha(\lambda)}{d\lambda} = \lim_{E\to 0+} \frac{1}{\pi}\mathrm{Im}\big(m_\alpha(\lambda+iE)\big) =: \frac{1}{\pi}\mathrm{Im}\big(m_\alpha(\lambda+i0)\big),$$

and may be thought of as a local probability density for the energy of the system.

Once we have a family of operators $H_\alpha$ depending on a parameter $\alpha \in \mathbb{R}$, it is quite natural to ask what happens when $\alpha$ varies. Are all $H_\alpha$ the same at least in some sense? How do properties or objects associated with $H_\alpha$ behave as we change $\alpha$? H. Weyl in [41, 42] (*cf.* also [3]) proved that the essential spectrum of $H_\alpha$, denoted by $\sigma_{ess}$, that is the set of points of accumulation of the spectrum, is independent of $\alpha$ and that point spectra corresponding to two different boundary conditions do not intersect. In the same paper Weyl states that he could not be sure that a similar stability result would hold for the continuous spectra.

It was N. Aronszajn who solved this problem in [1]. Since the methods used in his paper were the key to later developments, I shall try to explain their main aspects.

First note that the following relation holds for $m$:

$$m_\alpha(z) = \frac{m_\beta(z)\cos(\alpha-\beta) - \sin(\alpha-\beta)}{m_\beta(z)\sin(\alpha-\beta) + \cos(\alpha-\beta)}\,. \tag{5}$$

Once we have (5), we would like to know how the various spectral measures $\rho_\alpha$ are related, since spectral information about $H_\alpha$ is contained in the corresponding $\rho_\alpha$.

In [1] minimal supports $M_{ac}, M_s, M_{sc}$ and $M_p$ of the absolutely continuous, singular, singular continuous and point parts of $\rho_\alpha$, denoted respectively by $\rho_\alpha^{ac}, \rho_\alpha^s, \rho_\alpha^{sc}$ and $\rho_\alpha^p$, are given as follows

$$\begin{aligned}
M_{ac}^\alpha &= \bigl\{x \in E \mid 0 < \operatorname{Im}\bigl(m_\alpha(x+i0)\bigr) < \infty\bigr\}\,,\\
M_s^\alpha &= \bigl\{x \in E \mid \operatorname{Im}\bigl(m_\alpha(x+i0)\bigr) = \infty\bigr\}\,,\\
M_{sc}^\alpha &= \bigl\{x \in E \mid \operatorname{Im}\bigl(m_\alpha(x+i0)\bigr) = \infty,\ \rho_\alpha\{x\} = 0\bigr\}\,,\\
M_p^\alpha &= \bigl\{x \in E \mid \operatorname{Im}\bigl(m_\alpha(x+i0)\bigr) = \infty,\ \rho_\alpha\{x\} > 0\bigr\}\,,
\end{aligned}$$

where

$$\operatorname{Im}\bigl(m_\alpha(x+i0)\bigr) := \lim_{y\downarrow 0} \operatorname{Im}\bigl(m_\alpha(x+iy)\bigr)$$

and $E = \{x \mid \operatorname{Im}\bigl(m_\alpha(x+i0)\bigr)$ exists$\}$. Using (5) one observes that

$$M_{ac}^\alpha = M_{ac}^\beta, \text{ and } M_s^\alpha \cap M_s^\beta = \varnothing \text{ for } \alpha \neq \beta.$$

This implies that the absolutely continuous parts of $\rho_\alpha$ are equivalent for all $\alpha$, whereas their singular parts are mutually singular.

I should mention a remarkable paper due to D. Gilbert and D. Pearson [17] where the notion of subordinate solution is introduced and the above-mentioned supports are related to the behavior of solutions of the Schrödinger equation near the end points of the interval. See the contribution of D. Gilbert in this volume [16].

Since we have stability for the absolutely continuous part, the natural questions which arise concern the behavior of the singular part. For the singular part in the complement of the essential spectrum the following result holds (see [19], [20], [3] or [14, Theorem 2.5.3]).

**Theorem 1.** *Let the open interval $I$ be a gap in $\sigma_{ess}$ and let $\lambda$ be a point in $I$. Then there is a unique $\alpha$ such that $\lambda \in \sigma_{\alpha\mathrm{d}}$. Writing $\alpha$ as $\alpha(\lambda)$, $\alpha$ can be taken to be a continuous increasing function of $\lambda$ in $I$.*

Here $\sigma_{\alpha\mathrm{d}}$ denotes the set of isolated eigenvalues of $H_\alpha$. In particular, from this result it follows that $\complement\sigma_{ess} \subset S_0$ where $S_0$ is the interior of the set

$$S = \{\lambda \in \mathbb{R} \mid \exists u \text{ a solution of } \mathcal{L}u = \lambda u \text{ such that } \int_0^\infty |u(t)|^2 dt < \infty\}.$$

The problem whether the sets $\complement\sigma_{ess}$ and $S_0$ are always equal was first studied in [20] and it was shown in [6], [32] that equality does not always hold. However, if the spectrum is a perfect set, that is, equal to the set of its limit points, then $S_0 = \complement\sigma_{ess}$. See [6].

## 3. Embedded singular spectrum

Now we turn to the study of the singular part which is embedded in the essential spectrum. Since the supports of the singular parts corresponding to different boundary conditions are mutually disjoint, we cannot expect much stability of this part. We already saw that the way isolated eigenvalues move when the boundary condition varies is very smooth. No such smoothness occurs for the embedded singular part.

Eigenvalues embedded in the essential spectrum may "live" only in a set of first category in the sense of Baire. In fact this is a general statement when we talk of supports of the various spectral measures $\rho_\alpha$. Let $I$ be an interval such that the spectrum is essentially dense in $I$, meaning that $\rho_0(J) > 0$ for every subinterval $J$ of $I$. Then there exists a set $F$ of first category which supports each of the measures $\rho_\alpha$ (not just the point part), i.e., such that $\rho_\alpha(I \backslash F) = 0$ for all $\alpha$. See [11].

A basic tool for the understanding of the behavior of embedded eigenvalues is the following theorem of Aronszajn [1].

**Theorem 2.** *Consider the Sturm-Liouville equation* (1) *and two different boundary conditions corresponding to $\alpha \neq \beta \bmod \pi$. In order that $\xi$ be in the point spectrum relative to the boundary condition $\beta$, it is necessary and sufficient that $\int_{\mathbb{R}} (\lambda - \xi)^{-2} d\rho_\alpha(\lambda) < \infty$ and that $m_\alpha(\xi) + \cot(\beta - \alpha) = 0$.*

Let us define $G(\xi) = \int_{\mathbb{R}} (\lambda - \xi)^{-2} d\rho_0(\lambda)$. In [10] it was proven that $\{y \mid G(y) = \infty\}$ is a dense $G_\delta$ set in supp $(d\rho_0)$, the support of $d\rho_0$. (Remember that a $G_\delta$ is a set which is a countable intersection of open sets.)

If we assume that an interval is contained in the spectrum, then the complement of $\{y \mid G(y) = \infty\}$ cannot contain this interval; moreover the support of the point part embedded in the spectrum has to be small in Baire sense, that is of first category.

Now we can use properties of $m_0$ to map this set of first category in the spectrum to a set of first category in the boundary conditions to obtain the following theorem [10].

**Theorem 3.** *The set $\{\alpha \mid H_\alpha$ has no eigenvalues in the spectrum of $H_0\}$ is a dense $G_\delta$ in $[0, \pi]$.*

Therefore the set of $\alpha$ for which $H_\alpha$ may have embedded eigenvalues is of first category in $[0, \pi]$. This theorem tells us that dense point spectra are very unstable and that even a very small perturbation of the boundary condition will make the whole point part disappear.

In some sense the essential spectrum prevents the existence of point spectra for many boundary conditions. An example where the above applies is given by the operator $H = d^2/dx^2 + \cos(\sqrt{x})$ on $L^2(0,\infty)$.

It was shown that for any boundary condition $\alpha$ the spectrum of the operator is absolutely continuous on $(1,\infty)$ [37] and that $H_\alpha$ has pure point spectrum in [-1,1] for a.e. $\alpha$ [24]. From what was mentioned above, it follows that for a dense $G_\delta$ of $\alpha, H_\alpha$ has only singular continuous spectrum in $[-1,1]$. An open problem is to exhibit a $\alpha$ where this happens.

In 1993, N. Makarov [27] made the conjecture that

$$|\{\alpha \mid H_\alpha \text{ has only p.p. spectrum}\}| \cdot |\{\alpha \mid H_\alpha \text{ has only s.c. spectrum}\}| = 0\,;$$

here and in the sequel, $|\cdot|$ denotes Lebesgue measure. As far as I know this remains an open question. In this context, natural questions arise about the possibility of coexistence of different types of spectra for large sets of boundary conditions $\alpha$, in particular the coexistence of absolutely continuous and singular spectra. It is known that if the potential $v$ is in $L^2$, then there can be singular spectrum at positive energies only for a set of boundary conditions of measure zero. This follows from a result of [4] which states that the support of the singular part in this case has Lebesgue measure zero and formula (7) below.

As mentioned above, the singular spectrum may be very unstable. Nevertheless, the property of having singular spectra for a set of boundary conditions of positive measure is preserved under $L^1$ perturbations to the potential; see [25], [12]. It is also worth mentioning that the exact Hausdorff dimension of the spectral measures may be the same for all boundary conditions, at least in the discrete case. See Theorem 4.3 in the contribution of Y. Last to this volume [26].

## 4. Sketch of a result on coexistence

It is possible to have absolutely continuous spectrum for all boundary conditions and singular spectrum for some boundary conditions. In [33] an example was constructed where for a set of boundary conditions of positive measure there is singular spectrum supported on a Cantor type set. This construction can be modified to have singular spectrum supported on a dense set (see [40]). In this example there is not much information about the set of boundary conditions with mixed spectra other than that this set is of positive measure. In what follows I shall sketch a result (see [13]) which gives a clearer idea of this set.

The following equality holds

$$\int_\alpha^\beta \rho_\theta(A)d\theta = \frac{1}{\pi}\int_A \arg\left[\frac{\cos\beta + \sin\beta\, m_0(\lambda+i0)}{\cos\alpha + \sin\alpha\, m_0(\lambda+i0)}\right] d\lambda, \tag{6}$$

which is a generalization of the well-known result, (see [36]):

$$\int_0^\pi \rho_\theta(A)d\theta = |A|. \tag{7}$$

Let us define $\Lambda_M := \{\lambda \mid \operatorname{Im}\big(m_0(\lambda+i0)\big) > M\}$. Then we have the following bound:

**Lemma 1.**

$$\int_\alpha^\beta \rho_\theta\Big(I\cap\Lambda_M\Big)d\theta \le \frac{2}{\pi}\arctan\Big(\frac{1}{2M}(\cot\alpha-\cot\beta)\Big)\Big|I\cap\Lambda_M\Big|.$$

*Proof.* The statement of the lemma follows if we put together equality (6) and the definition of $\Lambda_M$. Observe that the transformation

$$w = Tz = \frac{z\sin\beta+\cos\beta}{z\sin\alpha+\cos\alpha}$$

maps the half-plane $\operatorname{Im} z > M$ onto the disk (see the figure)

$$\left(x-\frac{\sin\beta}{\sin\alpha}\right)^2+\left(y-\frac{\sin(\beta-\alpha)}{2M\sin^2\alpha}\right)^2<\left(\frac{\sin(\beta-\alpha)}{2M\sin^2\alpha}\right)^2;$$

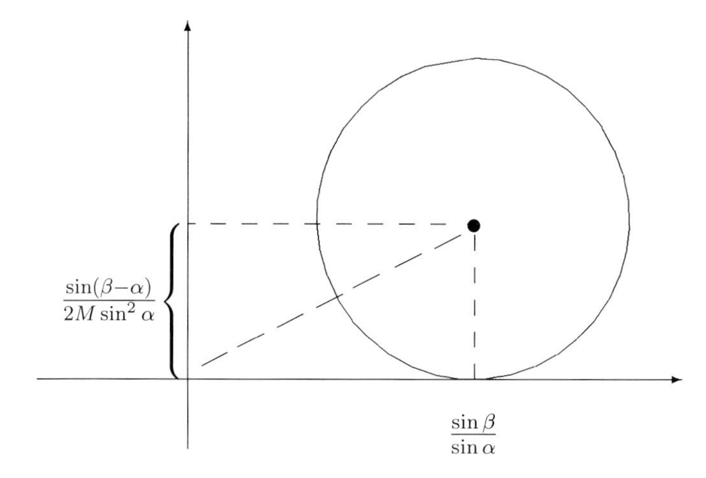


therefore $\operatorname{Im}\big(m_0(\lambda+i0)\big) > M$ implies that

$$\arg T(m_0(\lambda+i0)) \le 2\arctan\Big(\frac{1}{2M}(\cot\alpha-\cot\beta)\Big).$$

Using this and (6) we obtain

$$\begin{aligned}\int_\alpha^\beta \rho_\theta\Big(I\cap\Lambda_M\Big) &= \frac{1}{\pi}\int_{I\cap\Lambda_M} \arg\Big(T(m_0(\lambda+i0)\Big)d\lambda \\ &\le \frac{2}{\pi}\int_{I\cap\Lambda_M} \arctan\Big(\frac{1}{2M}(\cot\alpha-\cot\beta)\Big)d\lambda \\ &= \frac{2}{\pi}\arctan\Big(\frac{1}{2M}(\cot\alpha-\cot\beta)\Big)\Big|\Lambda_M\cap I\Big|.\end{aligned}$$

□

We also need a bound from below, for which the following theorem [34] will be useful.

**Theorem 4.** *Let $\lambda_i, \lambda_j$ be eigenvalues of*

$$-y''(x) + v(x)y(x) = \lambda y(x) \qquad x \in [0, N],$$
$$y(0)\cos\alpha - y'(0)\sin\alpha = 0,$$
$$y(N)\cos\beta - y'(N)\sin\beta = 0.$$

*Let $d\tilde{\rho}$ be the spectral measure of this problem. Then*

$$\tilde{\rho}\left((\lambda_i, \lambda_j)\right) = \min_{\rho \in M_N} \rho\left((\lambda_i, \lambda_j)\right).$$

Here $M_N$ is a family of measures which contains the spectral measure of the half-line problem. The bound from below that we need is given by the following lemma.

**Lemma 2.** *Let $N$ be a positive real number and assume that $v(x) = 0$ for all $x \in [0, N]$. Set $I = \left(\left(\frac{\pi k}{N}\right)^2, \left(\frac{\pi(k+2)}{N}\right)^2\right)$ where $k \in \mathbb{N}$; then*

$$\int_\alpha^\beta \rho_\theta(I)d\theta \geq \frac{2\pi(k+1)}{N^2} \int_{\frac{N}{\pi(k+1)}\cot\beta}^{\frac{N}{\pi(k+1)}\cot\alpha} \frac{dx}{1+x^2}.$$

*Proof.* The function

$$\psi_\alpha\left(x, \frac{\pi k}{N}\right) = \frac{1}{2}\left[\sin\alpha + \frac{N}{i\pi k}\cos\alpha\right] \mathrm{e}^{\frac{i\pi k}{N}x} + \frac{1}{2}\left[\sin\alpha - \frac{N}{i\pi k}\cos\alpha\right] \mathrm{e}^{\frac{-i\pi k}{N}x}$$

is a solution of

$$-\psi''(x) = \left(\frac{\pi k}{N}\right)^2 \psi(x),$$
$$\psi(0)\cos\alpha - \psi'(0)\sin\alpha = 0,$$
$$\psi(N)\cos\alpha - \psi'(N)\sin\alpha = 0.$$

Observe that we have the same eigenvalues $\left(\frac{\pi k}{N}\right)^2$ for all $\alpha \in [0, \pi)$. Since

$$\|\psi_\alpha\|^2 = \int_0^N |\psi_\alpha(x)|^2 dx = \frac{(\sin\alpha)^2}{2}N + \frac{1}{2}N\left(\frac{N\cos\alpha}{\pi k}\right)^2$$

we get from Theorem 4:

$$\rho_\theta(I) \geq \tilde{\rho}(I) = \left\|\psi_\theta\left(x, \frac{\pi(k+1)}{N}\right)\right\|^{-2},$$

$$\int_\alpha^\beta \rho_\theta(I)d\theta \geq \int_\alpha^\beta \|\psi_\theta\|^{-2}d\theta = \frac{2\pi(k+1)}{N^2} \int_{\frac{N}{\pi(k+1)}\cot\beta}^{\frac{N}{\pi(k+1)}\cot\alpha} (1+x^2)^{-1}dx. \qquad \square$$

If we put together the bounds from above and below we obtain the following result:

**Theorem 5.** *Let $N \in \mathbb{R}^+, k \in \mathbb{N}$, and assume that $v(x) = 0$ for all $x \in [0, N]$. Set $I = \left(\left(\frac{\pi k}{N}\right)^2, \left(\frac{\pi(k+2)}{N}\right)^2\right)$ and let $0 < \alpha < \beta < \pi$. If the following inequality is satisfied:*

$$\frac{2\pi(k+1)}{N^2} \int\limits_{\frac{N}{\pi(k+1)}\cot\beta}^{\frac{N}{\pi(k+1)}\cot\alpha} \frac{dx}{1+x^2} > \frac{2|\Lambda_M \cap I|}{\pi} \int\limits_0^{\frac{1}{2M}(\cot\alpha - \cot\beta)} \frac{dx}{1+x^2},$$

*then*

$$\int_\alpha^\beta \rho_\theta(I \cap \complement\Lambda_M) d\theta > 0.$$

*Proof.* The upper and lower bounds given by Lemmas 1 and 2, together with the hypotheses of the theorem imply

$$\int_\alpha^\beta \rho_\theta(I) d\theta > \int_\alpha^\beta \rho_\theta(I \cap \Lambda_M) d\theta.$$

Since

$$\int_\alpha^\beta \rho_\theta(I) d\theta = \int_\alpha^\beta \rho_\theta(I \cap \complement\Lambda_M) d\theta + \int_\alpha^\beta \rho_\theta(I \cap \Lambda_M) d\theta,$$

we have

$$\int_\alpha^\beta \rho_\theta(I \cap \complement\Lambda_M) d\theta > 0.$$

□

The theorem above can be used to analyze the set of boundary conditions $\theta$ which give rise to some singular spectrum. For this purpose we only need to consider the case $M = 0$, since $\Lambda_0$ happens to be a support for the absolutely continuous part of the spectral measure. Let us consider some examples.

**Example 1.** If we choose in the theorem above the parameters $k = 1$, $N = 2\pi$ and $M = 0$, then we obtain that the condition

$$\beta - \alpha > \pi|\Lambda_0 \cap I|,$$

where $I = \left(\frac{1}{4}, \frac{9}{4}\right)$, implies

$$\int_\alpha^\beta \rho_\theta\left(I \cap \complement\Lambda_0\right) d\theta > 0.$$

Since $\complement\Lambda_0$ is a support of the singular part we get the existence of a set $B \subset (\alpha, \beta)$ of positive measure such that for $\theta \in B$ the operator $H_0$ has some singular spectrum in $I$. Changing the parameters $k$ and $N$, we can get similar statements for other intervals.

**Example 2.** The theorem can also be applied to the examples constructed by Remling. He considers in [33] potentials of the form

$$v(x) = \sum_{n=1}^{\infty} g_n v_n(x - a_n),$$

where $g_n > 0, v_n \in L^1[-B_n, B_n], v_n(x) = \chi_{(-B_n,B_n)}(x)w(x),$

$$w(x) = \int_F \cos(2kx)\, dk.$$

The intervals $[a_n - B_n, a_n + B_n]$ are assumed to be disjoint and the set $F$, which appears in the definition of the barriers $w$, is Cantor type in an interval $[a, b]$ with Lebesgue measure any positive number less than $b - a$. The function $\chi_{(-B_n,B_n)}$ denotes the characteristic function of the set $(-B_n, B_n)$.

Let $L_n = a_n - B_n - a_{n-1} - B_{n-1}$ with $a_0 = B_0 = 0$. In [33], under minor assumptions on $F$ the following was proved:

**Theorem 6.** *Let $g_n = n^{-1/2}, B_n = n^\beta$ with $(2 - 4/\gamma)^{-1} < \beta < \gamma/8$, where $\gamma > 6$, and assume $n^{\beta/2\gamma} L_{n-1}/L_n \to 0$ as $n \to \infty$. Then the half-line Schrödinger operators $H_\alpha$ with potential $v$ given as above satisfy $\sigma_{ac}(H_\alpha) = \sigma_{ess}(H_\alpha) = [0, \infty), \sigma_d(H_\alpha) \cap (0, \infty) = \varnothing$, and $\sigma_{sc}(H_\alpha) \cap (0, \infty) \neq \varnothing$ for a set of boundary conditions $\alpha$ of positive measure.*

In proving this result it is shown that the singular part of the spectral measures $\rho_\alpha$ corresponding to $H_\alpha$ are supported on $F^2 = \{k^2 : k \in F\}$. In this theorem the potential can be chosen to be zero in an interval $[0, N]$.

We can apply Theorem 5 as we did in Example 1 taking $k = 1, N = 2\pi$ and $M = 0$. The requirement to have singular continuous spectrum for a set of boundary conditions of positive measure in $(\alpha, \beta)$ is

$$\beta - \alpha > \pi|\Lambda_0 \cap I| = |(\complement F^2) \cap I|,$$

where $I = \left(\frac{1}{4}, \frac{9}{4}\right)$. Observe that we can control the measure of $F^2$ and therefore, by modifying $|F|$, we can choose the length of $(\alpha, \beta)$.

## 5. Coexistence for all boundary conditions

As mentioned above, it is possible to have absolutely continuous spectrum and singular spectrum even for *all* boundary conditions. In fact very simple spectral measures generate this coexistence [7]. Consider, for example, a fixed interval $I$ and take a set $E \subset I$ such that for every subinterval $J \subset I$ we have

$$0 < |E \cap J| < |J|,$$

where $|\cdot|$ denotes the Lebesgue measure, that is, $E$ and $\complement E$ are essentially dense in $I$. We need this property of $E$ in order to get genuinely mixed spectra.

Let

$$u(x) = \begin{cases} 1 & x \in E \\ 0 & x \notin E \end{cases}$$

and define $d\mu_{\pi/2} = u\,dx$. We have to define $\mu_{\pi/2}$ outside $I$ so that

$$\int_{\mathbb{R}} \frac{d\mu_{\pi/2}(t)}{1+t^2} < \infty$$

and the necessary decay conditions required by the Gelfand-Levitan inverse theorem [31, Chapter VIII] are satisfied. The measure $\mu_{\pi/2}$ will be the spectral measure of a Sturm-Liouville operator $H_{\pi/2}$ and we denote by $\mu_\beta$ the spectral measures of $H_\beta$, where $\mu_\beta^s$ and $\mu_\beta^{ac}$ stand for their singular and absolutely continuous components respectively.

The family of measures $\mu_\beta$ so generated have the following properties:

**Theorem 7.**

a) $\mu_\beta^s(J) > 0$ *for every subinterval* $J \subset I$, $\beta \in \left(-\frac{\pi}{2}, \frac{\pi}{2}\right)$.
b) $\mu_\beta^{ac}(J) > 0$ *for every subinterval* $J \subset I$, $\beta \in \left(-\frac{\pi}{2}, \frac{\pi}{2}\right]$.

Before we prove this theorem we need a preliminary result. For real $\theta$ consider the family of functions

$$f_\theta(z) = \frac{\cos\theta + z\sin\theta}{\sin\theta - z\cos\theta}\,. \tag{8}$$

Observe that using (5) we have $f_\theta(m_{\pi/2}(z)) = m_\theta(z)$. For each $\theta$, $f_\theta$ is an analytic function that maps the upper half-plane into itself. We shall refer to such functions as Pick functions (also known as Nevanlinna or Herglotz functions). Given a Pick function $F$ it has an integral representation of the form

$$F(z) = a + bz + \int_{\mathbb{R}} \left( \frac{1}{t-z} - \frac{t}{1+t^2} \right) d\mu(t). \tag{9}$$

The integral on the right-hand side is the Cauchy integral of $\mu$ in the upper half-plane. The Weyl functions $m_\alpha$ are Pick functions and the spectral measure $\rho_\alpha$ is the measure that appears in the integral representation of these functions.

Let $L$ be a Pick function such that $0 \le \operatorname{Im} L(z) \le \pi$. For any $\alpha \in \mathbb{R}$, set

$$M_\alpha(z) := f_\alpha\big(L(z)\big) \quad \text{and} \quad N_\alpha(z) := f_\alpha\big(\exp L(z)\big).$$

Both $M_\alpha$ and $N_\alpha$ are Pick functions admitting representations similar to (9). We denote by $\mu_\alpha$ and $\nu_\alpha$ the associated measures that appear in the integral representations. The singular parts of these measures $\nu_\alpha^s$ and $\mu_\alpha^s$ satisfy the following relation:

**Lemma 3.** *Define a function* $\alpha : (-\frac{\pi}{2}, \frac{\pi}{2}) \to (0, \frac{\pi}{2})$ *by*

$$\alpha(\beta) = \arctan\left(\exp\tan\beta\right) \quad \textit{with } \beta \in \left(-\tfrac{\pi}{2}, \tfrac{\pi}{2}\right).$$

*Then*

$$\mu_\beta^s = \alpha'(\beta)\nu_{\alpha(\beta)}^s.$$

*Sketch of proof.* First we show that

$$\lim_{y\downarrow 0}\frac{\operatorname{Im} M_\beta(x+iy)}{\operatorname{Im} N_\alpha(x+iy)} = \alpha'(\beta) \quad \text{for } \mu_\beta^s\text{-a.e. } x. \tag{10}$$

To see this, using the definition of $M_\beta$ and $N_\alpha$ we have

$$\lim_{y\downarrow 0}\frac{\operatorname{Im} M_\beta(z)}{\operatorname{Im} N_\alpha(z)} = \lim_{y\downarrow 0}\frac{\operatorname{Im} L(z)}{\operatorname{Im}\exp L(z)}\left|\frac{\sin\alpha-\exp L(z)\cos\alpha}{\sin\beta - L(z)\cos\beta}\right|^2 .$$

From the definition of $\alpha$

$$\frac{\sin\alpha-\exp L(z)\cos\alpha}{\sin\beta - L(z)\cos\beta} = \frac{\cos\alpha}{\cos\beta}\cdot\frac{\exp(\tan\beta)-\exp L(z)}{\tan\beta - L(z)} . \tag{11}$$

It is well known that for $\mu_\beta^s$-a.e. $x$ the Cauchy integral of $\mu_\beta$ at $x+i\varepsilon$ tends to infinity as $\varepsilon\to 0$. Therefore $M_\beta(x+i\varepsilon)\longrightarrow\infty$ as $\varepsilon\downarrow 0$. The formula for $f_\beta$ and the definition of $M_\beta$ now imply that for $\mu_\beta^s$-a.e. $x$, $L(x+i\varepsilon)\overset{\varepsilon\downarrow 0}{\longrightarrow}\tan\beta$. Hence the expression in the right-hand side of (11) tends to

$$\frac{\cos\alpha}{\cos\beta}\exp(\tan\beta) \quad \text{when } \varepsilon\downarrow 0$$

and we obtain

$$\left|\frac{\sin\alpha-\exp L(z)\cos\alpha}{\sin\beta - L(z)\cos\beta}\right|^2 \overset{\varepsilon\downarrow 0}{\longrightarrow}\left(\frac{\cos\alpha}{\cos\beta}\right)^2(\exp\tan\beta)^2 \quad \text{for } \mu_\beta^s\text{-a.e. } x. \tag{12}$$

Once (10) is obtained, it follows that

$$\left|\frac{\operatorname{Im} M_\beta(x+iy)}{\operatorname{Im} N_\alpha(x+iy)} - \frac{\mu_\beta(x-y,\,x+y)}{\nu_\alpha(x-y,\,x+y)}\right| \overset{y\downarrow 0}{\longrightarrow} 0$$

for $\mu_\beta^s$-a.e. $x$ and from here the assertion of the lemma follows. □

*Proof of Theorem* 7. Let

$$F_{\nu_{\pi/2}}(z) := \exp\left(C + \int_{\mathbb{R}}\left(\frac{1}{\lambda - z} - \frac{\lambda}{\lambda^2+1}\right)d\mu_{\pi/2}(\lambda)\right), \tag{13}$$

where $C$ is chosen so that

$$C + \int_{\mathbb{R}}\left(\frac{1}{\lambda - z} - \frac{\lambda}{\lambda^2+1}\right)d\mu_{\pi/2}(\lambda) = m_{\pi/2}(z),$$

that is, $\log F_{\nu_{\pi/2}}$ is the Weyl function of some Sturm-Liouville operator with boundary condition $\alpha = \frac{\pi}{2}$. $F_{\nu_{\pi/2}}$ is a Pick function. Then recalling the definition of $\mu_{\pi/2}$ we get

$$u(x) = \frac{1}{\pi}\arg F_{\nu_{\pi/2}}(x+i0),$$

where arg stands for the principal branch of argument, taking values in $(-\pi,\pi]$. Therefore

$$\operatorname{Im} F_{\nu_{\pi/2}}(x+i0) = 0 \text{ for a.e. } x\in I.$$

Since the support of the absolutely continuous part of $\nu_\alpha$ is the set

$$\{x \mid \operatorname{Im} F_{\nu_{\pi/2}}(x+i0) > 0\},$$

it follows that $\nu_\alpha$ is purely singular in $I$ for every $\alpha \in (-\frac{\pi}{2}, \frac{\pi}{2})$.

Given an interval $J \subset I$ assume that $\nu_{\pi/2}(J) = 0$. Then $F_{\nu_{\pi/2}}(z)$ can be extended analytically across $J$ and from (13) the same follows for $m_{\pi/2}(z)$. Since this implies $\mu_{\pi/2}(J) = 0$, we get a contradiction to the construction of $\mu_{\pi/2}$. Hence $\nu_{\pi/2}(J) > 0$ for every $J \subset I$. Therefore $\nu_\alpha^s(J) > 0$ for every $\alpha \in (-\frac{\pi}{2}, \frac{\pi}{2})$ (see for instance [14]).

Now to obtain part a) in the theorem we just recall Lemma 3, and we have for every Borel set $A$ and every $\beta \in (-\frac{\pi}{2}, \frac{\pi}{2})$:

$$\mu_\beta^s(A) = \alpha'(\beta)\nu_{\alpha(\beta)}^s(A).$$

Part b) follows from the well-known stability of the absolutely continuous part. □

## 6. Some inverse spectral theory

Let us now refer briefly to the inverse spectral theory of regular Sturm-Liouville operators in relation to dependence on the boundary conditions. We shall consider the operator $-d^2/dx^2 + q$ in $L^2(0,1)$ with boundary conditions

$$\begin{aligned} y(0)\cos\alpha - y'(0)\sin\alpha &= 0, \\ y(1)\cos\theta - y'(1)\sin\theta &= 0, \end{aligned} \tag{14}$$

where $q \in L^1(0,1)$. We denote this operator again by $H_\alpha$ (we shall mostly consider various values of $\alpha$ for fixed $\theta$). It was G. Borg in 1946 [2] who proved the following:

**Theorem 8.** *The spectra of $H_\alpha$ for two values of $\alpha$ uniquely determine $q$.*

This statement should be interpreted as follows. Denote by $\sigma(q;\alpha)$ the spectrum of $-d^2/dx^2 + q$, with boundary conditions (14), where $\alpha \in [0,\pi)$. Let $q_1, q_2 \in L^1(0,1)$ be such that $\sigma(q_1;\alpha_1) = \sigma(q_2;\alpha_1)$ and $\sigma(q_1;\alpha_2) = \sigma(q_2;\alpha_2)$ for some $\alpha_1, \alpha_2$ ($\alpha_1 \neq \alpha_2$). Then $q_1(x) = q_2(x)$ for a.e. $x \in [0,1]$.

If we know more about the potential, then we need less information about the spectra as the following theorem of Hochstadt-Lieberman [21] states:

**Theorem 9.** *The spectrum of $H_\alpha$ for one value of $\alpha$ together with the values of $q$ on $[0,\frac{1}{2}]$ uniquely determine $q$ on $[\frac{1}{2},1]$.*

For interesting generalizations of these theorems to the case of matrix equations, see [28] as well as the contribution by M. Malamud [29] in this volume.

It is remarkable that we need exactly half of the potential $q$, since the knowledge of $q$ on $[0,\frac{1}{2}-\varepsilon]$ for any $\varepsilon > 0$ and the spectrum for one boundary condition is not enough to reconstruct $q$. This can be seen for instance for Dirichlet boundary conditions if we observe that in that case reflecting $q$ at the point $\frac{1}{2}$ does not change the spectrum and then take

$$q(x) = \begin{cases} c & \text{if } x \in [0, \frac{1}{2} - \varepsilon], \\ v(x) \neq c & \text{if } x \in (\frac{1}{2} - \varepsilon, \frac{1}{2}), \\ c & \text{if } x \in [\frac{1}{2}, 1], \end{cases}$$

so that $q$ and its reflection will coincide in $[0, \frac{1}{2} - \varepsilon]$ but will differ on $(\frac{1}{2} - \varepsilon, \frac{1}{2}) \cup (\frac{1}{2}, \frac{1}{2} + \varepsilon)$. In [15] the following was proved (we refer to that reference for a precise definition of the meaning of "half the spectrum"; for example it suffices to enumerate the eigenvalues in increasing order and take the lowest two eigenvalues and then every second one):

**Theorem 10.** *Half the spectrum of one $H_\alpha$ together with the values of $q$ on $[0, \frac{3}{4}]$ uniquely determine $q$.*

In [8] results of the following kind are presented.

**Theorem 11.** *Let $\alpha_1, \alpha_2, \alpha_3 \in [0, \pi)$ and denote by $\sigma_j$ the spectrum of $H_{\alpha_j}, j = 1, 2, 3$. Assume $S_j \subset \sigma_j$, and suppose that for all sufficiently large $\lambda_0 > 0$ we have*

$$\begin{aligned} &\#\{\lambda \in \{S_1 \cup S_2 \cup S_3\} \text{ with } \lambda \leq \lambda_0\} \\ &\qquad \geq \frac{2}{3}\#\{\lambda \in \{\sigma_1 \cup \sigma_2 \cup \sigma_3\} \text{ with } \lambda \leq \lambda_0\} - 1. \end{aligned}$$

*Then $q$ is uniquely determined a.e. on $[0, 1]$.*

In particular, two thirds of three spectra determine $q$. Another result (involving the values $\alpha = \pi/2$ and $\alpha = 0$) is the following:

**Theorem 12.** *Let $\sigma_N$ and $\sigma_D$ be the eigenvalues of $H_{\pi/2}$ and $H_0$ respectively. Let $S_N \subset \sigma_N, S_D \subset \sigma_D$. Fix $a \in (0, 1)$. Suppose that for all $\lambda_0 > 0$ sufficiently large we have*

$$\begin{aligned} &\#\{\lambda \in \{S_N \cup S_D\} \text{ with } \lambda \leq \lambda_0\} \\ &\qquad \geq (1 - a)\#\{\lambda \in \{\sigma_N \cup \sigma_D\} \text{ with } \lambda \leq \lambda_0\}. \end{aligned}$$

*Then $S_N, S_D$ and $q$ on $[0, a]$ uniquely determine $q$ a.e. on $[0, 1]$.*

For example if $a = \frac{1}{4}$, then knowing $q$ on $[0, \frac{1}{4}]$, all the Neumann eigenvalues and half the Dirichlet eigenvalues, uniquely determine $q$ a.e. on $[0, 1]$. Generalizations of some of the above results can be found in a paper by M. Horváth [22].

The strategy to prove the theorems just mentioned is to use the fundamental result of Marchenko, see Theorem 13 below, and to prove a general theorem that knowing $m$ at points $\lambda_0, \lambda_1, \ldots$ determines $m$ as long as $\{\lambda_n\}$ has enough density.

Since we are considering the regular case, we define the Weyl function $m$ as

$$m_\theta(z) = \frac{u_\theta'(0, z)}{u_\theta(0, z)}, \quad z \in \mathbb{C},$$

where $u_\theta(x,z)$ solves $-u''(x,z)+q(x)u(x,z)=zu(x,z)$ with boundary condition (14) at $x=1$.

**Theorem 13.** [30] *The Weyl m-function uniquely determines $q$ a.e. in $[0,1]$.*

Another result that should be possible to prove for Sturm-Liouville operators (I know the proof only for Dirac systems [9]) is the following statement in which, for $q \in L^2(0,1)$ and $\theta \in [0,\pi)$, we denote by $\mu_m(q;\theta)$ $(m=1,2,3,\dots)$ the eigenvalues of $H=-d^2/dx^2+q$ in $L^2(0,1)$ taken in increasing order, with boundary conditions (14) in the case $\alpha=0$:

*Let $a \in [0,1]$ and let $l,k$ be positive integers satisfying $\frac{1}{l}+\frac{1}{k} \geq 2a$. For some $q_1,q_2 \in L^2(0,1)$ and $\theta_1,\theta_2 \in [0,\pi)$ with $\theta_1 \neq \theta_2$, suppose that $\mu_{ln}(q_1;\theta_1) = \mu_{ln}(q_2;\theta_1)$ and $\mu_{kn}(q_1;\theta_2)=\mu_{kn}(q_2;\theta_2)$ for each $n \in \mathbb{N}$. Suppose moreover that $q_1(x)=q_2(x)$ a.e. on the interval $[a,1]$. Then $q_1(x)=q_2(x)$ a.e. in $[0,1]$.*

Thus the sets $\{\mu_{ln}(q;\theta_1) \mid n \in \mathbb{N}\}$ and $\{\mu_{kn}(q;\theta_2) \mid n \in \mathbb{N}\}$ together with $q|_{[a,1]}$ uniquely determine $q$. The result should also hold if we allow $l=\infty$ or $k=\infty$, where for example in the case $l=\infty$, the set $\{\mu_{kn}(q;\theta) \mid n \in \mathbb{N}\}$ for some $\theta$, together with $q|_{[a,1]}$, uniquely determine $q$ a.e. in $[0,1]$.

For particular values of $a,k,l$, one would then obtain:

- Borg type theorem: two spectra uniquely determine $q$ on $[0,1]$ ($a=1$, $l=k=1$).
- Hochstadt-Lieberman type theorem: one spectrum and $q$ on $[1/2,1]$ uniquely determine $q$ on $[0,1]$ ($a=1/2$, $l=1$, $k=\infty$).

Actually the above result would include many more general cases such as, for example:

- half of one spectrum and $q$ on $[1/4,1]$ uniquely determine $q$ on $[0,1]$ ($a=1/4$, $l=2$, $k=\infty$).
- half of two spectra and $q$ on $[1/2,1]$ uniquely determine $q$ on $[0,1]$ ($a=1/2$, $l=k=2$).

In results of the above kind involving spectra with different boundary conditions at one endpoint, it is important to take note of the endpoint at which the boundary condition may vary, relative to that at which the potential is given. As an example, consider the following situation: suppose that $q_1(x)=q_2(x)$ a.e. on $[0,1/2]$ and that $\mu_n(q_1;\theta_1)=\mu_n(q_2;\theta_2)$ for all $n \in \mathbb{N}$ and some $\theta_1,\theta_2 \in [0,\pi)$. A result of Hald [18] implies that $q_1(x)=q_2(x)$ a.e. on $[0,1]$ (and hence $\theta_1=\theta_2$). A similar uniqueness result would not, however, hold if the boundary condition at $x=0$, rather than at $x=1$, was varied. This may be shown as follows (see [5]).

Let $p$ be an arbitrary but fixed point in the open interval (0,1). For $x \in [p,1]$ consider the problem

$$-y''(x)+v(x)y(x)=\mu y(x) \tag{15}$$

with boundary conditions

$$\begin{aligned} y(p) - y'(p) &= 0, \\ y(1) &= 0, \end{aligned}$$

where $v$ is an arbitrary integrable function. We denote by $\{\mu_i\}_{i=0}^{\infty}$ the eigenvalues and by $\{\phi_i\}_{i=0}^{\infty}$ the corresponding eigenfunctions.

Now consider the potential $q$ defined for $x \in [0,1]$ as follows:

$$q(x) = \begin{cases} 1 + \mu_0 & \text{if } x \in [0,p] \\ v(x) & \text{if } x \in (p,1] \end{cases}$$

and the problem

$$-y''(x) + q(x)y(x) = \lambda y(x), \qquad x \in [0,1] \tag{16}$$

with boundary conditions

$$\begin{aligned} y(0) - y'(0) &= 0, \\ y(1) &= 0. \end{aligned} \tag{17}$$

Let us define

$$g_0(x) = \begin{cases} \sqrt{2}\, e^x & \text{for } x \in [0,p], \\ \phi_0(x) & \text{for } x \in (p,1], \end{cases}$$

where we have normalized $\phi_0$ in such a way that $\phi_0(p) = \sqrt{2}\, e^p$ holds. The function $g_0$ is continuously differentiable, solves (16), (17) for $\lambda = \mu_0$, and since $\mu_0$ is the first eigenvalue of (15), $g_0$ does not have any zeros in $[0,1)$. Define

$$q_1(x) := q(x) - 2\frac{d^2}{dx^2} \log\left(1 + \int_0^x g_0^2(s)ds\right).$$

It is easy to see that $q_1$ is integrable on $[0,1]$. The following result is proved in [5]:

**Theorem 14.** *The problems*

$$\begin{aligned} -y'' + q_1(x)y &= \lambda y \qquad x \in [0,1], \\ y(0) + y'(0) &= 0, \\ y(1) &= 0, \end{aligned}$$

*and*

$$\begin{aligned} -y'' + q(x)y &= \lambda y \qquad x \in [0,1], \\ y(0) - y'(0) &= 0, \\ y(1) &= 0, \end{aligned}$$

*have the same spectrum* $\{\lambda_i\}_{i=1}^{\infty}$. *In addition to this:* $q_1(x) = q(x)$ *for* $x \in [0,p]$ *and* $q_1 \neq q$ *as elements of* $L^1(0,1)$.

**Acknowledgements**

I thank Professors Werner Amrein, Andreas Hinz, David Pearson and the other organizers for a very interesting Conference on the occasion of the 200$^{\text{th}}$ anniversary of the birth of Charles François Sturm, for suggesting to me to write this paper and for all the work and patience in the careful revision of this contribution. I would like to thank Prof. Hubert Kalf for providing me with valuable information regarding the study of singular continuous spectra, Olga Tchebotareva for pertinent comments and Francine Gennai–Nicole for her help and kindness during my stay in Geneva.

## References

[1] N. Aronszajn, On a problem of Weyl in the theory of singular Sturm–Liouville equations, Amer. J. Math. **79** (1957), 597–610.

[2] G. Borg, Eine Umkehrung der Sturm–Liouvilleschen Eigenwertaufgabe, Acta Math. **78** (1946), 1–96.

[3] E.A. Coddington and N. Levinson, On the nature of the spectrum of singular second-order linear differential equations, Canad. J. Math. **3** (1951), 335–338.

[4] P. Deift and R. Killip, On the absolutely continuous spectrum of one-dimensional Schrödinger operators with square summable potentials, Comm. Math. Phys. **203** (1999), 341–347.

[5] R. del Rio, On Boundary Conditions of An Inverse Sturm–Liouville Problem, SIAM Journal on Applied Mathematics **50** (1990), 1745–1751.

[6] R. del Rio, On a problem of P. Hartman and A. Wintner, Memorias del XXVII Congreso Nacional de la Sociedad Matemática Mexicana, Aportaciones Matemáticas, Serie Comunicaciones **16** (1995), 119–123.

[7] R. del Rio, S. Fuentes and A. Poltoratski, Families of Spectral Measures with mixed types, Oper. Theory Adv. Appl. **132** (2002), 131–140.

[8] R. del Rio, F. Gesztesy and B. Simon, Inverse Spectral Analysis with Partial Information on the Potential III. Updating boundary conditions, International Mathematical Research Notes **15** (1997), 751–758.

[9] R. del Rio and B. Grébert, Inverse Spectral Results for AKNS Systems with Partial Information on the Potentials, Mathematical Physics, Analysis and Geometry **4** (2001), 229–244.

[10] R. del Rio, N. Makarov and B. Simon, Operators with Singular Continuous Spectrum II: Rank One Operators, Comm. Math. Phys. **165** (1994), 59–67.

[11] R. del Rio and A. Poltoratski, Spectral Measures and Category, Oper. Theory Adv. Appl. **108** (1999), 149–159.

[12] R. del Rio, B. Simon and G. Stolz, Stability of Spectral Types of Sturm-Liouville Operators, Math. Res. Let. **1** (1994), 437–450.

[13] R. del Rio and O. Tchebotareva, Boundary conditions of Sturm-Liouville operators with mixed spectra, Math. Anal. Appl. **288** (2003), 518–529.

[14] M.S.P. Eastham and H. Kalf, *Schrödinger-type operators with continuous spectra*, Pitman Advanced Publishing Program, Boston, 1982.

[15] F. Gesztesy and B. Simon, Inverse spectral analysis with partial information on the potential, II. The case of discrete spectrum, Trans. Amer. Math. Soc. **352** (2000), 2765–2787.

[16] D. Gilbert, Asymptotic Methods in the Spectral Analysis of Sturm-Liouville Operators, in this volume.

[17] D. Gilbert and D. Pearson, On subordinacy and analysis of the spectrum of one-dimensional Schrödinger operators, J. Math. Anal. Appl. **128** (1987), 30–56.

[18] O.H. Hald, Inverse eigenvalue problems for the mantle, Geophys. J. Royal Astron. Soc. **62** (1980), 41–48.

[19] P. Hartman and A. Wintner, An oscillation theorem for continuous spectra, Proc. Nat. Acad. Sci. USA **33** (1947), 376–379.

[20] P. Hartman and A. Wintner, A separation theorem for continuous spectra, Amer. J. Math. **71** (1949), 650–662.

[21] H. Hochstadt and B. Lieberman, An inverse Sturm-Liouville problem with mixed given data, SIAM J. Appl. Math. **34** (1978), 676–680.

[22] M. Horváth, On the inverse spectral theory of Schrödinger and Dirac operators, Trans. Amer. Math. Soc. **353** (2001), 4155–4171.

[23] T. Kato, Perturbation of continuous spectra by trace class operators, Proc. Japan Acad. **33** (1957), 260–264.

[24] W. Kirsch, S. Molchanov and L. Pastur, One-dimensional Schrödinger operators with high potential barriers, Oper. Theory Adv. Appl. **57** (1992), 163–170.

[25] A. Kiselev, Y. Last and B. Simon, Stability of singular spectral types under decaying perturbations, J. Funct. Anal. **198** (2003), 1–27.

[26] Y. Last, Spectral Theory of Sturm-Liouville Operators on Infinite Intervals: A Review of Recent Developments, in this volume.

[27] N. Makarov, Personal communication.

[28] M. Malamud, Uniqueness questions in inverse problems for systems of differential equations on a finite interval, Trans. Moscow Math. Soc. **60** (1999), 173–224.

[29] M. Malamud, Uniqueness of the Matrix Sturm-Liouville Equation given a Part of the Monodromy Matrix, and Borg Type Results, in this volume.

[30] V.A. Marchenko, Some questions in the theory of one-dimensional linear differential operators of the second order, I, Trudy Moskov. Mat. Obšč. **1** (1952), 327–420 (Russian); Amer. Math. Soc. Transl. **101** (1973), 1–104.

[31] M. Naimark, *Linear Differential Operators II*, Frederick Ungar Publishing Co., New York, 1968.

[32] C. Remling, Essential spectrum and $L_2$-solutions of one-dimensional Schrödinger operators, Proceedings of the Amer. Math. Soc. **124** (1996), 2097–2100.

[33] C. Remling, Embedded singular continuous spectrum for one-dimensional Schrödinger operators, Trans. Amer. Math. Soc. **351** (1999), 2479–2497.

[34] C. Remling, Universal bounds on spectral measures of one-dimensional Schrödinger operators, J. Reine Angew. Math **564** (2003), 105–117.

[35] M. Rosenblum, Perturbation of the continuous spectrum and unitary equivalence, Pacific J. Math. **7** (1957), 997–1010.

[36] B. Simon, Spectral analysis of rank one perturbations and applications, CMR Proc. Lecture Notes **8** (1995), 109–149.

[37] G. Stolz, Bounded solutions and absolute continuity of Sturm-Liouville operators, J. Math. Anal. Appl. **169** (1992), 210–228.

[38] C. Sturm, Mémoire sur les Équations différentielles linéaires du second ordre, J. Math. Pures Appl. **1** (1836), 106–186.

[39] C. Sturm, Mémoire sur une classe d'Équations à différences partielles, J. Math. Pures Appl. **1** (1836), 373–444.

[40] O. Tchebotareva, *Schrödinger Operators and Spectral Types*, PhD Thesis, UNAM, México, 2004.

[41] H. Weyl, Über gewöhnliche Differentialgleichungen mit Singularitäten und die zugehörigen Entwicklungen willkürlicher Funktionen, Math. Ann. **68** (1910), 220–269.

[42] H. Weyl, Über beschränkte quadratische Formen, deren Differenz vollstetig ist, Rend. Circ. Mat. Palermo **27** (1909), 373–392.

Rafael del Río
Instituto de Investigaciones en Matemáticas Aplicadas y en Sistemas
Universidad Nacional Autónoma de México
Apartado Postal 20-726
México D.F. 01000
México
e-mail: `delrio@servidor.unam.mx`

*W.O. Amrein, A.M. Hinz, D.B. Pearson*
Sturm-Liouville Theory: Past and Present, 237–270


# Uniqueness of the Matrix Sturm-Liouville Equation given a Part of the Monodromy Matrix, and Borg Type Results

Mark M. Malamud

**Abstract.** Uniqueness of the matrix Sturm-Liouville equation is investigated, given a part of its monodromy matrix. Generalizations of Borg's theorem and the Hochstadt-Lieberman result for the matrix Sturm-Liouville equation are presented.

**Mathematics Subject Classification (2000).** 34B10, 47E05.

**Keywords.** Inverse Sturm-Liouville problems, monodromy matrix, Borg type results, Hochstadt-Lieberman type results.

## 1. Introduction

Consider the scalar Sturm-Liouville differential equation

$$-y'' + q(x)y = zy \tag{1.1}$$

on the interval $[0, 1]$, with real potential $q = \overline{q} \in L^1[0, 1]$ and spectral parameter $z$. Denote by $S(q; h_0, h_1)$ the spectrum of equation (1.1) subject to the boundary conditions

$$y'(0) - h_0 y(0) = 0, \quad y'(1) - h_1 y(1) = 0, \tag{1.2}$$

where $h_j \in \mathbb{R} \cup \{\infty\}, j \in \{0, 1\}$ (with $h_j = \infty$ shorthand for the boundary condition $y(j) = 0$).

Alongside equation (1.1) we consider similar equations with real $\widetilde{q} \in L^1[0, 1]$ in place of $q$. The starting point of this paper is summarized by the following classical results of Borg [4] and Hochstadt-Lieberman [25].

**Theorem 1.1.** [4] *Suppose that $S(q; h_0, h_1) = S(\widetilde{q}; h_0, h_1)$ and that $S(q; h_0, h_2) = S(\widetilde{q}; h_0, h_2)$ for some $h_0, h_1, h_2$ with $h_1 \neq h_2$. Then $q(x) = \widetilde{q}(x)$ for almost all $x \in [0, 1]$, that is, two spectra $S(q; h_0, h_1)$ and $S(q; h_0, h_2)$ uniquely determine $q$.*

**Theorem 1.2.** [25] *Let $S(q; h_0, h_1) = S(\widetilde{q}; h_0, h_1)$ for some $h_0$ and $h_1$, and let $q(x) = \widetilde{q}(x)$ for almost all $x \in [0, 1/2]$. Then $q(x) = \widetilde{q}(x)$ for almost all $x \in [0, 1]$.*

We shall use the following notations. $\mathbb{C}^{n\times n}$ stands for the set of $n \times n$ matrices with complex entries, $I_n$ is the identity matrix in $\mathbb{C}^{n\times n}$, and $\mathbb{O}_n := 0 \cdot I_n$; $\operatorname{diag}(\lambda_1, \dots, \lambda_n)$ stands for the diagonal matrix in $\mathbb{C}^{n\times n}$ with entries $\lambda_1, \dots, \lambda_n$; $\operatorname{col}(a_1, \dots, a_n)$ stands for a column-vector with entries $a_1, \dots, a_n$. For a function $F$ of two real variables, we denote by $D_j F$ its derivative with respect to the $j$th argument ($j \in \{1, 2\}$).

The main purpose of the paper is to consider the matrix Sturm-Liouville equation

$$-y'' + Q(x)y = \lambda^2 y, \qquad y = \operatorname{col}(y_1, \dots, y_n) \tag{1.3}$$

with a potential matrix $Q(\cdot) \in L^1[0, 1] \otimes \mathbb{C}^{n\times n}$, not necessarily self-adjoint. We present generalizations of Theorems 1.1 and 1.2 to the case of the matrix equation (1.3). Both generalizations are essentially based on results on unique recovery of the equation (1.3) (the potential matrix $Q$) from a part of its monodromy matrix.

Define $n \times n$ matrix solutions $C(x, \lambda)$ and $S(x, \lambda)$ of (1.3) obeying the initial conditions

$$C(0, \lambda) = S'(0, \lambda) = I_n, \quad S(0, \lambda) = C'(0, \lambda) = \mathbb{O}_n. \tag{1.4}$$

Denote by $W(\lambda)$ the (canonical) monodromy matrix of equation (1.3),

$$W(\lambda) := \begin{pmatrix} C(1, \lambda) & S(1, \lambda) \\ C'(1, \lambda) & S'(1, \lambda) \end{pmatrix}. \tag{1.5}$$

Here, as in the sequel, prime denotes differentiation with respect to $x$.

The paper is organized as follows. In Section 2 we recall (with proofs) some basic facts on existence of triangular transformation operators for equation (1.3) and some properties of monodromy matrices.

In Section 3 we show that equation (1.3) is uniquely determined by one of the columns of the monodromy matrix (Proposition 3.1) as well as by its first row $\big(C(1, \lambda)\ S(1, \lambda)\big)$ (Propositions 3.5 and 3.6). Note that in the case of self-adjoint potential matrix $Q(\cdot) = Q(\cdot)^*$ Proposition 3.6 has recently been proved by R. Carlson [9] who used quite different methods for this purpose.

Both results are obtained by the method proposed by the author in [35] for the proof of Borg's theorem as well as its generalizations for ODEs of order $n$, and even nonintegral order $n - \epsilon$ (see also [38, 36, 37] for further applications). Namely, using triangular transformation operators (see (2.18)) we reduce both problems to an investigation of uniqueness for certain Goursat type problems for the hyperbolic equation

$$D_x^2 R(x, t) - D_t^2 R(x, t) = [\widetilde{Q}(x) - Q(t)] R(x, t), \tag{1.6}$$

in the triangle $\triangle = \{(u, v) : 0 \le v \le u \le 2 - v\}$. Here $R(\cdot, \cdot)$ is the kernel of the transformation operator. Note also that equation (1.3) is not uniquely determined by the second row $\big(C'(1, \lambda)\ S'(1, \lambda)\big)$ of the monodromy matrix $W(\lambda)$ nor by either of its diagonals.

In Section 4 we improve the second result (Proposition 3.6) by showing that equation (1.3) is uniquely determined by the meromorphic matrix function $S(1,\lambda)^{-1}C(1,\lambda)$ (see Theorem 4.1). Based on this result we prove an analogue of Borg's Theorem 1.1. Namely, it is shown (see Theorem 4.3) that the potential matrix $Q$ is uniquely determined by $n^2+1$ spectra of the respective (Sturm-Liouville type) boundary value problems for equation (1.3). Moreover, the number of spectra can be reduced to $n(n+1)/2+1$ if $Q=Q^*$.

In Section 5 we show that equation (1.3) is uniquely determined by the Weyl function $C'(1,\lambda)C^{-1}(1,\lambda)$. This result yields one more generalization of Borg's Theorem 1.1 (see Theorem 5.3).

In Section 6, Theorem 1.2 of Hochstadt-Lieberman is generalized to the case of the matrix equation (1.3). Namely we show that equation (1.3) is uniquely determined by one of the entries of the monodromy matrix if $Q(\cdot)$ is known on half of the segment $[0,1]$. The proof is similar to that of Propositions 3.1 and 3.5 and is based on reduction of the problem to a Goursat type problem for equation (1.6).

Note in conclusion that all uniqueness results involving monodromy matrices remain valid (without changes in the proofs) for operator-valued potentials $Q$.

## 2. Preliminaries

### 2.1. Transformation operators

Consider the matrix Sturm-Liouville equation

$$-y''+Q(x)y=\lambda^2 y, \qquad y=\mathrm{col}(y_1,y_2,\dots,y_n), \qquad x\in[0,1] \tag{2.1}$$

with an $n\times n$ matrix potential $Q(\cdot)\in L^1[0,1]\otimes\mathbb{C}^{n\times n}$, which is not assumed to be self-adjoint. Denote by $Y(x,\lambda)$ the $n\times n$ matrix solution of (2.1) obeying the initial conditions

$$Y(0,\lambda)=I_n, \qquad Y'(0,\lambda)=H \quad (\in\mathbb{C}^{n\times n}). \tag{2.2}$$

The following proposition is well known (see [32, 41, 42]).

**Proposition 2.1.** *Suppose that the potential matrix $Q$ is summable on $[0,1]$. Then there exists an $n\times n$-matrix kernel $K(\cdot,\cdot)\in C(\Omega)\otimes\mathbb{C}^{n\times n}$, where $\Omega=\{(x,t): 0\le t\le x\le 1\}$, such that*

$$Y(x,\lambda)=\cos(\lambda x)\cdot I_n+\int_0^x K(x,t)\cos(\lambda t)dt =: (I+K)\cos(\lambda x). \tag{2.3}$$

*Moreover, $K(\cdot,\cdot)$ belongs to $W_1^1(\Omega)\otimes\mathbb{C}^{n\times n}$, where $W_1^1(\Omega)$ stands for the Sobolev space, and satisfies (in the sense of distributions) the following boundary value problem:*

$$D_x^2K(x,t)-Q(x)K(x,t)=D_t^2K(x,t), \tag{2.4}$$

$$K(x,x)=H+\frac{1}{2}\int_0^x Q(s)ds, \quad x\in[0,1], \tag{2.5}$$

$$D_tK(x,t)|_{t=0}=0. \tag{2.6}$$

*If additionally $Q(\cdot) \in C^1[0,1] \otimes \mathbb{C}^{n\times n}$, then $K(\cdot,\cdot) \in C^2(\Omega) \otimes \mathbb{C}^{n\times n}$ and is a classical solution of the problem* (2.4)–(2.6).

*Proof.* (i) Assume that formula (2.3) is valid and the kernel $K(\cdot,\cdot) \in C^2(\Omega)\otimes\mathbb{C}^{n\times n}$. Substituting the expression (2.3) into (2.1) we derive after simple transformations that $K(\cdot,\cdot)$ is a solution of the problem (2.4)–(2.6) in $\Omega$. Conversely, to prove formula (2.3) with smooth $K(\cdot,\cdot)\big(\in C^2(\Omega) \otimes \mathbb{C}^{n\times n}\big)$ it suffices to show that the problem (2.4)–(2.6) has a (unique) solution. In order to prove this statement we impose the additional condition

$$K(x, 1-x) = 0, \qquad x \in [0,1] \tag{2.7}$$

and consider alongside problem (2.4)–(2.6) the Goursat problem (2.4), (2.5), (2.7). First we show that the Goursat problem has a unique solution in $\Omega$. Setting $u = x+t, v = x-t$ and $P_1(u,v) = K\big((u+v)/2, (u-v)/2\big)$ we rewrite system (2.4), (2.5),(2.7) in the equivalent form

$$4D_uD_vP_1(u,v) = Q((u+v)/2)P_1(u,v), \qquad (u,v) \in \Delta, \tag{2.8}$$

$$4P_1(u,0) = 4H + \int_0^u Q(s/2)ds, \qquad u \in [0,2], \tag{2.9}$$

$$P_1(1,v) = 0, \qquad v \in [0,1], \tag{2.10}$$

where $\triangle := \{(u,v) : 0 \le v \le u \le 2-v\}$. Integrating the system (2.8) along the characteristics and taking conditions (2.9),(2.10) into account we arrive at the following integral equation:

$$P_1(u,v) = H + \frac{1}{4}\int_0^u Q(s/2)ds - \frac{1}{4}\int_0^v d\mu \int_u^1 Q(\frac{\xi+\mu}{2})P_1(\xi,\mu)d\xi. \tag{2.11}$$

This is a Volterra type integral equation of the second kind. The existence and uniqueness of a solution of (2.11) can easily be proved by the method of successive approximations.

Let $P_1(\cdot,\cdot)$ be the (unique) solution of (2.11). It is clear that $P_1(\cdot,\cdot)$ is also the unique solution of the problem (2.8)–(2.10).

(ii) We put $K_1(x,t) := P_1(x+t, x-t)$. It follows from (2.11) that if $Q \in C^1[0,1] \otimes \mathbb{C}^{n\times n}$, then the derivatives $D_u^2 P_1(u,v)$ and $D_v^2 P_1(u,v)$ exist and are continuous in $\Delta$. Therefore in this case $K_1(\cdot,\cdot) \in C^2(\Omega) \otimes \mathbb{C}^{n\times n}$ and satisfies equation (2.4) and conditions (2.5) and (2.7).

Next we show that for any matrix function $\Phi \in C^2[0,1]\otimes\mathbb{C}^{n\times n}$ with $\Phi(0) = \mathbb{O}_n$ the function

$$K(x,t) = K_1(x,t) + \Phi(x-t) + \int_t^x K_1(x,s)\Phi(s-t)ds \tag{2.12}$$

satisfies the problem (2.4),(2.5). Indeed $K(x,x) = K_1(x,x) + \Phi(0) = K_1(x,x)$, from which (2.5) follows. Further, denote by $L$ the operator in (2.4), that is $L :=$

$D_x^2 - D_t^2 - Q(x)$. Applying the operator $L$ to (2.12) and taking (2.5) into account we find

$$LK = LK_1 - Q(x)\Phi(x-t) + \frac{dK_1(x,x)}{dx}\Phi(x-t) + K_1(x,x)\Phi'(x-t)$$

$$+ [D_x K_1(x,s)]_{s=x} \cdot \Phi(x-t) - K_1(x,t)\Phi'(0) + \int_t^x [(D_x^2 - Q(x))K_1(x,s)]\Phi(s-t)ds$$

$$- \int_t^x K_1(x,s)\Phi''(s-t)ds = -Q(x)\Phi(x-t) + \frac{dK_1(x,x)}{dx}\Phi(x-t)$$

$$+ [(D_x + D_t)K_1(x,t)]|_{t=x} \cdot \Phi(x-t) + \int_t^x [(LK_1)(x,s)]\Phi(s-t)ds = 0.$$

Thus every $K(\cdot,\cdot)$ of the form (2.12) satisfies equation (2.4) and condition (2.5). It remains to choose $\Phi(\in C^2[0,1] \otimes \mathbb{C}^{n\times n})$ so that $K(\cdot,\cdot)$ will satisfy condition (2.6). Starting with (2.12) and setting $g(x) := D_t K_1(x,t)|_{t=0}$, we can determine $\Phi$ from the following equation

$$\Phi'(x) + \int_0^x K_1(x,s)\Phi'(s)ds = g(x). \tag{2.13}$$

This is a Volterra integral equation of the second kind with respect to the unknown matrix function $\Phi_1 := \Phi'$. Therefore it has a unique solution $\Phi_1 \in C^1[0,1] \otimes \mathbb{C}^{n\times n}$, since $K_1(\cdot,\cdot) \in C^1(\Omega) \otimes \mathbb{C}^{n\times n}$ and $g \in C^1[0,1] \otimes \mathbb{C}^{n\times n}$. The required matrix function $\Phi \in C^2[0,1] \otimes \mathbb{C}^{n\times n}$ is then given by $\Phi(x) := \int_0^x \Phi_1(s)ds$. Thus the kernel $K(\cdot,\cdot)$ determined by (2.12), (2.13), satisfies the boundary value problem (2.4)–(2.6), hence the representation (2.3) holds with $K(\cdot,\cdot) \in C^2(\Omega) \otimes \mathbb{C}^{n\times n}$.

(iii) Let now $Q \in L^1[0,1] \otimes \mathbb{C}^{n\times n}$. Then the solution $P_1(\cdot,\cdot)$ of (2.11) is absolutely continuous with respect to $u$ (resp. $v$) for every fixed $v \in [0,1]$ (resp. $u \in [0,2]$). Moreover, it follows from (2.11) that $D_2 P_1(\cdot,v)$ is absolutely continuous for every fixed $v \in [0,1]$. Therefore $P_1(\cdot,\cdot)$ satisfies equation (2.8) for a.e. $(u,v) \in \triangle$ and the conditions (2.9)–(2.10) in the classical sense. Now equation (2.13) has a unique solution $\Phi_1 \in L^1[0,1] \otimes \mathbb{C}^{n\times n}$, since $K_1(\cdot,\cdot) \in C(\Omega) \otimes \mathbb{C}^{n\times n}$ and $g \in L^1[0,1] \otimes \mathbb{C}^{n\times n}$. Hence $\Phi(x) := \int_0^x \Phi_1(s)ds$ belongs to $W_1^1[0,1] \otimes \mathbb{C}^{n\times n}$. The corresponding kernel $K(\cdot,\cdot)$ determined by (2.12) has the required properties. Indeed, choosing a sequence $Q_m \in C^1[0,1] \otimes \mathbb{C}^{n\times n}$ approaching $Q$ in $L^1[0,1] \otimes \mathbb{C}^{n\times n}$, we easily find that the corresponding sequence of kernels $K_m(\cdot,\cdot)$ approaches the kernel $K(\cdot,\cdot)$ in the Sobolev space $W_1^1(\Omega) \otimes \mathbb{C}^{n\times n}$. □

**Remark 2.2.** It follows from (2.11) and (2.12) that $K(\cdot,t)$ (resp. $K(x,\cdot)$ ) is absolutely continuous for every fixed $t \in [0,1]$ (respectively $x \in [0,1]$) if $Q \in L^1[0,1] \otimes \mathbb{C}^{n\times n}$. Nevertheless the kernel $K(\cdot,\cdot)$ (resp. $K_1(\cdot,\cdot)$) is a solution of the problem (2.4)–(2.6) (resp. (2.4), (2.5), (2.7)) only in the sense of distributions, since the classical derivatives $D_x^2 K(x,t)$ and $D_t^2 K(x,t)$ do not exist in general if $Q \notin C^1[0,1] \otimes \mathbb{C}^{n\times n}$.

However, we emphasize that $P(u,v) := K\big((u+v/2),(u-v)/2\big)$, as well as $P_1(u,v)$, satisfy equation (2.8) for a.e. $(u,v) \in \triangle$ if $Q \in L^1[0,1] \otimes \mathbb{C}^{n\times n}$. We shall systematically use this fact in the sequel.

The following proposition can be proved in just the same way as Proposition 2.1.

**Proposition 2.3.** *Let $S(\cdot,\lambda)$ be the $n\times n$ matrix solution of (2.1) obeying $S(0,\lambda) = \mathbb{O}_n$, $S'(0,\lambda) = I_n$. Then there exists an $n\times n$-matrix kernel $K_2(\cdot,\cdot)$, such that*

$$\lambda S(x,\lambda) = \sin(\lambda x) I_n + \int_0^x K_2(x,t)\sin(\lambda t)dt =: (I + K_2)\sin(\lambda x). \tag{2.14}$$

*Moreover, $K_2(\cdot,\cdot)$ belongs to $W_1^1(\Omega) \otimes \mathbb{C}^{n\times n}$ and satisfies (in the sense of distributions) the boundary value problem (2.4)–(2.5) with $H = \mathbb{O}_n$ and the further condition*

$$K(x,0) = 0, \quad x \in [0,1], \tag{2.15}$$

*in place of (2.6). If additionally $Q(\cdot) \in C^1[0,1] \otimes \mathbb{C}^{n\times n}$, then $K_2(\cdot,\cdot) \in C^2(\Omega) \otimes \mathbb{C}^{n\times n}$ and is a classical solution of the problem (2.4), (2.5), (2.15).*

Alongside equation (2.1) consider the equation

$$-y'' + \widetilde{Q}(x)y = \lambda^2 y, \quad y = \mathrm{col}(y_1, y_2, \dots, y_n), \quad x \in [0,1], \tag{2.16}$$

with summable $n\times n$ matrix potential $\widetilde{Q}(\cdot)$. Denote by $\widetilde{Y}(\cdot,\cdot)$ the $n\times n$ matrix solution of (2.16) obeying the initial conditions

$$\widetilde{Y}(0,\lambda) = I_n, \qquad \widetilde{Y}'(0,\lambda) = \widetilde{H} \ (\in \mathbb{C}^{n\times n}). \tag{2.17}$$

**Proposition 2.4.** *Suppose that the potential matrices $Q$ and $\widetilde{Q}$ are summable on $[0,1]$. Then there exists an $n\times n$-matrix kernel $R(\cdot,\cdot) \in C(\Omega) \otimes \mathbb{C}^{n\times n}$, where $\Omega = \{(x,t) : 0 \le t \le x \le 1\}$, such that*

$$\widetilde{Y}(x,\lambda) = Y(x,\lambda) + \int_0^x R(x,t)Y(t,\lambda)dt =: (I+R)Y(x,\lambda). \tag{2.18}$$

*Moreover, $R(\cdot,\cdot)$ belongs to $W_1^1(\Omega) \otimes \mathbb{C}^{n\times n}$ and satisfies (in the sense of distributions) the following boundary value problem:*

$$D_x^2 R(x,t) - \widetilde{Q}(x)R(x,t) = D_t^2 R(x,t) - Q(t)R(x,t), \tag{2.19}$$

$$R(x,x) = \widetilde{H} - H + \frac{1}{2}\int_0^x [\widetilde{Q}(s) - Q(s)]ds, \tag{2.20}$$

$$[D_t R(x,t) - HR(x,t)]|_{t=0} = 0. \tag{2.21}$$

*If additionally $Q(\cdot), \widetilde{Q}(\cdot) \in C^1[0,1] \otimes \mathbb{C}^{n\times n}$, then $R(\cdot,\cdot) \in C^2(\Omega) \otimes \mathbb{C}^{n\times n}$ and $R(\cdot,\cdot)$ is a classical solution of the problem (2.19)–(2.21).*

The proof is similar to that of Proposition 2.1. On the other hand, starting with representations (2.3) for $\widetilde{Y}(\cdot,\cdot)$ and $Y(\cdot,\cdot)$ with kernels $\widetilde{K}(\cdot,\cdot)$ and $K(\cdot,\cdot)$, respectively, we immediately arrive at the representation (2.18) with $I+R = (I+\widetilde{K})(I+K)^{-1} = (I+\widetilde{K})(I+K_1)$, where $K_1$ is the Volterra operator with the $n\times n$-matrix kernel $K_1(\cdot,\cdot)$ satisfying

$$K_1(x,t)+K(x,t)+\int_t^x K_1(x,s)K(s,t)ds = 0. \tag{2.22}$$

Let $C(\cdot,\lambda)$ and $S(\cdot,\lambda)$ be the $n\times n$ matrix solutions of (2.1) obeying the initial conditions

$$C(0,\lambda)=S'(0,\lambda)=I_n, \qquad C'(0,\lambda)=S(0,\lambda)=\mathbb{O}_n. \tag{2.23}$$

Finally we consider representations (2.3) and (2.14) for equation (2.1) with a constant potential matrix $Q(x)=\mu^2\cdot I_n,\ \mu\in\mathbb{C}$. As usual, $J_1(\cdot)$ stands for the Bessel function

$$J_1(z)=\sum_{k=0}^{\infty}\frac{(-1)^k}{(k+1)!k!}\left(\frac{z}{2}\right)^{2k+1}. \tag{2.24}$$

**Lemma 2.5.** *The following representations hold:*

$$\cos(\sqrt{\lambda^2+\mu^2}x)\cdot I_n = \cos(\lambda x)\cdot I_n + \int_0^x P_1(x,t;\mu)\cos(\lambda t)dt, \tag{2.25}$$

$$\frac{\sin(\sqrt{\lambda^2+\mu^2}x)}{\sqrt{\lambda^2+\mu^2}}\cdot I_n = \frac{\sin(\lambda x)}{\lambda}\cdot I_n + \int_0^x P_2(x,t;\mu)\frac{\sin(\lambda t)}{\lambda}dt, \tag{2.26}$$

*where* $P_j(x,t;\mu):=p_j(x,t;\mu)\cdot I_n,\ j\in\{1,2\}$, *and*

$$p_1(x,t;\mu)=-x\mu\frac{J_1(\mu\sqrt{x^2-t^2})}{\sqrt{x^2-t^2}}, \quad p_2(x,t;\mu)=-t\mu\frac{J_1(\mu\sqrt{x^2-t^2})}{\sqrt{x^2-t^2}}. \tag{2.27}$$

*Proof.* Note that $C(x,\lambda)=\cos(\sqrt{\lambda^2+\mu^2}x)\cdot I_n$ and that $S(x,\lambda)=\frac{\sin(\sqrt{\lambda^2+\mu^2}x)}{\sqrt{\lambda^2+\mu^2}}\cdot I_n$ are the solutions of equation (2.1) with $Q(x)=-\mu^2\cdot I_n$ obeying (2.23). By Proposition 2.1, $C(x,\lambda)$ admits representation (2.3), where the kernel $P_1(x,t;\mu)$ solves problem (2.4)–(2.6) with $Q(x)=-\mu^2\cdot I_n$ and $H=\mathbb{O}_n$. It is easily seen that this solution is

$$P_1(x,t;\mu)=\frac{x}{2}\sum_{k=1}^{\infty}(-1)^k\mu^{2k}\frac{(x^2-t^2)^{k-1}}{4^{k-1}k!(k-1)!}\cdot I_n. \tag{2.28}$$

Combining (2.28) with (2.24), we arrive at the representation (2.25) with $P_1(x,t;\mu)$ as in (2.27).

Similarly, using Proposition 2.3 in place of Proposition 2.1, we obtain (2.26) with $P_2(x,t;\mu)$ as in (2.27). □

### 2.2. The monodromy matrix identities

Alongside equation (2.1) we consider the equation

$$-y'' + Q(x)^* y = \lambda^2 y, \qquad y = \operatorname{col}(y_1, y_2, \dots, y_n), \tag{2.29}$$

where $Q(x)^*$ is the adjoint matrix of $Q(x)$. Denote also by $C_*(\cdot,\lambda)$ and $S_*(\cdot,\lambda)$ the $n \times n$ matrix solutions of (2.29) obeying the initial conditions (2.23).

Next we denote by $W(\lambda)$ and $W_*(\lambda)$ the canonical monodromy matrices of equations (2.1) and (2.29) respectively, that is

$$W(\lambda) := \begin{pmatrix} C(1,\lambda) & S(1,\lambda) \\ C'(1,\lambda) & S'(1,\lambda) \end{pmatrix}, \quad W_*(\lambda) := \begin{pmatrix} C_*(1,\lambda) & S_*(1,\lambda) \\ C'_*(1,\lambda) & S'_*(1,\lambda) \end{pmatrix}. \tag{2.30}$$

**Lemma 2.6.** *The following identity holds*

$$W(\lambda) J W_*(\overline{\lambda})^* = J := \begin{pmatrix} 0 & I \\ -I & 0 \end{pmatrix}, \qquad \lambda \in \mathbb{C}. \tag{2.31}$$

*In particular,*

$$C(1,\lambda) S_*(1,\overline{\lambda})^* = S(1,\lambda) C_*(1,\overline{\lambda})^*, \quad \lambda \in \mathbb{C}. \tag{2.32}$$

*Proof.* First we prove the identity $W_*(\overline{\lambda})^* J W(\lambda) = J$. This is equivalent to the following three identities

$$C_*(1,\overline{\lambda})^* C'(1,\lambda) = C'_*(1,\overline{\lambda})^* C(1,\lambda), \quad S_*(1,\overline{\lambda})^* S'(1,\lambda) = S'_*(1,\overline{\lambda})^* S(1,\lambda), \tag{2.33}$$

and

$$C_*(1,\overline{\lambda})^* S'(1,\lambda) - C'_*(1,\overline{\lambda})^* S(1,\lambda) = I. \tag{2.34}$$

Let us prove (2.34). If follows from (2.1) and (2.29) that

$$-C_*(x,\overline{\lambda})^* S''(x,\lambda) + C_*(x,\overline{\lambda})^* Q(x) S(x,\lambda) = \lambda^2 C_*(x,\overline{\lambda})^* S(x,\lambda),$$

$$-C''_*(x,\overline{\lambda})^* S(x,\lambda) + C_*(x,\overline{\lambda})^* Q(x) S(x,\lambda) = \lambda^2 C_*(x,\overline{\lambda})^* S(x,\lambda).$$

Subtracting we obtain $C_*(x,\overline{\lambda})^* S'(x,\lambda) - C'_*(x,\overline{\lambda})^* S(x,\lambda) = const$. Substituting $x = 0$ and then $x = 1$ we get the required result. The identities (2.33) are proved similarly.

Thus, $W_*(\overline{\lambda})^* J W(\lambda) = J$. It follows that both matrices $W(\lambda)$ and $W_*(\lambda)$ are invertible. Taking the inverse we get $W^{-1}(\lambda) J \big(W_*(\overline{\lambda})^*\big)^{-1} = J$. Hence one has $W(\lambda) J W_*(\overline{\lambda})^* = J$. □

In the sequel we need also the following straightforward lemma.

**Lemma 2.7.** *Let $Y$ be the $n \times n$ matrix solution of* (2.1) *obeying $Y(0,\lambda) = A$, $Y'(0,\lambda) = B$. Suppose also that $Y_1$ is the $n \times n$ matrix solution of* (2.29) *obeying $Y_1(0,\lambda) = A_1$, $Y_1'(0,\lambda) = B_1$. If $A_1^* B = B_1^* A$ then the following identity holds:*

$$Y_1(x,\overline{\lambda})^* Y'(x,\lambda) = Y_1'(x,\overline{\lambda})^* Y(x,\lambda), \qquad \lambda \in \mathbb{C}. \tag{2.35}$$

## 3. Uniqueness of the potential matrix given a part of the monodromy matrix

### 3.1. Uniqueness of the potential matrix given a column of the monodromy matrix

First we show that the potential matrix is uniquely determined by a column of the monodromy matrix.

**Proposition 3.1.** *Let $Y(x,\lambda)$ (resp. $\widetilde{Y}(x,\lambda)$) be the $n \times n$ matrix solution of the problem* (2.1)–(2.2) (*resp.* (2.16)–(2.17)). *If*

$$Y(1,\lambda) = \widetilde{Y}(1,\lambda) \quad \textit{and} \quad Y'(1,\lambda) = \widetilde{Y}'(1,\lambda) \quad \textit{for} \quad \lambda \in \mathbb{C},$$

*then $H = \widetilde{H}$ and $Q(x) = \widetilde{Q}(x)$ for a.e. $x \in [0,1]$.*

*Proof.* (i) By Proposition 2.4, the solutions $\widetilde{Y}(x,\lambda)$ and $Y(x,\lambda)$ are connected by formula (2.18) where the kernel $R(\cdot,\cdot)$ satisfies (in the sense of distributions) the boundary value problem (2.19) – (2.21). After making the substitutions $u = x+t, v = x-t$ we see that $P(u,v) := R\big((u+v)/2, (u-v)/2\big)$ is a classical solution of the problem

$$4D_u D_v P(u,v) = [\widetilde{Q}((u+v)/2) - Q((u-v)/2)]P(u,v), \tag{3.1}$$

$$[P(u,v)H - (D_u - D_v)P(u,v)]|_{u=v} = 0, \tag{3.2}$$

$$4P(u,0) = 4(H - \widetilde{H}) + \int_0^u [\widetilde{Q}(s/2) - Q(s/2)]ds. \tag{3.3}$$

Since $\widetilde{Y}(1,\lambda) = Y(1,\lambda)$ and $\widetilde{Y}'(1,\lambda) = Y(1,\lambda)$ we easily get from (2.18) and (2.3) that $R(1,t) = D_x R(1,t)\big(:= [D_x R(x,t)]|_{x=1}\big) = 0$. Therefore

$$P(u,2-u) = 0, \qquad D_u P(u,v)|_{v=2-u} = D_v P(u,v)|_{v=2-u} = 0. \tag{3.4}$$

(ii) We will now show that the problem (3.1), (3.2), (3.4) has only the trivial solution $P(u,v) = 0$ in the triangle $\Delta = \{(u,v) : \ 0 \le v \le u \le 2-v\}$.

For the triangle $\Delta_1 = \{(u,v) : 1 \le u \le 2-v, v \ge 0\}$ this follows since this region is characteristic, and problem (3.1), (3.4) is a Cauchy problem in $\Delta_1$ with zero data on the non-characteristic base $u+v=2$. Furthermore, this is easily derived by integrating equation (3.1) along characteristics and taking (3.4) into account. Letting $Q(u,v) := \widetilde{Q}((u+v)/2) - Q((u-v)/2)$, we arrive at the following Volterra type equation in $\Delta_1$:

$$P(u,v) = \frac{1}{4}\int_u^{2-v}\int_v^{2-\alpha} Q(\alpha,\beta)P(\alpha,\beta)d\beta d\alpha,$$

which has only the trivial solution $P(u,v) = 0, \ (u,v) \in \Delta_1$. Hence

$$P(1,v) = 0, \qquad D_v P(1,v) = 0. \tag{3.5}$$

(iii) Turning now to the triangle $\Delta_2 = \{(u,v) : \ 0 \le v \le u \le 1\}$, we use the condition (3.2) in the form

$$\frac{dP(u,u)}{du} - HP(u,u) = 2D_2 P(u,u)\big(:= 2D_v P(u,v)|_{v=u}\big).$$

A solution of this equation obeying the condition $P(1,1)=0$ has the form

$$P(u,u)=2\int_1^u e^{H(u-\alpha)}D_2P(\alpha,\alpha)d\alpha. \tag{3.6}$$

Integrating (3.1) and taking (3.5) into account, we then obtain

$$-D_vP(u,v)=\frac{1}{4}\int_u^1 Q(\alpha,v)P(\alpha,v)d\alpha. \tag{3.7}$$

Integrating (3.7) with respect to the second variable from $v$ to $u$ and taking (3.6) into account, we obtain

$$P(u,v)=\frac{1}{4}\int_u^1\int_v^u Q(\alpha,\beta)P(\alpha,\beta)d\beta d\alpha \\ +\frac{1}{2}\int_u^1\int_u^\alpha e^{H(u-\beta)}Q(\alpha,\beta)P(\alpha,\beta)d\beta d\alpha.$$

Hence, observing that $v\le u\le\alpha\le 1$, we derive the inequality

$$\|P(u,v)\|\le C\int_u^1\int_v^\alpha \|Q(\alpha,\beta)\|\cdot\|P(\alpha,\beta)\|d\beta d\alpha, \tag{3.8}$$

where $\|P(\cdot,\cdot)\|$ stands for the matrix norm of $P(\cdot,\cdot)$.

By Gronwall's inequality we conclude from (3.8) that $P(u,v)=0$ for $(u,v)\in\Delta_2$. Thus $P(u,v)=0$ for $(u,v)\in\Delta$.

It follows now from (3.3) that $\widetilde{Q}(x)=Q(x)$ for a.e. $x\in[0,1]$. □

Consider the scalar case, $n=1$. In this case we write $q$, $h(\in\mathbb{C})$ and $y$ in place of $Q$, $H$ and $Y$, respectively. Similarly, we write $\widetilde{q}$, $\widetilde{h}$ and $\widetilde{y}$ in place of $\widetilde{Q}$, $\widetilde{H}$ and $\widetilde{Y}$, respectively.

**Corollary 3.2.** *Let $n=1$ and $y'(1,\lambda)y(1,\lambda)^{-1}=\widetilde{y}'(1,\lambda)\widetilde{y}(1,\lambda)^{-1}$ for $\lambda\in G$ ($G$ an open set in $\mathbb{C}$). Then $\widetilde{h}=h$ and $\widetilde{q}(x)=q(x)$ for a.e. $x\in[0,1]$.*

*Proof.* It follows from the uniqueness theorem for the equation (2.1) that the (scalar) entire functions $y(1,\lambda)$ and $y'(1,\lambda)$ have no common zeros. Therefore the sets of zeros of the entire functions $y(1,\lambda)$ and $\widetilde{y}(1,\lambda)$ (resp. $y'(1,\lambda)$ and $\widetilde{y}'(1,\lambda)$) coincide, taking account of multiplicities, with the set of zeros (resp. poles) of the meromorphic function $M(\lambda):=y'(1,\lambda)y(1,\lambda)^{-1}\ \big(=\widetilde{y}'(1,\lambda)\widetilde{y}(1,\lambda)^{-1}\big)$. It follows from (2.3) that both $y(1,\lambda)$ and $\widetilde{y}(1,\lambda)$ are (scalar) even entire functions of exponential type one. Therefore $y(1,\lambda)$ (resp. $\widetilde{y}(1,\lambda)$) is uniquely (up to a constant factor) determined by its zeros, hence $\widetilde{y}(1,\lambda)=ay(1,\lambda)$. It follows from (2.3) that both $y(1,\lambda)$ and $\widetilde{y}(1,\lambda)$) approach $\cos(\lambda)$ when $\lambda\to\infty$ in every sector $S_\epsilon:=\{\lambda:\arg\lambda\in(\epsilon,\pi-\epsilon),\ \epsilon>0\}$. Hence $a=1$, $\widetilde{y}(1,\lambda)=y(1,\lambda)$ and $\widetilde{y}'(1,\lambda)=y'(1,\lambda)$. By Proposition 3.1 we get $\widetilde{h}=h$ and $\widetilde{q}(x)=q(x)$ for a.e. $x\in[0,1]$. □

Denote by $S(q; h_0, h_1)$ the spectrum (that is, the set of the eigenvalues $\lambda^2$, counting their multiplicities) of the problem

$$-y'' + q(x)y = \lambda^2 y, \quad y'(0) - h_0 y(0) = 0,\ y'(1) - h_1 y(1) = 0, \tag{3.9}$$

where $h_0, h_1 \in \mathbb{C} \cup \{\infty\}$, with $h_0 = \infty$ (resp. $h_1 = \infty$) shorthand for the Dirichlet condition $y(0) = 0$ (resp. $y(1) = 0$).

**Corollary 3.3.** *Suppose that $h_0, h_j, \widetilde{h}_j \in \mathbb{C} \cup \{\infty\}$, $j \in \{1,2\}$, $h_1 \neq h_2$ and*

$$S(q; h_0, h_j) = S(\widetilde{q}; \widetilde{h}_0, \widetilde{h}_j), \qquad j \in \{1,2\}. \tag{3.10}$$

*Then $h_0 = \widetilde{h}_0$, $h_j = \widetilde{h}_j$ for $j \in \{1,2\}$ and $q(x) = \widetilde{q}(x)$ for a.e. $x \in [0,1]$.*

*Proof.* We confine ourselves to the case $|h_1| + |h_2| < \infty$. The other cases can be treated in just the same way. It is easily seen that the spectrum $S(q; h_0, h_j)$ (resp. $S(\widetilde{q}; \widetilde{h}_0, \widetilde{h}_j)$) coincides with the set of squares of zeros, counting multiplicities, of the entire function $\varphi_j(\lambda) := y'(1,\lambda) + h_j y(1,\lambda)$ (resp. $\widetilde{\varphi}_j(\lambda) := \widetilde{y}'(1,\lambda) + \widetilde{h}_j \widetilde{y}(1,\lambda)$). Since $\varphi_j$ (resp. $\widetilde{\varphi}_j$), being an even entire function of exponential type one, is uniquely determined (up to a constant factor) by its zeros (cf. the proof of Corollary 3.2), we get from (3.10) that $\widetilde{\varphi}_j(\lambda) = c_j \varphi_j(\lambda)$, $j \in \{1,2\}$. On the other hand by (2.3) both $\varphi_j$ and $\widetilde{\varphi}_j$ approach $\lambda \sin(\lambda)$ as $\lambda$ tends to infinity in every sector $S_\epsilon := \{\lambda : \arg\lambda \in (\epsilon, \pi - \epsilon)\}$. Hence $c_j = 1$, $j \in \{1,2\}$, and $y(1,\lambda) = a\widetilde{y}(1,\lambda)$ where $a = (h_1 - h_2)^{-1}(\widetilde{h}_1 - \widetilde{h}_2)$. It follows from (2.3) that $a = 1$.

Therefore the equation $\widetilde{\varphi}_1(\lambda) = \varphi_1(\lambda)$ becomes $\widetilde{y}'(1,\lambda) - y'(1,\lambda) = (h_1 - \widetilde{h}_1)y(1,\lambda)$. Note however, that (2.18) yields $\widetilde{y}'(1,\lambda) - y'(1,\lambda) \to 0$ as $\lambda = \overline{\lambda} \to \infty$. Combining these relations we get $\widetilde{h}_1 = h_1$ and $\widetilde{y}'(1,\lambda) = y'(1,\lambda)$. It remains to apply Proposition 3.1. □

**Remark 3.4.** The proof of Proposition 3.1 is borrowed from [35], where it was stated (and proved) only in the scalar case. Corollary 3.3 generalizes classical results of G. Borg [4] in the self-adjoint case, and the non-self-adjoint case is already implicit in the work of V.A. Marchenko [40]. Corollary 3.2 has also been proved in [5] and [40] even for the equation on the half-line. Note that Corollary 3.2 is implied by Corollary 3.3 with $h_1 = \infty$ and $h_2 = 0$. Different proofs of Borg's result can be found in [27, 31, 32, 34, 41, 45].

A complete solution to the classical two-spectra inverse problem for the scalar Sturm-Liouville equation with real potential has been obtained by Levitan and Gasymov [33]. Namely, they have obtained necessary and sufficient conditions for two sequences $\{\lambda_k\}_1^\infty$ and $\{\mu_k\}_1^\infty$ of real numbers to be the spectra of two boundary value problems (3.9) with fixed $h = \overline{h}$ and different $h_1 = \overline{h}_1 \neq h_2 = \overline{h}_2$.

### 3.2. Uniqueness of the Sturm-Liouville equation given the first row of its monodromy matrix

**Proposition 3.5.** *For $j \in \{1,2\}$ let $Y_j(x,\lambda)$ (resp. $\widetilde{Y}_j(x,\lambda)$) be the solution of the equation (2.1) (resp. (2.16)) obeying the initial conditions*

$$Y_j(0,\lambda) = I_n, \quad Y_j'(0,\lambda) = H_j \qquad (\textit{resp. } \widetilde{Y}_j(0,\lambda) = I_n, \quad \widetilde{Y}_j'(0,\lambda) = \widetilde{H}_j). \tag{3.11}$$

*Suppose that* $0 \in \rho(H_1 - H_2)$, *where* $\rho(\cdot)$ *denotes the resolvent set, and that*

$$Y_j(1,\lambda) = \widetilde{Y}_j(1,\lambda), \qquad \lambda \in \mathbb{C}, \qquad j \in \{1,2\}. \tag{3.12}$$

*Then* $H_j = \widetilde{H}_j$, $j \in \{1,2\}$ *and* $Q(x) = \widetilde{Q}(x)$ *for a.e.* $x \in [0,1]$.

*Proof.* (i) By Proposition 2.4 the solutions $\widetilde{Y}_j(x,\lambda)$ and $Y_j(x,\lambda)$ are connected by means of the triangular transformation operator (see (2.18))

$$\widetilde{Y}_j(x,\lambda) = Y_j(x,\lambda) + \int_0^x R_j(x,t) Y_j(t,\lambda)dt, \quad j \in \{1,2\}. \tag{3.13}$$

Moreover, $R_j \in C(\Omega) \otimes \mathbb{C}^{n\times n}$ and $R_j(x,\cdot)$ (resp. $R_j(\cdot,t)$) is absolutely continuous for every $x \in [0,1]$ (resp. $t \in [0,1]$) and satisfies in $\Omega = \{(x,t) : 0 \le t \le x \le 1\}$ (in the sense of distributions) the boundary value problem (2.19)–(2.21) with $H_j$ and $\widetilde{H}_j$ in place of $H$ and $\widetilde{H}$, respectively. Furthermore, if we define $P_j(u,v) := R_j\big((u+v)/2, (u-v)/2\big)$, $j \in \{1,2\}$, then $P_j$ is a classical solution of (3.1).

It follows from (3.12) and (3.13) that $\int_0^1 R_j(1,t)Y_j(t,\lambda)dt = 0,\ j \in \{1,2\}$. On the other hand, by Proposition 2.1, $Y_j$ admits a representation (2.3) with $K_j(\cdot,\cdot)$ in place of $K(\cdot,\cdot)$. Therefore

$$\begin{aligned} 0 = \int_0^1 R_j(1,t)Y_j(t,\lambda)dt &= \int_0^1 R_j(1,t)\big[\cos(\lambda t) + \int_0^t K_j(t,s)\cos(\lambda s)ds\big]dt \\ &= \int_0^1 \big[R_j(1,s) + \int_s^1 R_j(1,t)K_j(t,s)dt\big]\cos(\lambda s)ds, \qquad j \in \{1,2\}. \end{aligned}$$

Since the cosine-Fourier transform is injective, we arrive at the following matrix Volterra type integral equation of the second kind

$$R_j(1,s) + \int_s^1 R_j(1,t)K_j(t,s)dt = 0, \qquad j \in \{1,2\}.$$

Hence

$$R_1(1,t) = R_2(1,t) = 0, \qquad t \in [0,1]. \tag{3.14}$$

Setting $R := R_2 - R_1$ and taking (3.14) into account we derive from (2.19), (2.21), after making the substitutions $u = x + t$, $v = x - t$, that the function $P(u,v) := R\big((u+v)/2, (u-v)/2\big)$ is a (classical) solution of the problem

$$4D_uD_vP(u,v) = [\widetilde{Q}((u+v)/2) - Q((u-v)/2)]P(u,v), \tag{3.15}$$

$$P(u,0) = 0, \ \ u \in [0,2], \qquad P(u,2-u) = 0, \ \ u \in [1,2]. \tag{3.16}$$

in the triangle $\triangle := \{(u,v) : \ 0 \le v \le u \le 2 - v\}$.

We now prove that $P(u,v) = 0$ for $(u,v) \in \triangle$.

(ii) First we show that

$$P(u,v) = 0, \qquad (u,v) \in \triangle_1 = \{(u,v) : \ 1 \le u \le 2 - v, v \ge 0\}. \tag{3.17}$$

Setting $Q(u,v) := \widetilde{Q}((u+v)/2) - Q((u-v)/2)$ and integrating equation (3.15) along characteristics while taking (3.16) into account, we arrive at the equation

$$-P(u,v) = \frac{1}{4}\int_u^{2-v}\int_0^v Q(\alpha,\beta)P(\alpha,\beta)d\beta d\alpha. \tag{3.18}$$

Note that equation (3.18) is a Volterra type integral equation of the second kind since $Q \in L^1(\triangle)\otimes\mathbb{C}^{n\times n}$. Therefore $P(u,v)=0$ for $(u,v)\in\Delta_1$.

(iii) In this step we establish that

$$P(u,v)=0 \quad \text{for} \quad (u,v)\in\triangle_2 = \{(u,v): 0\le v\le u\le 1\}. \tag{3.19}$$

Since $P(u,v)=0$ in $\triangle_1$ we have in particular that $P(1,v)=0$. Therefore $P(\cdot,\cdot)$ is a (generalized) solution in $\triangle_2$ of the equation (3.15) satisfying the boundary conditions

$$P(u,0)=0, \quad u\in[0,1], \qquad P(1,v)=0, \quad v\in[0,1]. \tag{3.20}$$

Integrating equation (3.15) along characteristics and taking (3.20) into account we arrive at the equation

$$-P(u,v) = \frac{1}{4}\int_u^1\int_0^v Q(\alpha,\beta)P(\alpha,\beta)d\beta d\alpha. \tag{3.21}$$

This being a Volterra type equation of the second kind, there is only the trivial solution, and hence (3.19) holds. Thus $R_1(x,t)=R_2(x,t)$ for $(x,t)\in\Omega$.

(iv) Finally we prove that $R_1(x,t)=0$ for $(x,t)\in\Omega$. Recall that $R_j(\cdot,\cdot)$ is a generalized solution of equation (2.19) obeying conditions (2.20)–(2.21), with $H_j$ and $\widetilde{H}_j$ $(j\in\{1,2\})$ in place of $H$ and $\widetilde{H}$, respectively. Since $R_1(x,t) = R_2(x,t)$, $(x,t)\in\Omega$, we have

$$[D_tR_1(x,t) - H_jR_1(x,t)]|_{t=0} = 0, \quad j\in\{1,2\}.$$

Since $0\in\rho(H_1-H_2)$, this is equivalent to

$$R_1(x,0)=0, \qquad [D_tR_1(x,t)]|_{t=0}=0. \tag{3.22}$$

After making the substitutions $u=x+t$, $v=x-t$ and taking (3.14) into account we see that $P_1(u,v) := R_1\big((u+v)/2,(u-v)/2\big)$ is a (classical) solution of the problem

$$4D_uD_vP_1(u,v) = Q(u,v)P_1(u,v), \tag{3.23}$$

$$P_1(u,u)=0, \quad D_uP_1(u,v)|_{v=u} = D_vP_1(u,v)|_{v=u}=0, \tag{3.24}$$

$$P_1(u,2-u)=0, \qquad u\in[1,2]. \tag{3.25}$$

This problem has only the trivial solution in $\triangle$. Indeed, for the triangle $\triangle_2 = \{(u,v): 0\le v\le u\le 1\}$ this follows since the non-characteristic Cauchy problem (3.23), (3.24) is reduced to the following Volterra-type equation in $\triangle_2$:

$$P_1(u,v) = -\frac{1}{4}\int_v^u\int_v^\alpha Q(\alpha,\beta)P_1(\alpha,\beta)d\beta d\alpha, \tag{3.26}$$

which has only the trivial solution $P_1(u,v)=0$ for $(u,v)\in\triangle_2$.

Turning to the triangle $\triangle_1 = \{(u,v) : 1 \le u \le 2-v, v \ge 0\}$ we consider the problem (3.23), (3.25) and use the further condition $P_1(1,v) = 0$. Every solution of this problem is also a solution of the following Volterra-type equation in $\triangle_1$ :

$$P_1(u,v) = -\frac{1}{4}\int_v^{2-u}\int_1^u Q(\alpha,\beta)P_1(\alpha,\beta)d\alpha d\beta \tag{3.27}$$

and conversely. This equation has only the trivial solution $P_1(u,v) = 0$ for $(u,v) \in \triangle_1$. Thus $P_2(u,v) = P_1(u,v) = 0$ for $(u,v) \in \triangle$.

It now follows from (2.20) that $H_j = \widetilde{H}_j$, $j \in \{1,2\}$, and $Q(x) = \widetilde{Q}(x)$ for a.e. $x \in [0,1]$. □

Next we present an analogue of Proposition 3.5 for the canonical monodromy matrix. In order to state the corresponding result we denote by $C(x,\lambda)$ and $S(x,\lambda)$ the $n \times n$ matrix solutions of (2.1) obeying

$$C(0,\lambda) = S'(0,\lambda) = I_n, \qquad S(0,\lambda) = C'(0,\lambda) = \mathbb{O}_n. \tag{3.28}$$

Let also $\widetilde{C}(x,\lambda)$ and $\widetilde{S}(x,\lambda)$ be the solutions of (2.16) obeying the same initial conditions (3.28).

**Proposition 3.6.** *Let* $\widetilde{C}(1,\lambda) = C(1,\lambda)$ *and* $\widetilde{S}(1,\lambda) = S(1,\lambda)$ *for* $\lambda \in \mathbb{C}$. *Then* $\widetilde{Q}(x) = Q(x)$ *for a.e.* $x \in [0,1]$.

*Proof.* By Proposition 2.4 the solutions $\widetilde{C}(x,\lambda)$ and $C(x,\lambda)$ are related by (2.18), where the kernel $R_1(\cdot,\cdot)$ is a generalized solution (in the sense of distributions) of the problem (2.19)–(2.21) with $\widetilde{H} = H = \mathbb{O}_n$. Moreover, the solutions $\widetilde{S}(x,\lambda)$ and $S(x,\lambda)$ are related by the equality

$$\widetilde{S}(x,\lambda) = S(x,\lambda) + \int_0^x R_2(x,t)S(t,\lambda)dt, \tag{3.29}$$

in which the kernel $R_2(\cdot,\cdot)$ satisfies equation (2.19) and condition (2.20) as well as

$$R_2(x,0) = 0, \quad x \in [0,1], \tag{3.30}$$

in place of (2.21). The rest of the proof is similar to that of Proposition 3.5. □

**Corollary 3.7.** *Let* $n = 1$ *and* $\widetilde{S}(1,\lambda)^{-1}\widetilde{C}(1,\lambda) = S(1,\lambda)^{-1}C(1,\lambda)$ *for* $\lambda \in G$ *(where* $G$ *is an open set in* $\mathbb{C}$*). Then* $\widetilde{Q}(x) = Q(x)$ *for a.e.* $x \in [0,1]$.

*Proof.* The proof is similar to that of Corollary 3.2. It is only necessary to note, following from the uniqueness theorem for equation (2.1), that the (scalar) entire functions $C(1,\lambda)$ and $S(1,\lambda)$ have no common zeros. □

**Remark 3.8.** Proposition 3.6 has recently been proved by R. Carlson [9] in the self-adjoint case ($Q = Q^*$). The main idea of his proof has been borrowed from Levinson [31] who found an alternative method for establishing Borg's result [4]. Our proof of Propositions 3.5 and 3.6 essentially differs from that in [9] and is close to that of Proposition 3.1.

## 4. Borg type results

### 4.1. Uniqueness of the potential matrix given $S(1,\lambda)^{-1}C(1,\lambda)$

Here we generalize Corollary 3.3 to the case of matrix equations.

**Theorem 4.1.** *Let $C(x,\lambda)$, $S(x,\lambda)$, $\widetilde{C}(x,\lambda)$ and $\widetilde{S}(x,\lambda)$ be defined as in Subsection 3.2 and $G$ an open set in $\mathbb{C}$. If*

$$S(1,\lambda)^{-1}C(1,\lambda)=\widetilde{S}(1,\lambda)^{-1}\widetilde{C}(1,\lambda) \qquad \lambda\in G, \tag{4.1}$$

*then $Q(x)=\widetilde{Q}(x)$ for a.e. $x\in[0,1]$.*

*Proof.* (i) As in Section 2 we denote by $C_*(\cdot,\lambda)$ and $S_*(\cdot,\lambda)$ the $n\times n$ matrix solutions of (2.29) obeying the initial conditions

$$C_*(0,\lambda)=S_*'(0,\lambda)=I_n, \qquad S_*(0,\lambda)=C_*'(0,\lambda)=\mathbb{O}_n. \tag{4.2}$$

By Lemma 2.6, $C(1,\lambda)S_*(1,\overline{\lambda})^*=S(1,\lambda)C_*(1,\overline{\lambda})^*$ (see identity (2.32)), so that we can rewrite (4.1) in the form

$$C_*(1,\overline{\lambda})^*\big(S_*(1,\overline{\lambda})^*\big)^{-1}=\widetilde{S}(1,\lambda)^{-1}\widetilde{C}(1,\lambda), \qquad \lambda\in G. \tag{4.3}$$

Since $C_*(1,\overline{\lambda})^*$, $S_*(1,\overline{\lambda})^*$, $\widetilde{S}(1,\lambda)^{-1}$ and $\widetilde{C}(1,\lambda)$ are entire matrix functions, equality (4.3) is equivalent to

$$\widetilde{S}(1,\lambda)C_*(1,\overline{\lambda})^*=\widetilde{C}(1,\lambda)S_*(1,\overline{\lambda})^*, \qquad \lambda\in\mathbb{C}\,. \tag{4.4}$$

By Proposition 2.4 the solutions $\widetilde{C}(\cdot,\lambda)$ and $C(\cdot,\lambda)$ are connected by means of (2.18) with the Volterra operator $R_1$, that is $\widetilde{C}(x,\lambda)=(I+R_1)C(x,\lambda)$. Substituting this representation for $\widetilde{C}$ and the representation (3.29) for $\widetilde{S}$ in (4.4), and setting $R_j(t):=R_j(1,t)$ ($j\in\{1,2\}$), we easily find

$$\int_0^1 R_2(t)S(t,\lambda)dt\cdot C_*(1,\overline{\lambda})^*=\int_0^1 R_1(t)C(t,\lambda)dt\cdot S_*(1,\overline{\lambda})^*. \tag{4.5}$$

By Proposition 2.1, $C_*(t,\lambda)$ and $S_*(t,\lambda)$ admit the representations

$$\begin{aligned} C_*(t,\lambda)&=\cos(\lambda t)I_n+\int_0^t K_{*1}(t,s)\cos(\lambda s)ds,\\ \lambda S_*(t,\lambda)&=\sin(\lambda t)I_n+\int_0^t K_{*2}(t,s)\sin(\lambda s)ds. \end{aligned} \tag{4.6}$$

Moreover, both $C(t,\lambda)$ and $S(t,\lambda)$ admit similar representations with kernels $K_1(t,s)$ and $K_2(t,s)$ in place of $K_{*1}(t,s)$ and $K_{*2}(t,s)$ respectively. Setting

$$\Phi_j(t):=R_j(t)+\int_t^1 R_j(s)K_j(s,t)ds, \quad j\in\{1,2\} \tag{4.7}$$

we easily find

$$\int_0^1 R_1(t)C(t,\lambda)dt = \int_0^1 \Phi_1(t)\cos(\lambda t)dt =: a(\lambda), \tag{4.8}$$

$$\lambda\int_0^1 R_2(t)S(t,\lambda)dt = \int_0^1 \Phi_2(t)\sin(\lambda t)dt =: b(\lambda). \tag{4.9}$$

Finally, letting

$$k_j(t) := K_{*j}(1,t)^*, \qquad j\in\{1,2\} \tag{4.10}$$

and taking (4.6)–(4.10) into account we rewrite (4.5) as

$$a(\lambda)\cdot[\sin(\lambda)I_n + \int_0^1 k_2(t)\sin(\lambda t)dt] = b(\lambda)\cdot[\cos(\lambda)I_n + \int_0^1 k_1(t)\cos(\lambda s)ds]. \tag{4.11}$$

Let $A(\cdot)$ and $B(\cdot)$ be any summable $n\times n$ matrix functions. Using the simple identities

$$\int_0^1 e^{i\lambda t}A(t)dt\cdot\int_0^1 e^{-i\lambda t}B(t)dt = \int_0^1 e^{i\lambda t}dt\int_t^1 A(s)B(s-t)ds$$
$$+\int_0^1 e^{-i\lambda t}dt\int_0^{1-t} A(s)B(t+s)ds,$$

$$\int_0^1 e^{i\lambda t}A(t)dt\cdot\int_0^1 e^{i\lambda t}B(t)dt = \int_0^2 e^{i\lambda t}dt\int_0^t A(s)B(t-s)ds, \tag{4.12}$$

and the corresponding identities with $\lambda$ replaced by $-\lambda$, we can rewrite (4.11) in the form

$$\int_0^2 e^{i\lambda t}\Psi_1(t)dt = \int_0^2 \Psi_2(t)e^{-i\lambda t}dt,$$

with some summable $n\times n$ matrix functions $\Psi_1(\cdot)$ and $\Psi_2(\cdot)$. This identity yields

$$\Psi_1(t)=0,\quad t\in[0,2], \qquad \Psi_2(t)=0,\quad t\in[0,2]. \tag{4.13}$$

The precise calculation of $\Psi_j(\cdot)$, $j\in\{1,2\}$, shows that the first (as well as the second) equation in (4.13) is equivalent to the following system of two integral equations:

$$\Phi_1(t-1)-\Phi_2(t-1) = \int_0^t [\Phi_2(s)k_1(t-s)-\Phi_1(s)k_2(t-s)]ds,\ t\in[1,2], \tag{4.14}$$

$$\Phi_1(1-t)+\Phi_2(1-t)+\int_0^1[\Phi_1(s)\widetilde{k}_2(t,s)-\Phi_2(s)\widetilde{k}_1(t,s)]ds$$
$$+\int_0^{1-t}[\Phi_1(s)k_2(t+s)+\Phi_2(s)k_1(t+s)]ds = 0, \quad t\in[0,1]. \tag{4.15}$$

Here $\widetilde{k}_j(t,s)$, $j\in\{1,2\}$ is determined by

$$\widetilde{k}_j(t,s) = \begin{cases} k_j(t-s), & 0\le s\le t,\\ (-1)^{j+1}k_j(s-t), & t\le s\le 1.\end{cases} \tag{4.16}$$

In the sequel we assume the matrix functions $\Phi_j$ and $k_j$ defined by (4.7) and (4.10) to be extended by zero outside the segment $[0,1]$. Taking this agreement into account and setting $t' = t-1$ in (4.14) and $t' = 1-t$ in (4.15), we rewrite the system (4.14)–(4.15) as

$$\Phi_1(t) - \Phi_2(t) + \int_t^1 [\Phi_1(s)k_2(t+1-s) - \Phi_2(s)k_1(t+1-s)]ds = 0,$$

$$\Phi_1(t) + \Phi_2(t) + \int_0^1 [\Phi_1(s)\widetilde{k}_2(1-t,s) - \Phi_2(s)\widetilde{k}_1(1-t,s)]ds$$
$$+ \int_0^t [\Phi_1(s)k_2(1-t+s) + \Phi_2(s)k_1(1-t+s)]ds = 0, \qquad (4.17)$$

where $t \in [0,1]$.

(ii) In this step we show that system (4.17) has only the trivial solution $\Phi_1 = \Phi_2 = 0$.

Consider equation (2.1) (resp. (2.16)) with a potential matrix $Q_\mu := Q - \mu^2 \cdot I_n$ (resp. $\widetilde{Q}_\mu := \widetilde{Q} - \mu^2 \cdot I_n$) in place of $Q$ (resp. $\widetilde{Q}$), where $\mu \in \mathbb{C}$. Denote by $C_\mu(x,\lambda)$ and $S_\mu(x,\lambda)$ the $n \times n$-matrix solution of equation (2.1) (resp. (2.16)) subject to (3.28), with $Q_\mu$ in place of $Q$. Similarly $\widetilde{C}_\mu(x,\lambda)$ and $\widetilde{S}_\mu(x,\lambda)$ (resp. $C_{*\mu}(x,\lambda)$ and $S_{*\mu}(x,\lambda)$) stand for such $n \times n$-matrix solutions of equation (2.16) (resp. (2.29)) with the potential $\widetilde{Q}_\mu$ (resp. $Q_\mu^* := Q^* - \overline{\mu}^2 \cdot I_n$) in place of $Q$ (resp. $Q^*$).

It is clear that $\widetilde{C}_\mu = (I + R_1)C_\mu$ and $\widetilde{S}_\mu = (I + R_2)S_\mu$. On the other hand, combining representations (2.25) and (2.26) with (4.6) we arrive at the following representations for the $n \times n$-matrix solutions $C_{*\mu}$ and $S_{*\mu}$ of (2.29):

$$C_{*\mu}(x,\lambda) = (I + K_{*1})\cos\left(\sqrt{\lambda^2 + \overline{\mu}^2}x\right) = (I + K_{*1})(I + P_{1\overline{\mu}})\cos(\lambda x),$$
$$S_{*\mu}(x,\lambda) = (I + K_{*2})\frac{\sin(\sqrt{\lambda^2 + \overline{\mu}^2}x)}{\sqrt{\lambda^2 + \overline{\mu}^2}} = (I + K_{*2})(I + P_{2\overline{\mu}})\frac{\sin(\lambda x)}{\lambda}. \qquad (4.18)$$

Here $P_{1\mu}$ and $P_{2\mu}$ stand for the Volterra operators having the representations (2.25) and (2.26)) respectively, with kernels $P_j(x,t;\mu) = p_j(x,t;\mu) \cdot I_n$, $j \in \{1,2\}$ determined by (2.27). Setting

$$K_{*j}(x,t;\mu) := K_{*j}(x,t) + P_j(x,t;\overline{\mu}) + \int_t^x K_{*j}(x,s)P_j(s,t;\overline{\mu})ds,$$
$$K_j(x,t;\mu) := K_j(x,t) + P_j(x,t;\mu) + \int_t^x K_j(x,s)P_j(s,t;\mu)ds, \ j \in \{1,2\}, \qquad (4.19)$$

and denoting by $K_{*j\mu}$ the Volterra operator with kernel $K_{*j}(x,t;\mu)$, $j \in \{1,2\}$, we rewrite (4.18) as

$$C_{*\mu}(x,\lambda) = (I + K_{*1\mu})\cos(\lambda x), \quad \lambda S_{*\mu}(x,\lambda) = (I + K_{*2\mu})\sin(\lambda x). \qquad (4.20)$$

Next we set $k_j(t;\mu) := K_{*j}(1,t;\mu)^*,\ j\in\{1,2\}$ and

$$\Phi_j(t;\mu) := R_j(t) + \int_t^1 R_j(s)K_j(s,t;\mu)ds, \quad j\in\{1,2\}. \tag{4.21}$$

Since $S_\mu(1,\lambda) = S(1,\sqrt{\lambda^2+\mu^2})$ and $C_\mu(1,\lambda) = C(1,\sqrt{\lambda^2+\mu^2})$, the identity (4.1) yields

$$S_\mu(1,\lambda)^{-1}C_\mu(1,\lambda) = \widetilde{S}_\mu(1,\lambda)^{-1}\widetilde{C}_\mu(1,\lambda).$$

Starting with this identity instead of (4.1) and repeating the above arguments we arrive at the system (4.17) with $\Phi_j(t;\mu)$ and $k_j(t,s;\mu)$ in place of $\Phi_j(t)$ and $k_j(t,s),\ j\in\{1,2\}$.

According to the well-known asymptotic behavior of the Bessel function $J_1(\mu)$, we easily get from (2.27) the following asymptotic formulas

$$p_1(x,t;\mu) = -x\sqrt{\mu}\frac{\cos(\mu\sqrt{x^2-t^2}-\frac{3}{4}\pi)}{\sqrt{\pi}(x^2-t^2)^{3/4}}[1+o(1)] + O\big(\frac{\sqrt{\mu}}{(x^2-t^2)^{3/4}}\big), \tag{4.22}$$

$$p_2(x,t;\mu) = -t\sqrt{\mu}\frac{\cos(\mu\sqrt{x^2-t^2}-\frac{3}{4}\pi)}{\sqrt{\pi}(x^2-t^2)^{3/4}}[1+o(1)] + O\big(\frac{\sqrt{\mu}}{(x^2-t^2)^{3/4}}\big), \tag{4.23}$$

as $\mu\to\infty,\ |\arg\mu|\le\pi-\epsilon<\pi$.

It easily follows from (4.19) and (4.22)–(4.23) that

$$K_{*j}(x,t;i\mu) = K_{*j}(x,t;-i\mu) = P_j(x,t;i\mu)[1+o(1)], \text{ as } \mu\to+\infty. \tag{4.24}$$

Using (4.22)–(4.24) we obtain from first Volterra type equation (4.17) (with $\Phi_j(t;\mu)$ in place of $\Phi_j(t)$) that $\Phi_1(t;\mu) = \Phi_2(t;\mu)$. Now the second of equations (4.17) (with $\Phi_j(t;i\mu)$ in place of $\Phi_j(t)$) takes the form

$$\begin{aligned} -2\Phi_1(t;i\mu) &= \int_0^t \Phi_1(s;i\mu)G_1(1-t-s;i\mu)ds \\ &- \int_t^1 \Phi_1(s;i\mu)G_2(s+t-1;i\mu)ds + \int_0^t \Phi_1(s;i\mu)G_3(1-t+s;i\mu)ds, \end{aligned} \tag{4.25}$$

where $G_1(1-t-s;\mu) := k_2(1-t-s;\mu) - k_1(1-t-s;\mu)$, $G_2(s+t-1;\mu) := k_2(s+t-1;\mu)+k_1(s+t-1;\mu)$ and $G_3(1-t+s;\mu) := k_2(1-t+s;\mu)+k_1(1-t+s;\mu)$.

It follows from (4.22)–(4.24) that

$$\begin{aligned} G_1(1-t-s;i\mu) &= -\frac{\sqrt{\mu}(t+s)^{1/4}e^{i3\pi/4}}{\sqrt{\pi}(2-t-s)^{3/4}}e^{\mu\sqrt{(t+s)(2-t-s)}}\cdot I_n\big(1+o(1)\big), \\ G_3(1-t+s;i\mu) &= -\frac{\sqrt{\mu}(2-t+s)^{1/4}e^{i3\pi/4}}{\sqrt{\pi}(t-s)^{3/4}}e^{\mu\sqrt{(t-s)(2-t+s)}}\cdot I_n\big(1+o(1)\big), \end{aligned} \tag{4.26}$$

as $\mu\to+\infty$. Moreover, $G_2(s+t-1;i\mu)$ and $G_1(1-t-s;i\mu)$ have the same asymptotic behavior as $\mu\to+\infty$.

To prove that $(\varphi_{ij}(t;\mu))_{i,j=1}^n := \Phi_1(t;\mu) \equiv 0$, we suppose the contrary and choose any $t_0$ such that $\Phi_1(t_0;\mu) \not\equiv 0$. Then, it is clear from (4.21) and (4.22)–(4.24) that $\Phi_1(t_0;\mu)$ is a matrix function with entries being entire functions of exponential type and completely regular growth (see [30]). Suppose for definiteness that $\varphi_{12}(t_0;\mu)$ is one of the entries of maximal growth along the imaginary semi-axis $i\mathbb{R}_+$.

Setting $(g_{ij}(t;\mu))_{i,j=1}^n := G_1(t;\mu)$ and $(\psi_{ij}(t;\mu))_{i,j=1}^n := \Phi_1(t;\mu)G(1-2t;\mu)$, we get

$$\psi_{12}(t_0;i\mu) = \sum_{j=1}^n \varphi_{1j}(t_0;i\mu)g_{j2}(1-2t_0;i\mu). \tag{4.27}$$

It follows from the asymptotic formula (4.26) that

$$g_{2j}(1-t-s;i\mu)g_{22}^{-1}(1-t-s;i\mu) \to 0 \quad \text{as } \mu \to \infty, \ j \neq 2. \tag{4.28}$$

Combining (4.27) with (4.28) we obtain that

$$\psi_{12}(t_0;i\mu) = \phi_{12}(t_0;i\mu)g_{22}(1-2t_0;i\mu)[1+o(1)], \quad \text{as} \quad \mu \to +\infty.$$

Since both $\phi_{12}(t_0;\cdot)$ and $g_{22}(1-2t_0;\cdot)$ are entire functions of completely regular growth, the indicator diagram of the product $\phi_{12}(t_0;\cdot)g_{22}(1-2t_0;\cdot)$ is the sum of the indicator diagrams of its two factors. Hence and from (4.26) we get that $\psi_{12}(t_0;\cdot)$ is a (nonzero) entire function of exponential type with

$$\sigma\big(\psi_{12}(t_0;\cdot)\big) = \sigma\big(\phi_{12}(t_0;\cdot)\big) + 2\sqrt{t_0(1-t_0)}, \tag{4.29}$$

where $\sigma(f)$ stands for the type of the entire function $f$. Further, suppose first of all that $t_0 > 1/2$. It is easily seen that

$$\sigma\big(\psi_{12}(s;\cdot)\big) > \sigma\big(\phi_{12}(t_0;\cdot)\big) + 2\sqrt{t_0(1-t_0)}, \quad s \in [1-t_0, t_0). \tag{4.30}$$

Denoting by $I_j := I_j(t;i\mu)$ the $j$th summand in the right-hand side of (4.25), we derive from (4.30) that

$$\sigma\big(I_1(t_0;\cdot)\big) > \sigma\big(\phi_{12}(t_0;\cdot)\big) + 2\sqrt{t_0(1-t_0)}. \tag{4.31}$$

On the other hand, denoting by $a$ the smallest number in the interval $[0,1]$ such that $\operatorname{supp}\|R_1(\cdot)\| \subset [0,a]$, we find from (4.21) and (4.26) that

$$I_2(t;i\mu) = o\big(e^{\mu(\sqrt{a^2-t^2}+2\sqrt{t(1-t)})}\big), \quad \text{as} \quad \mu \to +\infty. \tag{4.32}$$

Combining (4.31) with (4.32) we obtain

$$\sigma\big(I_1(t_0;\cdot)\big) > \sigma\big(I_2(t_0;\cdot)\big). \tag{4.33}$$

Next, the inequality $(t+s)(2-t-s) > (t-s)(2-t+s)$ together with the asymptotic formulas (4.26) yield $\sigma\big(I_1(t_0;\cdot)\big) > \sigma\big(I_3(t_0;\cdot)\big)$. Combining this inequality with (4.33), yields $\sigma(I_1) > \max\{\sigma(I_2), \sigma(I_3)\}$. Hence $\sigma(I_1+I_2+I_3) = \sigma(I_1)$. It follows now from (4.25) and (4.31) that

$$\sigma\big(\Phi_1(t_0;\cdot)\big) = \sigma\big(I_1(t_0;\cdot)\big) > \sigma\big(\varphi_{12}(t_0;\cdot)\big) = \sigma\big(\Phi_1(t_0;\cdot)\big). \tag{4.34}$$

This inequality contradicts the assumption $\Phi_1(t_0;\mu) \neq 0$. Thus, $\Phi_1(t;\mu) = 0$ for $t \in [1/2, 1]$. The case $t \in (0, 1/2)$ is considered similarly.

Thus, $\Phi_1(t;\mu) = \Phi_2(t;\mu) \equiv 0$. It follows now either from (4.7) or from (4.21) that $R_1(t) = R_2(t) = 0$, $t \in [0,1]$. In turn the integral representations (2.18) and (3.29) yield $\widetilde{C}(1,\lambda) = C(1,\lambda)$ and $\widetilde{S}(1,\lambda) = S(1,\lambda)$. To complete the proof it remains only to apply Proposition 3.6. □

**Remark 4.2.** We show that, for $Q$ small enough, step (ii) of the proof can be considerably shortened. Taking the sum and difference of equations (4.17), we arrive at the following equivalent system of integral equations

$$-2\Phi_j(t) = \int_0^1 [\Phi_1(s)G_{2j}(t,s) + \Phi_2(s)G_{2j-1}(t,s)]ds, \quad j \in \{1,2\}, \tag{4.35}$$

with some kernels $G_j(t,s)$, $j \in \{1,2,3,4\}$.

Denote by $T$ the integral operator with $2n \times 2n$ matrix kernel $G(t,s) := \begin{pmatrix} G_2 & G_1 \\ G_4 & G_3 \end{pmatrix}$ acting in $C[0,1] \otimes \mathbb{C}^{2n\times 2n}$ and put $\Phi := \mathrm{col}(\Phi_1, \Phi_2)$. Then system (4.35) can be rewritten as $T\Phi = -2\Phi$.

The required statement is obvious if $\|T\| < 2$. In turn, this bound holds provided we assume $\|Q\|_{L^1\otimes\mathbb{C}^{n\times n}} < \epsilon$ with $\epsilon$ small enough. Thus, in this case system (4.35) has only the trivial solution $\Phi_1 = \Phi_2 = 0$.

### 4.2. Borg type results

Consider equation (2.1) subject to the boundary conditions

$$y'(0,\lambda) - Hy(0,\lambda) = 0, \qquad y(1,\lambda) = 0, \quad H \in \mathbb{C}^{n\times n} \tag{4.36}$$

and denote by $S(Q;H,D)$ the spectrum of the problem (2.1), (4.36).

Recall that, by definition, $S(Q;H,D)$ is the set of eigenvalues of the problem (2.1), (4.36), that is the set of values $\lambda_0^2$ such that (2.1), (4.36) has a nontrivial solution $y(x,\lambda_0)$. The root subspace $\mathfrak{N}(\lambda_0^2)$ is the set of eigenfunctions and associated functions corresponding to the eigenvalue $\lambda_0^2$.

Denote also by $S(Q;D,D)$ the spectrum of the Dirichlet problem

$$y(0,\lambda) = 0, \qquad y(1,\lambda) = 0, \tag{4.37}$$

for equation (2.1).

**Theorem 4.3.** *Let $Q, \widetilde{Q} \in L^1[0,1] \otimes \mathbb{C}^{n\times n}$ and $H_j \in \mathbb{C}^{n\times n}$, $j \in \{1,\dots,n^2\}$. Suppose that the matrices $H_j$, $j \in \{1,\dots,n^2\}$, are linearly independent and that the equations* (2.1) *and* (2.16) *have $n^2+1$ equal spectra:*

$$S(Q;D,D) = S(\widetilde{Q};D,D), \quad S(Q;H_j,D) = S(\widetilde{Q};H_j,D), \quad j \in \{1,\dots,n^2\}. \tag{4.38}$$

*Then $Q(x) = \widetilde{Q}(x)$ for almost all $x \in [0,1]$.*

*Proof.* (i) Denote by $Y_j(x,\lambda)$ (resp. $\widetilde{Y}_j(x,\lambda)$) the $n\times n$-matrix solution of equation (2.1) (resp.(2.16)) satisfying the initial conditions

$$\widetilde{Y}_j(0,\lambda)=Y_j(0,\lambda)=I_n,\quad \widetilde{Y}_j'(0,\lambda)=Y_j'(0,\lambda)=H_j,\quad j\in\{1,\dots,n^2\}. \tag{4.39}$$

As above, we denote by $S(x,\lambda)$ and $C(x,\lambda)$ the $n\times n$-matrix solutions of equation (2.1) obeying the initial conditions (3.28). It is clear that

$$Y_j(x,\lambda)=C(x,\lambda)+S(x,\lambda)H_j,\qquad j\in\{1,\dots,n^2\}. \tag{4.40}$$

Note that every determinant

$$F_j(\lambda):=\det\bigl(C(1,\lambda)+S(1,\lambda)H_j\bigr),\qquad j\in\{1,\dots,n^2\}, \tag{4.41}$$

is an even entire function because so are $C(1,\lambda)$ and $S(1,\lambda)$.

It is easily seen that the spectrum $S(Q;H_j,D)$ of the problem (2.1), (4.36) coincides with the set of squares of roots of the determinant $F_j(\cdot)$. Moreover, the dimension of the root subspace $\mathfrak{N}(\mu_0)$ corresponding to the eigenvalue $\mu_0=\lambda_0^2$ equals the multiplicity of the root $\lambda_0$ of the entire function $F_j(\lambda)$. Note that multiplicities of the roots $\pm\lambda_0$ coincide since $F_j$ is an even function.

Similarly, the spectrum $S(\widetilde{Q};H_j,D)$ of the problem (2.16), (4.39) coincides (taking multiplicities into account) with the roots of the entire function

$$\widetilde{F}_j(\lambda):=\det\bigl(\widetilde{C}(1,\lambda)+\widetilde{S}(1,\lambda)H_j\bigr),\quad j\in\{1,\dots,n^2\}. \tag{4.42}$$

Hence the equality of the spectra (4.38) is equivalent to the equality of the zeros (counted with multiplicities) of the entire functions $F_j$ and $\widetilde{F}_j$, $j\in\{1,\dots,n^2\}$.

By Proposition 2.1 the solutions $Y_j(x,\lambda)$ and $\widetilde{Y}_j(x,\lambda)$ admit the representations (2.3) with kernels $K_j(x,t)$ and $\widetilde{K}_j(x,t)$, respectively. Therefore both $F_j$ and $\widetilde{F}_j$, of the form (4.41) and (4.42) respectively, are even entire functions of exponential type. So $F_j$ (resp. $\widetilde{F}_j$) is uniquely (to within a constant factor) determined by its zeros. Since they have the same zeros (taking account of multiplicities), they can differ only by a constant factor, $\widetilde{F}_j(\lambda)=a_jF_j(\lambda)$. It also follows from (2.3), (4.41) and (4.42) that $a_j=1$ for $j\in\{1,\dots,n^2\}$. Similarly, the first equality in (4.38) yields $\det\widetilde{S}(1,\lambda)=\det S(1,\lambda)$.

(ii) For simplicity, we consider the boundary conditions (4.36) with special matrices $H_j$. Namely, we choose the basis $E_{ij}=(\delta_{ki}\delta_{mj})_{k,m=1}^n$ $(i,j\in\{1,\dots,n\})$ in $\mathbb{C}^{n\times n}$ and consider equalities (4.38) with $H_j$ replaced by $E_{ij}$. Using definitions (4.41) and (4.42) and setting

$$M_0(\lambda):=S(1,\lambda)^{-1}C(1,\lambda)\qquad\text{and}\qquad \widetilde{M}_0(\lambda):=\widetilde{S}(1,\lambda)^{-1}\widetilde{C}(1,\lambda)$$

we obtain

$$\begin{aligned}F_{ij}(\lambda)&:=\det\bigl(C(1,\lambda)+S(1,\lambda)E_{ij}\bigr)=\det S(1,\lambda)\cdot\det\bigl(M_0(\lambda)+E_{ij}\bigr)\\&=\det S(1,\lambda)[m(\lambda)+A_{ij}(\lambda)]=\det\widetilde{S}(1,\lambda)[\widetilde{m}(\lambda)+\widetilde{A}_{ij}(\lambda)],\end{aligned} \tag{4.43}$$

where $A_{ij}(\lambda)$ (resp. $\widetilde{A}_{ij}(\lambda)$) is a cofactor of the element $a_{ij}(\lambda)$ (resp. $\widetilde{a}_{ij}(\lambda)$) of the matrix $M_0(\lambda)$ (resp. $\widetilde{M}_0(\lambda)$), $m(\lambda) := \det M_0(\lambda)$ and $\widetilde{m}(\lambda) := \det \widetilde{M}_0(\lambda)$. Since $\det S(1,\lambda) = \det \widetilde{S}(1,\lambda)$ we get from (4.43) that

$$b_{ij}(\lambda) := A_{ij}(\lambda) + m(\lambda) = \widetilde{A}_{ij}(\lambda) + \widetilde{m}(\lambda), \quad i,j \in \{1,\dots,n^2\}. \tag{4.44}$$

Setting

$$A(\lambda) := \big(A_{ij}(\lambda)\big)_{i,j=1}^n, \qquad B(\lambda) := \big(b_{ij}(\lambda)\big)_{i,j=1}^n \qquad \text{and} \qquad E = \sum_{i,j} E_{ij}$$

and noting that $\det\big(A(\lambda)\big) = m(\lambda)^{n-1}$, we derive from (4.44) that

$$m(\lambda)^{n-1} = \det\big(B(\lambda) - m(\lambda)E\big) = \det\big(B(\lambda) - \widetilde{m}(\lambda)E\big) = \widetilde{m}(\lambda)^{n-1}.$$

It follows that both $m(\lambda)$ and $\widetilde{m}(\lambda)$ satisfy the equation

$$z^{n-1} = b_0(\lambda) - b_1(\lambda)z, \tag{4.45}$$

where $b_0(\lambda) := \det B(\lambda)$, $b_1(\lambda) := \sum_{1\le j\le n} \det B_j(\lambda)$ and $B_j$ is the $n\times n$-matrix obtained from $B(\lambda)$ by replacing the $j$th column by the column $\mathrm{col}(1,1,\dots,1)$.

Since both $m(\cdot)$ and $\widetilde{m}(\cdot)$ are the roots of equation (4.45) we have

$$\big(\widetilde{m}(\lambda) - m(\lambda)\big)\left[\sum_{k=0}^{n-2} \widetilde{m}(\lambda)^k m(\lambda)^{n-2-k} + b_1(\lambda)\right] = 0. \tag{4.46}$$

On the other hand, it follows from (2.3) and (2.14) that $M_0(-iy) = iI_n + O_n(-iy)$ and $\widetilde{M}_0(-iy) = iI_n + \widetilde{O}_n(-iy)$, where $O_n(\lambda) = \big(O_{ij}(\lambda)\big)_{i,j=1}^n$ and where $\widetilde{O}_n(\lambda) = \big(\widetilde{O}_{ij}(\lambda)\big)_{i,j=1}^n$ are $n\times n$-matrix functions satisfying $O_{ij}(-iy) = o(1)$ and $\widetilde{O}_{ij}(-iy) = o(1)$ as $y \to +\infty$. Hence $m(-iy) = i^n + o(1)$, $\widetilde{m}(-iy) = i^n + o(1)$ as $y \to +\infty$. These relations imply

$$\sum_{k=0}^{n-2} \widetilde{m}(-iy)^k m(-iy)^{n-2-k} = (n-1)i^{n-2} + o(1),\ b_1(-iy) = n\cdot i^{(n-1)^2} + o(1) \tag{4.47}$$

as $y \to +\infty$. In turn, by (4.47), the expression in the square brackets of (4.46) differs from zero for $\lambda = -iy$ with $y$ large enough. Therefore we have from (4.46) that $m(\lambda) = \widetilde{m}(\lambda)$. Combining this equality with (4.44) yields $A_{ij}(\lambda) = \widetilde{A}_{ij}(\lambda)$ for $i,j \in \{1,\dots,n^2\}$. Hence $M_0(\lambda) = \widetilde{M}_0(\lambda)$. To complete the proof it suffices to apply Theorem 4.1. □

**Corollary 4.4.** *Let $Q(x) = Q(x)^*$ and $\widetilde{Q}(x) = \widetilde{Q}(x)^*$ for a.e. $x \in [0,1]$. Suppose that $S(Q;D,D) = S(\widetilde{Q};D,D)$ and*

$$S(Q;E_{ij},D) = S(\widetilde{Q};E_{ij},D), \qquad 1 \le i \le j \le n. \tag{4.48}$$

*Then $Q(x) = \widetilde{Q}(x)$ for a.e. $x \in [0,1]$.*

*Proof.* Since $Q(x) = Q(x)^*$ we have $S(x,\lambda) = S_*(x,\lambda)$ and $C(x,\lambda) = C_*(x,\lambda)$. Therefore, by (2.32), $M_0(\lambda) := S(1,\lambda)^{-1}C(1,\lambda) = M_0(\overline{\lambda})^*$. Hence $M_0(\lambda) = M_0(\lambda)^*$ and $\widetilde{M_0}(\lambda) = \widetilde{M_0}(\lambda)^*$ for $\lambda \in \mathbb{R}$. Starting with (4.48) and repeating the arguments of the proof of Theorem 4.3, we obtain

$$b_{ij}(\lambda) := A_{ij}(\lambda) + m(\lambda) = \widetilde{A}_{ij}(\lambda) + \widetilde{m}(\lambda), \quad 1 \le i \le j \le n. \tag{4.49}$$

Combining (4.49) with the identities $A_{ij}(\lambda) = \overline{A_{ji}(\lambda)}$, $\widetilde{A}_{ij}(\lambda) = \overline{\widetilde{A}_{ji}(\lambda)}$, $\lambda \in \mathbb{R}$, we arrive at (4.44). Hence $M_0(\lambda) = \widetilde{M_0}(\lambda)$ for $\lambda \in \mathbb{R}$ and $Q(x) = \widetilde{Q}(x)$ for a.e. $x \in [0,1]$ by Theorem 4.1. □

## 5. Borg type results: second approach

### 5.1. Uniqueness of the potential matrix given the Weyl function

**Theorem 5.1.** *Let $Y$ (resp. $\widetilde{Y}$) be the $n \times n$ matrix solutions of* (2.1) (*resp.* (2.16)) *obeying the initial conditions* (2.2) (*resp.* (2.17)), *and $G$ an open set in $\mathbb{C}$. If*

$$Y'(1,\lambda) \cdot Y(1,\lambda)^{-1} = \widetilde{Y}'(1,\lambda) \cdot \widetilde{Y}(1,\lambda)^{-1} \qquad \forall \lambda \in G, \tag{5.1}$$

*then $H = \widetilde{H}$ and $Q(x) = \widetilde{Q}(x)$ for a.e. $x \in [0,1]$.*

*Proof.* Denote by $Y_*(\cdot,\cdot)$ the $n \times n$ matrix solution of (2.29) obeying the initial conditions $Y_*(0,\lambda) = I_n$, $Y_*'(0,\lambda) = H^*$ and note that both $Y$ and $Y_*$ satisfy the conditions of Lemma 2.7. Therefore in view of (2.35) identity (5.1) may be rewritten as

$$\left(Y_*(1,\overline{\lambda})^*\right)^{-1} \cdot Y_*'(1,\overline{\lambda})^* = \widetilde{Y}'(1,\lambda) \cdot \widetilde{Y}(1,\lambda)^{-1}, \qquad \lambda \in G. \tag{5.2}$$

Hence

$$Y_*'(1,\overline{\lambda})^* \cdot \widetilde{Y}(1,\lambda) = Y_*(1,\overline{\lambda})^* \cdot \widetilde{Y}'(1,\lambda), \qquad \lambda \in \mathbb{C}. \tag{5.3}$$

By Proposition 2.1, the solutions $Y(x,\lambda)$ and $\widetilde{Y}(x,\lambda)$ are connected by means of formula (2.18). Let us set

$$R_0(t) := R(1,t) \quad \text{and} \quad R_1(t) := D_x R(x,t)|_{x=1}. \tag{5.4}$$

Substituting (2.18) into (5.3) and taking into account identity (2.35) as well as the notation (5.4) we arrive at

$$Y_*'(1,\overline{\lambda})^* \int_0^1 R_0(t)Y(t,\lambda)dt = Y_*(1,\overline{\lambda})^* \int_0^1 R_1(t)Y(t,\lambda)dt + Y_*(1,\overline{\lambda})^* R(1,1)Y(1,\lambda). \tag{5.5}$$

By the Riemann-Lebesgue lemma both the first and the second term in (5.5) tend to zero as $\lambda \to \pm\infty$, hence $R_0(1) = R(1,1) = 0$.

Consider the representation (2.3) for $Y(x,\lambda)$ with kernel $K(x,t)$, and put

$$\Phi_j(t) := R_j(t) + \int_t^1 R_j(s)K(s,t)ds, \qquad j \in \{0,1\}. \tag{5.6}$$

Combining (2.3) with (5.6) and noting that $\Phi_0 \in AC[0,1]$ we easily find

$$\int_0^1 R_0(t)Y(t,\lambda)dt = \int_0^1 \Phi_0(t)\cos(\lambda t)dt = -\int_0^1 \Phi_0'(t)\frac{\sin(\lambda t)}{\lambda}dt =: a(\lambda),$$

$$\int_0^1 R_1(t)Y(t,\lambda)dt = \int_0^1 \Phi_1(t)\cos(\lambda t)dt =: b(\lambda). \tag{5.7}$$

Taking (5.7) into account and using $R(1,1) = 0$, we rewrite (5.5) as

$$Y_*'(1,\overline{\lambda})^* \cdot a(\lambda) = Y_*(1,\overline{\lambda})^* \cdot b(\lambda). \tag{5.8}$$

According to Proposition 2.1, $Y_*(x,\lambda)$ admits a representation similar to that for $Y$ (cf. (2.3)), so that we have

$$Y_*(x,\lambda) = \cos(\lambda x) \cdot I_n + \int_0^x K_*(x,t)\cos(\lambda t)dt, \tag{5.9}$$

where the $n \times n$-matrix kernel $K_*(x,t)$ is absolutely continuous with respect to $x$ (respectively $t$) for every fixed $t \in [0,1]$ (respectively $x \in [0,1]$).

Let us write

$$k_0(t) := K_*(1,t)^* \quad \text{and} \quad k_1(t) := D_x K_*(x,t)^*|_{x=1} \tag{5.10}$$

and assume for brevity that $K_*(1,1) = 0$. Substituting (5.9) into (5.8) and taking (5.10) into account we arrive at the identity

$$\left[-\lambda\sin(\lambda)I_n + \int_0^1 k_1(t)\cos(\lambda t)dt\right] \cdot a(\lambda) = \left[\cos(\lambda)I_n + \int_0^1 k_0(t)\cos(\lambda t)dt\right] \cdot b(\lambda). \tag{5.11}$$

Repeating the arguments of the proof of Theorem 4.1 and setting

$$\widetilde{k}_j(t,s) := \begin{cases} k_j(1-t-s), & 0 \le s+t \le 1 \\ k_j(t+s-1), & 1 \le s+t \le 2. \end{cases}$$

for $j \in \{0,1\}$, we arrive at the following system of integral equations:

$$\Phi_0'(t) + \Phi_1(t) - \int_t^1 [k_0(1+t-s)\Phi_1(s) + k_1(1+t-s)\Phi_0(s)]ds = 0,$$

$$-\Phi_0'(t) + \Phi_1(t) + \int_0^1 [\widetilde{k}_0(t,s)\Phi_1(s) - \widetilde{k}_1(t,s)\Phi_0(s)]ds +$$

$$\int_0^t [k_0(1-t+s)\Phi_1(s) - k_1(1-t+s)\Phi_0(s)]ds = 0, \quad t \in [0,1]. \tag{5.12}$$

Starting with the system (5.12) and following the proof of Theorem 4.1 we obtain $\Phi_0'(t) = \Phi_1(t) = 0$. Since $\Phi_0(0) = 0$ we get $\Phi_0(t) = \Phi_1(t) = 0$. It follows from (5.6) that $R_0(t) = R_1(t) = 0$ for $t \in [0,1]$. In turn, due to (2.18), we have $Y(1,\lambda) = \widetilde{Y}(1,\lambda)$ and $Y'(1,\lambda) = \widetilde{Y}'(1,\lambda)$ for $\lambda \in \mathbb{C}$. To complete the proof it suffices to apply Proposition 3.1. □

**Remark 5.2.** Theorem 5.1 extends to the matrix case the classical result due to Borg and Marchenko, which states that the half-line $m$-function uniquely determines the corresponding real potential coefficient. Another proof of the Borg-Marchenko result has recently been proposed by B. Simon [44] (see also the paper of Gesztesy and Simon [22]). As mentioned in [44], the proof remains valid for matrix Sturm-Liouville operators with self-adjoint $Q(\cdot) = Q(\cdot)^*$. Note, however, that the proofs contained in [44] and [22] cannot be extended to the non-self-adjoint case. Nevertheless Theorem 5.1 seems to be known implicitly for non-self-adjoint matrix functions $Q(\cdot)$ as well. Although I was not able to find this result in the literature, it can apparently be extracted from the result on unique recovery of the potential matrix from the generalized spectral function (see [42]); however, the proof given here seems to be simpler.

### 5.2. A further Borg type result

Consider equation (2.1) subject to the boundary conditions

$$y'(0,\lambda) - Hy(0,\lambda) = 0, \qquad y'(1,\lambda) + H_j y(1,\lambda) = 0, \quad H, H_j \in \mathbb{C}^{n\times n}, \tag{5.13}$$

and denote by $S(Q; H, H_j)$ the spectrum of the problem (2.1), (5.13). As in Section 4 we denote by $S(Q; D, D)$ the spectrum of the Dirichlet problem (2.1), (4.37).

**Theorem 5.3.** *Let $Q, \widetilde{Q} \in L^1[0,1] \otimes \mathbb{C}^{n\times n}$ and $H_j \in \mathbb{C}^{n\times n}$, $j \in \{1, \dots, n^2\}$. Suppose that the matrices $H_j$, $j \in \{1, \dots, n^2\}$, are linearly independent and that equations (2.1) and (2.16) have $n^2 + 1$ equal spectra*

$$S(Q; D, D) = S(\widetilde{Q}; D, D), \quad S(Q; H, H_j) = S(\widetilde{Q}; H, H_j), \quad j \in \{1, \dots, n^2\}. \tag{5.14}$$

*Then $Q(x) = \widetilde{Q}(x)$ for almost all $x \in [0,1]$.*

The proof is similar to that of Theorem 4.3. It is only necessary to use Theorem 5.1 in place of Theorem 4.1 and the asymptotic behavior

$$M(\lambda) := Y'(1,\lambda)Y(1,\lambda)^{-1} = i\sqrt{\lambda}\big((I_n + o(1)\big), \tag{5.15}$$

as $|\lambda| \to \infty$ with $\arg(\lambda) \in (-\pi/2 + \epsilon, -\epsilon)$, $\epsilon > 0$, in place of the asymptotic formulas for $M_0(\lambda) = S(1,\lambda)^{-1}C(1,\lambda)$.

Note also that another proof of Theorem 5.3 has been previously obtained by M. Lesch (private communication). While his proof is based on results from [6, 7], he also used algebraic formulas similar to (4.43) and (4.44).

### 5.3. Dirac-type operators

Let $B = \operatorname{diag}(\lambda_1 I_n, \lambda_2 I_n)$ ($\in \mathbb{R}^{2n\times 2n}$) be a diagonal $2n \times 2n$ matrix and let $Q(\cdot)$ and $\widetilde{Q}(\cdot)$ be $2n \times 2n$ potential matrices having block representations (with zero diagonals)

$$Q(x) = \big(Q_{ij}(x)\big)_{i,j=1}^2, \quad \widetilde{Q}(x) = \big(\widetilde{Q}_{ij}(x)\big)_{i,j=1}^2, \; Q_{ii}(x) = \widetilde{Q}_{ii}(x) = 0, \tag{5.16}$$

with respect to the decomposition $\mathbb{C}^{2n} = \mathbb{C}^n \oplus \mathbb{C}^n$.

Here we present an analogue of Theorem 5.1 for the following Dirac type system of ordinary differential equations:

$$B\frac{dy(x,\lambda)}{dx} + Q(x)y(x,\lambda) = \lambda y(x,\lambda), \quad y = \operatorname{col}(y_1,\dots,y_{2n}). \tag{5.17}$$

To state the corresponding result, consider alongside equation (5.17) the analogous equation

$$B\frac{dy(x,\lambda)}{dx} + \widetilde{Q}(x)y(x,\lambda) = \lambda y(x,\lambda), \quad y = \operatorname{col}(y_1,\dots,y_{2n}). \tag{5.18}$$

with a potential matrix $\widetilde{Q}(\cdot)$ of the form (5.16).

**Theorem 5.4.** [38] *Let* $B = \operatorname{diag}(\lambda_1 I_n, \lambda_2 I_n)$ *where* $\lambda_1 > 0$, $\lambda_2 < 0$, *and let* $Q$ *and* $\widetilde{Q}$ *be potential matrices of the form* (5.16) *belonging to* $L^\infty(\Omega) \otimes \mathbb{C}^{2n\times 2n}$. *Let*

$$Y(x,\lambda) = \begin{pmatrix} Y_1(x,\lambda) \\ Y_2(x,\lambda) \end{pmatrix} \quad \text{and} \quad \widetilde{Y}(x,\lambda) = \begin{pmatrix} \widetilde{Y}_1(x,\lambda) \\ \widetilde{Y}_2(x,\lambda) \end{pmatrix} \tag{5.19}$$

*be* $2n \times n$ *matrix solutions of equations* (5.17), (5.18), *respectively, obeying the initial conditions*

$$Y(0,\lambda) = \widetilde{Y}(0,\lambda) = \operatorname{col}(I_n, A_1), \quad A_1 \in \mathbb{C}^{n\times n}, \quad \det A_1 \neq 0. \tag{5.20}$$

*If in some open set* $G \subset \mathbb{C}$,

$$Y_1(1,\lambda)Y_2(1,\lambda)^{-1} = \widetilde{Y}_1(1,\lambda)\widetilde{Y}_2(1,\lambda)^{-1}, \qquad \lambda \in G, \tag{5.21}$$

*then* $Q(x) = \widetilde{Q}(x)$ *for almost all* $x \in [0,1]$.

**Remark 5.5.** Local uniqueness results for Dirac-type operators in terms of exponentially small differences of Weyl-Titchmarsh matrices with applications to other Borg-type uniqueness theorems appeared in [11]. Analogous results for Schrödinger and Jacobi operators can be found in [3], [12], [18], and [23].

Note in conclusion that different uniqueness results concerning first-order systems of ODE's as well as canonical (Hamiltonian) systems can be found in [1], [2], [8], [9], [10]–[15], [17], [18], [22], [29], [36]–[38], [43] (see also the references cited therein).

Next we present a generalization of Theorem 4.1 to the case of Dirac type systems (5.17).

**Theorem 5.6.** *Let* $B = \operatorname{diag}(\lambda_1 I_n, \lambda_2 I_n)$ *where* $\lambda_1 > 0$, $\lambda_2 < 0$, *and let* $Q$ *and* $\widetilde{Q}$ *be potential matrices of the form* (5.16) *belonging to* $L^\infty(\Omega) \otimes \mathbb{C}^{2n\times 2n}$. *Let* $Y$ *and* $\widetilde{Y}$ *be as in Theorem* 5.4 *and*

$$Z(x,\lambda) = \begin{pmatrix} Z_1(x,\lambda) \\ Z_2(x,\lambda) \end{pmatrix}, \qquad \widetilde{Z}(x,\lambda) = \begin{pmatrix} \widetilde{Z}_1(x,\lambda) \\ \widetilde{Z}_2(x,\lambda) \end{pmatrix}, \tag{5.22}$$

*be* $2n \times n$ *matrix solutions of equations* (5.17), (5.18), *respectively, obeying the initial conditions*

$$Z(0,\lambda) = \widetilde{Z}(0,\lambda) = \operatorname{col}(I_n, A_2), \quad A_2 \in \mathbb{C}^{n\times n}, \quad \det A_2 \neq 0. \tag{5.23}$$

*Assume that* $0 \in \rho(A_1 - A_2)$ *and that, for all* $\lambda$ *in some open set* $G \subset \mathbb{C}$*:*

$$Y_1(1,\lambda)Z_1(1,\lambda)^{-1} = \widetilde{Y}_1(1,\lambda)\widetilde{Z}_1(1,\lambda)^{-1}. \tag{5.24}$$

*Then* $Q(x) = \widetilde{Q}(x)$ *for almost all* $x \in [0,1]$.

The proof of Theorem 5.6 will be published elsewhere [39]. We note only that it is based on the following analogue of Proposition 3.5: a potential matrix of equation (5.17) is uniquely determined by the row $\big(Y_1(1,\lambda)\ Z_1(1,\lambda)\big)$ of the corresponding monodromy matrix.

## 6. A generalization of the Hochstadt-Lieberman result

### 6.1. The case of Sturm-Liouville operators

Here we obtain an extension of the Hochstadt-Lieberman theorem [25] to the matrix Sturm-Liouville operator.

Consider two Sturm-Liouville equations

$$-y'' + Q(x)y = \lambda^2 y, \qquad -\widetilde{y}'' + \widetilde{Q}(x)\widetilde{y} = \lambda^2 \widetilde{y} \tag{6.1}$$

with $n \times n$ potential matrices $Q(x)$ and $\widetilde{Q}(x)$, respectively; in general, they are non-self-adjoint.

First we show that the $n \times n$ potential matrix $Q$ is uniquely determined by a quarter of the monodromy matrix $W(\lambda)$ if $Q(\cdot)$ is known on half of an interval. Our proof is similar to that of Proposition 3.1.

**Theorem 6.1.** [38] *Let* $Q, \widetilde{Q} \in L^1[0,1] \otimes \mathbb{C}^{n\times n}$ *and let* $Q(x) = \widetilde{Q}(x)$ *for almost all* $x \in [1/2, 1]$. *Let* $Y(x,\lambda)$ *and* $\widetilde{Y}(x,\lambda)$ *be* $n \times n$ *matrix solutions of the Cauchy problems*

$$Y(0,\lambda) = \widetilde{Y}(0,\lambda) = I_n, \qquad Y'(0,\lambda) = H_1,\ \widetilde{Y}'(0,\lambda) = \widetilde{H}_1 \tag{6.2}$$

*for the first and second equation* (6.1), *respectively. If*

$$Y'(1,\lambda) + H_2 Y(1,\lambda) = \widetilde{Y}'(1,\lambda) + H_2 \widetilde{Y}(1,\lambda), \qquad \lambda \in \mathbb{C}, \tag{6.3}$$

*for some matrix* $H_2 \in \mathbb{C}^{n\times n}$, *then* $H_1 = \widetilde{H}_1$ *and* $Q(x) = \widetilde{Q}(x)$ *for almost all* $x \in [0,1]$.

*Proof.* (i) By Proposition 2.4 the solutions $\widetilde{Y}(x,\lambda)$ and $Y(x,\lambda)$ of the Cauchy problems (6.1), (6.2) are related by

$$\widetilde{Y}(x,\lambda) = Y(x,\lambda) + \int_0^x R(x,t)Y(t,\lambda)\,dt, \tag{6.4}$$

where $R(\cdot,\cdot) \in W_1^1(\Omega) \otimes \mathbb{C}^{n\times n}$.

Moreover, the $n\times n$ matrix kernel $R(\cdot,\cdot)$ is a generalized solution (in the sense of distributions) of the Goursat problem (2.19)–(2.21). Furthermore, $P(u,v) := R\big((u+v)/2, (u-v)/2\big)$ is a classical solution of the Goursat problem (3.1)–(3.3) in the triangle $\triangle = \{(u,v) : 0 \le v \le u \le 2 - v\}$.

(ii) First we show that $P(u,v)=0$ in the triangle $\triangle_1 := \{(u,v) : 1 \le u \le 2-v, v \ge 0\}$. To show this, we observe that, by (6.4), the equality (6.3) is equivalent to the following two equations:

$$[D_x R(x,t) + H_2 R(x,t)]|_{x=1} = 0 \quad \text{and} \quad R(1,1) = 0,$$

which in turn are equivalent to the similar equations for $P(u,v)$:

$$\big[(D_u + D_v)P(u,v) + H_2 P(u,v)\big]\big|_{v=2-u} = 0, \quad u \in [1,2], \quad P(2,0) = 0. \tag{6.5}$$

By rewriting these equalities as

$$\frac{dP(u,2-u)}{du} - H_2 P(u,2-u) = 2D_1 P(u,2-u), \quad P(2,0) = 0,$$

and integrating, we obtain

$$P(u,2-u) = 2\int_2^u e^{H_2(u-\alpha)} D_1 P(\alpha,2-\alpha)\,d\alpha, \quad u \in [1,2]. \tag{6.6}$$

Since $Q(x) = \widetilde{Q}(x)$ for almost all $x \in [1/2,1]$, equality (3.3) yields

$$D_u P(u,0) = 0, \qquad u \in [1,2]. \tag{6.7}$$

Taking (6.7) into account, we integrate (3.1) with respect to the second variable. Thus we obtain the equation

$$D_\alpha P(\alpha,v) = \frac{1}{4}\int_0^v Q(\alpha,\beta)P(\alpha,\beta)\,d\beta, \quad \alpha \in [1,2], \tag{6.8}$$

where $Q(\alpha,\beta) := \widetilde{Q}((\alpha+\beta)/2) - Q((\alpha-\beta)/2)$. Integrating equation (6.8) with respect to $\alpha$ and using (6.6), we obtain the following equation for $P(u,v)$ in $\triangle_1$:

$$\begin{aligned} P(u,v) = -\frac{1}{4}\int_u^{2-v}\int_0^v Q(\alpha,\beta)P(\alpha,\beta)\,d\beta\,d\alpha \\ -\frac{1}{2}\int_{2-v}^2 e^{H_2(2-v-\alpha)}\int_0^{2-\alpha} Q(\alpha,\beta)P(\alpha,\beta)\,d\beta\,d\alpha. \end{aligned} \tag{6.9}$$

Observing that $v \ge 2-\alpha$, it follows from (6.9) that

$$\|P(u,v)\| \le C_1 \int_u^2 \int_0^v \|Q(\alpha,\beta)\| \cdot \|P(\alpha,\beta)\|\,d\beta\,d\alpha, \qquad (u,v) \in \triangle_1, \tag{6.10}$$

where $C_1 = \big(1/4 + 1/2\max_{u\in[0,1]} \|\exp(-H_2 u)\|\big)$.

Note that the operator $T:\ C(\triangle_1)\otimes\mathbb{C}^{n\times n} \to C(\triangle_1)\otimes\mathbb{C}^{n\times n}$, given by

$$T:\ P(u,v) \mapsto \int_u^2\int_0^v Q(\alpha,\beta)P(\alpha,\beta)\,d\beta\,d\alpha,$$

is a Volterra operator, since the kernel $Q(\cdot,\cdot)$ can be approximated in $L^1(\triangle_1)\otimes\mathbb{C}^{n\times n}$ by bounded kernels. Therefore (6.10) yields the required equality $P(u,v)=0$ for $(u,v) \in \triangle_1$.

(iii) It remains to show that $P(u,v) = 0$ in the triangle $\triangle_2 = \{(u,v) : 0 \le v \le u \le 1\}$. From the previous step, we have

$$P(1,v) = D_v P(1,v) = 0, \qquad v \in [0,1]. \tag{6.11}$$

Adding condition (3.2) to (6.11) we obtain a Goursat problem in $\triangle_2$ for the equation (3.1). The required assertion $P(u,v) = 0$ in $\triangle_2$ has already been proved in step (iii) of the proof of Proposition 3.1.

Thus $P(u,v) = 0$ everywhere in the triangle $\triangle$. Therefore $P(u,0) = 0$ for $u \in [0,2]$, and by condition (3.3) we have $H_1 = \widetilde{H}_1$ and $\widetilde{Q}(x) = Q(x)$ for almost all $x \in [0,1]$. □

**Remark 6.2.** (a) The conclusion of Theorem 6.1 remains valid if condition (6.3) is replaced by the condition $Y(1,\lambda) = \widetilde{Y}(1,\lambda)$. The proof follows the same arguments, but with some simplifications. Indeed, we now obtain the condition $P(u, 2-u) = 0$ for $u \in [1,2]$ in place of (6.5), and we find

$$P(u,v) = \frac{1}{4} \int_u^{2-v} \int_0^v Q(\alpha,\beta) P(\alpha,\beta)\, d\beta\, d\alpha$$

in place of (6.9). The proof then runs as before.

(b) The case of the alternative initial conditions $Y(0,\lambda) = \widetilde{Y}(0,\lambda) = \mathbb{O}_n$, $Y'(0,\lambda) = H_1$, $\widetilde{Y}'(0,\lambda) = \widetilde{H}_1$ can be treated in a similar manner.

Let us briefly consider the scalar ($n = 1$) Sturm-Liouville equation. Denote by $S(q; h_1, h_2)$ the spectrum of the problem with separated boundary conditions for the first equation (6.1):

$$-y'' + q(x)y = \lambda^2 y, \quad y'(0) - h_1 y(0) = 0, \quad y'(1) + h_2 y(1) = 0. \tag{6.12}$$

**Corollary 6.3.** *Let $n = 1$ and suppose that $q, \tilde{q}$ are complex-valued summable potentials on $[0,1]$ satisfying the condition $q(x) = \widetilde{q}(x)$ for almost all $x \in [1/2, 1]$. If, moreover, $S(q; h_1, h_2) = S(\widetilde{q}; \widetilde{h}_1, h_2)$, then $h_1 = \widetilde{h}_1$ and $q(x) = \widetilde{q}(x)$ for almost all $x \in [0,1]$.*

*Proof.* Since the entire function $y'(1,\lambda) + h_2 y(1,\lambda)$ is uniquely determined by its zeros, i.e., by the spectrum $S(q; h_1, h_2)$, the proof follows by an application of Theorem 6.1. □

**Corollary 6.4.** *Let $n = 1$ and $q(x) = \widetilde{q}(x)$ for almost all $x \in [0, 1/2]$. If, moreover, $S(q; h_1, h_2) = S(\widetilde{q}; h_1, \widetilde{h}_2)$, then $h_2 = \widetilde{h}_2$ and $q(x) = \widetilde{q}(x)$ for almost all $x \in [0,1]$.*

**Remark 6.5.** Corollary 6.3 was first obtained by Hochstadt and Lieberman [25] under the assumption $h_1 = \widetilde{h}_1$, and in full generality (for $h_1 \ne \widetilde{h}_1$) by other methods in [24] and [45].

The conditions of Corollary 6.3 are sharp. Namely:

(i) It has been shown in [24], [13] that one cannot take different boundary conditions at the point $x_0 = 1$; i.e., the implication

$$S(q; h_1, h_2) = S(\widetilde{q}; h_1, \widetilde{h}_2) \Longrightarrow q(x) = \widetilde{q}(x) \text{ for almost all } x \in [0,1] \tag{6.13}$$

does not hold, in general, without the further assumption $h_2 = \widetilde{h}_2$;

(ii) The interval $[1/2, 1]$ cannot be replaced by a smaller interval (see [21] or the article by R. del Río in this volume);

(iii) Under the assumption $q,\ \widetilde{q} \in C[0,1]$, the requirement of equality of the spectra can be replaced by the requirement of equality of all but one corresponding eigenvalue (see [45]). Contrary to an assertion in [45], the condition of continuity of potentials is essential for this result (see [21] or [24]).

Note also that Gesztesy and Simon [19]–[22] and Gesztesy, Simon and del Rio [14], [15] have developed an original and promising approach to Hochstadt-Lieberman type results, as well as to other inverse problems (e. g., on a half-line or a line) with partial information on the potential. This enabled them to obtain a number of results on uniqueness of the potential given the potential on the interval $[1/2 - \varepsilon, 1]$ and some part of the spectrum.

Typical results obtained by Gesztesy and Simon are the following:

**Theorem 6.6.** [21] *Let $L = -d^2/dx^2 + q$ in $L^2[0,1]$ with boundary conditions $y'(0) - h_1 y(0) = 0$, $y'(1) + h_2 y(1) = 0$ and $h_1, h_2 \in \mathbb{R}$. Suppose $q$ is of class $C^{2k}$ on the interval $(\frac{1}{2} - \varepsilon, \frac{1}{2} + \varepsilon)$, for some $k = 0, 1, 2\dots$, and for some $\varepsilon > 0$. Then $q$ on $[0, 1/2]$, $h_1$, and all except $(k+1)$ eigenvalues of $L$ uniquely determine $h_2$ and $q$ on the whole of $[0,1]$.*

**Theorem 6.7.** [21] *Let $L = -d^2/dx^2 + q$ in $L^2[0,1]$ with boundary conditions $y'(0) - h_1 y(0) = 0$, $y'(1) + h_2 y(1) = 0$ and $h_1, h_2 \in \mathbb{R}$. Then $q$ on $[0, (1+\alpha)/2]$ for some $\alpha \in (0,1)$, $h_1$, and a subset $S \subseteq \sigma(L)$ of all the eigenvalues $\sigma(L)$ of $L$, satisfying*

$$\#\{\lambda \in S \mid \lambda \leq \lambda_0\} \geq (1-\alpha)\#\{\lambda \in \sigma(L) \mid \lambda \leq \lambda_0\} + (\alpha/2) \tag{6.14}$$

*for all sufficiently large $\lambda_0 \in \mathbb{R}$, uniquely determine $h_2$ and $q$ on the whole of $[0,1]$.*

**Remark 6.8.**

(i) As a typical example, knowing slightly more than half the eigenvalues and knowing $q$ on $[0, 3/4]$ determines $q$ uniquely on the whole of $[0,1]$.

(ii) As in the case $\alpha = 0$, there is an extension of the same type as Theorem 6.6. Explicitly, if $q$ is assumed to be $C^{2k}$ near $x = (1+\alpha)/2$, one only needs

$$\#\{\lambda \in S \mid \lambda \leq \lambda_0\} \geq (1-\alpha)\#\{\lambda \in \sigma(L) \mid \lambda \leq \lambda_0\} + (\alpha/2) - (k+1)$$

instead of (6.14).

(iii) The case $k = 0$ of Theorem 6.6 is due to Hald [24].

For further discussion of results of this kind, see the article of R. del Rio in this volume. Note also that there are no known generalizations of Theorems 6.6 and 6.7 to the vector case.

## 6.2. The case of Dirac-type operators

In this subsection we present an analogue of the Hochstadt-Lieberman theorem [25] for the system (5.17).

**Theorem 6.9.** [38] *Let* $B = \mathrm{diag}(\lambda_1 I_n, \lambda_2 I_n)$ *where* $\lambda_1 > 0,\ \lambda_2 < 0$, *and let* $Q$ *and* $\widetilde{Q}$ *be potential matrices of the form* (5.16), *which belong to* $L^\infty(\Omega) \otimes \mathbb{C}^{2n\times 2n}$. *Let*

$$Y(x,\lambda) = \begin{pmatrix} Y_1(x,\lambda) \\ Y_2(x,\lambda) \end{pmatrix} \quad \text{and} \quad \widetilde{Y}(x,\lambda) = \begin{pmatrix} \widetilde{Y}_1(x,\lambda) \\ \widetilde{Y}_2(x,\lambda) \end{pmatrix} \tag{6.15}$$

*be* $2n \times n$ *matrix solutions of equations* (5.17), (5.18), *respectively, obeying the initial conditions*

$$Y(0,\lambda) = \widetilde{Y}(0,\lambda) = \mathrm{col}(A_1, A_2), \quad \det A_j \neq 0,\ j \in \{1,2\}. \tag{6.16}$$

*Assume in addition that:*

1) *the equality*

$$Y_1(1,\lambda) - HY_2(1,\lambda) = \widetilde{Y}_1(1,\lambda) - H\widetilde{Y}_2(1,\lambda), \qquad \lambda \in \mathbb{C}, \tag{6.17}$$

*holds, where* $H \in \mathbb{C}^{n\times n}$ *and* $\det H \neq 0$;

2) $Q(x) = \widetilde{Q}(x)$ *for almost all* $x \in [x_0, 1]$, *where*

$$x_0 = \min\{|\lambda_1|(|\lambda_1| + |\lambda_2|)^{-1}, |\lambda_2|(|\lambda_1| + |\lambda_2|)^{-1}\}. \tag{6.18}$$

*Then* $Q(x) = \widetilde{Q}(x)$ *for almost all* $x \in [0,1]$.

As a corollary, we state the corresponding result for the Dirac system.

**Corollary 6.10.** [38] *Under the assumptions of Theorem* 6.9, *let* $\lambda_1 = 1,\ \lambda_2 = -1$. *Moreover, let condition* (6.17) *be fulfilled and* $Q(x) = \widetilde{Q}(x)$ *for almost all* $x \in [1/2, 1]$. *Then* $Q(x) = \widetilde{Q}(x)$ *for almost all* $x \in [0,1]$.

Passing to the scalar case, $n = 1$, denote by $S(Q; h, h_1)$ the spectrum (taking account of multiplicities) of equation (5.17) subject to the boundary conditions

$$y_2(0,\lambda) - hy_1(0,\lambda) = 0, \quad y_2(1,\lambda) - h_1 y_1(1,\lambda) = 0. \tag{6.19}$$

**Corollary 6.11.** *For* $n = 1$, *let* $Q(x)$ *and* $\widetilde{Q}(x)$ *be complex-valued summable* $2 \times 2$ *potential matrices of the form* (5.16), *and let* $h, h_1 \in \mathbb{C} \setminus \{0\}$. *If* $Q(x) = \widetilde{Q}(x)$ *for almost all* $x \in [x_0, 1]$, *where* $x_0$ *is defined by* (6.18), *and* $S(Q; h, h_1) = S(\widetilde{Q}; h, h_1)$, *then* $Q(x) = \widetilde{Q}(x)$ *for almost all* $x \in [0,1]$.

*Proof.* Let $y = \mathrm{col}(y_1, y_2)$ be the solution of equation (5.17) obeying the initial conditions $y_1(0,\lambda) = 1,\ y_2(0,\lambda) = h$. It is easily seen (compare the proof of Corollary 3.3) that the entire function $F(\lambda) := y_2(1,\lambda) - h_1 y_1(1,\lambda)$ is uniquely determined by the spectrum $S(Q; h, h_1)$ of the boundary value problem (5.17), (6.19). Therefore the required conclusion follows directly from Theorem 6.9. □

**Remark 6.12.** In the special case of a $2 \times 2$ Dirac system, that is the system (5.17) with $B = \text{diag}(1, -1)$, Corollary 6.11 has independently been discovered by R. del Rio and B. Grebert [16] and M. Horváth [26]. Their proofs differ from that proposed in [38]. Moreover, these authors have generalized Theorems 6.6 and 6.7 to the case of $2 \times 2$ Dirac systems.

**Acknowledgments.** I would like to thank Fritz Gesztesy for helpful discussions and hints regarding the literature. It is also a pleasure for me to thank W. Amrein and D. Pearson for the opportunity to give a talk at the Sturm Colloquium in Geneva and for stimulating the writing of this paper. I would also like to thank W. Amrein, A. Hinz and D. Pearson for helpful discussions and correspondence allowing to improve the paper.

## References

[1] D.Z. Arov and H. Dym, $J$-inner matrix functions, interpolation and inverse problems for canonical systems, II: the inverse monodromy problem, Integral Equations Operator Theory **36** (2000), 11–70.

[2] D.Z. Arov and H. Dym, $J$-inner matrix functions, interpolation and inverse problems for canonical systems, III: more on the inverse monodromy problem, Integral Equations Operator Theory **36** (2000), 127–181.

[3] C. Bennewitz, A proof of the local Borg-Marchenko theorem, Comm. Math. Phys. **218** (2001), 131–132.

[4] G. Borg, Eine Umkehrung der Sturm-Liouvilleschen Eigenwertaufgabe, Acta Math. **78** (1946), 1–96.

[5] G. Borg, Uniqueness theorems in the spectral theory of $y'' + (\lambda - q(x))y = 0$, in *Proc. 11th Scandinavian Congress of Math.*, 276–287, Johan Grundt Tanums Forlag, Oslo, 1952.

[6] D. Burghelea, L. Friedlander and T. Kappeler, On the determinant of elliptic differential and finite difference operators in vector bundles over $S^1$, Comm. Math. Phys. **138** (1991), 1–18.

[7] D. Burghelea, L. Friedlander and T. Kappeler, Regularized determinants for pseudodifferential operators in vector bundles over $S^1$, Integral Equations Operator Theory **16** (1993), 456–513.

[8] R. Carlson, Large eigenvalues and trace formulas for matrix Sturm-Liouville problems, SIAM J. Math. Anal. **30** (1999), 949–962.

[9] R. Carlson, An Inverse Problem for the Matrix Schrödinger Equation, J. Math. Anal. Appl. **267** (2002), 564–575.

[10] S. Clark and F. Gesztesy, Weyl-Titchmarsh $M$-function asymptotics for matrix-valued Schrödinger operators, Proc. London Math. Soc. **82** (2001), 701–724.

[11] S. Clark and F. Gesztesy, Weyl-Titchmarsh $M$-function asymptotics, local uniqueness results, trace formulas, and Borg-type theorems for Dirac operators, Trans. Amer. Math. Soc. **354** (2002), 3475–3534.

[12] S. Clark, F. Gesztesy, H. Holden and B. M. Levitan, Borg-type theorems for matrix-valued Schrödinger operators, J. Differential Equations **167** (2000), 181–210.

[13] R. del Rio Castillo, On boundary conditions of an inverse Sturm-Liouville problem, SIAM J. Appl. Math. **50** (1990), 1745–1751.

[14] R. del Rio, F. Gesztesy and B. Simon, Inverse spectral analysis with partial information on the potential, III: Updating boundary conditions, Internat. Math. Res. Notices **15** (1997), 751–758.

[15] R. del Rio, F. Gesztesy and B. Simon, Corrections and Addendum to "Inverse spectral analysis with partial information on the potential, III: Updating boundary conditions", Internat. Math. Res. Notices **11** (1999), 623–625.

[16] R. del Rio and B. Grebert, Inverse Spectral Results for AKNS Systems with Partial Information on the Potentials, Math. Phys. Anal. Geom. **4** (2001), 229–244.

[17] B. Després, The Borg theorem for the vectorial Hill's equation, Inverse Problems **11** (1995), 97–121.

[18] F. Gesztesy, A. Kiselev and K. Makarov, Uniqueness results for matrix-valued Schrödinger, Jacobi, and Dirac-type operators, Math. Nachr. **239/240** (2002), 103–145.

[19] F. Gesztesy and B. Simon, Uniqueness theorems in inverse spectral theory for one-dimensional Schrödinger operators, Trans. Amer. Math. Soc. **348** (1996), 349–373.

[20] F. Gesztesy and B. Simon, Inverse spectral analysis with partial information on the potential, I: The case of an a.c. component in the spectrum, Helv. Phys. Acta **70** (1997), 66–71.

[21] F. Gesztesy and B. Simon, Inverse spectral analysis with partial information on the potential, II: The case of discrete spectrum, Trans. Amer. Math. Soc. **352** (2000), 2765–2787.

[22] F. Gesztesy and B. Simon, A new approach to inverse spectral theory, II: General real potentials and the connection to the spectral measure, Ann. Math. **152** (2000), 593–643.

[23] F. Gesztesy and B. Simon, On local Borg-Marchenko uniqueness results, Comm. Math. Phys. **211** (2000), 273–287.

[24] O.H. Hald, Inverse eigenvalue problems for the mantle, Geophys. J. Royal Astron. Soc. **62** (1980), 41–48.

[25] H. Hochstadt and B. Lieberman, An inverse Sturm-Liouville problem with mixed given data, SIAM J. Appl. Math. **34** (1978), 676–680.

[26] M. Horváth, On the inverse spectral theory of Schrödinger and Dirac operators, Trans. Amer. Math. Soc. **353** (2001), 4155–4171.

[27] K. Iwasaki, Inverse problem for Sturm-Liouville and Hill equations, Ann. Mat. Pura Appl. Ser. 4, **149** (1987), 185–206.

[28] A.B. Khaled, Problème inverse de Sturm-Liouville associé à un opérateur différentiel singulier, C. R. Acad. Sci. Paris Sér. I Math. **299** (1984), 221–224.

[29] M. Lesch and M. Malamud, The inverse spectral problem for first-order systems on the half-line, in *Differential operators and related topics, Vol. I*, 199–238, Oper. Theory Adv. Appl. **117**, Birkhäuser, Basel, 2000.

[30] B.Ya. Levin, *Lectures on Entire Functions*, American Mathematical Society, Providence, Rhode Island, 1996.

[31] N. Levinson, The inverse Sturm-Liouville problem, Mat. Tidsskr. B. **1949** (1949), 25–30.

[32] B.M. Levitan, *Inverse Sturm-Liouville Problems*, VNU Science Press, Utrecht, 1987.

[33] B.M. Levitan and M.G. Gasymov, Determination of a differential equation by two of its spectra, Russ. Math. Surv. **19** no. 2 (1964), 1–63.

[34] B.M. Levitan and I.S. Sargsyan, *Sturm-Liouville and Dirac operators*, Kluwer, Dordrecht, 1991.

[35] M.M. Malamud, Similarity of Volterra operators and related questions of the theory of differential equations of fractional order, Trans. Moscow Math. Soc. **55** (1994), 57–122.

[36] M.M. Malamud, A connection between the potential matrix of the Dirac system and its Wronskian, Dokl. Math. **52** (1995), 296–299.

[37] M.M. Malamud, Borg type theorems for first-order systems on a finite interval, Funct. Anal. Appl. **33** (1999), 64–68.

[38] M.M. Malamud, Uniqueness questions in inverse problems for systems of differential equations on a finite interval, Trans. Moscow Math. Soc. **60** (1999), 204–262.

[39] M.M. Malamud, Borg type results for Dirac systems on a finite interval, Russ. J. Math. Phys., submitted.

[40] V.A. Marchenko, Some questions in the theory of one-dimensional linear differential operators of the second order, I, Trudy Moscow. Mat. Obsc. **1** (1952), 327–420 (Russian); English transl. in Amer. Math. Soc. Transl. (2) **101** (1973), 1–104.

[41] V.A. Marchenko, *Sturm-Liouville Operators and Applications*, Birkhäuser, Basel, 1986.

[42] F.S. Rofe-Beketov, Expansions in eigenfunctions of infinite systems of differential equations in non-selfadjoint and selfadjoint cases, Mat. Sb. **51** (1960), 293–342.

[43] L.A. Sakhnovich, *Spectral Theory of Canonical Differential Systems, Method of Operator Identities*, Birkhäuser, Basel, 1999.

[44] B. Simon, A new approach to inverse spectral theory, I: Fundamental formalism, Ann. Math. **150** (1999), 1029–1057.

[45] T. Suzuki, Inverse problems for heat equations on compact intervals and on circles, I, J. Math. Soc. Japan **38** (1986), 39–65.

Mark M. Malamud
Department of Mathematics
Donetsk National University
Universitetskaya Str. 24
Donetsk 83055
Ukraine
e-mail: mdm@dc.donetsk.ua

*W.O. Amrein, A.M. Hinz, D.B. Pearson*
Sturm-Liouville Theory: Past and Present, 271–331


# A Catalogue of Sturm-Liouville Differential Equations

W. Norrie Everitt

*Dedicated to all scientists who, down the long years,*
*have contributed to Sturm-Liouville theory.*

**Abstract.** This catalogue commences with sections devoted to a brief summary of Sturm-Liouville theory including some details of differential expressions and equations, Hilbert function spaces, differential operators, classification of interval endpoints, boundary condition functions and the Liouville transform.

There follows a collection of more than 50 examples of Sturm-Liouville differential equations; many of these examples are connected with well-known special functions, and with problems in mathematical physics and applied mathematics.

For most of these examples the interval endpoints are classified within the relevant Hilbert function space, and boundary condition functions are given to determine the domains of the relevant differential operators. In many cases the spectra of these operators are given.

The author is indebted to many colleagues who have responded to requests for examples and who checked successive drafts of the catalogue.

**Mathematics Subject Classification (2000).** Primary; 34B24, 34B20, 34B30: Secondary; 34L05, 34A30, 34A25.

**Keywords.** Sturm-Liouville differential equations; special functions; spectral theory.

## Contents

## 1. Introduction

The idea for this paper follows from the conference entitled:

**Bicentenaire de Charles François Sturm**

held at the University of Geneva, Switzerland from 15 to 19 September 2003. One of the main interests for this meeting involved the development of the theory of Sturm-Liouville differential equations. This theory began with the original work of Sturm from 1829 to 1836 and was then followed by the short but significant joint paper of Sturm and Liouville in 1837, on second-order linear ordinary differential equations with an eigenvalue parameter. Details for the 1837 paper are given as reference [78] in this paper; for a complete set of historical references see the historical survey paper [58] of Lützen.

This present catalogue of examples of Sturm-Liouville differential equations is based on four main sources:

1. The list of 32 examples prepared by Bailey, Everitt and Zettl in the year 2001 for the final version of the computer program SLEIGN2; this list is to be found within the LaTeX file xamples.tex contained in the package associated with the publication [11, Data base file xamples.tex]; all these 32 examples are contained within this catalogue.
2. A selection from the set of 59 examples prepared by Pryce and published in 1993 in the text [69, Appendix B.2]; see also [70].
3. A selection from the set of 217 examples prepared by Pruess, Fulton and Xie in the report [68].
4. A selection drawn up from a general appeal, made in October 2003, for examples but with the request relayed in the following terms: examples to be included should satisfy one or more of the criteria
   (i) The solutions of the differential equation are given explicitly in terms of special functions; see for example Abramowitz and Stegun [1], the Erdélyi *et al.* Bateman volumes [27], the recent text of Slavyanov and Lay [77] and the earlier text of Bell [16].
   (ii) Examples with special connections to applied mathematics and mathematical physics.
   (iii) Examples with special connections to numerical analysis; see the work of Zettl [82] and [83].

The overall aim was to be content with about 50 examples, as now to be seen in the list given below.

The naming of these examples of Sturm-Liouville differential equations is somewhat arbitrary; where named special functions are concerned the chosen name is clear; in certain other cases the name has been chosen to reflect one or more of the authors concerned.

## 2. Notations

The real and complex fields are represented by $\mathbb{R}$ and $\mathbb{C}$ respectively; a general interval of $\mathbb{R}$ is represented by $I$; compact and open intervals of $\mathbb{R}$ are represented by $[a, b]$ and $(a, b)$ respectively. The prime symbol $'$ denotes classical differentiation on the real line $\mathbb{R}$.

Lebesgue integration on $\mathbb{R}$ is denoted by $L$, and $L^1(I)$ denotes the Lebesgue integration space of complex-valued functions defined on the interval $I$. The local integration space $L^1_{\mathrm{loc}}(I)$ is the set of all complex-valued functions on $I$ which are Lebesgue integrable on all compact sub-intervals $[a, b] \subseteq I$; if $I$ is compact then $L^1(I) \equiv L^1_{\mathrm{loc}}(I)$.

Absolute continuity, with respect to Lebesgue measure, is denoted by $AC$; the space of all complex-valued functions defined on $I$ which are absolutely continuous on all compact sub-intervals of $I$, is denoted by $AC_{\mathrm{loc}}(I)$.

A weight function $w$ on $I$ is a Lebesgue measurable function $w : I \to \mathbb{R}$ satisfying $w(x) > 0$ for almost all $x \in I$.

Given an interval $I$ and a weight function $w$ the space $L^2(I; w)$ is defined as the set of all complex-valued, Lebesgue measurable functions $f : I \to \mathbb{C}$ such that

$$\int_I |f(x)|^2 \, w(x) \, dx < +\infty.$$

Taking equivalent classes into account $L^2(I; w)$ is a Hilbert function space with inner product

$$(f, g)_w := \int_I f(x)\overline{g}(x)w(x) \, dx \text{ for all } f, g \in L^2(I; w).$$

## 3. Sturm-Liouville differential expressions and equations

Given the interval $(a, b)$, then a set of Sturm-Liouville coefficients $\{p, q, w\}$ has to satisfy the minimal conditions

$$\begin{array}{ll} \text{(i)} & p, q, w : (a, b) \to \mathbb{R} \\ \text{(ii)} & p^{-1}, q, w \in L^1_{\mathrm{loc}}(a, b) \\ \text{(iii)} & w \text{ is a weight function on } (a, b). \end{array}$$

Note that in general there is no sign restriction on the leading coefficient $p$.

Given the interval $(a, b)$ and the set of Sturm-Liouville coefficients $\{p, q, w\}$ the associated Sturm-Liouville differential expression $M(p, q) \equiv M[\cdot]$ is the linear operator defined by

$$\begin{array}{ll} \text{(i)} & \text{domain } D(M) := \{f : (a, b) \to \mathbb{C} : f, pf' \in AC_{\mathrm{loc}}(a, b)\} \\ \text{(ii)} & \left\{ \begin{array}{l} M[f](x) := -(p(x)f(x)')' + q(x)f(x) \text{ for all } f \in D(M) \\ \text{and almost all } x \in (a, b). \end{array} \right. \end{array}$$

We note that $M[f] \in L^1_{\mathrm{loc}}(I)$ for all $f \in D(M)$; it is shown in [63, Chapter V, Section 17] that $D(M)$ is dense in the Banach space $L^1(a, b)$.

Given the interval $(a, b)$ and the set of Sturm-Liouville coefficients $\{p, q, w\}$ the associated Sturm-Liouville differential equation is the second-order linear ordinary differential equation

$$M[y](x) \equiv -(p(x)y'(x))' + q(x)y(x) = \lambda w(x)y(x) \text{ for all } x \in (a,b),$$

where $\lambda \in \mathbb{C}$ is a complex-valued spectral parameter.

The above minimal conditions on the set of coefficients $\{p, q, w\}$ imply that the Sturm-Liouville differential equation has a solution to any initial value problem at a point $c \in (a,b)$; see the existence theorem in [63, Chapter V, Section 15], *i.e.*, given two complex numbers $\xi, \eta \in \mathbb{C}$ and any value of the parameter $\lambda \in \mathbb{C}$, there exists a unique solution of the differential equation, say $y(\cdot, \lambda) : (a,b) \to \mathbb{C}$, with the properties:

(i) $y(\cdot,\lambda)$ and $(py')(\cdot,\lambda) \in AC_{\text{loc}}(a,b)$
(ii) $y(c,\lambda) = \xi$ and $(py')(c,\lambda) = \eta$
(iii) $y(x,\cdot)$ and $(py')(x,\cdot)$ are holomorphic on $\mathbb{C}$.

## 4. Operator theory

Full details of the following quoted operator theoretic results are to be found in [63, Chapter V, Section 17] and [34, Sections I, IV and V].

The Green's formula for the differential expression $M$ is, for any compact interval $[\alpha, \beta] \subset (a,b)$,

$$\int_\alpha^\beta \left\{\overline{g}(x)M[f](x) - f(x)\overline{M[g]}(x)\right\} dx = [f,g](\beta) - [f,g](\alpha) \text{ for all } f,g \in D(M),$$

where the symplectic form $[\cdot,\cdot](\cdot) : D(M) \times D(M) \times (a,b) \to \mathbb{C}$ is defined by

$$[f,g](x) := f(x)(p\overline{g}')(x) - (pf')(x)\overline{g}(x).$$

Incorporating now the weight function $w$ and the Hilbert function space $L^2((a,b);w)$, the maximal operator $T_1$ generated from $M$ is defined by

(i) $T_1 : D(T_1) \subset L^2((a,b);w) \to L^2((a,b);w)$
(ii) $D(T_1) := \{f \in D(M) : f, w^{-1}M[f] \in L^2((a,b);w)\}$
(iii) $T_1 f := w^{-1}M[f]$ for all $f \in D(T_1)$.

We note that, from the Green's formula, the symplectic form of $M$ has the property that, for all $f,g \in D(T_1)$, the following limits

$$[f,g](a) := \lim_{x \to a^+} [f,g](x) \text{ and } [f,g](b) := \lim_{x \to b^-} [f,g](x)$$

both exist and are finite in $\mathbb{C}$.

The minimal operator $T_0$ generated by $M$ is defined by

(i) $T_0 : D(T_0) \subset L^2((a,b);w) \to L^2((a,b);w)$
(ii) $D(T_0) := \{f \in D(T_1) : [f,g](b) = [f,g](a) = 0$ for all $g \in D(T_1)$
(iii) $T_0 f := w^{-1}M[f]$ for all $f \in D(T_0)$.

With these definitions the following properties hold for $T_0$ and $T_1$, and their adjoint operators

(i) $T_0 \subseteq T_1$

(ii) $T_0$ is closed and symmetric in $L^2((a,b);w)$
(iii) $T_0^* = T_1$ and $T_1^* = T_0$
(iv) $T_1$ is closed in $L^2((a,b);w)$
(v) $T_0$ has equal deficiency indices $(d,d)$ with $0 \leq d \leq 2$.
Self-adjoint extensions $T$ of $T_0$ exist and satisfy

$$T_0 \subseteq T \subseteq T_1$$

where the domain $D(T)$ is determined, as a restriction of the domain $D(T_1)$, by applying symmetric boundary conditions to the elements of the maximal domain $D(T_1)$.

## 5. Endpoint classification

Suppose given the interval $(a,b)$ and the set of coefficients $\{p,q,w\}$.

We now give the classification of the endpoints $a$ and $b$ of the differential equation valid under the coefficient conditions in Section 3 above; details are given for the endpoint $a$ but there is a similar classification scheme for the endpoint $b$.

Throughout this classification scheme let $c \in (a,b)$; however the classification that emerges is independent of the choice of the point $c$.

Additionally the scheme involves a choice of $\lambda$ but again the classification can be shown to be independent of this spectral parameter.

**Regular**
The endpoint $a$ is *regular* (notation R) if

$$\begin{cases} \text{(i)} & a \in \mathbb{R}, \textit{i.e.}, \ a > -\infty, \text{ and} \\ \text{(ii)} & p^{-1}, q, w \in L^1(a,c]. \end{cases}$$

**Singular**
The endpoint $a$ is *singular* (notation S) if it is not R, *i.e.*,

$$\begin{cases} \text{(i)} & \text{either } a = -\infty \\ \text{(ii)} & \text{or } a \in \mathbb{R} \ \text{ but } \ \int_a^c \{|p(x)|^{-1} + |q(x)| + w(x)\} dx = +\infty. \end{cases}$$

If $a$ is S then there are two main classification sub-cases as follows:

**Limit-point.** The endpoint $a$ is *limit-point* (notation LP) if $a$ is S and for some $\lambda \in \mathbb{C}$ there exists at least one solution $y(\cdot,\lambda)$ of the differential equation such that

$$\int_a^c w(x)\,|y(x,\lambda)|^2 \, dx = +\infty.$$

**Limit-circle.** The endpoint $a$ is *limit-circle* (notation LC) if $a$ is S and for some $\lambda \in \mathbb{C}$ all solutions $y(\cdot,\lambda)$ of the differential equation satisfy

$$\int_a^c w(x)\,|y(x,\lambda)|^2 \, dx < +\infty.$$

The LC classification has two sub-cases:

1. **Limit-circle non-oscillatory**

   The endpoint $a$ is *limit-circle non-oscillatory* (notation LCNO) if there exists a point $d \in (a, c)$, a real value $\lambda \in \mathbb{R}$ and a solution $y(\cdot, \lambda)$ with the property
$$y(x, \lambda) > 0 \text{ for all } x \in (a, d).$$

2. **Limit-circle oscillatory**

   The endpoint $a$ is *limit-circle oscillatory* (notation LCO) if
$$\left\{ \begin{array}{ll} \text{(i)} & \text{for any } \lambda \in \mathbb{R} \text{ and any non-null solution } y(\cdot, \lambda), \\ \text{(ii)} & \text{for any } d \in (a, c], \\ \text{(iii)} & \text{there exists a point } \xi \in (a, d] \text{ such that } y(\xi, \lambda) = 0. \end{array} \right.$$

**Remark 5.1.**

1. We stress the point made above that although the spectral parameter $\lambda$ is involved in the endpoint classification it can be shown that this classification is independent of $\lambda$ and depends only on the interval $(a, b)$ and the set of coefficients $\{p, q, w\}$.
2. When $a$ is R the initial value problem, as detailed above at the end of Section 3, can be solved at the endpoint $a$ to give a unique solution on the interval $[a, b)$.
3. The classification $a$ is R can be considered as a special case of the LCNO classification at $a$.
4. When $a$ is R or LCNO the differential equation, for any *real* value $\lambda \in \mathbb{R}$, has two linearly independent solutions $u(\cdot, \lambda), v(\cdot, \lambda) : (a, b) \to \mathbb{R}$, such that for some $c \in (a, b)$
$$\left\{ \begin{array}{ll} \text{(i)} & u(x, \lambda) > 0 \text{ and } v(x, \lambda) > 0 \text{ for all } x \in (a, c), \\ \text{(ii)} & \lim_{x \to a+} u(x, \lambda)/v(x, \lambda) = 0, \\ \text{(iii)} & \int_a^c \{p(x)u(x, \lambda)^2\}^{-1} dx = +\infty \text{ and } \int_a^c \{p(x)v(x, \lambda)^2\}^{-1} dx < +\infty. \end{array} \right.$$
   The solution $u(\cdot, \lambda)$ is unique, up to scalar multiples, and is called the *principal* solution of the differential equation for this value of the parameter $\lambda$. The solution $v(\cdot, \lambda)$ is called a *non-principal* solution, noting that this solution is not unique.

   In particular when $a$ is R, a principal solution $u(\cdot, \lambda)$ is determined by the initial conditions
$$u(a, \lambda) = 0 \quad \text{and} \quad (pu')(a) \neq 0.$$
5. When $a$ is LCO then for any real $\lambda \in \mathbb{R}$ all solutions of the differential equation have infinitely many zeros in any right-neighborhood $(a, d)$ of $a$.
6. All the above remarks apply equally well, with change of notation, to the classification cases of endpoint $b$; note that the classification of $a$ and of $b$ are independent of each other.

## 6. Endpoint boundary condition functions

Given the interval $(a, b)$ of $\mathbb{R}$ and a set of coefficients $\{p, q, w\}$, we can create a Sturm-Liouville differential equation with classified endpoints.

There is a very complete account of separated and coupled boundary conditions for the associated Sturm-Liouville boundary value problems, in the paper [10, Section 5].

Here, for use in cataloguing the Sturm-Liouville examples, we give information concerning the use of boundary condition functions at any endpoint in the LC classification. The use of these boundary condition functions takes the same form in both LCNO and LCO cases.

Let $a$ be R; then a separated boundary condition at this endpoint, for a solution $y$ of the Sturm-Liouville differential equation $M[y] = \lambda wy$ on $(a, b)$, takes the form, where $A_1, A_2 \in \mathbb{R}$ with $A_1^2 + A_2^2 > 0$,

$$A_1 y(a) + A_2(py')(a) = 0.$$

If $b$ is R then there is a similar form for a separated boundary condition

$$B_1 y(b) + B_2(py')(b) = 0.$$

Let $a$ be LC; then a separated boundary condition at this endpoint, for a solution $y \in D(T_1)$ of the Sturm-Liouville differential equation $M[y] = \lambda wy$ on $(a, b)$, takes the form,

$$A_1[y, u](a) + A_2[y, v](a) = 0$$

where

(i) $A_1, A_2 \in \mathbb{R}$ with $A_1^2 + A_2^2 > 0$
(ii) $u, v : (a, b) \to \mathbb{R}$
(iii) $u, v \in D(T_1)$
(iv) $[u, v](a) \neq 0$.

Such pairs $\{u, v\}$ of elements from the maximal domain $D(T_1)$ always exist under the LC classification on the endpoint $a$, see [10, Section 5].

If $b$ is LC then there is a similar form for a separated boundary condition involving a pair $\{u, v\}$ of boundary condition functions, in general a different pair from the pair required for the endpoint $a$, to give

$$B_1[y, u](b) + B_2[y, v](b) = 0.$$

For any given particular Sturm-Liouville differential equation the search for pairs of such boundary condition functions may start with a study of the solutions of the differential equation $M[y] = \lambda wy$ on $(a, b)$, and also with a direct search within the elements of the maximal domain $D(T_1)$.

For the examples given in the catalogue a suitable choice of these boundary condition functions is given, for endpoints in the LC case.

**Remark 6.1.** In practice it is sufficient to determine the pair $\{u, v\}$ in a neighborhood $(a, c]$ of $a$, or $[c, b)$ of $b$, so that they are locally in the maximal domain $D(T_1)$; this practice is adopted in many of the examples given in this catalogue.

## 7. The Liouville transformation

The named Liouville transformation, see [30, Section 4.3] and [17, Chapter 10, Section 10] for details, of the general Sturm-Liouville differential equation

$$-(p(x)y'(x))' + q(x)y(x) = \lambda w(x)y(x) \text{ for all } x \in (a,b)$$

provides a means, under additional conditions on the coefficients $\{p,q,w\}$, to yield a simpler Sturm-Liouville form of the differential equation

$$-Y''(X) + Q(X)Y(X) = \lambda Y(X) \text{ for all } X \in (A,B).$$

The minimal additional conditions required, see [30, Section 4.3], are

(i) $p$ and $p' \in AC_{\text{loc}}(a,b)$, and $p(x) > 0$ for all $x \in (a,b)$
(ii) $w$ and $w' \in AC_{\text{loc}}(a,b)$, and $w(x) > 0$ for all $x \in (a,b)$.

The Liouville transformation changes the variables $x$ and $y$ to $X$ and $Y$ as follows, see [30, Section 4.3]:

(i) For $k \in (a,b)$ and $K \in \mathbb{R}$ the mapping $X(\cdot) : (a,b) \to (A,B)$ defines a new independent variable $X(\cdot)$ by

$$X(x) = l(x) := K + \int_k^x \{w(t)/p(t)\}^{1/2}\,dt \text{ for all } x \in (a,b)$$

$$A := K - \int_a^k \{w(t)/p(t)\}^{1/2}\,dt \text{ and } B := K + \int_k^b \{w(t)/p(t)\}^{1/2}\,dt$$

where $-\infty \le A < B \le +\infty$; there is then an inverse mapping $L(\cdot) : (A,B) \to (a,b)$.

(ii) Define the new dependent variable $Y(\cdot)$ by

$$\begin{aligned} Y(X) &:= \{p(x)w(x)\}^{1/4} y(x) \text{ for all } x \in (a,b) \\ &:= \{p(L(X))w(L(X))\}^{1/4} y(L(X)) \text{ for all } X \in (A,B). \end{aligned}$$

The new coefficient $Q$ is given by

$$Q(X) = w(x)^{-1}q(x) - \{w(x)^{-3}p(x)\}^{1/4}(p(x)(\{p(x)w(x)\}^{-1/4})')' \text{ for all } x \in (a,b).$$

An example of this Liouville transformation is worked in Section 11 for one form of the Bessel equation.

## 8. Fourier equation

This is the classical Sturm-Liouville differential equation, see [79, Chapter I, and Chapter IV, Section 4.1],

$$-y''(x) = \lambda y(x) \text{ for all } x \in (-\infty,+\infty)$$

with solutions

$$\cos(x\sqrt{\lambda}) \text{ and } \sin(x\sqrt{\lambda}).$$

Endpoint classification in $L^2(-\infty, +\infty)$:

| Endpoint | Classification |
|---|---|
| $-\infty$ | LP |
| 0 | R |
| $+\infty$ | LP |

This is a simple constant coefficient equation; for any self-adjoint boundary value problem on a compact interval the eigenvalues can be characterized in terms of the solutions of a transcendental equation involving only trigonometric functions.

For a study of boundary value problems on the half-line $[0, \infty)$ or the whole line $(-\infty, \infty)$ see [79, Chapter IV, Section 4.1] and [2, Volume **II**, Appendix 2, Section 132, Part 2].

## 9. Hypergeometric equation

The standard form for this differential equation is, see [46, Chapter 4, Section 3], [81, Chapter XIV, Section 14.2], [16, Chapter 9, Section 9.2], [27, Chapter II, Section 2.1.1], [1, Chapter 15, Section 15.5] and [79, Chapter IV, Sections 4.18 to 4.20],

$$z(1-z)y''(z) + [c - (a+b+1)z]y'(z) - aby(z) = 0 \text{ for all } z \in \mathbb{C}$$

where, in general $a, b, c \in \mathbb{C}$. In terms of the hypergeometric function ${}_2F_1$, solutions of this equation are, with certain restrictions on the parameters and the independent variable $z$,

$${}_2F_1(a, b; c; z) \text{ and } z^{1-c}{}_2F_1(a+1-c, b+1-c; 2-c; z).$$

For consideration of this hypergeometric equation in Sturm-Liouville form we replace the variable $z$ by the real variable $x \in (0,1)$. Thereafter, on multiplying by the factor $x^\alpha(1-x)^\beta$ and rearranging the terms gives the Sturm-Liouville equation, for all $\alpha, \beta \in \mathbb{R}$,

$$-\left(x^{\alpha+1}(1-x)^{\beta+1}y'(x)\right)' = \lambda x^\alpha(1-x)^\beta y(x) \text{ for all } x \in (0,1).$$

In this form the relationship between the parameters $\{a, b, c\}$ and $\{\alpha, \beta, \lambda\}$ is

$$c = \alpha + 1 \qquad a + b = \alpha + \beta + 1 \qquad ab = -\lambda;$$

these equations can be solved for $\{a, b, c\}$ in terms of $\{\alpha, \beta, \lambda\}$ as in [79, Chapter IV, Section 4.18].

Given $\alpha, \beta \in \mathbb{R}$ and $\lambda \in \mathbb{C}$ the solutions of this Sturm-Liouville equation can then be represented in terms of the hypergeometric function ${}_2F_1$, as above.

For the case when $\lambda = 0$ the general solution of this differential equation takes the form, for $c \in (0,1)$,

$$y(x) = k\int_c^x \frac{1}{t^{\alpha+1}(1-t)^{\beta+1}}\,dt + l \text{ for all } x \in (0,1)$$

where the numbers $k, l \in \mathbb{C}$. From this representation it may be shown that the following classifications, in the space $L^2((0,1); x^\alpha(1-x)^\beta)$, of the endpoints 0 and 1 hold:

| Endpoint | Parameters $\alpha, \beta$ | Classification |
|---|---|---|
| 0 | For $\alpha \in (-1, 0)$ and all $\beta \in \mathbb{R}$ | R |
| 0 | For $\alpha \in [0, 1)$ and all $\beta \in \mathbb{R}$ | LCNO |
| 0 | For $\alpha \in (-\infty, -1] \cup [1, \infty)$ and all $\beta \in \mathbb{R}$ | LP |
| 1 | For $\beta \in (-1, 0)$ and all $\alpha \in \mathbb{R}$ | R |
| 1 | For $\beta \in [0, 1)$ and all $\alpha \in \mathbb{R}$ | LCNO |
| 1 | For $\beta \in (-\infty, -1] \cup [1, \infty)$ and all $\alpha \in \mathbb{R}$ | LP |

For the endpoint 0, for $\alpha \in [0, 1)$ and for all $\beta \in \mathbb{R}$ the LCNO boundary condition functions $u, v$ take the form, for all $x \in (0, 1)$,

| Parameter | $u$ | $v$ |
|---|---|---|
| $\alpha = 0$ | 1 | $\ln(x)$ |
| $\alpha \in (0, 1)$ | 1 | $x^{-\alpha}$ |

For the endpoint 1, for $\beta \in [0, 1)$ and for all $\alpha \in \mathbb{R}$ the LCNO boundary condition functions $u, v$ take the form, for all $x \in (0, 1)$,

| Parameter | $u$ | $v$ |
|---|---|---|
| $\beta = 0$ | 1 | $\ln(1-x)$ |
| $\beta \in (0, 1)$ | 1 | $(1-x)^{-\beta}$ |

Another form of the hypergeometric differential equation is obtained if in the original equation above the independent variable $z$ is replaced by $-z$ to give

$$z(1+z)y''(z) + [c + (a+b+1)z]y'(z) + aby(z) = 0 \text{ for all } z \in \mathbb{C}$$

with general solutions

$${}_2F_1(a, b; c; -z) \text{ and } z^{1-c}{}_2F_1(a+1-c, b+1-c; 2-c; -z);$$

see the account in [79, Chapter IV, Section 4.18].

For consideration of this hypergeometric equation in Sturm-Liouville form we replace the variable $z$ by the real variable $x \in (0, \infty)$. Thereafter, on multiplying by the factor $x^\alpha(1+x)^\beta$ and rearranging the terms gives the Sturm-Liouville equation, for all $\alpha, \beta \in \mathbb{R}$,

$$-\left(x^{\alpha+1}(1+x)^{\beta+1}y'(x)\right)' = \lambda x^\alpha(1+x)^\beta y(x) \text{ for all } x \in (0, \infty).$$

In this form the relationship between the parameters $\{a, b, c\}$ and $\{\alpha, \beta, \lambda\}$ is

$$c = \alpha + 1 \qquad a + b = \alpha + \beta + 1 \qquad ab = \lambda;$$

these equations can be solved for $\{a, b, c\}$ in terms of $\{\alpha, \beta, \lambda\}$ as in [79, Chapter IV, Section 4.18].

Given $\alpha, \beta \in \mathbb{R}$ and $\lambda \in \mathbb{C}$ the solutions of this Sturm-Liouville equation can then be represented in terms of the hypergeometric function ${}_2F_1$, as above.

For the case when $\lambda = 0$ the general solution of this differential equation takes the form, for $c \in (0, \infty)$,

$$y(x) = k \int_c^x \frac{1}{t^{\alpha+1}(1+t)^{\beta+1}}\, dt + l \text{ for all } x \in (0, \infty)$$

where the numbers $k, l \in \mathbb{C}$. From this representation it may be shown that the following classifications, in the space $L^2((0, \infty); x^\alpha(1+x)^\beta)$, of the endpoints 0 and $+\infty$ hold:

| Endpoint | Parameters $\alpha, \beta$ | Classification |
|---|---|---|
| 0 | For $\alpha \in (-1, 0)$ and all $\beta \in \mathbb{R}$ | R |
| 0 | For $\alpha \in [0, 1)$ and all $\beta \in \mathbb{R}$ | LCNO |
| 0 | For $\alpha \in (-\infty, -1] \cup [1, \infty)$ and all $\beta \in \mathbb{R}$ | LP |
| $+\infty$ | For all $\alpha, \beta \in \mathbb{R}$ | LP |

For the endpoint 0, for $\alpha \in [0, 1)$ and for all $\beta \in \mathbb{R}$ the LCNO boundary condition functions $u, v$ take the form, for all $x \in (0, 1)$,

| Parameter | $u$ | $v$ |
|---|---|---|
| $\alpha = 0$ | 1 | $\ln(x)$ |
| $\alpha \in (0, 1)$ | 1 | $x^{-\alpha}$ |

The spectral properties of these hypergeometric Sturm-Liouville differential equations seem not to have been studied in detail; however there are a number of very interesting special cases, together with their spectral properties, considered in [79, Chapter IV, Sections 4.18 to 4.20].

## 10. Kummer equation

The Kummer differential equation is a special case of the confluent hypergeometric differential equation

$$zw''(z) + (b - z)y'(z) - aw(z) = 0 \text{ for } z \in \mathbb{C}.$$

Taking the parameter $b \in \mathbb{R}$ to be real-valued, putting $\lambda = -a \in \mathbb{C}$, replacing the independent variable $z$ by $x \in \mathbb{R}$, and then writing the resulting differential equation in Lagrange symmetric form, gives the Sturm-Liouville differential equation

$$-(x^b \exp(-x)y'(x))' = \lambda x^{b-1} \exp(-x)y(x) \text{ for all } x \in (0, \infty).$$

Solutions of this Sturm-Liouville differential equation are given in the form, using the Kummer functions $M$ and $U$,

$$M(-\lambda, b, x) \text{ and } U(-\lambda, b, x) \text{ for all } x \in (0, \infty), b \in \mathbb{R} \text{ and } \lambda \in \mathbb{C};$$

see [1, Chapter 13, Section 13.1].

Endpoint classification in $L^2((0,+\infty);x^{b-1}\exp(-x))$:

| Endpoint | Parameter $b$ | Classification |
|---|---|---|
| 0 | For $b \leq 0$ | LP |
| 0 | For $0 < b < 1$ | R |
| 0 | For $1 \leq b < 2$ | LCNO |
| 0 | For $b \geq 2$ | LP |
| $\infty$ | For all $b \in \mathbb{R}$ | LP |

For the endpoint 0 and then for $b \in (0,2)$, the LCNO boundary condition functions $u, v$ take the form, for all $x \in (0,\infty)$,

| Parameter | $u$ | $v$ |
|---|---|---|
| $b = 1$ | 1 | $\ln(x)$ |
| $1 < b < 2$ | 1 | $x^{1-b}$ |

See also the Laguerre differential equation given in Section 27 below.

## 11. Bessel equation

The Bessel differential equation has many different forms, see [80, Chapter IV], [1, Chapters 9 and 10], [27, Volume II, Chapter VII], [46, Chapter 8], [16, Chapter 4]; see in particular [52, Part C, Section 2.162].

One elegant Sturm-Liouville form, see [33, Section 1], of this differential equation is, where the parameter $\alpha \in \mathbb{R}$,

$$-(x^{2\alpha+1}y'(x))' = \lambda x^{2\alpha+1}y(x) \text{ for all } x \in (0,\infty).$$

Solutions of this differential equation are, for all $\alpha \in \mathbb{R}$,

$$x^{-\alpha}J_\alpha(x\sqrt{\lambda}) \text{ and } x^{-\alpha}Y_\alpha(x\sqrt{\lambda}) \text{ for all } x \in (0,\infty)$$

where $J_\alpha$ and $Y_\alpha$ are the classical Bessel functions, and the power $x^{-\alpha}$ is defined by $x^{-\alpha} := \exp(-\alpha\ln(x))$ for all $x \in (0,\infty)$.

For the case when $\lambda = 0$ the general solution of this differential equation takes the form, for $c \in (0,\infty)$,

$$y(x) = k\int_c^x \frac{1}{t^{2\alpha+1}}\,dt + l \text{ for all } x \in (0,\infty)$$

where the numbers $k, l \in \mathbb{C}$. From this representation it may be shown that the following classifications, in the space $L^2((0,\infty);x^{2\alpha+1})$, of the endpoints 0 and $+\infty$ hold:

| Endpoint | Parameter $\alpha$ | Classification |
|---|---|---|
| 0 | For $\alpha \in (-1,1)$ | LCNO |
| 0 | For $\alpha \in (-\infty,-1] \cup [1,\infty)$ | LP |
| $\infty$ | For all $\alpha \in \mathbb{R}$ | LP |

For the endpoint 0 and then for $\alpha \in (-1,1)$, the LCNO boundary condition functions $u, v$ take the form, for all $x \in (0,\infty)$,

| Parameter | $u$ | $v$ |
|---|---|---|
| $\alpha = 0$ | 1 | $\ln(x)$ |
| $\alpha \in (-1,0) \cup (0,1)$ | 1 | $x^{-2\alpha}$ |

As an example of the Liouville transformation, see Section 7 above, let $k = 1$ and $K = 1$ to give, for the form of the Bessel differential equation above,

$$X(x) = 1 + \int_1^x dt = x \text{ for all } x \in (0,\infty);$$

a computation then shows that

$$Q(X) = (\alpha^2 - 1/4)x^{-2} = (\alpha^2 - 1/4)X^{-2} \text{ for all } X \in (0,\infty).$$

Thus the Liouville form of this Bessel differential equation is

$$-Y''(X) + (\alpha^2 - 1/4)X^{-2}Y(X) = \lambda Y(X) \text{ for all } X \in (0,\infty),$$

where we can now take the parameter $\alpha \in [0,\infty)$.

## 12. Bessel equation: Liouville form

In the Liouville normal form, see Sections 7 and 11 above, the Bessel differential equation appears as

$$-y''(x) + \left(\nu^2 - 1/4\right)x^{-2}y(x) = \lambda y(x) \text{ for all } x \in (0,+\infty),$$

with the parameter $\nu \in [0,+\infty)$; this differential equation is extensively studied in [79, Chapter IV, Sections 4.8 to 4.15]; see also [2, Volume II, Appendix 2, Section 132, Part 5]. In this form the equation has solutions

$$x^{1/2}J_\nu(x\sqrt{\lambda}) \text{ and } x^{1/2}Y_\nu(x\sqrt{\lambda}).$$

Endpoint classification in $L^2(0,+\infty)$:

| Endpoint | Parameter $\nu$ | Classification |
|---|---|---|
| 0 | For $\nu = 1/2$ | R |
| 0 | For all $\nu \in [0,1)$ but $\nu \neq 1/2$ | LCNO |
| 0 | For all $\nu \in [1,\infty)$ | LP |
| $+\infty$ | For all $\nu \in [0,\infty)$ | LP |

For endpoint 0 and $\nu \in [0,1)$ but $\nu \neq 1/2$, the LCNO boundary condition functions $u, v$ are determined by, for all $x \in (0,+\infty)$,

| Parameter | $u$ | $v$ |
|---|---|---|
| $\nu \in (0,1)$ but $\nu \neq 1/2$ | $x^{\nu+1/2}$ | $x^{-\nu+1/2}$ |
| $\nu = 0$ | $x^{1/2}$ | $x^{1/2}\ln(x)$ |

(a) Problems on $(0,1]$ with $y(1)=0$:
For $0 \le \nu < 1, \nu \ne \frac{1}{2}$ : the Friedrichs case: $A_1 = 1, A_2 = 0$ yields the classical Fourier-Bessel series; here $\lambda_n = j_{\nu,n}^2$ where $\{j_{\nu,n} : n = 0,1,2,\ldots\}$ are the zeros (positive) of the Bessel function $J_\nu(\cdot)$.
For $\nu \ge 1$; LP at 0 so that there is a unique boundary value problem with $\lambda_n = j_{\nu,n}^2$ as before.

(b) Problems on $[1,\infty)$ all have continuous spectrum on $[0,\infty)$:
For Dirichlet and Neumann boundary conditions there are no eigenvalues.
For $A_1 = A_2 = 1$ at 1 there is one isolated negative eigenvalue.

(c) Problems on $(0,\infty)$ all have continuous spectrum on $[0,\infty)$:
For $\nu \ge 1$ there are no eigenvalues.
For $0 \le \nu < 1$ the Friedrichs case is given by $A_1 = 1, A_2 = 0$; there are no eigenvalues.
For $\nu = 0.45$ and $A_1 = 10, A_2 = -1$ there is one isolated eigenvalue near to the value $-175.57$.

One of the interesting features of this Liouville form of the Bessel equation is that it is possible to choose purely imaginary values of the order $\nu$ of the Bessel function solutions. If $\nu = ik$, with $k \in \mathbb{R}$, then the Liouville form of the equation becomes

$$-y''(x) - \left(k^2 + 1/4\right) x^{-2} y(x) = \lambda y(x) \text{ for all } x \in (0,+\infty)$$

with solutions

$$x^{1/2} J_{ik}(x\sqrt{\lambda}) \text{ and } x^{1/2} Y_{ik}(x\sqrt{\lambda}).$$

This differential equation is considered below in Section 44 under the name of the Rellich equation.

## 13. Bessel equation: form 1

This special case of the Bessel equation is

$$-y''(x) - xy(x) = \lambda y(x) \text{ for all } x \in [0,\infty).$$

This differential equation has explicit solutions in terms of Bessel functions of order $1/3$; see [1, Chapter 10, Section 10.4], [28, Section 3], [29, Section 4] and [79, Chapter IV, Section 4.13].

Endpoint classification in $L^2(0,+\infty)$:

| Endpoint | Classification |
|---|---|
| 0 | R |
| $+\infty$ | LP |

## 14. Bessel equation: form 2

This special case of the Bessel equation is

$$-(x^{\beta}y'(x))' = \lambda x^{\alpha}y(x) \text{ for all } x \in (0, \infty)$$

with the parameters $\alpha > -1$ and $\beta < 1$. This differential equation has solutions of the form, see [52, Section C, Equation 2.162 (1a)] and [35, Section 2.3],

$$y(x, \lambda) = x^{\frac{1}{2}(1-\beta)} Z_{\nu}\left(k^{-1}x^{k}\sqrt{\lambda}\right) \text{ for all } x \in (0, \infty) \text{ and all } \lambda \in \mathbb{C}$$

where the real parameters $\nu$ and $k$ are defined by

$$\nu := (1 - \beta)/(\alpha - \beta + 2) \text{ and } k := \tfrac{1}{2}(\alpha - \beta + 2),$$

and $Z_{\nu}$ is any Bessel function, $J_{\nu}, Y_{\nu}, H_{\nu}^{(1)}, H_{\nu}^{(2)}$, of order $\nu$. A calculation shows that with the given restrictions on $\alpha$ and $\beta$ we have

$$0 < \nu < 1 \text{ and } k > 0.$$

Endpoint classification in $L^2((0, +\infty); x^{\alpha})$, for all $\alpha$ and $\beta$ as above,

| Endpoint | Classification |
|---|---|
| 0 | R |
| $+\infty$ | LP |

## 15. Bessel equation: form 3

This special case of the Bessel equation is

$$-(x^{\tau}y'(x))' = \lambda y(x) \text{ for all } x \in [1, \infty).$$

where the real parameter $\tau \in (-\infty, \infty)$. This differential equation has solutions of the form, see [52, Section C, Equation 2.162] and [32, Section 5],

$$y(x, \lambda) = x^{\frac{1}{2}(1-\tau)} Z_{v}\left(2(2-\tau)^{-1}x^{\frac{1}{2}(2-\tau)}\sqrt{\lambda}\right) \text{ for all } x \in [1, \infty) \text{ and all } \lambda \in \mathbb{C}$$

where the real parameter $\nu$ is defined by

$$\nu := (1 - \tau)/(2 - \tau) \ ,$$

and where $Z_{\nu}$ is any Bessel function, $J_{\nu}, Y_{\nu}, H_{\nu}^{(1)}, H_{\nu}^{(2)}$, of order $\nu$. The case when $\tau = 2$ requires special attention; the solutions can then be expressed in elementary terms.

Endpoint classification in $L^2(1, \infty)$, for all $\tau \in (-\infty, \infty)$,

| Endpoint | Classification |
|---|---|
| 1 | R |
| $+\infty$ | LP |

## 16. Bessel equation: form 4

This special case of the Bessel equation is, with $a > 0$,

$$-y''(x) + (\nu^2 - 1/4)x^{-2}y(x) = \lambda y(x) \text{ for all } x \in [a, \infty).$$

This equation is a special case of the Liouville form of the Bessel differential equation, see Section 12 above, with the parameter $\nu \geq 0$ and considered on the interval $[a, \infty)$ to avoid the singularity at the endpoint 0. The reason for this choice of endpoint is to relate to the Weber integral transform as considered in [79, Chapter IV, Section 4.10]. As in Section 12 above this equation has solutions

$$x^{1/2}J_\nu(x\sqrt{\lambda}) \text{ and } x^{1/2}Y_\nu(x\sqrt{\lambda}) \text{ for all } x \in [a, \infty)$$

but now the endpoint classification in $L^2[a, \infty)$ is

| Endpoint | Classification |
|---|---|
| $a$ | R |
| $+\infty$ | LP |

## 17. Bessel equation: modified form

The modified Bessel functions, notation $I_\nu$ and $K_\nu$, are best defined, on the real line $\mathbb{R}$, in terms of the classical Bessel functions $J_\nu$ and $Y_\nu$ by, see [16, Chapter 4, Section 4.7],

$$I_\nu(x) := i^{-\nu}J_\nu(ix) \text{ and } K_\nu(x) := \frac{\pi}{2}i^{\nu+1}\{J_\nu(ix) + iY_\nu(ix)\} \text{ for all } x \in \mathbb{R}.$$

The properties of these special functions are considered in [16, Chapter 4, Section 4.7 to 4.9].

With careful attention to the branch definition of the powers of the factors $i^\nu$ it may be shown that

$$I_\nu(\cdot) : \mathbb{R} \to \mathbb{R} \text{ and } K_\nu(\cdot) : \mathbb{R} \to \mathbb{R}.$$

The functions $I_\nu(x\sqrt{\lambda})$ and $K_\nu(x\sqrt{\lambda})$ form an independent basis of solutions for the differential equation

$$(xy'(x))' - \nu^2 x^{-1}y(x) = \lambda xy(x) \text{ for all } x \in (0, \infty)$$

and have properties similar to the classical Bessel functions $J_\nu(x\sqrt{\lambda})$ and $Y_\nu(x\sqrt{\lambda})$, respectively, when $x \in \mathbb{R}$ and $\lambda \in \mathbb{C}$.

If the Liouville transformation is applied to this last equation, or if in the Bessel Liouville differential equation of Section 12 above, the formal transformation $x \longmapsto ix$ is applied, then the resulting differential equation has the form

$$y''(x) - \left(\nu^2 - 1/4\right)x^{-2}y(x) = \lambda y(x) \text{ for all } x \in (0, \infty).$$

This gives one interesting property of the Liouville form of the differential equation for the modified Bessel functions: in the standard Sturm-Liouville form given in

Section 3 above the leading coefficient $p$ has to be taken as negative-valued on the interval $(0, \infty)$, *i.e.*,

$$p(x) = -1 \qquad q(x) = -\left(\nu^2 - 1/4\right) x^{-2} \qquad w(x) = 1 \text{ for all } x \in (0, \infty).$$

The independent solutions of this Liouville form are

$$x^{1/2} I_\nu(x\sqrt{\lambda}) \text{ and } x^{1/2} K_\nu(x\sqrt{\lambda}) \text{ for all } x \in (0, \infty).$$

Endpoint classification in $L^2(0, +\infty)$:

| Endpoint | Parameter $\nu$ | Classification |
|---|---|---|
| 0 | For $\nu = 1/2$ | R |
| 0 | For all $\nu \in [0, 1)$ but $\nu \neq 1/2$ | LCNO |
| 0 | For all $\nu \in [1, \infty)$ | LP |
| $+\infty$ | For all $\nu \in [0, \infty)$ | LP |

For endpoint 0 and $\nu \in [0, 1)$ but $\nu \neq 1/2$, the LCNO boundary condition functions $u, v$ are determined, for all $x \in (0, +\infty)$, by

| Parameter | $u$ | $v$ |
|---|---|---|
| $\nu \in (0, 1)$ but $\nu \neq 1/2$ | $x^{\nu+1/2}$ | $x^{-\nu+1/2}$ |
| $\nu = 0$ | $x^{1/2}$ | $x^{1/2} \ln(x)$ |

## 18. Airy equation

The Airy differential equation, in Sturm-Liouville form, is

$$-y''(x) + xy(x) = \lambda y(x) \text{ for all } x \in \mathbb{R}.$$

The solutions of this equation can be expressed in terms of the Bessel functions $J_{1/3}$ and $J_{-1/3}$, or in terms of the Airy functions $\mathrm{Ai}(\cdot)$ and $\mathrm{Bi}(\cdot)$. For a detailed study of the properties of these functions see [1, Chapter 10, Section 10.4]; see also the results in [29, Section 5].

Endpoint classification in $L^2(-\infty, \infty)$:

| Endpoint | Classification |
|---|---|
| $-\infty$ | LP |
| $+\infty$ | LP |

The spectrum of the boundary value problem on the interval $(-\infty, \infty)$ has no eigenvalues and is continuous on the real line in $\mathbb{C}$; the spectrum for any problem on the interval $[0, \infty)$ is discrete.

## 19. Legendre equation

The standard form for this differential equation is, see [81, Chapter XV],

$$-\left(\left(1-x^2\right)y'(x)\right)'+\tfrac{1}{4}y(x)=\lambda y(x) \text{ for all } x\in(-1,+1);$$

see also [1, Chapter 8], [27, Volume I, Chapter III], [16, Chapter 3] and [2, Volume II, Appendix 2, Section 132, Part 3].

Endpoint classification in $L^2(-1,+1)$:

| Endpoint | Classification |
|---|---|
| $-1$ | LCNO |
| $+1$ | LCNO |

For both endpoints the boundary condition functions $u, v$ are given by (note that $u$ and $v$ are solutions of the Legendre equation for $\lambda = 1/4$)

$$u(x)=1 \qquad v(x)=\frac{1}{2}\ln\left(\frac{1+x}{1-x}\right) \text{ for all } x\in(-1,+1).$$

(i) The Legendre polynomials are obtained by taking the principal (Friedrichs) boundary condition at both endpoints $\pm 1$: enter $A_1 = 1, A_2 = 0,\ B_1 = 1, B_2 = 0$; *i.e.*, take the boundary condition function $u$ at $\pm 1$; eigenvalues: $\lambda_n = (n+1/2)^2$ ; $n = 0, 1, 2, \ldots$; eigenfunctions: Legendre polynomials $P_n(x)$.

(ii) Enter $A_1 = 0,\ A_2 = 1,\ B_1 = 0,\ B_2 = 1$, *i.e.*, use the boundary condition function $v$ at $\pm 1$; eigenvalues: $\mu_n$; $n = 0, 1, 2, \ldots$ but no explicit formula is available; eigenfunctions are logarithmically unbounded at $\pm 1$.

(iii) Observe that $\mu_n < \lambda_n < \mu_{n+1}$; $n = 0, 1, 2, \ldots$.

The Liouville normal form of the Legendre differential equation is

$$-y''(x)+\tfrac{1}{4}\sec^2(x)y(x)=\lambda y(x) \text{ for all } x\in\left(-\tfrac{1}{2}\pi,\tfrac{1}{2}\pi\right);$$

this form of the equation is studied in detail in [79, Chapter IV, Sections 4.5 to 4.7].

## 20. Legendre equation: associated form

This Sturm-Liouville differential equation is an extension of the classical Legendre equation of Section 19:

$$-\left(\left(1-x^2\right)y'(x)\right)'+\frac{\mu^2}{1-x^2}y(x)=\lambda y(x) \text{ for all } x\in(-1,+1)$$

where the parameter $\mu \in [0, \infty)$; see [1, Chapter 8], [27, Volume I, Chapter III], [16, Chapter 3, Section 3.9] and [79, Chapter IV, Section 4.3].

Endpoint classification in $L^2(-1,+1)$:

| Endpoint | Parameter | Classification |
|---|---|---|
| $-1$ | $0 \leq \mu < 1$ | LCNO |
| $-1$ | $1 \leq \mu$ | LP |

| Endpoint | Parameter | Classification |
|---|---|---|
| $+1$ | $0 \leq \mu < 1$ | LCNO |
| $+1$ | $1 \leq \mu$ | LP |

For the endpoint $-1$ and for the LCNO cases the boundary condition functions $u, v$ are determined by

| Parameter | $u$ | $v$ |
|---|---|---|
| $\mu = 0$ | $1$ | $\ln\left(\dfrac{1+x}{1-x}\right)$ |
| $0 < \mu < 1$ | $(1-x^2)^{\mu/2}$ | $(1-x^2)^{-\mu/2}$ |

For the endpoint $+1$ and for the LCNO cases the boundary condition functions $u, v$ are determined by

| Parameter | $u$ | $v$ |
|---|---|---|
| $\mu = 0$ | $1$ | $\ln\left(\dfrac{1+x}{1-x}\right)$ |
| $0 < \mu < 1$ | $(1-x^2)^{\mu/2}$ | $(1-x^2)^{-\mu/2}$ |

If the spectral parameter $\lambda$ is written as $\lambda = \nu(\nu+1)$ then the solutions of this modified Legendre equation are the associated Legendre functions $P_\nu^\mu(x)$ and $Q_\nu^\mu(x)$ for $x \in (-1,+1)$; see [1, Chapter 8] and [16, Chapter 3, Section 3.9].

## 21. Hermite equation

The most elegant Sturm-Liouville form for this differential equation is

$$-(\exp(-x^2)y'(x))' = \lambda \exp(-x^2)y(x) \text{ for all } x \in (-\infty, \infty).$$

For all $n \in \mathbb{N}_0 = \{0, 1, 2, \ldots\}$ and for $\lambda = 2n + 1$ this equation has the Hermite polynomials $H_n$ for solutions. These polynomials are orthogonal and complete in the Hilbert function space $L^2((-\infty,\infty);\exp(-x^2))$.

Endpoint classification in $L^2((-\infty,\infty);\exp(-x^2))$:

| Endpoint | Classification |
|---|---|
| $-\infty$ | LP |
| $+\infty$ | LP |

## 22. Hermite equation: Liouville form

The Liouville transformation applied to the Hermite differential equation gives

$$-y''(x) + x^2 y(x) = \lambda y(x) \text{ for all } x \in (-\infty, +\infty).$$

For all $n \in \mathbb{N}_0 = \{0, 1, 2, \ldots\}$ and for $\lambda = 2n + 1$ this equation has the Hermite functions $\exp(-\frac{1}{2}x^2)H_n$ for solutions. These functions are orthogonal and complete in the Hilbert function space $L^2(-\infty, \infty)$.

Endpoint classification in $L^2(-\infty, +\infty)$:

| Endpoint | Classification |
|---|---|
| $-\infty$ | LP |
| $+\infty$ | LP |

For a classical treatment see [79, Chapter IV, Section 2].

This differential equation is also called the harmonic oscillator equation; see example 15 in the list to be found within the LaTeX file xamples.tex contained in the package associated with the publication [11, Data base file xamples.tex; example 15].

This differential equation is also considered under the name of the parabolic cylinder equation; see [1, Chapter 19].

## 23. Jacobi equation

The general form of the Jacobi differential equation is

$$-\left((1-x)^{\alpha+1}(1+x)^{\beta+1}y'(x)\right)' = \lambda(1-x)^{\alpha}(1+x)^{\beta}y(x) \text{ for all } x \in (-1, +1),$$

where the parameters $\alpha, \beta \in (-\infty, +\infty)$. Apart from an isomorphic transformation of the independent variable this differential equation coincides with the Sturm-Liouville form of the hypergeometric equation considered in Section 9 above.

Endpoint classification in the weighted space $L^2((-1, +1); (1-x)^{\alpha}(1+x)^{\beta}))$:

| Endpoint | Parameter | Classification |
|---|---|---|
| $-1$ | $\beta \le -1$ | LP |
| $-1$ | $-1 < \beta < 0$ | R |
| $-1$ | $0 \le \beta < 1$ | LCNO |
| $-1$ | $1 \le \beta$ | LP |

| Endpoint | Parameter | Classification |
|---|---|---|
| $+1$ | $\alpha \le -1$ | LP |
| $+1$ | $-1 < \alpha < 0$ | R |
| $+1$ | $0 \le \alpha < 1$ | LCNO |
| $+1$ | $1 \le \alpha$ | LP |

For the endpoint $-1$ and for the LCNO cases the boundary condition functions $u, v$ are determined by

| Parameter | $u$ | $v$ |
|---|---|---|
| $\beta = 0$ | 1 | $\ln\left(\dfrac{1+x}{1-x}\right)$ |
| $0 < \beta < 1$ | 1 | $(1+x)^{-\beta}$ |

For the endpoint $+1$ and for the LCNO cases the boundary condition functions $u, v$ are determined by

| Parameter | $u$ | $v$ |
|---|---|---|
| $\alpha = 0$ | 1 | $\ln\left(\dfrac{1+x}{1-x}\right)$ |
| $0 < \alpha < 1$ | 1 | $(1-x)^{-\alpha}$ |

To obtain the classical Jacobi orthogonal polynomials it is necessary to take $-1 < \alpha,\ \beta$; then note the required boundary conditions:

Endpoint $-1$:

| Parameter | Boundary condition |
|---|---|
| $-1 < \beta < 0$ | $(py')(-1) = 0$ or $[y, v](-1) = 0$ |
| $0 \leq \beta < 1$ | $[y, u](-1) = 0$ |

Endpoint $+1$:

| Parameter | Boundary condition |
|---|---|
| $-1 < \alpha < 0$ | $(py')(+1) = 0$ or $[y, v](+1) = 0$ |
| $0 \leq \alpha < 1$ | $[y, u](+1) = 0$ |

For the classical Jacobi orthogonal polynomials the eigenvalues are given by:

$$\lambda_n = n(n + \alpha + \beta + 1) \text{ for } n = 0, 1, 2, \ldots$$

and this explicit formula can be used to give an independent check on the accuracy of the results from the SLEIGN2 code.

It is interesting to note that the required boundary condition for these Jacobi polynomials is the Friedrichs condition in the LCNO cases.

In addition to the cases of the Jacobi equation mentioned in this section, there are other values of the parameters $\alpha$ and $\beta$ which lead to important Sturm-Liouville differential equations; see the paper [53] and the book [3].

## 24. Jacobi equation: Liouville form

The Liouville transformation applied to the Jacobi differential equation gives

$$-y''(x) + q(x)y(x) = \lambda y(x) \quad \text{for all} \quad x \in (-\pi/2, +\pi/2)$$

where the coefficient $q$ is given by, for all $x \in (-\pi/2, +\pi/2)$,

$$\begin{aligned} q(x) &= \frac{\beta^2 - 1/4}{4\tan^2((2x+\pi)/4)} + \frac{\alpha^2 - 1/4}{4\tan^2((2x-\pi)/4)} - \frac{4\alpha\beta + 4\beta + 4\alpha + 3}{8} \\ &= \frac{\beta^2 - 1/4}{4\sin^2((2x+\pi)/4)} + \frac{\alpha^2 - 1/4}{4\sin^2((2x-\pi)/4)} - \frac{(\alpha+\beta+1)^2}{4}. \end{aligned}$$

Here the parameters $\alpha, \beta \in (-\infty, +\infty)$.

Endpoint classification in the space $L^2(-\pi/2, +\pi/2)$:

| Endpoint | Parameter | Classification |
|---|---|---|
| $-\pi/2$ | $\beta \le -1$ | LP |
| $-\pi/2$ | $-1 < \beta < 1$ but $\beta^2 \neq 1/4$ | LCNO |
| $-\pi/2$ | $\beta^2 = 1/4$ | R |
| $-\pi/2$ | $1 \le \beta$ | LP |

| Endpoint | Parameter | Classification |
|---|---|---|
| $+\pi/2$ | $\alpha \le -1$ | LP |
| $+\pi/2$ | $-1 < \alpha < 1$ but $\alpha^2 \neq 1/4$ | LCNO |
| $+\pi/2$ | $\alpha^2 = 1/4$ | R |
| $+\pi/2$ | $1 \le \alpha$ | LP |

For the endpoint $-\pi/2$ and for LCNO cases the boundary condition functions $u, v$ are determined by, where $b(x) = \pi/2 + x$ for all $x \in (-\pi/2, +\pi/2)$,

| Parameter | $u$ | $v$ |
|---|---|---|
| $-1 < \beta < 0$ | $b(x)^{\frac{1}{2}-\beta}$ | $b(x)^{\frac{1}{2}+\beta}$ |
| $\beta = 0$ | $\sqrt{b(x)}$ | $\sqrt{b(x)}\ln(b(x))$ |
| $0 < \beta < 1$ | $b(x)^{\frac{1}{2}+\beta}$ | $b(x)^{\frac{1}{2}-\beta}$ |

For the endpoint $+\pi/2$ and for LCNO cases the boundary condition functions $u, v$ are determined by, where $a(x) = \pi/2 - x$ for all $x \in (-\pi/2, +\pi/2)$,

| Parameter | $u$ | $v$ |
|---|---|---|
| $-1 < \alpha < 0$ | $a(x)^{\frac{1}{2}-\alpha}$ | $a(x)^{\frac{1}{2}+\alpha}$ |
| $\alpha = 0$ | $\sqrt{a(x)}$ | $\sqrt{a(x)}\ln(a(x))$ |
| $0 < \alpha < 1$ | $a(x)^{\frac{1}{2}+\alpha}$ | $a(x)^{\frac{1}{2}-\alpha}$ |

The classical Jacobi orthogonal polynomials are produced only when both $\alpha, \beta > -1$. For $\alpha, \beta > +1$ the LP condition holds and no boundary condition is required to give the polynomials. If $-1 < \alpha, \beta < 1$ then the LCNO condition holds and boundary conditions are required to produce the Jacobi polynomials; these conditions are as follows:

Endpoint $-\pi/2$:

| Parameter | Boundary condition |
|---|---|
| $-1 < \beta < 0$ | $[y, v](-\pi/2) = 0$ |
| $0 \le \beta < 1$ | $[y, u](-\pi/2) = 0$ |

Endpoint $+\pi/2$:

| Parameter | Boundary condition |
|---|---|
| $-1 < \alpha < 0$ | $[y, v](+\pi/2) = 0$ |
| $0 \leq \alpha < 1$ | $[y, u](+\pi/2) = 0$ |

Recall from Section 23 for the classical orthogonal Jacobi polynomials the eigenvalues are given explicitly by:

$$\lambda_n = n(n + \alpha + \beta + 1) \text{ for } n = 0, 1, 2, \ldots$$

## 25. Jacobi function equation

This is another Jacobi differential equation which corresponds to the hypergeometric differential equation considered over the half-line $[0, \infty)$, see the second equation in Section 9 above, and the paper [37].

This equation is written in the form

$$-(\omega(x)y'(x))' - \rho^2\omega(x)y(x) = \lambda\omega(x)y(x) \text{ for all } x \in (0, \infty)$$

where

(i) $\alpha \geq \beta \geq -1/2$
(ii) $\rho = \alpha + \beta + 1$
(iii) $\omega(x) \equiv \omega(x)_{\alpha,\beta} = 2^{2\rho}(\sinh(x))^{2\alpha+1}(\cosh(x))^{2\beta+1}$ for all $x \in (0, \infty)$.

Endpoint classification, for all $\beta \in [-1/2, \infty)$, in $L^2((0, \infty); \omega)$:

| Endpoint | Parameter $\alpha$ | Classification |
|---|---|---|
| 0 | For $\alpha \in [-1/2, 0)$ | R |
| 0 | For $\alpha \in [0, 1)$ | LCNO |
| 0 | For $\alpha \in [1, \infty)$ | LP |
| $+\infty$ | For all $\alpha \in [-1/2, \infty)$ | LP |

For the endpoint 0, for $\alpha \in [0, 1)$ and for all $\beta \in [1/2, \infty)$ the LCNO boundary condition functions $u, v$ take the form, for all $x \in (0, 1)$,

| Parameter | $u$ | $v$ |
|---|---|---|
| $\alpha = 0$ | 1 | $\ln(x)$ |
| $\alpha \in (0, 1)$ | 1 | $x^{-2\alpha}$ |

## 26. Jacobi function equation: Liouville form

In the Liouville normal form, see Sections 7 and 11 above, the Jacobi function differential equation of Section 25 above appears as

$$-y''(x) + q(x)y(x) = \lambda y(x) \text{ for all } x \in (0, \infty),$$

where the coefficient $q$ is determined, again with $\alpha \geq \beta \geq -1/2$, by

$$q(x) = \frac{\alpha^2 - 1/4}{(\sinh(x))^2} - \frac{\beta^2 - 1/4}{(\cosh(x))^2} \text{ for all } x \in (0, \infty).$$

Endpoint classification, for all $\beta \in [-1/2, \infty)$, in $L^2(0, \infty)$:

| Endpoint | Parameter $\alpha$ | Classification |
|---|---|---|
| 0 | For $\alpha = -1/2$ | R |
| 0 | For $\alpha \in (-1/2, 1/2)$ | LCNO |
| 0 | For $\alpha = 1/2$ | R |
| 0 | For $\alpha \in (1/2, 1)$ | LCNO |
| 0 | For $\alpha \in [1, \infty)$ | LP |
| $+\infty$ | For all $\alpha \in [-1/2, \infty)$ | LP |

For the endpoint 0, for $\alpha \in [-1/2, 1)$ but $|\alpha| \neq 1/2$ and for all $\beta \in [-1/2, \infty)$ the LCNO boundary condition functions $u, v$ take the form, for all $x \in (0, 1)$,

| Parameter | $u$ | $v$ |
|---|---|---|
| $\alpha = 0$ | $x^{1/2}$ | $x^{1/2} \ln(x)$ |
| $\alpha \in [-1/2, 1)$ but $|\alpha| \neq 1/2$ | $x^{|\alpha|+1/2}$ | $x^{-|\alpha|+1/2}$ |

## 27. Laguerre equation

The general form of the Laguerre differential equation is

$$-(x^{\alpha+1} \exp(-x) y'(x))' = \lambda x^{\alpha} \exp(-x) y(x) \text{ for all } x \in (0, +\infty)$$

where the parameter $\alpha \in (-\infty, +\infty)$.

Endpoint classification in the weighted space $L^2((0, +\infty); x^{\alpha} \exp(-x))$:

| Endpoint | Parameter | Classification |
|---|---|---|
| 0 | $\alpha \leq -1$ | LP |
| 0 | $-1 < \alpha < 0$ | R |
| 0 | $0 \leq \alpha < 1$ | LCNO |
| 0 | $1 \leq \alpha$ | LP |
| $+\infty$ | $\alpha \in (-\infty, +\infty)$ | LP |

For these LCNO cases the boundary condition functions $u, v$ are given by:

| Endpoint | Parameter | $u$ | $v$ |
|---|---|---|---|
| 0 | $\alpha = 0$ | 1 | $\ln(x)$ |
| 0 | $0 < \alpha < 1$ | 1 | $x^{-\alpha}$ |

This is the classical form of the differential equation which for parameter $\alpha > -1$ produces the classical Laguerre polynomials as eigenfunctions; for the boundary condition $[y, 1](0) = 0$ at 0, when required, the eigenvalues are then (remarkably!) independent of $\alpha$ and given by $\lambda_n = n$ $(n = 0, 1, 2, \ldots)$; see [1, Chapter 22, Section 22.6].

See also the Kummer differential equation given in Section 10 above.

## 28. Laguerre equation: Liouville form

The Liouville transformation applied to the Laguerre differential equation gives

$$-y''(x) + \left(\frac{\alpha^2 - 1/4}{x^2} - \frac{\alpha+1}{2} + \frac{x^2}{16}\right) y(x) = \lambda y(x) \text{ for all } x \in (0, +\infty)$$

where the parameter $\alpha \in (-\infty, +\infty)$.

Endpoint classification in the space $L^2(0, +\infty)$:

| Endpoint | Parameter | Classification |
|---|---|---|
| 0 | $\alpha \leq -1$ | LP |
| 0 | $-1 < \alpha < 1$, but $\alpha^2 \neq 1/4$ | LCNO |
| 0 | $\alpha^2 = 1/4$ | R |
| 0 | $1 \leq \alpha$ | LP |
| $+\infty$ | $\alpha \in (-\infty, +\infty)$ | LP |

For these LCNO cases the boundary condition functions $u, v$ are given by:

| Endpoint | Parameter | $u$ | $v$ |
|---|---|---|---|
| 0 | $-1 < \alpha < 0$ but $\alpha \neq -1/2$ | $x^{\frac{1}{2}-\alpha}$ | $x^{\frac{1}{2}+\alpha}$ |
| 0 | $\alpha = -1/2$ | $x$ | 1 |
| 0 | $\alpha = 0$ | $x^{1/2}$ | $x^{1/2}\ln(x)$ |
| 0 | $0 < \alpha < 1$ but $\alpha \neq 1/2$ | $x^{\frac{1}{2}+\alpha}$ | $x^{\frac{1}{2}-\alpha}$ |
| 0 | $\alpha = 1/2$ | $x$ | 1 |

The Laguerre polynomials are produced as eigenfunctions only when $\alpha > -1$. For $\alpha \geq 1$ the LP condition holds at 0. For $0 \leq \alpha < 1$ the appropriate boundary condition is the Friedrichs condition: $[y, u](0) = 0$; for $-1 < \alpha < 0$ use the non-Friedrichs condition: $[y, v](0) = 0$. In all these cases $\lambda_n = n$ for $n = 0, 1, 2, \ldots$.

## 29. Heun equation

One Sturm-Liouville form of the general Heun differential equation is

$$-(py')' + qy = \lambda wy \text{ on } (0, 1)$$

where the coefficients $p, q, w$ are given explicitly, for all $x \in (0, 1)$, by

$$\begin{aligned} p(x) &= x^c(1-x)^d(x+s)^e \\ q(x) &= abx^c(1-x)^{d-1}(x+s)^{e-1} \\ w(x) &= x^{c-1}(1-x)^{d-1}(x+s)^{e-1}. \end{aligned}$$

The parameters $a, b, c, d, e$ and $s$ are all real numbers and satisfy the following two conditions

(i) $s > 0$ and $c \geq 1, d \geq 1, a \geq b$,

and

(ii) $a + b + 1 - c - d - e = 0$.

From these conditions it follows that

$$a \geq 1, b \geq 1, e \geq 1 \text{ and } a + b - d \geq 1.$$

The differential equation above is a special case of the general Heun equation

$$\frac{d^2w(z)}{dz^2} + \left(\frac{\gamma}{z} + \frac{\delta}{z-1} + \frac{\varepsilon}{z-a}\right)\frac{dw(z)}{dz} + \frac{\alpha\beta z - q}{z(z-1)(z-a)}w(z) = 0$$

with the general parameters $\alpha, \beta, \gamma, \delta, \varepsilon$ replaced by the real numbers $a, b, c, d, e$, $a$ replaced by $-s$, and $q$ replaced by the spectral parameter $\lambda$. For general information concerning the Heun equation see [27, Chapter XV, Section 15.3], [77] and the compendium [74]; for the special form of the Heun equation considered here, and for the connection with confluence of singularities and applications, see the recent paper [55].

We note that the coefficients of the Sturm-Liouville differential equation above satisfy the conditions

1. $q, w \in C[0,1]$ and $w(x) > 0$ for all $x \in (0,1)$
2. $p^{-1} \in L^1_{\text{loc}}(0,1), p(x) > 0$ for all $x \in (0,1)$
3. $p^{-1} \notin L^1(0,1/2]$ and $p^{-1} \notin L^1[1/2,1)$.

Thus both endpoints 0 and 1 are singular for the differential equation. Analysis shows that the endpoint classification for this equation is

| Endpoint | Parameter | Classification |
|---|---|---|
| 0 | $c \in [1,2)$ | LCNO |
| 0 | $c \in [2,+\infty)$ | LP |
| 1 | $d \in [1,2)$ | LCNO |
| 1 | $d \in [2,+\infty)$ | LP |

For the endpoint 0 and for LCNO cases the boundary condition functions $u, v$ are determined by:

| Parameter | $u$ | $v$ |
|---|---|---|
| $c = 1$ | 1 | $\ln(x)$ |
| $1 < c < 2$ | 1 | $x^{1-c}$ |

For the endpoint 1 and for LCNO cases the boundary condition functions $u, v$ are determined by:

| Parameter | $u$ | $v$ |
|---|---|---|
| $d = 1$ | 1 | $\ln(1-x)$ |
| $1 < d < 2$ | 1 | $(1-x)^{1-d}$ |

Further it may be shown that the spectrum of any self-adjoint problem on $(0,1)$, with the parameters $a, b, c, d, e$ and $s$ satisfying the above conditions, and considered in the space $L^2((0,1); w)$ with either separated or coupled boundary conditions, is bounded below and discrete. For the analytic properties, and proofs of the spectral properties of this Heun differential equation, see the paper [7].

## 30. Whittaker equation

The general form of the Whittaker differential equation is

$$-y''(x) + \left(\frac{1}{4} + \frac{k^2-1}{x^2}\right) y(x) = \lambda \frac{1}{x} y(x) \text{ for all } x \in (0, +\infty)$$

where the parameter $k \in [1, +\infty)$.

Endpoint classification in the space $L^2((0,+\infty); x^{-1})$, for all $k \in [1, +\infty)$:

| Endpoint | Classification |
|---|---|
| 0 | LP |
| $+\infty$ | LP |

This equation is studied in [49, Part II, Section 10], where there it is shown that the LP case holds at $+\infty$ and also at 0 for $k \geq 1$; the general properties of Whittaker functions are given in [1, Chapter 13, Section 13.1.31]. The spectrum of the boundary value problem on $(0, \infty)$ is discrete and is given explicitly by:

$$\lambda_n = n + (k+1)/2, \quad n = 0, 1, 2, 3, \ldots.$$

## 31. Lamé equation

This differential equation has many forms; there is an extensive literature devoted to the definition, theory and properties of this equation and the associated Lamé functions; see [81, Chapter XXIII, Section 23.4] and [27, Chapter XV, Section 15.2]. The Lamé equation is a special case of the Heun equation; see [27, Chapter XV, Section 15.3] and Section 29 above.

Here we consider two cases of the Lamé equation involving the Weierstrass doubly periodic elliptic function $\wp$, considered for the special case when the fundamental periods $2\omega_1$ and $2\omega_2$ of $\wp$ satisfy

$$\omega_1 \in (0, \infty) \text{ and } \omega_2 = i\chi \text{ where } \chi \in (0, \infty).$$

We note that the lattice of double poles for $\wp$ is rectangular with points $[2m\omega_1 + 2n\omega_2 : m, n \in \mathbb{Z}\}$ of $\mathbb{C}$.

For the general theory of the Weierstrass elliptic function $\wp$ see [22, Chapter XIII] and [81, Chapter XX].

1. Consider the Sturm-Liouville differential equation

$$-y''(x) + k\wp(x)y(x) = \lambda y(x) \text{ for all } x \in (0, 2\omega_1)$$

where $k$ is a real parameter, $k \in \mathbb{R}$. We note that $\wp(\cdot) : (0, 2\omega_1) \to \mathbb{R}$ and that $\wp(\cdot) \in L^1_{\text{loc}}(0, 2\omega_1)$; from [81, Chapter XX, Section 20.2] and [22, Chapter XIII, Section 13.4] it follows that

$$\wp(x) = x^{-2} + O(x^2) \text{ as } x \to 0^+ \text{ and}$$
$$\wp(x) = (2\omega_1 - x)^{-2} + O((2\omega_1 - x)^2) \text{ as } x \to 2\omega_1^-.$$

These order results for the coefficient $\wp$, at the endpoints of the interval $(0, 2\omega_1)$, taken together with the parameter $k$, allow of a comparison between this form of the Lamé equation and $(a)$ the Liouville Bessel equation of Section 12 when $k \in [-1/4, +\infty)$, and $(b)$ the Rellich equation of Section 44 when $k \in (-\infty, -1/4)$. This comparison leads to the following endpoint classification for this Lamé equation in the space $L^2(0, 2\omega_1)$ :

| Endpoint | Parameter $k$ | Classification |
|---|---|---|
| 0 | For $k \in (-\infty, -1/4)$ | LCO |
| 0 | For $k \in [-1/4, 0)$ | LCNO |
| 0 | $k = 0$ | R |
| 0 | For $k \in (0, 3/4)$ | LCNO |
| 0 | For $k \in [3/4, \infty)$ | LP |

and

| Endpoint | Parameter $k$ | Classification |
|---|---|---|
| $2\omega_1$ | For $k \in (-\infty, -1/4)$ | LCO |
| $2\omega_1$ | For $k \in [-1/4, 0)$ | LCNO |
| $2\omega_1$ | $k = 0$ | R |
| $2\omega_1$ | For $k \in (0, 3/4)$ | LCNO |
| $2\omega_1$ | For $k \in [3/4, \infty)$ | LP |

The boundary condition functions $u, v$ for the LCO and LCNO classifications at the endpoint zero can be copied from the corresponding cases for the Liouville Bessel equation in Section 12, and for the Rellich equation Section 44; similarly for the endpoint $2\omega_1$.

2. Consider the Sturm-Liouville differential equation

$$-y''(x) + k\wp(x + \omega_2)y(x) = \lambda y(x) \text{ for all } x \in (-\infty, +\infty)$$

where $k$ is a real parameter, $k \in \mathbb{R}$.

From the information about the fundamental periods $2\omega_1$ and $2\omega_2$ given in case 1 above, it follows that $\wp(\cdot + \omega_2)$ is real-valued, periodic with period $2\omega_1$, and real-analytic on $\mathbb{R}$, see [1, Chapter 18, Section 18.1]; see also the corresponding case for the algebro-geometric form 3 differential equation, in Subsection 40.3.

Endpoint classification in $L^2(-\infty, +\infty)$ for all $k \in (-\infty, 0) \cup (0, +\infty)$:

| Endpoint | Classification |
|---|---|
| $-\infty$ | LP |
| $+\infty$ | LP |

This differential equation is of the Mathieu type, see Section 32 below, and the general properties of Sturm-Liouville differential equations with periodic coefficients given in [26, Chapter 2].

## 32. Mathieu equation

The general form of the Mathieu differential equation is

$$-y''(x) + 2k\cos(2x)y(x) = \lambda y(x) \text{ for all } x \in (-\infty, +\infty),$$

where that parameter $k \in (-\infty, 0) \cup (0, +\infty)$.

Endpoint classification in $L^2(-\infty, +\infty)$, for all $k \in (-\infty, 0) \cup (0, +\infty)$:

| Endpoint | Classification |
|---|---|
| $-\infty$ | LP |
| $+\infty$ | LP |

The classical Mathieu equation has a celebrated history and voluminous literature; for the general properties of the Mathieu functions see [1, Chapter 20], [26, Chapter 2, Section 2.5], [27, Volume **III**, Chapter XVI, Section 16.2] and [81, Chapter XIX, Sections 19.1 and 19.2]. For the general properties of Sturm-Liouville differential equations with periodic coefficients see the text [26].

There are no eigenvalues for this problem on $(-\infty, +\infty)$. There may be one negative eigenvalue of the problem on $[0, \infty)$ depending on the boundary condition at the endpoint 0. The continuous (essential) spectrum is the same for the whole line or half-line problems and consists of an infinite number of disjoint closed intervals. The endpoints of these – and thus the spectrum of the problem – can be characterized in terms of periodic and semi-periodic eigenvalues of Sturm-Liouville problems on the compact interval $[0, 2\pi]$; these can be computed with SLEIGN2.

The spectrum depends quantitatively but not qualitatively upon the parameter $k$.

The above remarks also apply to the general Sturm-Liouville equation with periodic coefficients of the same period, the so-called Hill's equation.

Of special interest is the starting point of the continuous spectrum – this is also the oscillation number of the equation. For the Mathieu equation ($p = 1, q = \cos(2x), w = 1$) on both the whole line and the half-line it is approximately -0.378; this result may be obtained by computing the first eigenvalue $\lambda_0$ of the periodic problem on the interval $[0, 2\pi]$.

For extensions of this theory to Sturm-Liouville differential equations with almost periodic coefficients see the paper [59] and the text [66].

## 33. Bailey equation

The general form of the Bailey differential equation, see [11, Data base file xamples.tex; example 7], is

$$-(xy'(x))' - x^{-1}y(x) = \lambda y(x) \text{ for all } x \in (-\infty, 0) \cup (0, +\infty).$$

Endpoint classification in $L^2(-\infty,0)\cup L^2(0,+\infty)$:

| Endpoint | Classification |
|---|---|
| $-\infty$ | LP |
| $0-$ | LCO |
| $0+$ | LCO |
| $+\infty$ | LP |

For both endpoints $0-$ and $0+$:

$$u(x)=\cos(\ln(|x|)) \qquad v(x)=\sin(\ln(|x|)) \text{ for all } x\in(-\infty,0)\cup(0,+\infty).$$

This example is based on the earlier studied Sears-Titchmarsh equation; see Section 58 below.

For numerical results see [11, Data base file xamples.tex; example 7].

## 34. Behnke-Goerisch equation

The general form of the Behnke-Goerisch differential equation, see [11, Data base file xamples.tex; example 28], is

$$-y''(x)+k\cos^2(x)y(x)=\lambda y(x) \text{ for all } x\in(-\infty,+\infty)$$

where the parameter $k\in(-\infty,+\infty)$,

Endpoint classification in the space $L^2(-\infty,+\infty)$, for all $k\in(-\infty,+\infty)$:

| Endpoint | Classification |
|---|---|
| $-\infty$ | LP |
| $+\infty$ | LP |

This is a form of the Mathieu equation. In [15] the authors computed a number of Neumann eigenvalues of this equation, on certain compact intervals, using interval arithmetic to obtain rigorous bounds for the eigenvalues.

## 35. Boyd equation

The general form of the Boyd equation, see [11, Data base file xamples.tex; example 4], is

$$-y''(x)-x^{-1}y(x)=\lambda y(x) \text{ for all } x\in(-\infty,0)\cup(0,+\infty).$$

Endpoint classification in $L^2(-\infty,0)\cup L^2(0,+\infty)$:

| Endpoint | Classification |
|---|---|
| $-\infty$ | LP |
| $0-$ | LCNO |
| $0+$ | LCNO |
| $+\infty$ | LP |

For both endpoints $0-$ and $0+$

$$u(x)=x \qquad v(x)=x\ln(|x|) \text{ for all } x\in(-\infty,0)\cup(0,+\infty).$$

This equation arises in a model studying eddies in the atmosphere; see [18]. There is no explicit formula for the eigenvalues of any particular boundary condition; eigenfunctions can be given in terms of Whittaker functions; see [8, Example 3].

## 36. Boyd equation: regularized

The form of this regularized Boyd equation is

$$-(p(x)y'(x))' + q(x)y(x) = \lambda w(x)y(x) \text{ for all } x \in (-\infty, 0) \cup (0, +\infty)$$

where

$$p(x) = r(x)^2 \quad q(x) = -r(x)^2 \left(\ln(|x|)\right)^2 \quad w(x) = r(x)^2$$

with

$$r(x) = \exp\left(-(x\ln(|x|) - x)\right) \text{ for all } x \in (-\infty, 0) \cup (0, +\infty).$$

Endpoint classification in $L^2((-\infty, 0); w) \cup L^2((0, +\infty); w)$:

| Endpoint | Classification |
|---|---|
| $-\infty$ | LP |
| $0-$ | R |
| $0+$ | R |
| $+\infty$ | LP |

This is a regularized R form of the Boyd equation in Section 35; the LCNO singularity at zero has been made R but requiring the introduction of quasi-derivatives. There is a close relationship between these two forms of the Boyd equation; in particular they have the same eigenvalues – see [4]. For a general discussion of regularization using non-principal solutions see [65]. For numerical results see [8, Example 3].

## 37. Dunford-Schwartz equation

This differential equation is considered in detail in [25, Chapter VIII, Pages 1515–20];

$$-\left((1-x^2)y'(x)\right)' + \left(\frac{2\alpha^2}{(1+x)} + \frac{2\beta^2}{(1-x)}\right) y(x) = \lambda y(x) \text{ for all } x \in (-1, +1)$$

where the independent parameters $\alpha, \beta \in [0, +\infty)$.

Boundary value problems for this differential equation are discussed in [25, Chapter XIII, Section 8].

Endpoint classification in the space $L^2(-1, +1)$ for $-1$:

| Parameter | Classification |
|---|---|
| $0 \leq \alpha < 1/2$ | LCNO |
| $1/2 \leq \alpha$ | LP |

Endpoint classification in the space $L^2(-1,+1)$ for $+1$:

| Parameter | Classification |
|---|---|
| $0 \le \beta < 1/2$ | LCNO |
| $1/2 \le \beta$ | LP |

For the LCNO cases the boundary condition functions $u, v$ are given by

| Endpoint | Parameter | $u$ | $v$ |
|---|---|---|---|
| $-1$ | $\alpha = 0$ | $1$ | $\frac{1}{2}\ln\left(\frac{1+x}{1-x}\right)$ |
| $-1$ | $0 < \alpha < 1/2$ | $(1+x)^{\alpha}$ | $(1+x)^{-\alpha}$ |
| $+1$ | $\beta = 0$ | $1$ | $\frac{1}{2}\ln\left(\frac{1+x}{1-x}\right)$ |
| $+1$ | $0 < \beta < 1/2$ | $(1-x)^{\beta}$ | $(1-x)^{-\beta}$ |

Note that these $u$ and $v$ are not solutions of the differential equation but maximal domain functions.

In the case when $\alpha \in [0,1/2)$ and $\beta \in [0,1/2)$ it is shown in [25, Chapter XIII, Section 8, Page 1519] that the boundary value problem determined by the boundary conditions

$$[y,u](-1) = 0 = [y,u](1)$$

has a discrete spectrum with eigenvalues given by the explicit formula

$$\lambda_n = (n+\alpha+\beta+1)(n+\alpha+\beta) \text{ for } n = 0,1,2,\dots;$$

the eigenfunctions are determined in terms of the hypergeometric function ${}_2F_1$.

## 38. Dunford-Schwartz equation: modified

This modification of the Dunford-Schwartz equation replaces one of the LCNO singularities by a LCO singularity;

$$-\left((1-x^2)y'(x)\right)' + \left(\frac{-2\gamma^2}{(1+x)} + \frac{2\beta^2}{(1-x)}\right)y(x) = \lambda y(x) \text{ for all } x \in (-1,+1)$$

where the independent parameters $\gamma, \beta \in [0,+\infty)$.

Endpoint classification in the space $L^2(-1,+1)$ for $-1$:

| Parameter | Classification |
|---|---|
| $\gamma = 0$ | LCNO |
| $0 < \gamma$ | LCO |

Endpoint classification in the space $L^2(-1,+1)$ for $+1$:

| Parameter | Classification |
|---|---|
| $0 \le \beta < 1/2$ | LCNO |
| $1/2 \le \beta$ | LP |

For these LCNO/LCO cases the boundary condition functions $u, v$ are given by

| Endpoint | Parameter | $u$ | $v$ |
|---|---|---|---|
| $-1$ | $\gamma = 0$ | $1$ | $\frac{1}{2}\ln\left(\frac{1+x}{1-x}\right)$ |
| $-1$ | $0 < \gamma$ | $\cos(\gamma \ln(1+x))$ | $\sin(\gamma \ln(1+x))$ |
| $+1$ | $\beta = 0$ | $1$ | $\frac{1}{2}\ln\left(\frac{1+x}{1-x}\right)$ |
| $+1$ | $0 < \beta < 1/2$ | $(1-x)^{\beta}$ | $(1-x)^{-\beta}$ |

This is a modification of the Dunford-Schwartz equation above, see Section 37, which illustrates an LCNO/LCO mix obtained by replacing $\alpha$ with $i\gamma$; this changes the singularity at $-1$ from LCNO to LCO.

Again these $u$ and $v$ are not solutions of the differential equation but maximal domain functions.

## 39. Hydrogen atom equation

It is convenient to take this equation in two forms:

$$-y''(x) + (kx^{-1} + hx^{-2})y(x) = \lambda y(x) \text{ for all } x \in (0, +\infty) \tag{1}$$

where the two independent parameters $h \in [-1/4, +\infty)$ and $k \in \mathbb{R}$, and

$$-y''(x) + (kx^{-1} + hx^{-2} + 1)y(x) = \lambda y(x) \text{ for all } x \in (0, +\infty) \tag{2}$$

where the two independent parameters $h \in (-\infty, -1/4)$ and $k \in \mathbb{R}$.

Note that form (2) is introduced as a device to aid the numerical computations in the difficult LCO case; it forces the boundary value problem to have a non-negative eigenvalue.

Endpoint classification, for both forms (1) and (2), in $L^2(0, +\infty)$:

| Endpoint | Form | Parameters | Classification |
|---|---|---|---|
| 0 | 1 | $h = k = 0$ | R |
| 0 | 1 | $h = 0, k \in \mathbb{R} \setminus \{0\}$ | LCNO |
| 0 | 1 | $-1/4 \leq h < 3/4, h \neq 0, k \in \mathbb{R}$ | LCNO |
| 0 | 1 | $h \geq 3/4, k \in \mathbb{R}$ | LP |
| 0 | 2 | $h < -1/4, k \in \mathbb{R}$ | LCO |
| $+\infty$ | 1 and 2 | $h, k \in \mathbb{R}$ | LP |

This is the two parameter version of the classical one-dimensional equation for quantum modelling of the hydrogen atom; see [49, Section 10].

For form (1) and all $h, k$ there are no positive eigenvalues; form (2) is best considered in the single LCO case when some eigenvalues are positive; in form (1) there is a continuous spectrum on $[0, \infty)$; in form (2) there is a continuous spectrum on $[1, \infty)$.

If $k = 0$ and $h = \nu^2 - 1/4$ then form (1) reduces to the Liouville form of the Bessel equation, see Section 12 above.

### 39.1. Results for form 1

In all cases below $\rho$ is defined by

$$\rho := (h + 1/4)^{1/2} \text{ for all } h \geq -1/4.$$

(a) For $h \geq 3/4$ and $k \geq 0$ no boundary conditions are required; there is at most one negative eigenvalue and $\lambda = 0$ may be an eigenvalue; for $h \geq 3/4$ and $k < 0$ there are infinitely many negative eigenvalues given by

$$\lambda_n = \frac{-k^2}{(2n + 2\rho + 1)^2}, \ \rho = (h + 1/4)^{1/2} > 0, \ n = 0, 1, 2, 3, \dots$$

and $\lambda = 0$ is not an eigenvalue.

(b) For $h = 0$ and $k \in \mathbb{R} \setminus \{0\}$ a boundary condition is required at 0 for which

$$u(x) = x \qquad v(x) = 1 + k\, x \ln(x).$$

For some computed eigenvalues see [8] and [49, Section 10].

(c) For $-1/4 < h < 3/4$, *i.e.*, $0 < \rho < 1$, and $h \neq 0$, *i.e.*, $\rho \neq 1/2$, then a boundary condition is required at 0 for which, for all $x \in (0, +\infty)$,

$$u(x) = x^{\frac{1}{2}+\rho} \qquad v(x) = x^{\frac{1}{2}-\rho} + \frac{k}{1-2\rho} x^{\frac{3}{2}-\rho}.$$

The following results hold for the non-Friedrichs boundary condition $[y, v](0) = 0$, *i.e.*, $A_1 = 0, A_2 = 1$:

1. $k > 0, \ 0 < \rho < 1/2$ there are no negative eigenvalues
2. $k > 0, \ 1/2 < \rho < 1$ there is exactly one negative eigenvalue given by

$$\lambda_0 = \frac{-k^2}{(2\rho - 1)^2}$$

3. if $k < 0, \ 0 < \rho < 1/2$ there are infinitely many negative eigenvalues given by

$$\lambda_n = \frac{-k^2}{(2n - 2\rho + 1)^2}, \ n = 0, 1, 2, 3, \dots$$

4. if $k < 0, \ 1/2 < \rho < 1$ there are infinitely many negative eigenvalues given by

$$\lambda_n = \frac{-k^2}{(2n - 2\rho + 3)^2}, \ n = 0, 1, 2, 3, \dots$$

5. for $k = 0$ and $A_1 A_2 < 0$ there is exactly one negative eigenvalue given by:

$$\lambda_0 = \frac{4A_1 \Gamma(1 + \rho)}{A_2 \Gamma(1 - \rho)^{1/\rho}}.$$

(d) For $h = -1/4, \ k \in R$ , the LCNO classification at 0 prevails and a boundary condition is required for which, for all $x \in (0, +\infty)$,

$$u(x) = x^{1/2} + kx^{3/2} \qquad v(x) = 2x^{1/2} + \left(x^{1/2} + kx^{3/2}\right) \ln(x).$$

For $k = 0$ and $A_1 A_2 < 0$ there is exactly one negative eigenvalue given by:

$$\lambda_0 = -c \exp(2A_1/A_2), \quad c = 4\exp(4 - 2\gamma)$$

where $\gamma$ is Euler's constant: $\gamma = 0.5772156649\ldots$.

### 39.2. Results for form 2

For $h < -1/4$, $k \in R$, the equation is LCO at 0 (recall that we added 1 to the coefficient $q(\cdot)$ for this case, thus moving the start of the continuous spectrum from 0 to 1). On defining

$$\sigma := (-h - 1/4)^{1/2},$$

then, for all $x \in (0, +\infty)$,

$$u(x) = x^{1/2}\left[(1 - (4h)^{-1}kx)\cos(\sigma\ln(x)) + k\sigma x\sin(\sigma\ln(x))/2\right]$$
$$v(x) = x^{1/2}\left[(1 - (4h)^{-1}kx)\sin(\sigma\ln(x)) + k\sigma x\cos(\sigma\ln(x))/2\right];$$

(i) when $k = 0$ this equation reduces to the Rellich equation, see Section 44 below (but note that the notation is different)

(ii) when $k \neq 0$ explicit formulas for the eigenvalues are not available; however we report here on the qualitative properties of the spectrum for any boundary condition at 0:

($\alpha$) for all $k \in R$ there are infinitely many negative eigenvalues tending exponentially to $-\infty$

($\beta$) for $k > 0$ there are only a finite number of eigenvalues in any bounded interval, in particular they do not accumulate at 1

($\gamma$) for $k \leq 0$ the eigenvalues accumulate also at 1.

($\delta$) for $k = 0$ and $A_1 A_2 < 0$ there is exactly one negative eigenvalue given by:

$$\lambda_0 = \frac{4A_1\Gamma(1+\rho)}{A_2\Gamma(1-\rho)^{1/\rho}}.$$

Most of these results are due to Jörgens, see [49, Section 10]; a few new results were established by the authors of [11, Data base file xamples.tex; example 13].

## 40. Algebro-geometric equations

A potential $q$ of the one-dimensional Schrödinger equation

$$L[y](x) := -y''(x) + q(x)y(x) = \lambda y(x) \text{ for all } x \in I \subseteq \mathbb{R}$$

is called an algebro-geometric potential if there exists a linear ordinary differential expression $P$ of odd-order and leading coefficient 1, which commutes with $L$. There are deep relationships between algebro-geometric equations and the Korteweg-de Vries hierarchy of non-linear differential equations. An overview of these properties and results can be found in the survey article [44] which contains a substantial list of references.

The main structure and properties of the algebro-geometric equations can only be observed when the differential equations are considered in the complex plane, which would take the contents of this catalogue outside the environment of the Sturm-Liouville symmetric differential equations as described in Section 3 above.

However, three forms of algebro-geometric differential equations are given here; all three examples are Sturm-Liouville equations; two cases are related to other examples in this catalogue. However, all of these three examples have to be seen within the structure of algebro-geometric potentials and the relationships to non-linear differential equations.

### 40.1. Algebro-geometric form 1

Let $l \in \mathbb{N}_0$; then the differential equation is

$$-y''(x) + l(l+1)x^{-2}y(x) = \lambda y(x) \text{ for all } x \in (0, \infty).$$

This equation is a special case of:

(i) the hydrogen atom equation of Section 39 above, which gives the endpoint classification on $(0, \infty)$ for this example

(ii) the Liouville form of the Bessel differential equation, see Section 12 above, when the parameter $\nu = l + 1/2$; these cases of Bessel functions are named as the "spherical" Bessel functions; see [1, Chapter 10, Section 10.1] and [80, Chapter III, Section 3.41].

Endpoint classification in $L^2(0, +\infty)$:

| Endpoint | Parameter | Classification |
|---|---|---|
| 0 | $l = 0$ | R |
| 0 | $l \in \mathbb{N}$ | LP |
| $+\infty$ | $l \in \mathbb{N}_0$ | LP |

It is shown in [44] that this differential equation has two solutions of the form

$$y(x, \lambda) = \exp(ixs)\left(s^l + \sum_{j=0}^{l} a_j \frac{s^{l-j}}{x^j}\right) \text{ for all } x \in (0, \infty) \text{ and all } \lambda \in \mathbb{C},$$

where:

(i) $s^2 := \lambda$

(ii) the coefficients $\{a_j : j \in \mathbb{N}_0\}$ are determined by

$$a_0 = 1 \text{ and } a_{n+1} = i\frac{l(l+1) - n(n+1)}{2(n+1)} \text{ for all } n \in \mathbb{N}.$$

### 40.2. Algebro-geometric form 2

Let $g \in \mathbb{N}_0$; then the differential equation is

$$-y''(x) - \frac{g(g+1)}{\cosh(x)^2}y(x) = \lambda y(x) \text{ for all } x \in (-\infty, \infty);$$

this equation is a special case of:

(i) the hypergeometric differential equation, see Section 9 above but in particular [79, Chapter IV, Section 4.19].
(ii) the Liouville form of the Jacobi function differential equation, see Section 26 above, with the special case of $\alpha = -1/2$ and $\beta = g + 1/2$; here, the interval $(0, \infty)$ for the equation can be extended to $(-\infty, \infty)$ since the origin 0 is no longer a singular point of the equation when $\alpha = -1/2$.
Endpoint classification in $L^2(-\infty, +\infty)$:

| Endpoint | Parameter | Classification |
|---|---|---|
| $-\infty$ | $g \in \mathbb{N}_0$ | LP |
| $+\infty$ | $g \in \mathbb{N}_0$ | LP |

It is shown in [44] that this differential equation has two solutions of the form

$$y(x, \lambda) = \exp(ixs) \left( \sum_{n=0}^{g} a_n(s) \tanh(x)^n \right) \text{ for all } x \in (-\infty, \infty) \text{ and all } \lambda \in \mathbb{C},$$

where:
(i) $s^2 := \lambda$
(ii) the coefficients $\{a_n : n = 0, \ldots, g\}$ are determined by a five-term recurrence relation.

### 40.3. Algebro-geometric form 3

Let $g \in \mathbb{R}$; then this differential equation is a special case of Lamé's equation, see Section 31 above, and is given by

$$-y''(x) + g(g+1)\wp(x + \omega')y(x) = \lambda y(x) \text{ for all } x \in (-\infty, \infty),$$

where $\wp$ is the Weierstrass elliptic function with fundamental periods $2\omega$ and $2\omega'$, with $\omega$ real and $\omega'$ purely imaginary.

In this situation $\wp(\cdot + \omega')$ is real-valued, periodic with period $2\omega$, and real-analytic on $\mathbb{R}$, see [1, Chapter 18, Section 18.1].

Endpoint classification in the space $L^2(-\infty, \infty)$:

| Endpoint | Parameter | Classification |
|---|---|---|
| $-\infty$ | $g \in \mathbb{R}$ | LP |
| $+\infty$ | $g \in \mathbb{R}$ | LP |

When $g \in \mathbb{N}_0$, it is shown in [44] that this differential equation (which is an example of the general Lamé differential equation) has solutions of the form

$$y(\mathbf{a}, x) = \sigma(x + \omega')^{-g} \prod_{j=1}^{g} \sigma(x + \omega' - a_j) \exp\left( x \sum_{j=1}^{g} \zeta(a_j) \right) \text{ for all } x \in (-\infty, \infty),$$

where the vector $\mathbf{a} = (a_1, a_2, \ldots, a_g)$ has to satisfy the conditions

$$\sum_{\substack{j=1 \\ j \neq k}}^{g} (\zeta(a_j - a_k) - \zeta(a_j) + \zeta(a_k)) = 0 \text{ for } k = 1, 2, \cdots, g$$

and the spectral parameter $\lambda$ is then given by

$$\lambda = (1 - 2g) \sum_{j=1}^{g} \wp(a_j).$$

Here $\sigma$ and $\zeta$ are the Weierstrass-$\sigma$ and Weierstrass-$\zeta$ functions respectively, see [1, Chapter 18, Section 18.1].

The spectrum of the unique self-adjoint operator, in the Hilbert function space $L^2(-\infty, \infty)$, generated by this example of the Lamé differential equation, consists of $g+1$ disjoint intervals, one of which is a semi-axis; these are the spectral bands of this differential operator.

Note that $\mathbf{a}$ satisfies the constraints mentioned if and only if $-\mathbf{a}$ satisfies the same constraint, since $\zeta$ is an odd function; as $\wp$ is an even function these properties lead to the same value of $\lambda$. Both the functions $y(\mathbf{a}, \cdot)$ and $y(-\mathbf{a}, \cdot)$ do then satisfy the same differential equation; they are linearly independent except when $\lambda$ is one of the $2g + 1$ band edges.

For these results and additional examples of algebro-geometric differential equations see the survey paper [44].

### 40.4. Algebro-geometric form 4

This form is named as the $N$-soliton potential.

We introduce the $N \times N$ matrix, for $1 \leq j,\ k \leq N$ and all $x \in (-\infty, \infty)$,

$$C_N(x) = \big(c_j c_k (\kappa_j + \kappa_k)^{-1} \exp(-(\kappa_j + \kappa_k)x)\big)$$

with

$$c_j > 0, \kappa_j > 0, \kappa_j \neq \kappa_k \text{ for all } 1 \leq j, k \leq N \text{ with } j \neq k;$$

the $N$-soliton potential $q_N : (-\infty, \infty) \to \mathbb{R}$ is then defined by

$$q_N(x) := -2 \frac{d^2}{dx^2} \ln(\det(I_N + C_N(x))) \text{ for all } x \in (-\infty, \infty)$$

(with $I_N$ the identity matrix in $\mathbb{C}^N$). The corresponding Sturm-Liouville differential equation then reads

$$-y''(x) + q_N(x)y(x) = \lambda y(x) \text{ for all } x \in (-\infty, \infty) \text{ and } \lambda \in \mathbb{C}.$$

Since

$$q_N \in C^\infty(-\infty, \infty), \quad q_N(x) = O(\exp(-2\kappa_{j_0}|x|)) \text{ for } |x| \to \infty,$$

where $\kappa_{j_0} = \min_{1 \leq j \leq N}(\kappa_j)$, the endpoint classification in $L^2(-\infty, \infty)$ is

| Endpoint | Classification |
|---|---|
| $-\infty$ | LP |
| $+\infty$ | LP |

Defining

$$c_{N,j,+} := c_j, \quad c_{N,j,-} := c_j^{-1} \times \begin{cases} 2\kappa_1, & j = N = 1, \\ 2\kappa_j \prod_{k=1}^{N} \dfrac{\kappa_j + \kappa_k}{\kappa_j - \kappa_k} \text{ with } k \neq j, & 1 \leq j \leq N, \ N \geq 2 \end{cases}$$

two independent solutions of the differential equation, associated with $q_N$, are then given by

$$f_{N,\pm}(x,\lambda) := \left[1 - i \sum_{j=1}^{N} (\sqrt{\lambda} + i\kappa_j)^{-1} c_{N,j,\pm} \psi_{N,j}(x) \exp(\mp \kappa_j x)\right] \exp(\pm i \sqrt{\lambda} x),$$

for all $\lambda \in \mathbb{C}$ with $\operatorname{Im}(\sqrt{\lambda}) \geq 0$, and all $x \in (-\infty, \infty)$. Here $\{\psi_{N,j}(\cdot) : j = 1, 2, \ldots, N\}$ are given as follows; define the column vector

$$\Psi_N^0(x) := (c_1 \exp(-\kappa_1 x), \ldots, c_N \exp(-\kappa_N x))^\top,$$

and then $\Psi_N(\cdot)$ by

$$\Psi_N(x) := [I_N + C_N(x)]^{-1} \psi_N^0(x) \text{ for all } x \in (-\infty, \infty),$$

both for all $x \in (-\infty, \infty)$. Writing now

$$\Psi_N(x) = (\psi_{N,1}(x), \ldots, \psi_{N,N}(x))^\top$$

this defines the components $\{\psi_{N,j}(\cdot) : j = 1, 2, \ldots, N\}$ and completes the definition of the two solutions $f_{N,\pm}$.

Now let $H_N$ denote the (maximally defined) self-adjoint Schrödinger operator with potential $q_N$ in $L^2(-\infty, \infty)$. Then $\psi_{N,j} \in C^\infty(-\infty, \infty)$ are exponentially decaying eigenfunctions of $H_N$ as $|x| \to \infty$, corresponding to the negative eigenvalues $-\kappa_j^2$; thus

$$H_N \psi_{N,j} = -\kappa_j^2 \psi_{N,j}, \quad 1 \leq j \leq N.$$

Moreover, $H_N$ has spectrum

$$\{-\kappa_j^2 : 1 \leq j \leq N\} \cup [0, \infty)$$

and $q_N$ satisfies

$$q_N(x) = -4 \sum_{j=1}^{N} \kappa_j \psi_{N,j}(x)^2 < 0, \quad 0 < -q_N(x) \leq 2\hat{\kappa}^2,$$

where $\hat{\kappa} := \max_{1 \leq j \leq N}(\kappa_j)$ for all $x \in \mathbb{R}$.

The potentials $q_N$ are reflectionless since the corresponding $2 \times 2$ scattering matrix $S_N(\lambda)$ is of the form

$$S_N(\lambda) = \begin{pmatrix} T_N(\lambda) & R_N^r(\lambda) \\ R_N^\ell(\lambda) & T_N(\lambda) \end{pmatrix} \text{ for all } \lambda \geq 0,$$

with transmission coefficients given by

$$T_N(\lambda) = \prod_{j=1}^{N} \frac{\sqrt{\lambda} + i\kappa_j}{\sqrt{\lambda} - i\kappa_j} \text{ for all } \lambda \geq 0$$

and vanishing reflection coefficients from the right and left incidence

$$R_N^r(\lambda) = R_N^\ell(\lambda) = 0 \text{ for all } \lambda \geq 0.$$

Thus, the $N$-soliton potentials $q_N$ can be thought of as a particular construction of reflectionless potentials that add $N$ negative eigenvalues $-\kappa_j^2$, $1 \leq j \leq N$, to the spectrum of $H_0$, where $H_0 = -d^2/dx^2$ is the Schrödinger operator in $L^2(-\infty, \infty)$ associated with the trivial potential $q_0(x) = 0$ for all $x \in \mathbb{R}$, and spectrum $[0, \infty)$.

It can be shown that $q_N$ satisfies a particular $N$th stationary KdV equation, see [40, Section 1.3]. In addition, introducing an appropriate time-dependence in $c_j$ leads to KdV $N$-soliton potentials, see [40, Section 1.4].

We also note that $q_g(x) = -g(g+1)[\cosh(x)]^{-2}$, treated in Subsection 40.2 above, is a special case of $q_N$ for $N = g$ and a particular choice of $\kappa_j$ and $c_j$, $1 \leq j \leq N$.

Reflectionless potentials $q_N$ were first derived by Kay and Moses [51] (see also [23], [24], [39], and [41] for detailed discussions).

## 41. Bargmann potentials

Let $\varphi_0(x, \lambda) = s^{-1} \sin(sx)$ for $\lambda = s^2 \in \mathbb{C}$ and $x \in [0, \infty)$; then for $N \in \mathbb{N}$ introduce the $N \times N$ matrix

$$B_N(x) = (B_{N,j,k}(x)) \text{ for } 1 \leq j, k \leq N \text{ and all } x \in [0, \infty)$$

given by

$$\begin{aligned} B_{N,j,k}(x) &= \int_0^x C_j \varphi_0(t, -\gamma_j^2) \varphi_0(t, -\gamma_k^2) \, dt \\ &= C_j (2\gamma_j \gamma_k)^{-1} \begin{cases} (2\gamma_j)^{-1} \sinh(2\gamma_j x) - x, \\ \text{for } j = k, \\ (\gamma_j + \gamma_k)^{-1} \sinh((\gamma_j + \gamma_k)x) - (\gamma_j - \gamma_k)^{-1} \sinh((\gamma_j - \gamma_k)x), \\ \text{for } j \neq k, \end{cases} \end{aligned}$$

$C_j > 0$, $\gamma_j > 0$ for $1 \leq j, k \leq N$.

Bargmann potentials $q_N$ are then defined by

$$q_N(x) = -2 \frac{d^2}{dx^2} \ln(\det(I_N + B_N(x))) \text{ for all } x \in [0, \infty)$$

($I_N$ the identity matrix in $\mathbb{C}^N$), and the associated Sturm-Liouville differential equation reads

$$-y''(x) + q_N(x) y(x) = \lambda y(x) \text{ for all } x \in [0, \infty) \text{ and } \lambda \in \mathbb{C}.$$

It can be shown that

$$\int_0^\infty (1+x)|q_N(x)|\,dx < \infty.$$

Actually, much more detailed information can be obtained to give

$$q_N(x) \underset{x\downarrow 0}{=} -4\left(\sum_{j=1}^N C_j x\right) + o(x)$$

$$q_N(x) \underset{x\uparrow\infty}{=} -2C_{j_0}^{-1}(2\gamma_{j_0})^5 \exp(-2\gamma_{j_0}x)[1+o(1)],$$

where $\gamma_{j_0} = \min_{1\le j\le N}(\gamma_j)$ and $C_{j_0}$ is the corresponding normalization constant.

Hence the endpoint classification of this equation in $L^2(0,+\infty)$ is

| Endpoint | Classification |
|---|---|
| 0 | R |
| $+\infty$ | LP |

The regular solution $\varphi_N(\cdot,\lambda)$ associated with $q_N$ is then given by

$$\varphi(x,\lambda) = \frac{\det\begin{vmatrix} I_N + B_N(x) & \psi(x) \\ \beta(x,\lambda) & \varphi_0(x,\lambda)\end{vmatrix}}{\det(I_N + B_N(x))}, \quad x\in[0,\infty),\ \lambda\in\mathbb{C},$$

where the matrix in the numerator is obtained by adding to $I_N + B_N(x)$ the column $\psi$, the row $\beta$, and the last diagonal element $\varphi_0$. Here $\psi$ and $\chi$ are vectors with components $C_j\varphi_0(x,-\gamma_j^2)$ and $\varphi_0(x,-\gamma_j^2)$, respectively, and

$$\beta(x,\lambda) = \int_0^x \chi(t)^\top \varphi_0(t,\lambda)\,dt.$$

Similarly, the Jost solution $f(x,\lambda)$ corresponding to $q_N$ can be computed, but we omit the lengthy expression; the Jost function $F(s)$ associated with $q_N$ finally reads

$$F(s) = f(0,\lambda) = \prod_{j=1}^N \frac{s - i\gamma_j}{s + i\gamma_j} \text{ with } \lambda = s^2.$$

This shows that the Schrödinger operator $H_N$ in $L^2(0,+\infty)$ associated with $q_N$ and a Dirichlet boundary condition at $x=0$ has spectrum

$$\{-\gamma_j^2 : 1\le j\le N\}\cup[0,\infty).$$

Next we denote by $q_0$ the trivial potential $q_0(x) = 0$ for all $x\in[0,\infty)$, and by $H_0 = -d^2/dx^2$ the corresponding Schrödinger operator in $L^2(0,+\infty)$ with a Dirichlet boundary condition at $x=0$; the operator $H_0$ then has spectrum $[0,\infty)$.

In comparison with the trivial potential $q_0(x) = 0$, the Bargmann potential $q_N(x)$ is constructed such that the corresponding operator $H_N$ has $N$ additional strictly negative eigenvalues at $-\gamma_j^2$ for $1\le j\le N$. Put differently, $N$ negative eigenvalues $-\gamma_j^2$ have been added to the spectrum of $H_0$.

However, since $|F(s)| = 1$ for all $s \geq 0$, the spectral densities of $H_N$ and $H_0$ coincide for $\lambda \geq 0$ (recall $\lambda = s^2$). Explicitly, the spectral function $\rho_N(\lambda)$, for all $\lambda \in (-\infty, +\infty)$, of $H_N$ is of the form

$$\rho_N(\lambda) = \begin{cases} (2/3)\pi^{-1}\lambda^{3/2}, & \lambda \geq 0, \\ \sum_{j=1}^{N} C_j \theta(\lambda + \gamma_j^2), & \lambda < 0 \end{cases}$$

(here $\theta(t) = 1$ for $t > 0$, $\theta(t) = 0$ for $t < 0$), which should be compared with the spectral function $\rho_0(\lambda)$ of $H_0$,

$$\rho_0(\lambda) = \begin{cases} (2/3)\pi^{-1}\lambda^{3/2}, & \lambda \geq 0, \\ 0, & \lambda < 0. \end{cases}$$

For Bargmann's original work we refer to [13], [14]; more details on Bargmann potentials can be found in [21, Sections III.2, IV.1 and IV.3], and the references therein (see also [42, Section 11]).

## 42. Halvorsen equation

The Halvorsen differential equation exhibits the difficulties created at R endpoints, both analytically and numerically, in certain circumstances:

$$-y''(x) = \lambda x^{-4} \exp(-2/x) y(x) \text{ for all } x \in (0, +\infty).$$

The endpoint classification in the space $L^2((0, +\infty); x^{-4} \exp(-2/x))$:

| Endpoint | Classification |
|---|---|
| 0 | R |
| $+\infty$ | LCNO |

For the endpoints 0 and $+\infty$ in the R and LCNO classification the boundary condition functions $u, v$ are determined by

| Endpoint | $u$ | $v$ |
|---|---|---|
| 0 | $x$ | 1 |
| $+\infty$ | 1 | $x$ |

In this example the LC boundary condition form can be used at the R endpoint 0, with $u$ and $v$ as shown.

Since this equation is R at 0 and LCNO at $+\infty$ the spectrum is discrete and bounded below for all boundary conditions. However, this example illustrates that even a R endpoint can cause difficulties for computation; details of the computation of eigenvalues are given in [11, Data base file xamples.tex; example 3].

At 0, the principal boundary condition entry is $A_1 = 1,\ A_2 = 0$; at $\infty$ with $u(x) = 1,\ v(x) = x$ the principal boundary condition entry is also $A_1 = 1,\ A_2 = 0$, but note the interchange of the definitions of $u$ and $v$ at these two endpoints.

## 43. Jörgens equation

We have this example due to Jörgens [49]:

$$-y''(x) + (\exp(2x)/4 - k\exp(x))y(x) = \lambda y(x) \text{ for all } x \in (-\infty, +\infty)$$

where the parameter $k \in (-\infty, +\infty)$.

Endpoint classification in the space $L^2(-\infty, +\infty)$, for all $k \in (-\infty, +\infty)$:

| Endpoint | Classification |
|---|---|
| $-\infty$ | LP |
| $+\infty$ | LP |

This is a remarkable example from Jörgens; numerical results are given in [11, Data base file xamples.tex; example 27]. Details of this problem are given in [49, Part II, Section 10]. For all $k \in (-\infty, +\infty)$ the boundary value problem on the interval $(-\infty, +\infty)$ has a continuous spectrum on $[0, +\infty)$; for $k \leq 1/2$ there are no eigenvalues; for $h = 0, 1, 2, 3, \ldots$ and then $k$ chosen by $h < k - 1/2 \leq h + 1$, there are exactly $h+1$ eigenvalues and these are all below the continuous spectrum; these eigenvalues are given explicitly by

$$\lambda_n = -(k - 1/2 - n)^2, \quad n = 0, 1, 2, 3, \ldots, h.$$

## 44. Rellich equation

The Rellich differential equation is, where the parameter $K \in \mathbb{R}$,

$$-y''(x) + Kx^{-2}y(x) = \lambda y(x) \text{ for all } x \in (0, +\infty);$$

this equation has a long and interesting history as indicated in the references, see [72], [64], [54], [20] and [43].

Here, we consider the equation in the form, where the parameter $k \in (0, +\infty)$,

$$-y''(x) + (1 - (k^2 + 1/4)x^{-2})y(x) = \lambda y(x) \text{ for all } x \in (0, +\infty)$$

as discussed by Krall in the paper [54], and in view of the connection with the computer program SLEIGN2, see [8] and [10].

Endpoint classification, for all $k \in (0, +\infty)$, in the space $L^2(0, +\infty)$:

| Endpoint | Classification |
|---|---|
| 0 | LCO |
| $+\infty$ | LP |

This example should be seen as a special case of the Liouville form of the Bessel equation as discussed at the end of Section 12 above; solutions can be obtained in terms of the modified Bessel functions.

To help with the computations for this example the spectrum is translated by a term $+1$; this simple device is used for numerical convenience.

For problems with a boundary condition at the endpoint 0 there is a continuous spectrum on $[1, \infty)$ with a discrete (and simple) spectrum on $(-\infty, 1)$. This discrete spectrum has cluster points at both $-\infty$ and 1.

For the LCO endpoint at 0 the boundary condition functions are given by

$$u(x) = x^{1/2}\cos(k\ln(x)) \qquad v(x) = x^{1/2}\sin(k\ln(x)).$$

For the boundary value problem on $[0,\infty)$ with boundary condition $[y,u](0) = 0$ let the following conditions and notations apply:

(i) suppose $\Gamma(1+i) = \alpha + i\beta$ and $\mu > 0$ satisfies $\tan\big(\ln(\frac{1}{2}\mu)\big) = -\alpha/\beta$

(ii) $\theta = \mathrm{Im}(\log(\Gamma(1+i)))$

(iii) $\ln(\frac{1}{2}\mu) = \frac{1}{2}\pi + \theta + s\pi$ for $s = 0, \pm 1, \pm 2, \dots$

(iv) $\mu_s^2 = \big(2\exp(\theta + \frac{1}{2}\pi)\big)^2 \exp(2s\pi),\ \ s = 0, \pm 1, \pm 2, \dots$

then the eigenvalues are given explicitly by $\lambda_n = -\mu^2_{-(n+1)} + 1$ $(n = 0, \pm 1, \pm 2, \dots)$.

This problem creates major computational difficulties; see [11, Data base file xamples.tex; example 20]. The program SLEIGN2 can compute only six of these eigenvalues in a normal UNIX server, even in double precision, specifically $\lambda_{-3}$ to $\lambda_2$; other eigenvalues are, numerically, too close to 1 or too close to $-\infty$. Here we list these SLEIGN2 computed eigenvalues in double precision in a normal UNIX server and compare them with the same eigenvalues computed from the transcendental equation given above; for the problem on $(0,\infty)$ with $k = 1$ and $A_1 = 1,\ A_2 = 0$, the results are:

| Eigenvalue | eig from SLEIGN2 | eig from trans. equ. |
|---|---|---|
| $-3$ | $-276,562.5$ | $-14,519.130$ |
| $-2$ | $-27,114.48$ | $-27,114.67$ |
| $-1$ | $-49.62697$ | $-49.63318$ |
| 0 | 0.9054452 | 0.9054454 |
| 1 | 0.9998234 | 0.9998234 |
| 2 | 0.9999997 | 0.9999997 |

## 45. Laplace tidal wave equation

This differential equation is given by:

$$-(x^{-1}y'(x))' + \left(kx^{-2} + k^2x^{-1}\right)y(x) = \lambda y(x) \text{ for all } x \in (0,+\infty),$$

where the parameter $k \in (-\infty,0)\cup(0,+\infty)$. This equation has been studied by many authors, in particular by Homer in his doctoral thesis [47] where a detailed list of references is to be found.

Endpoint classification in $L^2(0,\infty)$:

| Endpoint | Classification |
|---|---|
| 0 | LCNO |
| $+\infty$ | LP |

For the endpoint 0:

$$u(x) = x^2 \qquad v(x) = x - k^{-1} \text{ for all } x \in (0,+\infty).$$

This equation is a particular case of the more general equation with this name; for details and references see [47].

There are no representations for solutions of this differential equation in terms of the well-known special functions. Thus to determine boundary conditions at the LCNO endpoint 0 use has to be made of maximal domain functions; see the $u$, $v$ functions given above. Numerical results for some boundary value problems and certain values of the parameter $k$ are given in [11, Data base file xamples.tex; example 8].

## 46. Latzko equation

This differential equation is given by

$$-((1-x^7)y'(x))' = \lambda x^7 y(x) \text{ for all } x \in (0,1].$$

Endpoint classification in $L^2(0,1]$:

| Endpoint | Classification |
|---|---|
| 0 | R |
| 1 | LCNO |

For the endpoint 1:

$$u(x) = 1 \qquad v(x) = -\ln(1-x) \text{ for all } x \in (0,1).$$

This differential equation has a long and celebrated history; in particular it has been studied by Fichera, see [36, Pages 43 to 45]. There is a LCNO singularity at the endpoint 1 which requires the use of maximal domain functions; see the $u$, $v$ functions given above. The endpoint 0 is R but there are computational difficulties in general when a weight has the property $w(0) = 0$.

This example is similar in some respects to the Legendre equation of Section 19 above.

For numerical results see [11, Data base file xamples.tex; example 7].

## 47. Littlewood-McLeod equation

This important example gives a Sturm-Liouville boundary value problem that has a discrete spectrum that is unbounded above and below; the differential equation is

$$-y''(x) + x\sin(x)y(x) = \lambda y(x) \text{ for all } x \in [0,+\infty).$$

Endpoint classification in the space $L^2(0,+\infty)$:

| Endpoint | Classification |
|---|---|
| 0 | R |
| $+\infty$ | LP |

This differential equation is an example of the LPO endpoint classification introduced in [69, Chapter 7], see item 6 of Remark 5.1 above.

The spectral analysis of this differential equation is considered in [56] and [61]; the equation is R at 0 and LP at $+\infty$. All self-adjoint operators in $L^2[0,\infty)$

have a simple, discrete spectrum $\{\lambda_n : n = 0, \pm 1, \pm 2, \ldots\}$ that is unbounded both above and below, *i.e.*,

$$\lim_{n \to -\infty} \lambda_n = -\infty \qquad \lim_{n \to +\infty} \lambda_n = +\infty.$$

Every eigenfunction has infinitely many zeros in $(0, \infty)$.

SLEIGN2, and all other codes, fail to compute the eigenvalues for this type of LP oscillatory problem. However there is qualitative information to be obtained by considering regular problems on $[0, X]$ with, say, Dirichlet boundary conditions $y(0) = y(X) = 0$.

## 48. Lohner equation

This Sturm-Liouville example was one of the first differential equations to be subjected to the Lohner code, see [57], for computing guaranteed numerical bounds for eigenvalues of boundary value problems, but see the earlier paper of Plum [71].

The differential equation is

$$-y''(x) - 1000xy(x) = \lambda y(x) \text{ for all } x \in (-\infty, +\infty).$$

Endpoint classification in the space $L^2(-\infty, +\infty)$:

| Endpoint | Classification |
|---|---|
| $-\infty$ | LP |
| $+\infty$ | LP |

In [57] Lohner computed the Dirichlet eigenvalues of certain regular problems on compact intervals, using interval arithmetic, and obtained rigorous bounds. In double precision SLEIGN2 computed eigenvalues to give numerical values that are in good agreement with these guaranteed bounds.

## 49. Pryce-Marletta equation

This differential equation presents a difficult problem for computational programs; it was devised by Pryce, see [69, Appendix B, Problem 60], and studied by Marletta [60]; the differential equation is:

$$-y''(x) + \frac{3(x-31)}{4(x+1)(x+4)^2} y(x) = \lambda y(x) \text{ for all } x \in [0, +\infty).$$

Endpoint classification in $L^2(0, +\infty)$:

| Endpoint | Classification |
|---|---|
| 0 | R |
| $+\infty$ | LP |

For this differential equation boundary value problems on the interval $[0, \infty)$ are considered. Since $q(x) \to 0$ as $x \to \infty$ the continuous spectrum consists of $[0, \infty)$ and every negative number is an eigenvalue for some boundary condition at 0.

For the boundary condition $A_1 = 5, A_2 = 8$ at the endpoint 0, there is a negative eigenvalue $\lambda_0$ near $-1.185$. However the equation with $\lambda = 0$ has a solution

$$y(x) = \frac{1-x^2}{(1+x/4)^{5/2}} \text{ for all } x \in [0,\infty),$$

that satisfies this boundary condition, which is not in $L^2(0,\infty)$ but is "nearly" in this space. This solution deceives most computer programs; however SLEIGN2 correctly reports that $\lambda_0$ is the only eigenvalue, and the start of the continuous spectrum at 0.

Additional details of this example are to be found in the Marletta certification report on SLEIGN [60].

## 50. Meissner equation

The Meissner equation has piecewise constant but discontinuous coefficients; it has a remarkable distribution of simple and double eigenvalues for periodic boundary conditions on the interval $(-1/2, 1/2)$: the differential equation is

$$-y''(x) = \lambda w(x)y(x) \text{ for all } x \in (-\infty, +\infty),$$

where the weight coefficient $w$ is defined by

$$\begin{aligned} w(x) &= 1 \text{ for all } x \in (-\infty, 0] \\ &= 9 \text{ for all } x \in (0, +\infty). \end{aligned}$$

Endpoint classification in the space $L^2(-\infty, +\infty)$:

| Endpoint | Classification |
|---|---|
| $-\infty$ | LP |
| $+\infty$ | LP |

This equation arose in a model of a one-dimensional crystal. For this constant coefficient equation with a weight function which has a jump discontinuity the eigenvalues can be characterized as roots of a transcendental equation involving only trigonometrical and inverse trigonometrical functions. There are infinitely many simple eigenvalues and infinitely many double ones for the periodic case; they are given by:

**Periodic boundary conditions** on $(-1/2, +1/2)$, *i.e.*,

$$y(-1/2) = y(+1/2) \qquad y'(-1/2) = y'(+1/2).$$

We have $\lambda_0 = 0$ and for $n = 0, 1, 2, \ldots$

$$\lambda_{4n+1} = (2n\pi + \alpha)^2; \quad \lambda_{4n+2} = (2(n+1)\pi - \alpha))^2;$$

$$\lambda_{4n+3} \;=\; \lambda_{4n+4} = (2(n+1)\pi))^2.$$

where $\alpha = \cos^{-1}(-7/8)$

**Semi-periodic boundary conditions** on $(-1/2,+1/2)$, *i.e.*,

$$y(-1/2) = -y(+1/2) \qquad y'(-1/2) = -y'(+1/2).$$

With $\beta = \cos^{-1}((1+\sqrt{33})/16)$ and $\gamma = \cos^{-1}((1-\sqrt{33})/16)$ these are all simple and given by, for $n = 0, 1, 2, \ldots$

$$\lambda_{4n} = (2n\pi + \beta)^2; \ \lambda_{4n+1} = (2n\pi + \gamma)^2;$$

$$\lambda_{4n+2} = (2(n+1)\pi - \gamma)^2; \ \lambda_{4n+3} = (2(n+1)\pi - \beta)^2.$$

For the general theory of periodic differential boundary value problems see [26]; for a special case with discontinuous coefficients see [45].

## 51. Morse equation

This differential equation has exponentially small and large coefficients; the differential equation is

$$-y''(x) + (9\exp(-2x) - 18\exp(-x))y(x) = \lambda y(x) \text{ for all } x \in (-\infty,+\infty).$$

Endpoint classification in the space $L^2(-\infty,+\infty)$:

| Endpoint | Classification |
|---|---|
| $-\infty$ | LP |
| $+\infty$ | LP |

This differential equation on the interval $(-\infty,\infty)$ is studied in [5, Example 6]; the spectrum has exactly three negative, simple eigenvalues, and a continuous spectrum on $[0,\infty)$; the eigenvalues are given explicitly by

$$\lambda_n = -(n-2.5)^2 \text{ for } n = 0, 1, 2.$$

## 52. Morse rotation equation

This differential equation is considered in [5] and is given as

$$-y''(x) + (2x^{-2} - 2000(2e(x) - e(x)^2))y(x) = \lambda y(x) \text{ for all } x \in (0,+\infty),$$

where

$$e(x) = \exp(-1.7(x-1.3)) \text{ for all } x \in (0,+\infty).$$

Endpoint classification in the space $L^2(0,+\infty)$

| Endpoint | Classification |
|---|---|
| 0 | LP |
| $+\infty$ | LP |

This classical problem on the interval $(0,\infty)$ has a continuous spectrum on $[0,\infty)$ and exactly 26 negative eigenvalues; it provides an invaluable numerical test for computer programs.

## 53. Brusencev/Rofe-Beketov equations

### 53.1. Example 1

The Sturm-Liouville differential equation

$$-(x^4 y'(x))' - 2x^2 y(x) = \lambda y(x) \text{ for all } x \in (0, \infty)$$

is considered in the paper [19]; this example provides a LC case with special properties.

Endpoint classification in $L^2(0, +\infty)$:

| Endpoint | Classification |
|---|---|
| 0 | LP |
| $+\infty$ | LCNO |

For the endpoint $+\infty$ in the LCNO classification the boundary condition functions $u, v$ are determined by

| Endpoint | $u$ | $v$ |
|---|---|---|
| $+\infty$ | $x^{-1}$ | $x^{-2}$ |

### 53.2. Example 2

The Sturm-Liouville differential equation

$$-y''(x) - \left(x^{10} + x^4 \operatorname{sign}(\sin(x))\right) y(x) = \lambda y(x) \text{ for all } x \in [0, \infty)$$

is considered in the paper [73]; this example provides a LC case with special properties.

Endpoint classification in $L^2(0, +\infty)$:

| Endpoint | Classification |
|---|---|
| 0 | R |
| $+\infty$ | LCO |

For the endpoint $+\infty$ in the LCO classification the boundary condition functions $u, v$ may be determined as the real and imaginary parts of the expression

$$x^{-5/2} \exp(ix^6/6) Y(x) \text{ for all } x \in [1, \infty),$$

where the function $Y(\cdot)$ is the solution of the integral equation, for $x \in [1, \infty)$,

$$Y(x) = 1 + \frac{i}{2} \int_x^\infty \left(t^{-1} \operatorname{sign}(\sin(t)) + \frac{35}{4} t^{-7}\right) \left[\exp\left(\frac{i}{3}(t^6 - x^6)\right) - 1\right] Y(t)\, dt.$$

The solution $Y(\cdot)$ of this integral equation may be obtained by the iteration method of successive approximations; in this process it has to be noted that the integrals concerned are only conditionally convergent.

## 54. Slavyanov equations

In the important text [77] the authors give a systematic presentation of a unified theory of special functions based on singularities of linear ordinary differential equations in the complex plane $\mathbb{C}$. In particular, in [77, Chapter 3], there is to be found an authoritative account of the definition and properties of the Heun differential equation.

In [77, Chapter 4] there is a chapter devoted to physical applications, including the use of the Heun differential equation, resulting from the application of separation techniques to boundary value problems for linear partial differential equations. From this chapter we have selected three examples of Sturm-Liouville differential equations; each equation contains a number of symbols denoting physical constants and parameters which are given here without explanation. To allow the quoted examples to be given in Sturm-Liouville form the notation for one of these parameters has been changed to play the role of the spectral parameter $\lambda \in \mathbb{C}$.

The resulting Sturm-Liouville examples given below have not yet been considered for their endpoint classification, nor for their boundary condition functions if required for LC endpoints.

### 54.1. Example 1

The hydrogen-molecule ion problem, see [77, Chapter 4, Section 4.1.3], gives the two differential equations:

$$-\left((1-\eta^2)Y'(\eta)\right)' + \left(n^2(1-\eta^2)^{-1} - \mu\right)Y(\eta) = \lambda\eta^2 Y(\eta) \text{ for all } \eta \in (-1,1)$$

and

$$-\left((1-\xi^2)X'(\xi)\right)' + \left(\kappa\xi + n^2(1-\xi^2) - \mu\right)X(\xi) = \lambda\xi^2 X(\xi) \text{ for all } \xi \in (1,\infty).$$

### 54.2. Example 2

The Teukolsky equations in astrophysics gives the equation, see [77, Chapter 4, Section 4.2.1]:

$$-((1-u^2)X'(u))' + \left(2 + (m-2u)^2(1-u^2)^{-1} - 4a\omega u - a^2\omega^2u^2\right)X(u) = \lambda X(u)$$

for all $u \in (-1,1)$.

### 54.3. Example 3

The theory of tunnelling in double-well potentials, see [77, Chapter 4, Section 4.4], gives the differential equation

$$-y''(x) + V(x)y(x) = \lambda y(x) \text{ for all } x \in (-\infty,\infty)$$

with the potential $V$ determined by

$$V(x) = -A(\operatorname{sech}^2(x+x_0) + \operatorname{sech}^2(x-x_0)) \text{ for all } x \in (-\infty,\infty);$$

here $A$ is a number and $x_0$ is a parameter.

## 55. Fuel cell equation

This Sturm-Liouville differential equation

$$-(xy'(x))' - x^3y(x) = \lambda xy(x) \text{ for all } x \in (0, b]$$

plays an important role in a fuel cell problem as discussed in the paper [6].

Endpoint classification in the space $L^2((0, b); x)$:

| Endpoint | Classification |
|---|---|
| 0 | LCNO |
| $b$ | R |

For the LCNO endpoint at 0 the $u, v$ boundary condition functions can be taken as, see [6, Section 8]:

$$u(x) = 1 \text{ and } v(x) = \ln(x) \text{ for all } x \in (0, b].$$

Various boundary value problems are considered in [6, Section 8]; the technical requirements of the fuel cell problem require a study of the analytic properties of these boundary value problems, as the endpoint $b$ tends to zero.

## 56. Shaw equation

This Sturm-Liouville differential equation is considered in the paper [76] and has the form

$$y''(x) - Q(x)y(x) = \lambda y(x) \text{ for all } x \in (0, \infty)$$

where

$$Q(x) = A - B\exp(-Cx) + Dx^{-2} \text{ for all } (0, \infty)$$

for positive real numbers $A, B, C, D$ with $D \geq 3/4$.

Endpoint classification in $L^2(0, +\infty)$:

| Endpoint | Classification |
|---|---|
| 0 | LP |
| $+\infty$ | LP |

In the paper [76] the following specific values for $A, B, C, D$ are used in connection with the chemical photodissociation of methyl iodide:

$$A = 19362.8662 \quad B = 19362.8662 \times 46.4857$$

$$C = 1.3 \quad D = 2.0.$$

## 57. Plum equation

This Sturm-Liouville equation is one of the first to be considered for numerical computation using interval arithmetic: the equation is

$$-y''(x) + 100\cos^2(x)y(x) = \lambda y(x) \text{ for all } x \in (-\infty, +\infty).$$

Endpoint classification in $L^2(-\infty, +\infty)$:

| Endpoint | Classification |
|---|---|
| $-\infty$ | LP |
| $+\infty$ | LP |

In [71] the first seven eigenvalues for periodic boundary conditions on the interval $[0, \pi]$, *i.e.*,

$$y(0) = y(\pi) \qquad y'(0) = y'(\pi),$$

are computed using a numerical homotopy method together with interval arithmetic; rigorous bounds for these seven eigenvalues are obtained.

## 58. Sears-Titchmarsh equation

This differential equation is considered in detail in [79, Chapter IV, Section 4.14] and [75]; the equation is

$$-y''(x) - \exp(2x)y(x) = \lambda y(x) \text{ for all } x \in (-\infty, \infty)$$

and has solutions of the form, using the Bessel function $J_\nu$ and writing $\sqrt{\lambda} = s = \sigma + it$,

$$y(x, \lambda) = J_{is}(\exp(x)) \text{ for all } x \in (-\infty, \infty);$$

in the space $L^2(-\infty, \infty)$ this equation is LP at $-\infty$ and is LCO at $+\infty$. This differential equation is then another example of equations derived from the original Bessel differential equation.

This Sears-Titchmarsh differential equation is the Liouville form, see Section 7 above, of the Sturm-Liouville equation

$$-(xy'(x))' - xy(x) = \lambda x^{-1} y(x) \text{ for all } x \in (0, +\infty).$$

In the space $L^2((0, \infty); x^{-1})$ this differential equation is LP at 0 and is LCO at $+\infty$.

Endpoint classification in $L^2((0, \infty); x^{-1})$:

| Endpoint | Classification |
|---|---|
| 0 | LP |
| $+\infty$ | LCO |

For the LCO endpoint $+\infty$ the boundary condition functions can be chosen as, for all $x \in (0, +\infty)$,

$$u(x) = x^{-1/2}(\cos(x) + \sin(x)) \quad v(x) = x^{-1/2}(\cos(x) - \sin(x)).$$

For details of boundary value problems for this Sturm-Liouville equation, on $[1,\infty)$ see [8, Example 4]. For problems on $[1,\infty)$ the spectrum is simple and discrete but unbounded both above and below, since the endpoint $+\infty$ is LCO.

Numerical results are given in [11, Data base file xamples.tex; example 6].

## 59. Zettl equation

This differential equation is closely linked to the classical Fourier equation 8;

$$-(x^{1/2}y'(x))' = \lambda x^{-1/2}y(x) \text{ for all } x \in (0,+\infty).$$

Endpoint classification in $L^2((0,+\infty);x^{-1/2})$

| Endpoint | Classification |
|---|---|
| 0 | R |
| $+\infty$ | LP |

This is a devised example to illustrate the computational difficulties of regular problems which have mild (integrable) singularities, in this example at the endpoint 0 of $(0,\infty)$.

The differential equation gives $p(0) = 0$ and $w(0) = \infty$ but nevertheless 0 is a regular endpoint in the Lebesgue integral sense; however this endpoint 0 does give difficulties in the computational sense.

The Liouville normal form of this equation is the Fourier equation, see Section 8 above; thus numerical results for this problem can be checked against numerical results from

(i) an R problem,
(ii) the roots of trigonometrical equations, and
(iii) as an LCNO problem (see below).

There are explicit solutions of this equation given by

$$\cos(2x^{1/2}\sqrt{\lambda}) \;\;;\;\; \sin(2x^{1/2}\sqrt{\lambda})/\sqrt{\lambda}.$$

If 0 is treated as an LCNO endpoint then $u$, $v$ boundary condition functions are

$$u(x) = 2x^{1/2} \qquad v(x) = 1.$$

The regular Dirichlet condition $y(0) = 0$ is equivalent to the singular condition $[y,u](0) = 0$. Similarly the regular Neumann condition $(py')(0) = 0$ is equivalent to the singular condition $[y,\ v](0) = 0$.

The following indicated boundary value problems have the given explicit formulae for the eigenvalues:

$$y(0) = 0 \text{ or } [y,u](0) = 0, \text{ and } y(1) = 0 \text{ gives}$$
$$\lambda_n = ((n+1)\pi)^2/4 \ (n = 0,1,\dots)$$

$$(py')(0) = 0 \text{ or } [y,v](0) = 0, \text{ and } (py')(1) = 0 \text{ gives}$$
$$\lambda_n = \left((n+\tfrac{1}{2})\pi\right)^2/4 \ (n = 0,1,\dots).$$

## 60. Remarks

1. The author has made use of an earlier collection of examples of Sturm-Liouville differential equations drawn up by Bailey, Everitt and Zettl, in connection with the development and testing of the computer program SLEIGN2; see [8] and [10].
2. The author has made use of major collections of Sturm-Liouville differential equations from Pryce [69] and [70], and from Fulton, Pruess and Xie [38] and [68].
3. This catalogue will continue to be developed; the author welcomes corrections to the present form, and information about additional examples to extend the scope, of the catalogue.

## 61. Acknowledgments

1. The author is grateful to Werner Amrein (University of Geneva, Switzerland), David Pearson (University of Hull, England, UK) and other colleagues who organized the Sturm meeting, held at the University of Geneva in September 2003, which brought together so many scientists working in and making application of Sturm-Liouville theory.
2. The author is grateful to many colleagues who commented on the earlier drafts and sent information concerning possible examples to include in this catalogue: Werner Amrein, Paul Bailey, Richard Cooper, Desmond Evans, Charles Fulton, Fritz Gesztesy, Don Hinton, Hubert Kalf, Lance Littlejohn, Clemens Markett, Marco Marletta, Lawrence Markus, David Pearson, Michael Plum, John Pryce, Fiodor Rofe-Beketov, Ken Shaw, Barry Simon, Sergei Slavyanov, Rudi Weikard, Tony Zettl.
3. The author is especially indebted to Barry Simon, Fritz Gesztesy, Clemens Markett, John Pryce, Fiodor Rofe-Beketov and Rudi Weikard who supplied detailed information for the examples in Sections 11, 24, 25, 26, 31, 40, 41, 49 and 53.
4. The author thanks the editors of the proceedings volume, for the coming publication of the Sturm-Liouville manuscripts, for checking the final draft of this catalogue; their careful scrutiny led to many improvements in the text, and to the correction of not a small number of errors.

## 62. The future

As mentioned above it is hoped to continue this catalogue as a database for Sturm-Liouville differential equations.

The main contributor to the assessment and extension of successive drafts of this catalogue is Fritz Gesztesy, who has agreed to join with the author in continuing to update the content of this database.

Together we ask that all proposals for enhancing and extending the catalogue be sent to both of us, if possible by e-mail and LaTeX file. The author's affiliation data is to be found at the end of this paper; the corresponding data for Fritz Gesztesy is:

Fritz Gesztesy
Department of Mathematics
University of Missouri
Columbia, MO 65211, USA

e-mail: `fritz@math.missouri.edu`
fax: ++ 1 573 882 1869

## References

[1] M. Abramowitz and I.A. Stegun, *Handbook of mathematical functions*, Dover Publications, Inc., New York, 1972.

[2] N.I. Akhiezer and I.M. Glazmann, *Theory of linear operators in Hilbert space: I and II*, Pitman and Scottish Academic Press, London and Edinburgh, 1981.

[3] R.A. Askey, T.H. Koornwinder and W. Schempp, Editors of *Special functions: group theoretical aspects and applications,* D. Reidel Publishing Co., Dordrecht, 1984.

[4] F.V. Atkinson, W.N. Everitt and A. Zettl, Regularization of a Sturm-Liouville problem with an interior singularity using quasi-derivatives, Diff. and Int. Equations **1** (1988), 213–222.

[5] P.B. Bailey, *SLEIGN: an eigenvalue-eigenfunction code for Sturm-Liouville problems*, Report Sand77-2044, Sandia National Laboratory, New Mexico, USA, 1978.

[6] P.B. Bailey, J. Billingham, R.J. Cooper, W.N. Everitt, A.C. King, Q. Kong, H. Wu and A. Zettl, Eigenvalue problems in fuel cell dynamics, Proc. Roy. Soc. London (A) **459** (2003), 241–261.

[7] P.B. Bailey, W.N. Everitt, D.B. Hinton and A. Zettl, Some spectral properties of the Heun differential equation, Operator Theory: Advances and Applications **132** (2002), 87–110.

[8] P.B. Bailey, W.N. Everitt and A. Zettl, Computing eigenvalues of singular Sturm-Liouville problems, Results in Mathematics **20** (1991), 391–423.

[9] P.B. Bailey, W.N. Everitt and A. Zettl, Regular and singular Sturm-Liouville problems with coupled boundary conditions, Proc. Royal Soc. Edinburgh (A) **126** (1996), 505–514.

[10] P.B. Bailey, W.N. Everitt and A. Zettl, The SLEIGN2 Sturm-Liouville code, ACM Trans. Math. Software **27** (2001), 143–192. This paper may also be downloaded as the LaTeX file bailey.tex from the web site: http://www.math.niu.edu/~zettl/SL2.

[11] P.B. Bailey, W.N. Everitt and A. Zettl, The SLEIGN2 database, Web site: http://www.math.niu.edu/~zettl/SL2.

[12] P.B. Bailey, W.N. Everitt, J. Weidmann and A. Zettl, Regular approximation of singular Sturm-Liouville problems, Results in Mathematics **23** (1993), 3–22.

[13] V. Bargmann, Remarks on the determination of a central field of force from the elastic scattering phase shifts, Phys. Rev. **75** (1949), 301–303.

[14] V. Bargmann, On the connection between phase shifts and scattering potential, Rev. Mod. Phys. **21** (1949), 488–493.

[15] H. Behnke and F. Goerisch, Inclusions for eigenvalues of self-adjoint problems, in *Topics in Validated Computation*, J. Herzberger (Hrsg.), North Holland Elsevier, Amsterdam, 1994.

[16] W.W. Bell, *Special functions for scientists and engineers*, Van Nostrand, London, 1968.

[17] G. Birkhoff and G.-C. Rota, *Ordinary differential equations*, Wiley, New York, 1989.

[18] J.P. Boyd, Sturm-Liouville eigenvalue problems with an interior pole, J. Math. Physics **22** (1981), 1575–1590.

[19] A.G. Brusencev and F.S. Rofe-Beketov, Conditions for the selfadjointness of strongly elliptic systems of arbitrary order, Mt. Sb. (N.S.) **95**(**137**) (1974), 108–129.

[20] W. Bulla and F. Gesztesy, Deficiency indices and singular boundary conditions in quantum mechanics, J. Math. Phys. **26** (1985), 2520–2528.

[21] K. Chadan and P.C. Sabatier, *Inverse Problems in Quantum Scattering Theory,* 2nd ed., Springer, New York, 1989.

[22] E.T. Copson, *Theory of functions of a complex variable*, Oxford University Press, Oxford, 1946.

[23] P.A. Deift, Applications of a commutation formula, Duke Math. J. **45** (1978), 267–310.

[24] P. Deift and E. Trubowitz, Inverse scattering on the line, Comm. Pure Appl. Math. **32** (1979), 121–251.

[25] N. Dunford and J.T. Schwartz, *Linear Operators, part II*, Interscience Publishers, New York, 1963.

[26] M.S.P. Eastham, *The spectral theory of periodic differential equations*, Scottish Academic Press, Edinburgh and London, 1973.

[27] A. Erdélyi, *Higher transcendental functions: I, II and III*, McGraw-Hill, New York, 1953.

[28] W. D. Evans and W.N. Everitt, On an inequality of Hardy-Littlewood type: I, Proc. Royal Soc. Edinburgh (A) **101** (1985), 131–140.

[29] W.D. Evans, W.N. Everitt, W.K. Hayman and D.S. Jones, Five integral inequalities: an inheritance from Hardy and Littlewood, Journal of Inequalities and Applications **2** (1998), 1–36.

[30] W.N. Everitt, On the transformation theory of ordinary second-order linear symmetric differential equations, Czechoslovak Mathematical Journal **32** (107) (1982), 275–306.

[31] W.N. Everitt, J. Gunson and A. Zettl, Some comments on Sturm-Liouville eigenvalue problems with interior singularities, J. Appl. Math. Phys. (ZAMP) **38** (1987), 813–838.

[32] W.N. Everitt and D.S. Jones, On an integral inequality, Proc. Royal Soc. London (A) **357** (1977), 271–288.

[33] W.N. Everitt and C. Markett, On a generalization of Bessel functions satisfying higher-order differential equations, Jour. Comp. Appl. Math. **54** (1994), 325–349.

[34] W.N. Everitt and L. Markus, *Boundary value problems and symplectic algebra for ordinary and quasi-differential operators*, Mathematical Surveys and Monographs **61**, American Mathematical Society, RI, USA, 1999.

[35] W.N. Everitt and A. Zettl, On a class of integral inequalities, J. London Math. Soc. (2) **17** (1978), 291–303.

[36] G. Fichera, *Numerical and quantitative analysis*, Pitman Press, London, 1978.

[37] M. Flendsted-Jensen and T. Koornwinder, The convolution structure for Jacobi function expansions, Ark. Mat. **11** (1973), 245–262.

[38] C.T. Fulton and S. Pruess, *Mathematical software for Sturm-Liouville problems*, NSF Final Report for Grants DMS88 and DMS88-00839, 1993.

[39] C.S. Gardner, J.M. Greene, M.D. Kruskal and R.M. Miura, Korteweg-deVries equation and generalizations, VI. Methods for exact solution, Comm. Pure Appl. Math. **27** (1974), 97–133.

[40] F. Gesztesy and H. Holden, *Soliton equations and their algebro-geometric solutions. Vol. I:* (1+1)*-dimensional continuous models*, Cambridge Studies in Advanced Mathematics **79**, Cambridge University Press, 2003.

[41] F. Gesztesy, W. Karwowski and Z. Zhao, Limits of soliton solutions, Duke Math. J. **68** (1992), 101–150.

[42] F. Gesztesy and B. Simon, A new approach to inverse spectral theory, II. General real potentials and the connection to the spectral measure, Ann. Math. **152** (2000), 593–643.

[43] F. Gesztesy and M. Ünal, Perturbative oscillation criteria and Hardy-type inequalities, Math. Nach. **189** (1998), 121–144.

[44] F. Gesztesy and R. Weikard, Elliptic algebro-geometric solutions of the KdV and AKNS hierarchies – an analytic approach, Bull. Amer. Math. Soc. **35** (1998), 271–317.

[45] H. Hochstadt, A special Hill's equation with discontinuous coefficients, Amer. Math. Monthly **70** (1963), 18–26.

[46] H. Hochstadt, *The functions of mathematical physics*, Wiley-Interscience, New York, 1971.

[47] M.S. Homer, Boundary value problems for the Laplace tidal wave equation, Proc. Roy. Soc. of London (A) **428** (1990), 157–180.

[48] E.L. Ince, *Ordinary differential equations*, Dover, New York, 1956.

[49] K. Jörgens, *Spectral theory of second-order ordinary differential operators*, Lecture Notes: Series no.2, Matematisk Institut, Aarhus Universitet, 1962/63.

[50] K. Jörgens and F. Rellich, *Eigenwerttheorie gewöhnlicher Differentialgleichungen*, Springer-Verlag, Heidelberg, 1976.

[51] I. Kay and H.E. Moses, Reflectionless transmission through dielectrics and scattering potentials, J. Appl. Phys. **27** (1956), 1503–1508.

[52] E. Kamke, *Differentialgleichungen: Lösungsmethoden und Lösungen: Gewöhnliche Differentialgleichungen*, $3^{rd}$ edition, Chelsea Publishing Company, New York, 1948.

[53] T.H. Koornwinder, Jacobi functions and analysis on noncompact semisimple Lie groups, in *Special functions: group theoretical aspects and applications*, 1–85, edited by R.A. Askey, T.H. Koornwinder and W. Schempp, D. Reidel Publishing Co., Dordrecht, 1984.

[54] A.M. Krall, Boundary value problems for an eigenvalue problem with a singular potential, J. Diff. Equations **45** (1982), 128–138.

[55] W. Lay and S. Yu. Slavyanov, Heun's equation with nearby singularities, Proc. R. Soc. Lond. A **455** (1999), 4347–4361.

[56] J.E. Littlewood, On linear differential equations of the second order with a strongly oscillating coefficient of $y$, J. London Math. Soc. **41** (1966), 627–638.

[57] R.J. Lohner, Verified solution of eigenvalue problems in ordinary differential equations, personal communication, 1995.

[58] J. Lützen, Sturm and Liouville's work on ordinary linear differential equations. The emergence of Sturm-Liouville theory, Arch. Hist. Exact Sci. **29** (1984), 309–376.

[59] L. Markus and R.A. Moore, Oscillation and disconjugacy for linear differential equations with almost periodic coefficients, Acta. Math. **96** (1956), 99–123.

[60] M. Marletta, Numerical tests of the SLEIGN software for Sturm-Liouville problems, ACM Trans. Math. Software **17** (1991), 501–503.

[61] J.B. McLeod, Some examples of wildly oscillating potentials, J. London Math. Soc. **43** (1968), 647–654.

[62] P.M. Morse, Diatomic molecules according to the wave mechanics II: Vibration levels, Phys. Rev. **34** (1929), 57–61.

[63] M.A. Naimark, *Linear differential operators: II*, Ungar Publishing Company, New York, 1968.

[64] H. Narnhofer, Quantum theory for $1/r^2$-potentials, Acta Phys. Austriaca **40** (1974), 306–322.

[65] H.-D. Niessen and A. Zettl, Singular Sturm-Liouville problems; the Friedrichs extension and comparison of eigenvalues, Proc. London Math. Soc. **64** (1992), 545–578.

[66] L. Pastur and A. Figotin, *Spectra of random and almost-periodic operators*, Springer, Berlin, 1992.

[67] E.G.P. Poole, *Introduction to the theory of linear differential equations*, Oxford University Press, 1936.

[68] S. Pruess, C.T. Fulton and Y. Xie, *Performance of the Sturm-Liouville software package SLEDGE*, Technical Report MCS-91-19, Department of Mathematical and Computer Sciences, Colorado School of Mines, USA, 1994.

[69] J.D. Pryce, *Numerical solution of Sturm-Liouville problems*, Oxford University Press, 1993.

[70] J.D. Pryce, A test package for Sturm-Liouville solvers, ACM Trans. Math. Software **25** (1999), 21–57.

[71] M. Plum, Eigenvalue inclusions for second-order ordinary differential operators by a numerical homotopy method, ZAMP **41** (1990), 205–226.

[72] F. Rellich, *Die zulässigen Randbedingungen bei den singulären Eigenwertproblemen der mathematischen Physik*, (Gewöhnliche Differentialgleichungen zweiter Ordnung), Math. Z. **49** (1944), 702–723.

[73] F.S. Rofe-Beketov, Non-semibounded differential operators. Teor. Funkciĭ Funkcional. Anal. i Priložen. Vyp. **2** (1966), 178–184.

[74] A. Ronveaux, *Heun differential equations*, Oxford University Press, 1995.

[75] D.B. Sears and E.C. Titchmarsh, Some eigenfunction formulae, Quart. J. Math. Oxford (2) **1** (1950), 165–175.

[76] J.K. Shaw, A.P. Baronavski and H.D. Ladouceur, Applications of the Walker method, in *Spectral Theory and Computational Methods of Sturm-Liouville problems*, 377–395, Lecture Notes in Pure and Applied Mathematics **191**, Marcel Dekker, Inc., New York, 1997.

[77] S. Yu. Slavyanov and W. Lay, *Special functions: a unified theory based on singularities*, Oxford University Press, 2000.

[78] C. Sturm and J. Liouville, Extrait d'un Mémoire sur le développement des fonctions en séries dont les différents termes sont assujettis à satisfaire à une même équation différentielle linéaire, contenant un paramètre variable, J. Math. Pures Appl. **2** (1837), 220–223.

[79] E.C. Titchmarsh, *Eigenfunction expansions associated with second-order differential equations I*, Clarendon Press, Oxford, 1962.

[80] G.N. Watson, *A treatise on the theory of Bessel functions*, Cambridge University Press, Cambridge, England, 1958.

[81] E.T. Whittaker and G.N. Watson, *Modern analysis*, Cambridge University Press, 1950.

[82] A. Zettl, Computing continuous spectrum, in *Trends and Developments in Ordinary Differential Equations*, 393–406, Y. Alavi and P. Hsieh editors, World Scientific, 1994.

[83] A. Zettl, Sturm-Liouville problems, in *Spectral Theory and Computational Methods of Sturm-Liouville problems*, 1–104, Lecture Notes in Pure and Applied Mathematics **191**, Marcel Dekker, Inc., New York, 1997.

W. Norrie Everitt
School of Mathematics and Statistics
University of Birmingham
Edgbaston
Birmingham B15 2TT, UK
e-mail: `w.n.everitt@bham.ac.uk`

# Index